New Frontiers in Oil and Gas Exploration

Congrui Jin • Gianluca Cusatis
Editors

New Frontiers in Oil and Gas Exploration

Springer

Editors
Congrui Jin
Department of Mechanical Engineering
Binghamton University
State University of New York
Binghamton, NY, USA

Gianluca Cusatis
Department of Civil and Environmental Engineering
Northwestern University
Evanston, IL, USA

ISBN 978-3-319-40122-5 ISBN 978-3-319-40124-9 (eBook)
DOI 10.1007/978-3-319-40124-9

Library of Congress Control Number: 2016949461

Printed on acid-free paper

This Springer imprint is published by Springer Nature
The registered company is Springer International Publishing AG Switzerland

Preface

The United States has seen a resurgence in petroleum production, especially when new discoveries of unconventional oil and gas sources have dramatically reshaped both the energy outlook and the future of chemicals production in this country. They have the potential to positively influence our overall trade balance and future economic growth rate for decades to come. Gas shale production sites throughout the United States are replacing coal power plant operations with environmental benefits associated with lower levels of life-cycle carbon dioxide emissions. However, production of gas from low permeability shale involves hydraulic fracturing of shale into a sufficiently dense system of cracks, and thus establishing fracture/joint connectivity to facilitate fluid and gas flow. Unfortunately, these procedures are not without adverse environmental consequences, such as potential contamination of fresh water resources, seismic activities resulting from the stimulation of the rock formation (induced seismicity), and the disposal of flow-back and production water (triggered seismicity). It is clear that mitigation of the environmental impact of unconventional resource development needs to be pursued by a variety of technological innovations aiming at optimizing hydraulic fracturing protocols to increase recovery efficiency above the current level and to reduce water usage, which, however, requires the fundamental understanding of the failure and flow phenomena occurring in the heterogeneous shale formations, deeply buried under the earth surface, which cannot be investigated easily through experimental or field observations. This book will highlight how this new domestic energy source may be utilized and managed and the projection for the long-lasting economic impact if the United States is to take full advantage of this new unique opportunity. With the technologies simply changing too fast, this book will provide the latest research works and findings in this field, with a focus on key practical issues, such as computational characterization of shale at multiple length scales, mechanical interactions of proppant and hydraulic fractures, and production analysis of a multi-fractured horizontal well, etc. This book will bring unique perspectives of knowledge and experience in dealing with many of the issues about oil and gas exploration. It contains both cutting-edge original research and comprehensive reviews addressing both theory and practice. This book focuses on key practical issues, and it is a practical reference for geoscientists and engineers in the petroleum and geothermal industries, and for research scientists interested in reservoir modeling and their application to the improvement of current design of hydraulic fractures.

Binghamton, NY, USA Congrui Jin
Evanston, IL, USA Gianluca Cusatis

Contents

Chapter 1
Understanding Asphaltene Aggregation and Precipitation Through Theoretical and Computational Studies

Cuiying Jian and Tian Tang

Abstract Asphaltenes are known to cause serious problems during the processing of petroleum compounds due to their aggregation and precipitation behaviors. Despite the significant amount of experimental works that have been performed, large debates still exist in literature. Parallel with experimental work, great efforts have been spent from theoretical and computational perspectives to predict asphaltene behaviors under given conditions, to provide atomic/molecular information on their aggregation as well as precipitation, and to further shed lights on existing debates. This chapter presents a detailed review of previous theoretical and computational works on asphaltene aggregation and precipitation. Theoretical models developed, systems simulated, and the key findings are summarized; and discrepancies among those works are highlighted.

Abbreviations

CCC	Critical cluster concentration
CMC	Critical micelle concentration
CNAC	Critical nanoaggregation concentration
CPA-EoS	Cubic-Plus-Association EoS
EoS	Equation of state
L^2MS	Laser desorption laser ionization mass spectrometry
MD	Molecular dynamics
MM	Molecular mechanics
MW	Molecular weight
NMR	Nuclear magnetic resonance
PA	Polyaromatic

C. Jian • T. Tang (✉)
Department of Mechanical Engineering, University of Alberta, Edmonton, AB, Canada, T6G 2G8
e-mail: tian.tang@ualberta.ca

C. Jin, G. Cusatis (eds.), *New Frontiers in Oil and Gas Exploration*,
DOI 10.1007/978-3-319-40124-9_1

PC-SAFT EoS	Perturbed-chain SAFT-EoS
PME	Potential of mean force
PR-EoS	Peng-Robinson EoS
QM	Quantum mechanics
RICO	Ruthenium-ion-catalyzed oxidation
SAFT EoS	Statistical Associating Fluid Theory EoS
SAFT-VR EoS	SAFT-EoS for potentials of variable range
SARA	Saturate aromatic, resin, and asphaltene sc-CO_2 supercritical carbon dioxide
XANES	X-ray absorption near-edge structure
XPS	X-ray photoelectron spectroscopy

1.1 Introduction

Petroleum resources are of great importance to daily life and economy all over the world [1]. To optimize petroleum processing and utilization, it's thus of great interest to understand the behaviors of petroleum components [2–4]. Crude oil can be categorized into saturate, aromatic, resin, and asphaltene (SARA) fraction groups [5, 6]. Asphaltenes, as the heaviest and most aromatic component, are operationally defined as a solubility class that are insoluble in n-alkanes but soluble in aromatic solvents [7–9]. They are known to be the "cholesterol" of petroleum due to their deleterious effects on oil processing, from extraction to refinery [10]. For instance, asphaltenes clog rock pores and production facilities, form deposit during transportation and storage, and deactivate catalysts, thus generating a large cost in petroleum industry [11–15]. Most of these serious problems can be traced to asphaltene aggregation and precipitation [16–18]. Therefore a considerable amount of effort, based on experimental or theoretical approaches, has been dedicated to investigate asphaltene aggregation and precipitation dynamics as well as the associated mechanisms and resultant structures. Recently, with the development of computational techniques, asphaltene investigations from simulation aspect have been emerging quickly.

Many detailed reviews exist for experimental studies, such as the works of Speight et al. [19], Strausz et al. [20, 21], and Mullins et al. [22]. However, there is a lack of review on theoretical and computational studies of asphaltenes. These studies are very useful in that they provide fundamental explanations to asphaltenes' behaviors observed in experiments. Therefore, this chapter aims at reviewing theoretical and computational works performed on asphaltenes. As these studies have been motivated by experimental observations, a brief summary on experimental studies is first given. This is followed by a detailed review on theoretical and computational works performed on asphaltene aggregation and precipitation. Finally, summaries and future perspectives are provided.

1.2 Experiment Studies

The behaviors of asphaltenes are strongly related to their chemical structures. Hence significant amount of experimental work has been performed on probing the molecular structures and functional groups of asphaltenes. Two types of molecular structures (island vs. archipelago, see a schematic representation in Fig. 1.1) have been proposed in literature for various asphaltene samples investigated, each supported by different experimental observations. Island-type, or continental-type, models were proposed to possess a center that has a large number of fused aromatic cores and a periphery characterized by aliphatic chains, while archipelago-type structures were believed to have small aromatic regions interconnected by aliphatic chains [23–25].

In the 1960s, Dickie and Yen [24] examined the molecular weight (MW) of asphaltene using a range of techniques including X-ray diffraction and scattering, mass spectrometer, gel permeation chromatography, vapor pressure osmometry, ultracentrifugation, and electron microscope. Through investigating the results of MW measurements, the individual asphaltene molecules were inferred to consist of condensed aromatic rings and peripheral aliphatic chains. Later on, various conceptual island-type models constructed from experimental data have been reported. For instance, structural parameters calculated for asphaltene extracted from a Lagunillas crude oil agreed well in most cases with the constructed hypothetical island molecular model; León et al. [18] determined the average molecular parameters (e.g., MW and number of aromatic rings) for four asphaltene samples using LECO CHNS 244 elemental analyzer model, Knauer vapor pressure osmometer, as

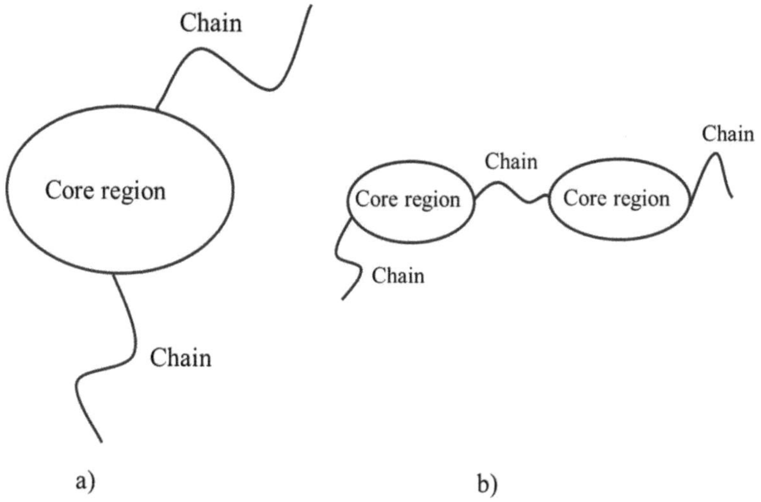

Fig. 1.1 Schematic representation of (**a**) island-type and (**b**) archipelago-type molecular structures

well as nuclear magnetic resonance (NMR) spectra. Based on these data, the average molecular models constructed all possessed a single condensed core.

More recently, in the work of Sabbah et al. [26] using laser desorption laser ionization mass spectrometry (L^2MS), the spectra of 1, 3, 6, 8-tetradecyl pyrene which possesses one-polyaromatic (PA) core shared similarities with those of asphaltenes, suggesting the existence of island-type asphaltene molecules. Through L^2MS, Zare and co-workers [27] compared the fragmentation behaviors of 23 model compounds with 2 petroleum asphaltene samples extracted from Middle Eastern black oils. The results showed that all model compounds which have one aromatic core, with or without pendant alkyl groups, exhibited similar behaviors to what was observed for the two asphaltene samples, whereas all model compounds which have more than one aromatic core showed energy-dependent fragmentation, different from the two asphaltene samples. Further, a series of work done by Mullins and co-workers, using mass spectral methods [28], diffusion measurements [29–31], and X-ray Raman [32], has widely popularized the island-type asphaltene molecular structure. This model was also supported by a sequence of work using other methods. For instance, the MW distribution of asphaltenes was measured by Pinkston et al. [33] through analyzing laser-induced acoustic desorption/electron ionization mass spectra, and by Qian et al. [34] using field desorption mass spectrometry and electrospray ionization mass spectrometry; the size of asphaltene monomers was determined by Lisitza et al. [35] through NMR diffusion measurements, and by Bouhadda et al. [36] employing Raman spectrometry and X-ray diffraction. These studies all provided evidence for the island-type asphaltene model.

The archipelago-type structure, on the other hand, was supported by other experimental evidences. Strausz and co-workers [37–39] showed that the mild thermolysis products of Athabasca asphaltenes mainly contained mono- to pentacyclic condensed aromatic molecules, which could not be formed from the cracking and rearrangement of large aromatic clusters under mild thermal cracking conditions. Therefore they proposed that small aromatic fragments should be present in the asphaltenes. Similarly, in the work of Calemma et al. [40], the asphaltenes from different feedstocks (i.e., vacuum residue, atmospheric residue and crude oil) were investigated by pyrolysis-gas chromatography/mass spectrometry, and through comparison with thermolysis reactions of model compounds, the asphaltene samples were considered to be of archipelago-type molecular structures. For asphaltene fraction of a Venezuela residue examined by Ferris et al. [41], as MW increased, the aromaticity calculated from NMR data only slightly increased, suggesting that asphaltene molecules were consisting of small units of similar aromatic-to-saturate ratios. Analysis based on ruthenium-ion-catalyzed oxidation (RICO) reaction of asphaltenes was also shown to support the existence of archipelago-type molecular structure. For instance, Su et al. [42] investigated the RICO products of an asphaltene sample from a vacuum residue of Arabian crude mixture, and the resolved aliphatic di- and polycarboxylic acids suggested the presence of aliphatic linkages in a single asphaltene molecule connecting more than two aromatic rings or hydroaromatics; RICO used in Murgich et al. [43], for

Athabasca sand oil, and in Artok et al. [44], for Arabian crude mixture, also confirmed the existence of sulfide links among fused rings within asphaltene molecules. Furthermore, Siskin et al. [45], characterized asphaltenes from several shot-coke- and sponge-coke-producing vacuum residue feeds by a combination of solid-state 13C NMR, X-ray photoelectron spectroscopy (XPS), and elemental abundance, and their average chemical structures were shown to have aliphatic bridges among fused rings within the molecular structure. This is consistent with the work of Gray and co-workers [46, 47], where it was proposed that to be consistent with asphaltene pyrolysis and coking behaviors, asphaltene chemical structures must consist of a variety of aromatic groups joined by bridges and substituted by aliphatic groups. Regarding the island-type structure concluded by Mullins et al. [29, 30, 48–51], Strausz et al. [20] commented that the fluorescence decay and depolarization kinetic time measurements used in their studies [29, 30, 48–51] were erroneous, which is still being debated [52].

While great controversy exists in the molecular architectures of asphaltenes, it is generally agreed that these molecules [30, 53–56] contain heteroatoms such as nitrogen, oxygen, sulfur, as well as various metals, e.g., iron, nickel, and vanadium, associated with different functional groups. For instance, Strausz and co-workers [23] proposed the presence of thiolane, thiane, thiophene, pyridine, carboxylic esters, and vanadium porphyrins in asphaltene samples from Alberta; X-ray absorption near-edge structure (XANES) spectroscopy and XPS by Mullins and co-workers [57, 58] as well as Kelemen et al. [59] showed that sulfur forms thiophene, sulfide, and sulfoxide in asphaltene molecules. Furthermore, XANES spectra also confirmed the existence of nitrogen in pyrrolic and pyridine [60]. In the work of Desando et al. [61], using NMR spectroscopy, two types of oxygen containing functionality (hydroxyl and carboxyl) have been detected for Athabasca oil sand asphaltene. Speight and co-workers [62], using infrared spectroscopy, inferred that for the asphaltene fraction of Athabasca bitumen, oxygen could also exist as sulfur-oxygen functions, or carbonyl groups, i.e., ketones, quinones, and ethers. The presence of carbonyl functional groups were also proposed by Ignasiak et al. [63] using the same technique. For metals, the fluorescence absorption spectra of asphaltenes, measured using CARY 500 UV−visible−NIR spectrometer by Groenzin et al. [30], showed similarity with octaethyl porphyrin, suggesting the asphaltene porphyrins were of the β-octa-alkyl type.

The complexity of asphaltene molecular structures has greatly impeded the clear understanding of asphaltene aggregation, leaving this as a highly debated area to date. The most populated theory to describe asphaltene aggregation is perhaps the Yen-Mullins model [22, 25, 64, 65] proposed by Yen and co-workers [65] and later advanced by Mullins and co-workers [25, 64]. In this model (shown in Fig. 1.2), it was proposed that the asphaltenes molecules first form nanoaggregates through the stacking of a small number ($\leq$10) of poly-aromatic cores, surrounded by side-chains, when its concentration reaches the critical nanoaggregation concentration (CNAC, 50–150 mg/L). These nanoaggregates further form clusters at the critical cluster concentration (CCC, 2–5 g/L), each containing less than ten nanoaggregates. Within this hierarchical description for asphaltene aggregation,

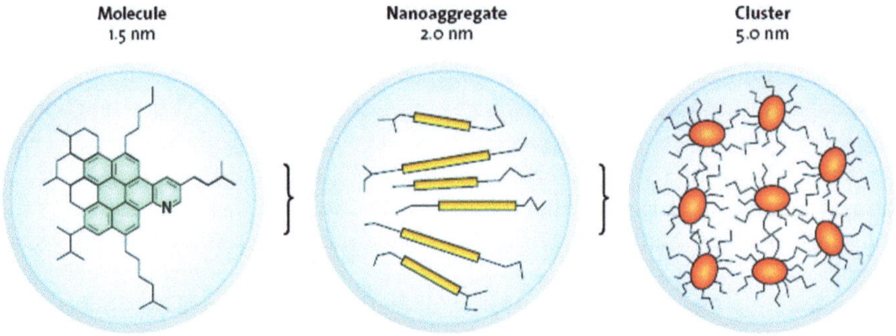

Fig. 1.2 Yen-Mullins model [22, 25, 64, 65] for asphaltene aggregation. Asphaltene molecules are proposed to be of island-type structures. When their concentration reaches CNAC, asphaltene molecules form nanoaggregates. At higher concentrations (above CCC), nanoaggregates further form clusters. Reprinted with permission from Mullins, O. C.; Sabbah, H.; Eyssautier, J.; Pomerantz, A. E.; Barré, L.; Andrews, A. B.; Ruiz-Morales, Y.; Mostowfi, F.; McFarlane, R.; Goual, L. Advances in asphaltene science and the Yen–Mullins model. *Energy Fuels* **2012**, 26, 3986–4003. Copyright 2012 American Chemical Society

the individual asphaltene molecules were proposed to be of island-type architectures. In good solvents such as toluene, this theory was supported by the CNAC and CCC detected using high-Q ultrasonic spectroscopy, direct-current electrical conductivity and fluorescence measurements [66, 67]. However, there are still ongoing debates even regarding the existence of CNAC and CCC, or the general critical micelle concentration (CMC) for asphaltenes in good solvents. Shown by Andersen et al. [68], no CMC was detected by calorimetric titration method for heptane-precipitated asphaltenes in pure toluene. Evdokimov et al. [69, 70] investigated the aggregation behaviors of asphaltene from Tatarstan virgin crude oils using UV/visible spectroscopy. In toluene, a gradual aggregation process was observed for these asphaltenes, and asphaltene monomers were only abundant when the concentrations were below 1–5 mg/L. These observations were distinct from the two-step aggregation process as described in the Yen-Mullins model [22, 25, 64, 65]. Similarly, in the work of Hoepfner et al. [71], as asphaltenes were diluted from 5 vol. % to 0.00125 vol. % (15 mg/L) in toluene, dissociation of asphaltene nanoaggregates was observed to occur gradually, and aggregates even persisted down to asphaltene concentrations as low as 15 mg/L. Very different aggregated structures of asphaltenes have also been reported. For instance, using the same technique of small angle neutron scattering, studies have reported different morphologies, ranging from prolate ellipsoid [72] and oblate cylinders [73] to thin discs [74].

Other components, especially resins, can also have strong effect on the stability of asphaltene monomers and aggregates in crude oil, and hence their final precipitation. There are two views on the effect of resins: the first proposes that resins form an adsorption layer around asphaltene particles which are responsible for the asphaltene stabilization [75, 76]; the second proposes resins and other crude oil components to act as a solvent which may solvate asphaltenes [77–79]. Each of these two views is supported by different experimental observations. For instance, the peptizing role of resins was suggested in the work of Soorghali et al. [75] using wettability measurements, while the latter view was supported by DC conductivity measurements in the work of Sedghi et al. [79]

Despite the existing debates on the asphaltene aggregation and precipitation, it is generally accepted that the interaction between PA cores (so-called $\pi - \pi$ interaction) is an important contributor to aggregation [22, 46, 80, 81], and hence precipitation. The stacking of PA cores has been widely observed [22, 46, 80, 81], although whether $\pi - \pi$ interaction was the dominating driving force for asphaltene aggregation [82–85] is still under debate. In this context, theoretical and the emerging computational approaches, based on the available experimental observation, can help with the prediction of asphaltene behaviors under given conditions, provide atomic/molecular information for the system studied, and further shed lights on resolving the existing debates.

1.3 Theoretical Studies

In this section, we provide a detailed review of the theoretical approaches used for investigating asphaltene aggregation and precipitation.

1.3.1 Theoretical Modeling of Asphaltene Aggregation

Based on Yen's model that describes asphaltene aggregates as stacked aromatic sheets surrounded by aliphatic side chains [22, 25, 64], Rogel [86] first proposed a molecular thermodynamic approach that predicted asphaltene aggregation behaviors in different solvents. An analytical expression was developed for the free energy of aggregation in benzene, toluene, cyclohexane, and n-decylbenzene, which included contributions from (1) transferring of the PA rings from the solvent into the aromatic core stacks, (2) mixing of the aliphatic chains with the solvent, (3) deformation of the aliphatic chains, (4) steric repulsion among the aliphatic chains, and (5) aggregate core-solvent interaction. The calculated free energy curves from the five contributions showed that the aggregation was mostly affected by the free energy of transferring the PA rings of asphaltenes from the solvent into the stacking structure. Later, Rogel [87] applied the same approach to compare the free energy of aggregation for asphaltene aggregates of different shapes in benzene, toluene, 1-methyl naphthalene, nitrobenzene, quinoline, and 1,2 dichlorobenzene. It was showed that cylindrical shape containing stacked aromatic cores was preferred over spherical shape and disks. According to this model, the existence of CMC for asphaltene solutions depended on the characteristics of the PA ring: asphaltenes with low aromaticities and aromatic condensations did not have an operational CMC. These results suggested [86, 87] that the interaction between PA cores was the main driving forces for asphaltene aggregation. The same approach was also applied by Rogel [88] to study the effect of resins on asphaltene aggregation. Resins were considered to be of similar chemical structures with asphaltenes but have a smaller PA core area. Consistent with experiments, Rogel's model predicted [88] that resins decreased the size of asphaltene aggregates and increased the CMC of the asphaltene solutions in toluene. The series of models proposed by Rogel were based on assuming island-type molecular structure for the asphaltenes. Contrarily, Agrawala et al. [89] proposed an asphaltene association model analogous to linear polymerization, where asphaltene molecules were assumed to contain multiple active sites capable of linking with each other. This oligomerization-like asphaltene association model was consistent with archipelago-type asphaltene molecular structure, and well fitted experimental molar mass data for asphaltenes in different solvents at different temperatures.

While the above works focused on a one-step aggregation scenario, several studies considered the aggregation process to be of multiple steps. In the work of Zhang et al. [90], a hindered stepwise aggregation model was developed to simulate asphaltene aggregation. In this model, it was assumed that the effective interaction/collision among asphaltene monomers promoted the initial aggregation, and the sequential collisions of monomers with smaller aggregates resulted in larger aggregates. The results showed that asphaltene molecules formed polymer aggregates, and the size of the aggregates increased significantly with its concentration in solution. Later, this model was used by the same group [91] to infer the aggregation morphologies of asphaltenes, and it was shown that the dominant aggregates were of 2–8 monomers with stacked aromatic cores. Acevedo et al. [92] investigated asphaltene aggregation using a mathematical model based on two consecutive equilibrium steps: $nA \overset{K_1}{\leftrightarrow} A_n$ and $mA + A_n \overset{K_2}{\leftrightarrow} A_{n+m}$, where n and m were number of moles; A, A_n, and A_{n+m} represented single molecules, n aggregates and $n+m$ aggregates of A, respectively; K_1 and K_2 were equilibrium constants. The results suggested that in toluene nanoaggregates were first formed by asphaltene fraction of extremely low solubility; asphaltene fraction of higher solubility would then keep the aggregates in solution and prevent phase separation.

1.3.2 Theoretical Modeling of Asphaltene Precipitation

Tremendous theoretical efforts have been spent on predicting asphaltene's precipitation behaviors. Depending on the way asphaltenes are assumed to be stabilized in crude oils, two types of theories have been developed, namely colloidal and solubility theories [93]. According to the colloidal theory, asphaltenes form insoluble solid particles in crude oil, which are stabilized by the presence of resins adsorbed onto their surface (see Fig. 1.3); detachment of the resins from the asphaltene particles leads to their aggregation and final precipitation [93]. Contrarily, the solubility theory adopts the view that asphaltene molecules are soluble in crude oil and precipitation will occur if its solubility drops below a certain threshold [94].

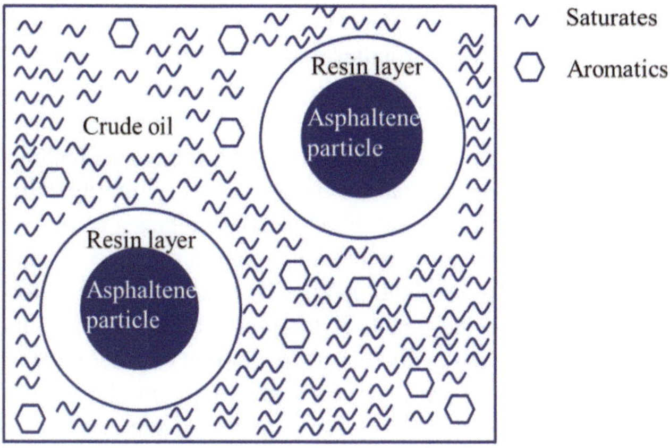

Fig. 1.3 Schematic representation of crude oil system based on colloidal theory

1.3.2.1 Models Based on Colloidal Theory

In 1987, Leontaritis and Mansoori proposed [95] a thermodynamic model for asphaltene precipitation based on colloidal theory. In this model, it was assumed that resin molecules were adsorbed onto asphaltene particles and the repulsive forces between these resins kept asphaltene aggregates from precipitation. The amount of adsorbed resin depended on its concentration in the crude oil phase. Therefore the concentration must be maintained above certain critical values to keep asphaltene particles stabilized. The corresponding critical resin chemical potential was calculated based on the Flory-Huggins statistical thermodynamic theory of polymer solutions [96, 97]. Adding precipitants affected resin concentration as well as its chemical potential. The amount of precipitants, needed to bring the resin chemical potential to its critical values, was calculated and defined as the onset condition for asphaltene precipitation. It was shown that the predicted onset agreed well with experimental ones. Later on, Firoozabadi and co-workers [98, 99] incorporated asphaltene aggregation model into precipitation. In their model, resins adhered to the surface of the asphaltene colloidal particles, and stabilized these particles in a way analogous to that of surfactant amphiphilic molecules in the theories of solubilization [100]. In their first work [98], the Gibbs free energy of micellization was derived by mainly considering lyophobic contributions gained from transferring asphaltene molecule from an infinitely diluted crude to a micelle. Adding a precipitant would change the concentration of asphaltene monomers and the associated chemical potential, hence the Gibbs free energy of micellization, leading to a new equilibrium. The monomeric asphaltene concentration in the crude oil phase at this new equilibrium state was calculated using the Peng-Robinson equation of state (PR-EoS) [101], which in turn determined the amount of precipitated asphaltenes. This model predicted that the amounts of asphaltenes precipitate decreased with increasing the chain length of n-alkanes added to the crude, which agreed well with experimental data. In their subsequent work, Pan and Firoozabadi

[99] evaluated the total Gibbs free energy of a system consisting of a bulk liquid phase and a precipitated phase, the latter assumed to be an ideal solid mixture of asphaltenes and resins. Then, the composition of the precipitated phase was determined by minimizing the total Gibbs free energy subject to mass balance equations. It was shown that [99] the predicted amount of precipitated asphaltenes by adding precipitants *n*-pentane, *n*-heptane, and *n*-decane was in good agreement with the experimentally measured data. Furthermore, when n-alkanes of short chains, such as propane and ethane were added to the crude, coprecipitation of asphaltenes and resins were predicted in this model [99]. Consistent with the work of Rogel et al. [88], it was also revealed that [99] for precipitant *n*-heptane and *n*-decane, increasing the resin concentration decreased the amount of precipitated asphaltenes. The performance of this model was also examined in the work of Assoori et al. [102], in which the effect of each micellization parameter on the asphaltene precipitation was demonstrated.

Also based on colloidal theory, Wu et al. [103] used pseudo-pure components to represent asphaltenes and resins, while all other components in the crude oil were modeled as a continuous solvent medium that affected interactions among asphaltene and resin molecules. While it was assumed that asphaltene molecules were stabilized by resins molecules in the crude oil, the precipitated asphaltene-rich phase was considered a liquid phase. At equilibrium, the chemical potentials and the osmotic pressures for each species (asphaltene or resins) were the same in both liquid (oil and precipitated) phases, and they were calculated from the Helmholtz free energy derived by a perturbation theory. Together with mass balance equations, the compositions of each liquid phase were obtained. The predicted fraction of precipitated asphaltenes was compared with experiments to study the effect of precipitants. It was shown that [103] when a large amount of precipitants were added, the calculated results were in line with the experiments. To address the effect of resins on the solubilization of asphaltene particles, Victorov et al. [104] formulated a thermodynamic model which explicitly considered the shapes of resin molecules as well as the deformation of the resin layer adsorbed onto the asphaltene aggregates. By incorporating the PR-EoS, the proposed model was applied to predict the amounts of asphaltene precipitates [104, 105].

1.3.2.2 Models Based on Solubility Theory

In this category, two main types of models have emerged for investigating the phase separation of asphaltenes, which is based on either regular solution theory or equation of state (EoS) methods. The works reviewed in this section are summarized in Fig. 1.4.

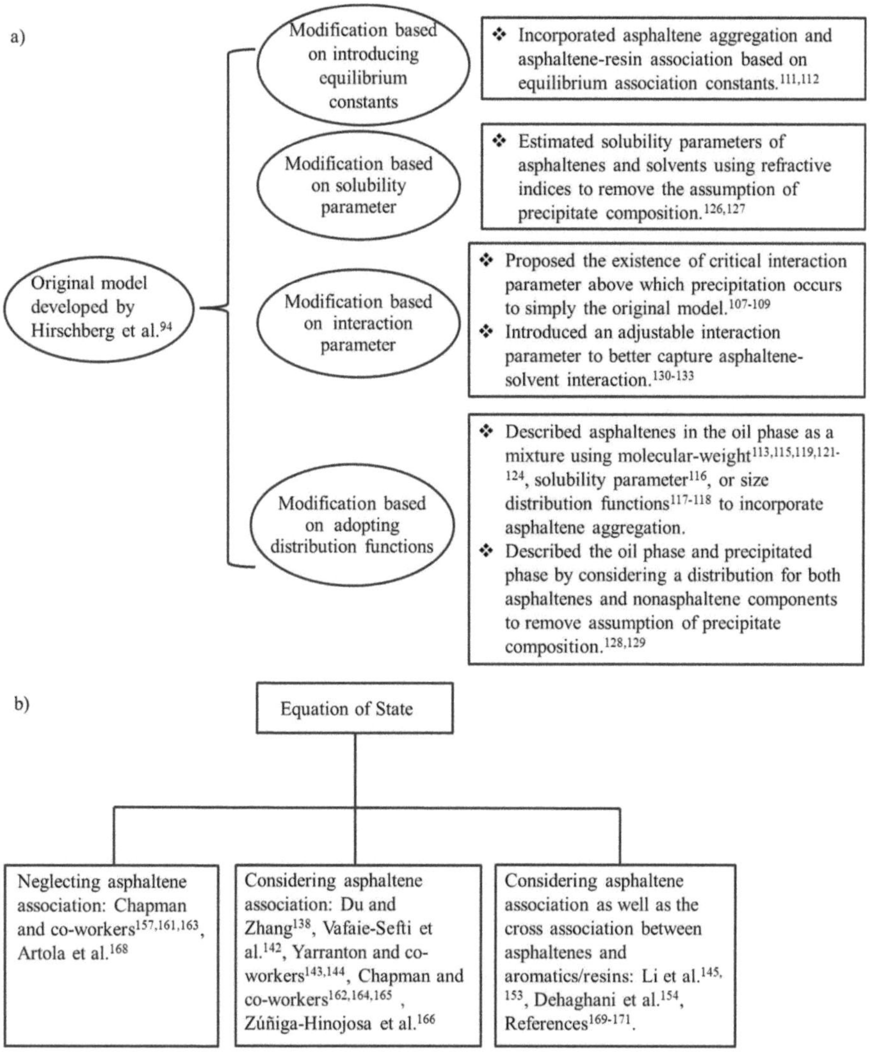

Fig. 1.4 Summary of theoretical works based on (**a**) regular solution theory and (**b**) equation of state

Models Based on Regular Solution Theory

Regular solution theory was first applied by Hirschberg et al. [94] to model asphaltene precipitation. Crude oil was described as a quasi-binary mixture of two pseudo-components with asphaltene as the solute and the remainder as the mixed solvent. Asphaltene precipitation occurred when its concentration exceeded the solubility limit. At equilibrium, the chemical potentials for each pseudocomponent must be the same in both the precipitated phase P and solvated phase S:

$$\mu_{\mathrm{a}}^{\mathrm{P}} = \mu_{\mathrm{a}}^{\mathrm{S}}, \tag{1.1}$$

$$\mu_{\mathrm{m}}^{\mathrm{P}} = \mu_{\mathrm{m}}^{\mathrm{S}}, \tag{1.2}$$

where subscripts a and m represent asphaltene and the mixed solvent, respectively. Following the Flory-Huggins theory [96, 97], the chemical potential for each of the two components is given by

$$\frac{\mu_{\mathrm{a}} - (\mu_{\mathrm{a}})_{\mathrm{ref}}}{RT} = \ln\varphi_{\mathrm{a}} + 1 - \frac{V_{\mathrm{a}}}{V} + \chi_{\mathrm{a}}, \tag{1.3}$$

$$\frac{\mu_{\mathrm{m}} - (\mu_{\mathrm{m}})_{\mathrm{ref}}}{RT} = \ln\varphi_{\mathrm{m}} + 1 - \frac{V_{\mathrm{m}}}{V} + \chi_{\mathrm{m}}, \tag{1.4}$$

where R is the universal gas constant; T is the absolute temperature; $(\mu_{\mathrm{a}})_{\mathrm{ref}}$ and $(\mu_{\mathrm{m}})_{\mathrm{ref}}$ were, respectively, the chemical potentials for pure liquids a and m (the reference state); χ_{a} and χ_{m} are the interaction parameters for a and m. φ_{a} and φ_{m} were the volume fractions of a and m in the crude oil; V_{a} and V_{m} were the molar volumes for a and m; they are related by

$$\varphi_{\mathrm{a}} = \frac{x_{\mathrm{a}} V_{\mathrm{a}}}{V}, \tag{1.5}$$

$$\varphi_{\mathrm{m}} = \frac{x_{\mathrm{m}} V_{\mathrm{m}}}{V}. \tag{1.6}$$

Here, x_{a} and x_{m} are the mole factions of components a and m. The molar volume, V, of the crude oil and the interaction parameter χ were calculated by using Hildebrand's approximation [106]:

$$V = x_{\mathrm{a}} V_{\mathrm{a}} + x_{\mathrm{m}} V_{\mathrm{m}}, \tag{1.7}$$

and

$$\chi_{\mathrm{a}} = \frac{V_{\mathrm{a}}}{RT}(\delta - \delta_{\mathrm{a}})^2, \tag{1.8}$$

$$\chi_{\mathrm{m}} = \frac{V_{\mathrm{m}}}{RT}(\delta - \delta_{\mathrm{m}})^2, \tag{1.9}$$

where δ_{a} and δ_{m} are the solubility parameters for components a and m, estimated by fitting the cohesive energy density to experimental data using EoS. The solubility parameter δ for the crude oil was given by

$$\delta = \varphi_a \delta_a + \varphi_m \delta_m. \tag{1.10}$$

Hirschberg et al. [94] assumed that the precipitation phase was pure asphaltenes and $\varphi_a{}^S \ll 1$. Therefore combing Eqs. (1.1), (1.3), and (1.8), the maximum volume fraction, $(\varphi_a{}^S)_{max}$, of asphaltenes soluble in the crude was given by

$$\left(\varphi_a{}^S\right)_{max} = \exp\left(-1 + \frac{V_a}{V_m} - \frac{V_a}{RT}(\delta_a - \delta_m)^2\right). \tag{1.11}$$

Because of its simplicity, this model has been widely adopted in the oil industry, and many adaptations have been performed.

Correra and Donaggio [107] simplified the model proposed by Hirschberg et al. [94] by using the interaction parameter χ. to predict whether asphaltene precipitation occurred or not. This new model was based on the assumption that asphaltene precipitation was ruled by the overall solvent power of the crude oil toward asphaltenes, which in turn could be predicted by the interaction parameters χ defined in Eqs. (1.8) and (1.9) as functions of the solubility parameters. According to this model, precipitation occurred when χ exceeded certain critical value χ_C. Correra et al. [108] showed that $\chi_C = 1$ when asphaltenes were considered to be at infinite dilution. In their subsequent work [109], the authors presented a detailed discussion on the determination of the parameters used in the proposed model.

While Hirschberg et al. [94] assumed asphaltene to be a homogenous component of petroleum crude, Burke et al. [110] applied this approach to predict the effect of composition on precipitate formation. As the model was only able to qualitatively match the experimental precipitation data, it was suggested that the aggregation of asphaltenes in crude oil might affect the performance of this model. Novosad and Costain [111] adopted a modified Hirschberg's model [94, 112] to study asphaltene precipitation, which incorporated both asphaltene aggregation and asphaltene-resin association through introducing equilibrium association constants. Correspondingly, the crude oil was described to consist of three components: asphaltene, resin, and solvent (oil components that were neither asphaltene nor resin). It was shown that the precipitation model, qualitatively and to some degree quantitatively, reproduced the trends in asphaltene behaviors.

Kawanaka et al. [113] applied the Scott-Magat [114] heterogeneous polymer solution theory, developed based on Flory-Huggins theory, to edict the onset point and amount of asphaltene precipitation from petroleum crude under the influence of precipitants. They considered asphaltenes to be a mixture of many similar polymeric molecules and used a continuous model in which a molecular-weight-distribution was assumed. For precipitants *n*-pentane, *n*-heptane, and *n*-decane, the predicted onset point and amount of precipitates were shown to be in good agreement with experimental data. The Scott-Magat [114] heterogeneous polymer solution theory was also used in the work of Manshad et al. [115], where equilibrium conditions were established by minimizing the total Gibbs free energy. In this work, Manshad et al. [115] compared the performance of three distribution functions for asphaltene molecular weight, namely, Gamma, Schultz-Zimm, and fractal distribution functions. It was shown that for predicting the amount of asphaltene precipitation, fractal distribution function had the lowest average absolute deviation and the best rate of convergence. To include the intermolecular interaction, Browarzik et al. [116] described the crude oils using continuous distribution functions for the solubility parameter. The crude oil was considered to consist of maltenes and asphaltenes, each described by a separate distribution function. The calculated precipitation point was found to be in excellent agreement with the experimental results. Later on, Browarzik et al. [117] proposed that the association of the asphaltenes led to a great number of aggregates with different size. Therefore a continuous size distribution function was used to describe asphaltenes to improve the performance of the proposed model for crude oils of high asphaltene content [117, 118].

Yarranton et al. [119] combined the material balance equation with the Flory-Huggins equation to predict the amount of asphaltene precipitated, similar to the approach developed by Nor-Azian and Adewumi [120]. In this model, the activity coefficients of asphaltene and solvent were used to develop the criterion for thermonamic equilibrium, which were related to the change of the chemical potentials expressed in Eqs. (1.3) and (1.4). The asphaltenes were considered as a class of PA hydrocarbons with randomly distributed functional groups. Their molar mass distribution was determined from experiments, and their molar volume as well as solubility parameter was correlated to the molar mass. The proposed method was tested on asphaltenes solvated in toluene with hexane being the precipitant. The predicted onset point and the amount of precipitate were shown to be in good agreement with experimental data. Alboudwarej et al. [121] extended this method to study the precipitation of asphaltene from mixtures of Western Canadian bitumen by *n*-alkanes. The bitumen was separated into SARA pseudo-fractions, and the molar mass distribution of asphaltenes was assumed to follow a Schultz-Zimm distribution function. Later, Akbarzadeh et al. [122] further tested this model on four international bitumen and heavy oil samples. It was shown that the predicted onset and amount of precipitation were in good agreement with the experiments in all cases. The same model was also applied by Wiehe et al. [123] to study the effect of *n*-alkane size on its amount to cause precipitation. From *n*-pentane to *n*-hexadecane, it was demonstrated that the model quantitatively described the precipitation of asphaltenes from the onset to full precipitation. In addition, application of this model to asphaltene precipitation from live oils was available [124].

While Hirschberg et al. [94] assumed that only asphaltene partitioned into the precipitated phase, in the model of Cimino et al. [125], the precipitated phase consisted of asphaltene and nonasphaltene components but the oil phase was free of asphaltene. To remove these two assumptions, Buckley and co-worker [126] used refractive indices to estimate solubility parameters of asphaltenes and solvents. This was based on their earlier work [127], where London dispersion forces were proposed to dominate aggregation and precipitation of asphaltenes. It was suggested [127] that the solubility parameter, hence the interaction parameter χ in

Eqs. (1.8) and (1.9), could be described by the refractive indices of asphaltene and the solvent. This approach was then adopted to develop a new asphaltene solubility model [126]. It was shown [126] that this model produces fairly good matches to experimental data for the onset conditions of asphaltene precipitation from seven crude oils using precipitants from pentane to pentadecane. Also in order to remove the two assumptions on the oil and precipitated phases, Mohammadi et al. [128] proposed a different approach, namely a monodisperse thermodynamic model, in which a distribution of asphaltenes and nonasphaltene components in the oil and precipitated phases was considered. It was further shown that the models developed by Hirschberg et al. [94] and Cimino et al. [125] were special cases of this new model. Later on, asphaltene aggregation was taken into account by assuming that the asphaltenes existed in solution in the form of linear hydrogen-like bonded polymers [129]. Compared with the case where the aggregation was ignored, the model provided a better agreement with experimental data.

In the original model of Hirschberg et al. [94], the solute-solvent interaction was captured through the solubility parameters derived from pure liquids, and there was no other adjustable parameters. To improve accuracy, Pazuki et al. [130] introduced an adjustable interaction parameter λ_{PS} to Eqs. (1.3) and (1.4). λ_{PS} was defined to depend on the ratio of molecular weight of asphaltene and solvent, and was obtained by optimization of an objective function on mass fractions of asphaltene precipitates. With this parameter, the model provided a good match with experimental data on asphaltene precipitation when precipitants n-pentane, n-hexane, and n-heptane were added to the crude oil sample from south west of Iran. Following a similar approach, in the work by Mofidi et al. [131], an adjustable interaction parameter was added to Eqs. (1.3) and (1.4) to calculate the activity coefficients, which was subsequently used to determine the change of Gibbs free energy. At thermodynamic equilibrium, the change of Gibbs free energy was zero. Therefore, the solubility parameters δ_a and δ_m were obtained through solving equilibrium equations instead of being quantities that must be prescribed a priori for asphaltene precipitation calculation. The adjustable interaction parameter depended on molecular weight of asphaltene and solvent, and a tuning procedure was used to determine its value. The predicted onset precipitation was shown to agree well with experimental data. In the work of Nourbakhsh et al. [132], this adjustable interaction parameter was further modified to take solvent ratio (volume of the solvent divided by the mass of the oil) into account. These above models did not separate precipitants from the crude oil and considered them as a single-phase solvent. In the work of Orangi et al. [133], a three-component system of asphaltene + solvent + precipitant was proposed. The predicted fractional precipitation agreed well with experimental data for the n-alkanes at high dilution ratios.

While majority of the above models focused on investigating asphaltene precipitation upon the addition of precipitants, the original model proposed by Hirschberg et al. [94] addressed the influence of temperature and pressure on asphaltene precipitation. Therefore theoretical models have also been developed to predict the effect of temperature and pressure. For instance, Orangi et al. [133] obtained a spinodal diagram (phase stability limits) of temperature (T) versus the volume faction of asphaltenes in the crude oil (φ_a), based on solving the phase stability limit (spinodal curve) equation

$$\frac{\partial \mu_a}{\partial \varphi_a} = 0, \tag{1.12}$$

where μ_a was obtained using Flory-Huggins theory as in Eqs. (1.3) and (1.4). Mousavi-Dehghani et al. [134] adopted Miller's combinatory entropy [135] to modify Flory-Huggins theory, and predicted the amount of asphaltene precipitation versus pressure and temperature. Furthermore, theoretical models developed based on the Flory-Huggins theory have also been employed to predict asphaltene precipitation when gas was injected, such as in the works of Hirschberg et al. [94], Mousavi-Dehghani et al. [134], and Yang et al. [136]

Models Based on Equation of State Methods

Traditional cubic EoS has been used to investigate the phase separations of crude oils where asphaltene precipitation occurs. In addition to providing critical parameters such as molar volumes and solubility parameter to other prediction models [94, 128, 137], they have also been used to directly predict the onset conditions of asphaltene precipitation and the amount of precipitates.

Du and Zhang [138] developed a thermodynamic model to predict asphaltene precipitation conditions, in which the PR-EoS was used to calculate the component fugacity coefficient and the phase compressibility. Asphaltene association was taking into account by adding a chemical contribution [139–141] (resulted from treating aggregates as "new chemical species") to the expression of fugacity coefficients, similar to the approach adopted in the work of Vafaie-Sefti et al. [142]. It was shown that the proposed model provided satisfactory predictions of asphaltene precipitation such as the onset point temperature of asphaltene precipitation with *n*-heptane injection. Assuming equilibrium between a primary liquid phase and a second dense asphaltene phase, Yarranton and co-workers [143] applied the PR-EoS to calculate the fugacity and K-values of each phase and further determine their composition. The bitumens and heavy oils were separated into four pseudocomponents corresponding to the SARA fractions. To incorporate the effects of asphaltene self-association, asphaltenes were divided into 30 pseudocomponent fractions of different molar mass based on a gamma distribution function. The model was tested on three Western Canadian oils (Athabasca, Cold Lake, and Lloydminster) and four other oils (Venezuela 1 and 2, Russia, and Indonesia). In all cases, the model successfully matched asphaltene yields for *n*-alkane-diluted bitumens at temperatures from 0 to 100 °C and pressures up to 7 MPa. Later on, an advanced PR-EoS was used by Yarranton and co-workers [144] to model asphaltene precipitation, in which a volume translation correction to better describe liquid densities was included. In this model, to fit both the saturation pressure and asphaltene onset condition, a temperature-dependent correlation was developed for the binary interaction parameters between solvent and bitumen pseudo-components.

Even though the above works considered asphaltene self-association, the cross association between asphaltenes and aromatics/resins were not included. To do so, Li et al. [145] applied a Cubic-Plus-Association EoS (CPA-EoS) model [146] to study asphaltene precipitation due to the injection of *n*-alkanes from model solution (asphaltene + toluene), and real heavy oils and bitumens (characterized by three pseudocomponents: saturates, aromatics/resins, and asphaltenes). In this model, the excess Helmholtz free energy A^{ex} was assumed to consist of two parts: the physical (non-associating) part describing molecular interactions such as short-range repulsions and dispersion attractions; and the association part describing the polar−polar cross-association of asphaltene and toluene (or aromatic/resin) molecules as well as self-association of asphaltenes. The physical and association contributions were, respectively, described by the PR-EoS and Wertheim thermodynamic perturbation theory [147–152]. It was shown that the proposed model reproduced most of the experiment results on the amount of asphaltene precipitation by different *n*-alkanes over a wide range of temperatures, pressures, and compositions. In their subsequent work [153], the application of this CPA-EoS was extended to seven live oils. Similarly in the work of Dehaghani et al. [154], a CPA-EoS was used to predict asphaltene deposition during gas injection. However, in this work, the physical compressibility factor was obtained from Soave-Redlich-Kwong EoS and the chemical part was from the generalized Anderko's model [139] that included the cross association between different components of a mixture.

Recently developed statistical associating fluid theory (SAFT) EoS by Chapman et al. [155, 156], in terms of residual Helmholtz free energy, has also been employed in predicting asphaltene phase behaviors. The original SAFT EoS modeled each molecule as a chain of freely jointed spherical segments connected through covalent bonds, and aggregates were formed by the association of chain-like molecules. Therefore, the residual Helmholtz free energy A^{res} of a mixture was described as

$$A^{res} = A^{seg} + A^{chain} + A^{assoc}, \tag{1.13}$$

where A^{seg}, A^{chain}, and A^{assoc}, respectively, represent contributions from the segment-segment interaction, i.e., Lennard-Jones (LJ) interaction, chain-forming covalent bonds, and the site-site-specific interactions among segments, for example, hydrogen bonding interactions. A^{seg} was assumed to be composed of two parts that, respectively, corresponded to the repulsive (hard sphere, A^{hs}) and attractive (dispersion, A^{disp}) parts of the potential:

$$A^{seg} = A^{hs} + A^{disp}. \tag{1.14}$$

In 2003, Chapman and co-workers [157] modeled asphaltene phase behavior with the perturbed-chain SAFT (PC-SAFT) EoS [158], in which the dispersion term was derived by applying a perturbation theory of Barker and Henderson [159, 160] to a hard-chain reference fluid. Molecular size interactions and non-polar van der Waals interactions were assumed to dominate asphaltene phase behaviors. Therefore, the SAFT association term (A^{assoc}), due to the presence of site-site specific interactions, was not considered. Three parameters were required by PC-SAFT EoS for each non-associating component of the mixture: the temperature-independent diameter of each molecular segment (σ), the number of segments per molecule (m), and the segment–segment dispersion energy (ε/k; ε is an LJ energy, d k is Boltzmann constant.). The PC-SAFT parameters for asphaltenes were obtained by fitting EoS results to experimental data from oil titrations with n-alkanes at ambient condition, while those for other subfractions in the live oil liquid phase (i.e., saturates, aromatics, and resins) were obtained from correlations with MW for each class of compound. It was shown [161, 162] that the PC-SAFT EoS properly predicted the effect of temperature, pressure, and composition on asphaltene precipitation behaviors. In the subsequent work of Chapman and co-workers [163], 3D asphaltene phase plots between pressure, temperature, and gas content were generated.

In a few of the works by Chapman and co-workers [157, 161, 163] reviewed above, asphaltenes were assumed to be monodisperse. In a separate work, the same research group [164] investigated the case where asphaltenes were considered as a polydisperse class of compounds with resins as their lower MW subfraction. Polydisperse asphaltenes were represented by four pseudocomponents in SAFT: $n - C_{15+}$, $n - C_{7-15}$, $n - C_{5-7}$, and the $n - C_{3-5}$ n-alkane-insoluble subfraction which corresponded to the resins subfraction. The SAFT parameters for each subfraction were obtained by fitting the experimental onset data at ambient conditions. In their recent work [165], a detailed approach was presented to reduce the number of fitting parameters when adopting the polydisperse asphaltene model. The analysis showed [164, 165] that the largest pseudocomponent ($n - C_{15+}$ insoluble fraction) precipitated first, followed by the precipitation of smaller pseudocomponents, consistent with the results reported in their separate work [162]. Compared with monodisperse asphaltene model, the amount of asphaltenes precipitated increased more gradually with precipitant (n-alkane) addition; as well polydisperse asphaltenes' re-dissolution started at lower precipitant volume fractions and with a higher re-dissolution amount. Adopting the gamma molar mass distribution function reported in the work of Sabbagh et al. [143], Zúñiga-Hinojosa et al. [166] considered polydisperse asphaltenes by dividing them into 30 subfractions To determine the PC-SAFT parameter, empirical correlations were proposed for σ, m, and ε/k as functions of the molar mass of each subfraction. In addition, unlike in the work of Chapman and co-workers [157, 161–165] where binary interaction parameters between different fractions in crude oil mixtures were assumed to be constants, Zúñiga-Hinojosa et al. [166] estimated the parameters for asphaltene-n-alkane binary interaction as a function of the n-alkane mass fraction. The results obtained showed [166] that the PC-SAFT EoS was able to predict the amount of asphaltene precipitates from n-alkane-diluted heavy oils and bitumens at different temperatures and pressures.

The SAFT EoS for potentials of variable range (SAFT-VR) [167] is the other emerging approach within the SAFT framework that has been used to model apshaltene precipitation behavior. Compared with the original SAFT model, the attractive potentials, and hence A^{disp} in Eq. (1.14), are allowed to have variable widths. Artola et al. [168] applied SAFT-VR EoS to investigate their model oil system, where asphaltenes were represented by polystyrene oligomers with similar MW and the lighter components in the crude oil by alkanes. The square-well (SW) potential was used to model segment-segment interaction in Eq. (1.13), and similar to the works of Chapman et al. [157, 161–165], A^{assoc} was not considered when calculating the residual Helmholtz free energy. Using this simplified approach, Artola et al. [168] generated constant-composition pressure–temperature-phase diagrams that incorporated both the bubble curve and the asphaltene precipitation boundary. Their analysis supported the view that asphaltene precipitation was initiated by a thermodynamic instability leading to liquid–liquid-phase separation. SAFT-VR EoS has also been applied in the case [169–171] where resins were believed to play important roles for asphaltene stabilization in crude oil, and its effect was taken into account by explicitly calculating A^{assoc}. However, unlike traditional colloidal theory for asphaltenes, these works [169–171] described asphaltene precipitation from crude oil solution as a liquid–liquid equilibrium process. Actually, the A^{assoc} calculated here was obtained from the Wertheim thermodynamic perturbation theory [147–152] as used in the original SAFT EoS [155, 156] and was similar to the one in the work of Li et al. [145, 153] where CPA-EoS was applied to model asphaltene precipitation.

Besides the colloidal and solubility models reviewed above, other approaches, such as the scaling equation [172–174], integral equation theory [175], and artificial neural network algorithm [176], support vector regression [177], have also been applied to predict asphaltene precipitation.

1.4 Computational Studies

While most theoretical works have been focusing on predicting asphaltene precipitation behaviors at macroscopic level, a large amount of computational studies aim at, from microscopic level, understanding the forces driving asphaltene aggregation as well as probing the aggregated structures. One of the major motivations behind these studies is that the extent of aggregation determines whether precipitation will occur. Furthermore, these computational studies can shed lights on the existing debates on the mechanisms and structures of asphaltene aggregation. Because these computational studies require knowledge of the exact chemical structures of simulated molecules, they were all based on model compounds as surrogates for real asphaltenes. According to the methods used, these works can be categorized into three groups: (1) studies using quantum mechanics (QM) approach; (2) studies using molecular mechanics (MM) and molecular dynamics (MD) approaches; and (3) studies using mesoscopic simulation techniques. The computational works reviewed here are listed in Table 1.1.

Table 1.1 Summary of computational works on aggregation: (a) QM studies, (b) MM studies, (c) MD studies, and (d) mesoscopic studies. In each table, for the works listed in the first column, the systems investigated are given in the second column, and the last three columns describe the key findings. Empty cells in the last three columns indicate no discussion on the corresponding topic from the cited work

References	Systems	Forces driving aggregations	Favorite configurations	Effect of solvent and/or external conditions
(a) QM studies (DFT approach)				
Alvarez-Ramirez et al. [178]	Island-type asphaltene dimers in vacuum	$\pi-\pi$ interaction	Face-to-face stacking of PA cores	
Moreira da Costa et al. [183]	Stacked di-, tri-, tetra-, and pentamer aggregates of island-type molecules in chloroform	$\pi-\pi$ interaction	Parallel stacking of PA cores	Chloroform weakened the stacking interaction, especially for the compounds with small PA cores.
Gray and co-workers [192, 195]	Archipelago-type dimers in chloroform	$\pi-\pi$ interaction	Parallel stacking of PA cores	
	One water molecule added to the above system	Hydrogen bonding to water introduced additional stability to the asphaltene aggregates	Parallel stacking of PA cores	Water enhanced the aggregation of model asphaltenes.
Carauta et al. [196]	Asphaltene dimers of island-type molecules in vacuum	No intermolecular hydrogen bond formed	Parallel stacking of PA cores	
(b) MM studies				
Brandt et al. [198]	Island-type asphaltene dimer in vacuum	$\pi-\pi$ interaction	Parallel stacking of PA cores	
Ortega-Rodriguez et al. [199]	Island-type asphaltene dimer in vacuum		Parallel stacking of PA cores	
	Island-type asphaltene dimer in mixture of heptane + pyridine		Parallel stacking of PA cores	Solvency power: pyridine > heptane
Pacheco-Sánchez et al. [200, 201]	Island-type asphaltene molecules in vacuum	$\pi-\pi$ interaction, which was influenced by the steric hindrance of the side chains	Face-to-face, offset π-stacking and T-shaped stacking of PA cores	

(continued)

Table 1.1 (continued)

References	Systems	Forces driving aggregations	Favorite configurations	Effect of solvent and/or external conditions
Murgich and co-workers [43, 203–205]	Island-type asphaltene molecules	$\pi - \pi$ interaction	Parallel stacking of PA cores	Solvency power: resin > toluene, n-octane
	Archipelago-type asphaltene molecules	H-bonding and bridging provided additional mechanisms driving aggregation	Complex shape including parallel stacking of PA cores, molecular tangling and inter aggregate links	Solvency power: resin > toluene, n-octane
Rogel [206]	Island-type asphaltene dimer in vacuum	$\pi - \pi$ interaction	Parallel stacking of PA cores	
(c) MD studies				
Pacheco-Sánchez et al. [207]	Ninety-six island-type asphaltene molecules in vacuum		Face-to-face, T-shaped and offset π-stacking of PA cores	High temperature could dissociate larger aggregates
Takanohashi et al. [208]	Two island-type and 1 archipelago-type molecules in vacuum	$\pi - \pi$ interaction as well as hydrogen bond		Compared with $\pi - \pi$ interactions, hydrogen bond between molecules was easily dissociated at high temperature
Takanohashi et al. [220]	Asphaltene aggregates in quinolone and toluene			Solvency power: quinolone > toluene
Boek, Headen and co-workers [209]	Six island-type asphaltene molecules in toluene or heptane		Parallel stacking of PA cores in heptane; less preference of parallel stacking in toluene	
Headen et al. [211]	Archipelago-type and island-type structures in toluene and heptane		Offset stacking	Solvency power: toluene > heptane. High temperature decreased the average aggregation time
	PMF calculations of asphaltene dimers in toluene and heptane			Similar values of free energy for dimer formation in toluene and heptane

(continued)

Table 1.1 (continued)

References	Systems	Forces driving aggregations	Favorite configurations	Effect of solvent and/or external conditions
Headen and Boek [222]	Asphaltene molecules in sc-CO_2 with and without limonene		Parallel stacking of PA cores	Adding limonene considerably decreased the size and stability of the aggregate
Rogel [210]	Two island-type molecules in mixed heptane and toluene		Parallel stacking of PA cores	Solvency power: toluene > heptane
Ungerer et al. [212]	Two island-type and one archipelago-type model asphaltenes in toluene and heptane		Significant parallel stacking of PA cores was observed for island-type molecules	Solvency power: toluene > heptane
Kuznicki et al. [213]	A mixture of archipelago-type and island-type asphaltene molecules in toluene and heptane	$\pi - \pi$ interaction	Parallel stacking of PA cores	Solvency power: toluene > heptane
Jian et al [214, 215, 224]	Island-type molecules in toluene, heptane and water	$\pi - \pi$ interaction and additional hydrophobic association in water	Parallel stacking of PA cores	Solvency power: toluene > heptane > water
Sedghi et al. [216]	Island-type molecules in heptane	$\pi - \pi$ interaction and additional aggregation driving forces introduced by hydroxyl groups in the aliphatic side chains	T-shaped configuration was predominant in the nanoaggregates formed by model asphaltene molecules without aliphatic chains while model compounds with several short side chains could only form parallel stacking configuration	Solvency power: toluene > heptane
	PMF calculations of asphaltene dimers in toluene and heptane	$\pi - \pi$ interaction		Lower association free energy (absolute value) of asphaltene in toluene compared with that in heptane

(continued)

Table 1.1 (continued)

References	Systems	Forces driving aggregations	Favorite configurations	Effect of solvent and/or external conditions
Teklebrhan et al. [217]	Island-type molecules in toluene and heptane	$\pi - \pi$ interaction, which was greatly influenced by the steric hindrance and polarity of the substituted side chains	PA core parallel stacking, T-shaped stacking, and tail-tail association	Solvency power: toluene > heptane
Carauta et al. [218]	Island-type asphaltene dimers in n-butane, isobutene, toluene, and n-heptane	$\pi - \pi$ interaction	Parallel stacking of PA cores	Solvency power: toluene > butane, isobutene >heptane. High temperature did not lower the stability of asphaltene dimers
Frigerio et al. [219]	Island-type asphaltenes in toluene, tetrahydrofurane, pentane, and heptane	$\pi - \pi$ interaction	Parallel stacking of PA cores	Solvency power: toluene, tetrahydrofurane > pentane and heptane
Pacheco-Sánchez et al. [221]	Asphaltene aggregates in toluene and pentane		Face-to-face stacking, offset π-stacking and T-shaped stacking of PA cores were all observed with offset π-stacking being the most common geometry	Solvency power: toluene > pentane
Hu et al. [223]	Asphaltene aggregates in CO_2	Pure CO_2 enhanced the hydrogen bond and the dipolar interaction between asphaltene molecules	Face-to-face stacking of PA cores in pure CO_2	CO_2 injection could enhance asphaltene aggregation in methane, ethane, and octane
Zhang et al. [225]	Model asphaltenes in asphalt mixtures	Core-core interaction, core-tail interaction, and	Parallel stacking at high temperature but	

(continued)

Table 1.1 (continued)

References	Systems	Forces driving aggregations	Favorite configurations	Effect of solvent and/or external conditions
		tail-tail interaction for model asphaltenes with long aliphatic side chains	perpendicular at low temperature; at intermediate temperatures, the most probable relative orientation between the cores near 40°	
		$\pi-\pi$ interaction for highly aromatic asphaltene molecules (with short aliphatic side chains)	Parallel stacking at low temperature and almost perpendicular at high temperature	
Aray et al. [227]	Solubility parameter calculation for asphaltenes and amphiphiles	$\pi-\pi$ interaction	Parallel stacking of PA cores in vacuum	Archipelago-type models had a large solubility in amphiphiles containing large aliphatic tails and no aromatic rings; island-type models showed high solubility in those containing aromatic rings and small aliphatic chains
(d) Mesoscopic studies				
Aguilera-Mercado et al. [229]	Asphaltene in crude oil mixture studied by CGMC method			Temperature and asphaltene concentration affected aggregation while resin concentration did not
Ortega-Rodriguez et al. [230]	Asphaltene in crude oil mixture studied by CGMD method			Solvency power: pyridine > heptane. Resins had stabilizing effect on asphaltene aggregates
Zhang et al. [231, 232]	Asphaltene in crude oil mixture studied by DPD program		Face-to-face stacking, offset π-stacking and T-shaped	

(continued)

Table 1.1 (continued)

References	Systems	Forces driving aggregations	Favorite configurations	Effect of solvent and/or external conditions
	integrated with a rotational algorithm [231]		stacking of PA cores	
	Asphaltene in crude oil mixture studied by DPD program integrated with quaternion method [232]		Face-to-face stacking, offset π-stacking and T-shaped stacking of PA cores	

1.4.1 Studies Using QM Approach

QM calculations provide numerical solutions to the Schrödinger equation and thus can accurately determine the interaction energy among atoms/molecules and further evaluate the stability of the formed structure.

Density functional theory (DFT) calculation has been used to investigate the driving forces and structure of asphaltene aggregation. Alvarez-Ramirez et al. [178] used two DFT approaches, namely Harris functional [179] with the Vosko, Wilk, Nusair exchange-correlation potential [180], and GGA Perdew-Wang 91 functional [181], to calculate the isotropic intermolecular potentials between asphaltene dimers in a vacuum, and a double-numerical basis set [182] was adopted in both cases. In this work, the Groenzin–Mullins molecular model [29, 30] was used as asphaltene surrogates; the interaction energy curves for different configurations (see a schematic representation in Fig. 1.5) of model asphaltene dimers, namely T-shaped, face-to-face, and edge-edge, were obtained by varying the minimum atomic distance between two asphaltene dimers. The results [178] showed that among the different configurations studied, the face-to-face orientation had the lowest energy, suggesting it as the most preferred configuration. Moreira da Costa et al. [183] compared the accuracy of different DFT approaches with respect to the highly accurate spin component scaled Møller–Plesset second-order perturbation theory [184, 185], and the results obtained using DFT with ωB97X-D functional [186, 187] was shown to be in closest agreement with those from the highly accurate method. Therefore this functional [188] with 6–31+G(d,p) basis set [189] was employed to evaluate the strength of π–π interaction for stacked di-, tri-, tetra-, and pentamer aggregates of fused polycyclic aromatic hydrocarbons, such as naphthalene, anthracene, phenanthrene, tetracene, pyrene, coronene, and benzene, in chloroform, and the solvent effect was taking into account using the conductor-like polarizable continuum model [190, 191]. The results suggested that [183] the polycyclic aromatic hydrocarbons containing more than four fused benzene rings would spontaneously form aggregates at 298 K while those containing fewer rings

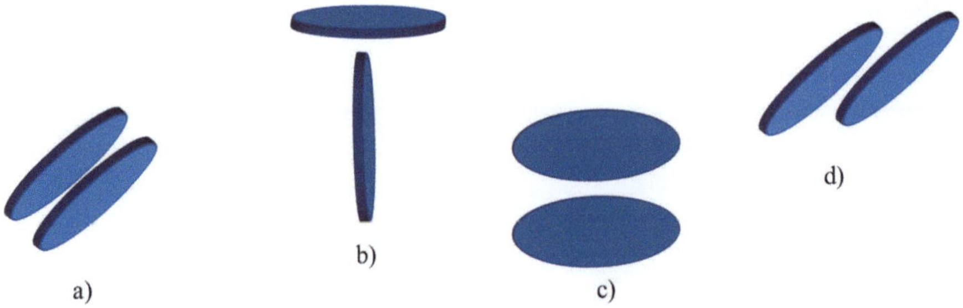

Fig. 1.5 A schematic representation of the possible aggregated geometries formed by two PA cores: (**a**) face-to-face stacking (PA cores parallel to each other with one directly on top of the other) (**b**) T-shaped stacking (PA cores perpendicular to each other), (**c**) edge-edge stacking (PA cores coplanar), (**d**) offset stacking (PA cores parallel with the two cores shifted laterally from directly on top of each other). Face-to-face stacking and offset stacking together define parallel stacking

couldn't, confirming the importance of aromatic rings for aggregation of asphaltene-like compounds.

Gray and co-workers [192] applied DFT with the ωB97Xd functional [193] and 6–31G(d,p) basis [189] set to investigate contributions of $\pi - \pi$ stacking and hydrogen-bonding interactions to the homodimerization of their ten asphaltene model compounds in chloroform, which was modeled using the polarizable continuum solvation method [194]. These model compounds contained a 2,2′-bipyridine moiety which was bonded to one (monosubstituted) or two (disubstituted) aromatic hydrocarbon moieties (phenyl, naphthyl, anthracyl, phenanthryl, and pyrenyl) through ethylene tethers. Consistent with the work [195] of the same authors performed on 4,4′-bis(2-pyren-1-yl-ethyl)-2,2′-bipyridine, it was shown that because of π–π stacking interactions, the homodimers of both mono- and disubstituted compounds were stable and had a negative Gibbs free energy change upon aggregation, demonstrating that aggregation was a spontaneous process. To explore the effect of water which may be present in solvents, one explicit water molecule was added to the simulated system. After optimization, this water molecule was found to form a bridge between the 2,2′-bipyridine moieties, and the Gibbs free energy of homodimers decreased, suggesting hydrogen bonding to water might bring additional stability to the asphaltene aggregates. However, Carauta et al. [196], using a hybrid methodology (QM:MM), did not observe the formation of intermolecular hydrogen bonds between their asphaltene dimers of island-type molecules in vacuum.

Regarding the effect of resins on asphaltene aggregation, DFT calculations have been applied to evaluate the interaction energy between resins and asphaltenes, which showed that stable structures of stacked PA cores were formed between model resin and asphaltene molecules [178, 197].

1.4.2 Studies Using MM and MD Approaches

MM and MD approaches are based on classical mechanics and thus less time consuming compared with QM approaches. In MM and MD calculations, the potential energy of the system studied is defined by a force field, which is a set of equations describing the bonded as well as nonbonded interactions between atoms. After an initial configuration is specified, MM adjusts the atomic positions through an iterative procedure until the minimum energy configuration is obtained. MD, on the other hand, evaluates forces on every atom and updates its velocity and position through the integration of Newton's equation of motion. Therefore standard MM and MD can provide energetic and structural information regarding the configurations and dynamic behaviors of asphaltene aggregation.

1.4.2.1 Studies Using MM Approach

Brandt et al. [198] assumed an island-type model structure for asphaltene molecules and constructed a dimer from two optimized unit sheets. The conformational search using MM calculations showed that the minimum energy structure was formed by parallel stacked PA cores. Ortega-Rodriguez et al. [199] applied MM to calculate the interaction energy in vacuum between two model asphaltene molecules as a function of their relative distance. These two molecules have an identical chemical structure, which consists of 630 atoms with an extended aromatic region and long aliphatic side chains. Similar to the work of Brandt et al. [198], they found that given the distance between the PA cores, the most stable molecular configuration of the dimers was of parallel stacked PA cores with opposite-oriented aliphatic chains. The interaction curve between the dimmers was then constructed for this relative orientation at different PA core separations, and the potential well was shown to be at ~0.52 nm, corresponding to the equilibrium distance between two parallel stacked PA cores. A stable aggregate of several stacking sheets was observed in the works of Pacheco-Sánchez et al. [200, 201], who investigated the aggregated structures of four island-type asphaltene molecular models in vacuum. However, T-shaped stacking was also reported [201].

MM calculations have also been performed to probe the effect of resins and solvents on asphaltene aggregation. For instance, in the work of Ortega-Rodriguez et al. [199] reviewed above, by adopting a representative molecular model for resins, the potential curve for asphaltene-resin interaction was constructed, and the potential well was shown to be at ~0.36 nm for the separation of aromatic regions of asphaltenes and resins. In the same work [199], the effect of the surrounding medium (heptane + pyridine) was taken into account by a mean-field approximation, where the Coulomb and London interaction terms of the asphaltene-asphaltene and asphaltene-resin potentials were scaled. It was shown that as the molar fraction of pyridine in the medium was increased, the equilibrium separations in both cases (asphaltene-asphaltene and asphaltene-resin) shifted to a larger distance and the depths of the potential wells decreased, indicating that pyridine tended to lower the dimer-binding energy and enhance asphaltene solubility. Murgich et al. [202] applied MM to study the interaction of two archipelago-type asphaltene molecules with a water molecule in toluene solution. Different initial configurations were constructed, each of which was energy minimized through MM calculations. It was shown that the strength of water-mediated hydrogen bonding between the two asphaltene molecules depended on the heteroatoms involved, and water molecules might generate additional mechanisms for aggregation of asphaltenes in toluene, similar to the observation of Gray and co-workers [192, 195].

Murgich and co-workers [43, 203–205] have also done a series of work to investigate asphaltene aggregation using MM calculations together with short MD simulations. For a specified starting configuration, the general procedure was to first do a minimization to relax any initial strain, followed by an MD equilibration period. Then MM minimization was performed again for the conformation obtained from MD. For island-type asphaltene molecules, it was found that the most stable configurations for asphaltene dimers, trimers, and tetramer were of parallel stacked aromatic cores; and the growth of aggregates was limited by the presence of alkyl and cycloalkyls groups in asphaltenes [203]. Analysis of the different molecular interactions confirmed the dominant role of $\pi - \pi$ interaction for asphaltene aggregates while the alkyl and cycloalkyls groups introduced steric interference to the aggregation [203]. However, for archipelago-type asphaltene molecules, the aggregates had more complex configurations, and H-bonding and bridging were proposed to be additional mechanisms driving aggregation [205]. Compared with *n*-octane and toluene, resins showed higher affinities to model asphaltenes constructed from the same oil [43, 203–205].Also, resins showed selectivity for the adsorption sites of the asphaltene, i.e., given the overall shape of asphaltene molecules and aggregates, only certain resins were allowed to adsorb [43, 203–205]. Using similar procedure (coupled MM and MD simulation), Rogel [206] built initial monomers and further constructed dimers for their model asphaltenes (island-type). It was shown in vacuum, dimers were mainly stabilized by the van der Waals interaction, i.e., $\pi - \pi$ interaction, which depended on the asphaltene structure. Adding solvents may only dissociate the asphaltene dimers of weak van der Waals interaction.

With the further development of computational capacities and MD efficiency, nowadays a more general application of MM method in studying asphaltene aggregation behaviors is to provide energy-minimized initial configurations for the subsequent MD runs.

1.4.2.2 Studies Using MD Approach

The emerging MD techniques have been widely used in probing the aggregation behaviors of asphaltenes in vacuum, organic and inorganic solvents.

Pacheco-Sánchez et al. [207] investigated the aggregation behaviors of the model asphaltene proposed by Groenzin and Mullins [29, 30] in vacuum. Initially, 96 identical model asphaltene molecules were set into an array with the neighboring sulfur-sulfur separation of ~4 nm. After energy minimization, MD simulations were conducted under 5 different temperatures: 298, 310, 340, 370, and 400 K. The final MD configurations showed that many aggregates were not in the stacking form and remained irregular shapes; for the stacked molecules, face-to-face, T-shaped, and offset π-stacked geometries formed by PA cores were all observed (see Fig. 1.5). Furthermore, as the temperature increased, dissociation of large aggregates into smaller ones was observed. Temperature-induced dissociation was also observed in the work of Takanohashi et al. [208], in which a mixture of asphaltene molecular models were adopted. These models included two island-type and one archipelago-type structures. A simulation cell containing three molecules, each representing one asphaltene molecular model, was constructed and simulated at 300 K, 523 K, and 673 K. It was shown that the hydrogen bond between asphaltene molecules dissociated at 523 K, while aromatic–aromatic stacking interactions, i.e., $\pi - \pi$ interactions, appeared to be stable at the same temperature. In addition, it was reported [208] that the removal of all aliphatic side chains and the replacement of heteroatoms with carbon atoms decreased the stability of stacking in the aggregates.

As asphaltenes are a solubility class which is soluble in aromatic solvents but insoluble in *n*-alkanes [7–9], aggregation behaviors of asphaltenes in toluene (representing aromatic solvents) and heptane (representing *n*-alkanes) have been extensively studied using MD approach. Boek, Headen, and co-workers [209] investigated the aggregation of six island-type model asphaltene molecules in toluene and heptane at 7 wt% [209]. In both solvents, no stable dimers living longer than roughly 10 ns were observed. It was suggested that in heptane, the preferred configuration between two nearest asphaltene molecules were parallel stacking, although non-parallel stacked PA cores were also reported, and in toluene, the preference of parallel stacking was less. For mixed heptane and toluene (heptol), Rogel [210] monitored the interaction energy between two island-type asphaltene monomers along the MD simulation course. The results showed the dimer formation in pure heptane, but not in heptol or in pure toluene.

In a separate work of Headen et al. [211],both island-type and archipelago structures were adopted, generated using a Monte Carlo method, and simulated in toluene and heptane, again at 7 wt%. Although similar to their previous study [209], no stable dimers/trimers were observed, aggregates lived longer in heptane. In addition, it was shown [211] that higher temperature decreased the average aggregation time. For the aggregated structures, offset stacking was reported to be the prevalent conformation in both solvents. Ungerer et al. [212] studied the aggregation behaviors of three model asphaltenes, two island types and one archipelago type, in toluene and heptane. In both solvents, significant aggregation was observed for larger island-type molecules while for smaller island-type molecules, the aggregation was stronger in heptane than in toluene. On the other hand, it was shown that the archipelago-type molecules did not evidently aggregate in heptane or toluene. Their results suggested that small island-type molecules could represent the average

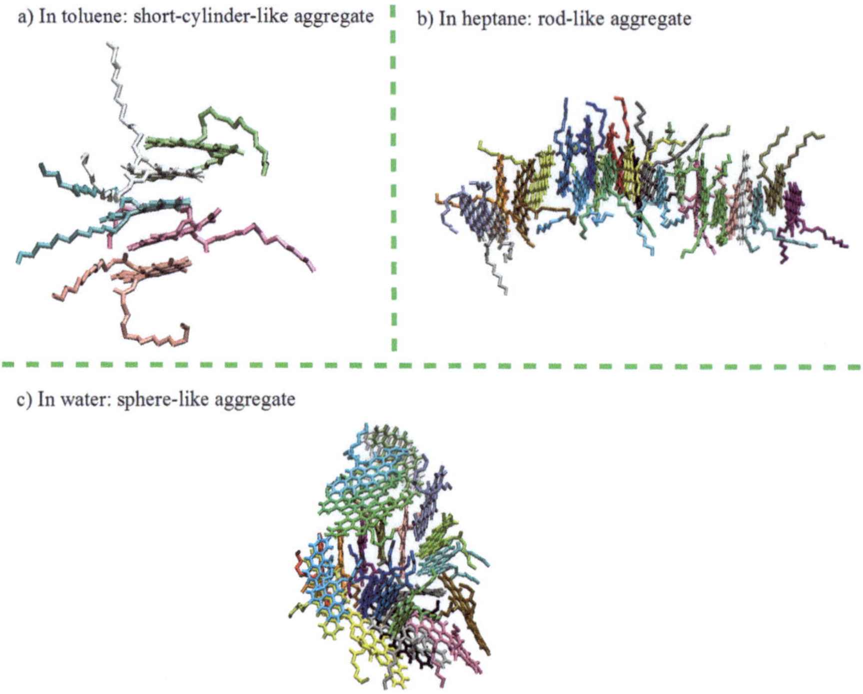

Fig. 1.6 Representative aggregates formed in the work of Jian et al. [214, 215, 224]: (**a**) in toluene, (**b**) in heptane, and (**c**) in water

structures of real asphaltenes. Aggregation of a mixture of archipelago-type and island-type asphaltene molecules in toluene and heptane were studied by Kuznicki et al. [213]. In their work, an ordered array of 12 island-type and 12 archipelago-type molecules were packed into a cubic box of edge 6.5 nm, followed by solvation. After 10 ns simulation, all 24 molecules formed a single aggregate in both solvents, and the stacking of PA cores was evident in both cases, suggesting the importance of $\pi - \pi$ interaction for aggregation. Furthermore, the stacking behaviors were reported [213] to be more pronounced in heptane compared with those in toluene.

In the work of Jian et al [214, 215], adopting an initial configuration described by Kuznicki et al. [213], long-time MD simulations were performed to investigate the aggregations of violanthrone-78 based model asphaltene molecules in toluene and heptane. Four asphaltene models were used, which had the same PA cores but different side-chain lengths. The results showed [214, 215] that distinct aggregates were formed in these two solvents (see Fig. 1.6): in toluene, the dominant short-cylinder-like aggregates were of several parallel stacked PA cores, consistent with the Yen-Mullins model [22, 25, 64]; in heptane, a one-dimensional rod-like structure was formed by almost all the 24 molecules in each system. Furthermore, independent of the side-chain lengths, the favorite configuration between two nearest asphaltene molecules was parallel stacking, confirming the importance of

$\pi - \pi$ interaction. Sedghi et al. [216] investigated the aggregation behaviors of 8 types of asphaltene-like compounds (island-type) in heptane. The concentrations of asphaltenes were kept constant at 7 wt% in all the simulations. Stable aggregates of different sizes were observed. It was proposed that T-shape configuration was predominant in the nanoaggregates formed by model asphaltene molecules without aliphatic chains while model compounds with several short aliphatic side chains could only form nanoaggregates with parallel configuration due to the steric hindrance between the side chains. Furthermore, the presence of hydroxyl groups in the aliphatic side chains was suggested [216] to help aggregation.

The effect of hydroxyl groups, present in carboxylic acid, on asphaltene aggregation was discussed in the work of Teklebrhan et al. [217], who studied the aggregation of five synthesized PA compounds in toluene and heptane. These five island-type models were of the same PA cores and differentiated by the structures of the substituent side chains. Twenty-four molecules of a single type of model compounds were dispersed in a cubic box of dimension 12 nm $\times$ 12 nm $\times$ 12 nm, and solvated by toluene or heptane. After 20 ns MD simulations, depending on the nature of the side chains attached, aggregation extents were shown to be different and aggregates of various configurations were observed including PA core parallel stacking, T-shaped stacking, and tail-tail association. For instance, the change from -COOH to -$COOCH_3$ functionality enhanced the nanoaggregate formation in toluene. Generally, it was shown that the toluene hindered the molecular association of model asphaltenes through weakening $\pi - \pi$ stacking, and aggregates grew to a much larger size in heptane than in toluene.

Besides toluene and heptane, the effects of other organic solvents have also been studied in literature. For instance, in the work of Carauta et al. [218], the effects of *n*-butane and isobutane on the dissociation of asphaltene dimers were compared with those of toluene and heptane, where model asphaltenes were of island-type architectures. It was shown that the ability of *n*-butane and isobutane to break asphaltene dimers fell in between that of toluene (most effective) and heptane (least effective). In contrast to the work of Headen et al. [211], increasing temperature did not lower the stability of asphaltene dimers in n-butane, isobutene, and toluene. Frigerio et al. [219] investigated the aggregation behaviors of their island-type asphaltenes in four solvents: toluene, tetrahydrofuran, pentane and heptane. A high concentration (20–30 wt%) was used, and the parallel stacking of PA cores were reported in all cases. Similar to the work of Jian et al. [215], a decamer of parallel stacked PA cores were observed for model asphaltene in pentane. In the work of Takanohashi et al. [220], the effects of solvent on the stability of the asphaltene aggregates were studied at 573 K. They observed that quinolone could disrupt some stacking interactions of model asphaltenes while 1-methylnaphthalene, having a similar structure to toluene, could not. On the other hand, Pacheco-Sánchez et al. [221] showed that toluene could intrigue the dissociation of asphaltenes while pentane could not.

In addition to these organic solvents, the aggregation behaviors of asphaltenes in inorganic solvents were also probed. Adopting previous simulation procedure [209, 211], Headen and Boek [222] studied the aggregation of asphaltenes in supercritical carbon dioxide (sc-CO_2) with and without limonene. At all temperatures (300, 350, and 400 K) and all pressures (100, 150 and 200 bar), an aggregate of six molecules, mainly by stacking of the aromatic cores, was quickly formed in pure sc-CO_2 with some structural fluctuations. The presence of limonene considerably decreased the size and stability of the aggregate, confirming the role of limonene as a dispersant. Hu et al. [223] investigated the effect of CO_2 on the interaction in asphaltene aggregates, and showed that pure CO_2 would enhance the hydrogen bond and the dipolar interaction between asphaltene molecules. In the work of Jian et al. [224], four types of VO-78-based model asphaltenes were simulated in pure water, which had the same PA core but different lengths of aliphatic side chains. Stable aggregates were found for all the four models studied, and the favorite configuration between two neighboring molecules was the parallel stacking of PA cores. However, compared with in toluene and heptane [214, 215], the stacking of PA cores inside each aggregate was found to be of short range and the final aggregated geometry was close to a sphere (see Fig. 1.6). More interestingly, a non-monotonic relation between the side-chain length and the size of the aggregate was observed: model asphaltenes with very long or very short side chains could fully aggregate while those with intermediate side chains could not. This suggested that the role of side chains was twofold in water. On one hand, they brought in steric hindrance to PA core stacking. On the other hand, very long side chains also introduced significant hydrophobic association which helped the aggregation.

Using MD simulations, Zhang et al. [225] investigated the molecular orientation of asphaltenes in asphalt mixtures, consisting of asphaltenes, resins and maltenes. In this work, two island-type models were adopted as asphaltene surrogates, and the resin and maltene were, respectively, represented by 1,7-dimethylnaphthalene and linear alkane n-C_{22}. It was shown that the relative orientation between two nearest asphaltene molecules was affected by their chemical structures and temperatures. For model asphaltene molecules with long aliphatic side chains, the PA cores preferred to be almost parallel at high temperature but perpendicular at low temperature, and at intermediate temperatures, the most probable relative orientation between the cores was near 40°. This was attributed to the combination of $\pi - \pi$ interaction from aromatic rings, $\pi - \theta$ interactions between the aromatic ring and aliphatic chains, and $\theta - \theta$ interaction between aliphatic chains. However for highly aromatic asphaltene molecules (with short aliphatic side chains), because of the dominant $\pi - \pi$ interaction, they preferred to stay almost parallel to each other at low temperatures and almost perpendicular at high temperatures.

MD simulations have also been conducted to calculate solubility parameters (δ) which are of great importance in solubility theory. In MD simulations, the cohesive energy E_{coh}, used to define solubility parameters, is calculated as the energy difference between per mole of materials in their condensed phase and in the ideal gas phase where all intermolecular interactions are eliminated. Then the solubility parameter δ is calculated using the following equation:

$$\delta = \left(\frac{E_{\text{coh}}}{V}\right)^{0.5}, \tag{1.15}$$

where V is the molar volume. Rogel [210] calculated the solubility parameters for the monomer, dimer, trimer, and tetramer of their model asphaltenes, each of which was treated as a "single molecule." It was showed [210] that aggregate of more molecules had a lower solubility parameter. Diallo et al. [226] investigated the effect of temperature on the solubility parameters of model asphaltenes and observed a linear decrease in δ as temperature increased. Aray et al. [227] investigated the solubility parameters of 13 island-type and 2 archipelago-type asphaltenes. In general, δ became larger as the number of aromatic rings forming the asphaltene cores increased or as the number and length of aliphatic chains decreased. In addition, the solubilities of these model asphaltenes in toluene, heptane, and amphiphiles (resins) were also studied using the Scatchard-Hildebrand and the Hansen sphere methodologies [228]. According to these theories, the maximum of solubility was observed when the solubility parameters of asphaltenes and solvents were identical. Aray et al. [227] showed that the increase of aromatic ring number decreased the asphaltene solubility in toluene and heptane. Further, archipelago-type models were observed to have a large solubility in amphiphiles containing large aliphatic tails and no aromatic rings while island-type models showed high solubility in those containing aromatic rings and small aliphatic chains [227].

Recently, MD techniques have been used to calculate the free energy of asphaltene association, which was defined as the difference in the potential of mean force (PMF) at infinite and equilibrium separations. Headen and co-workers [211] calculated the asphaltene-asphaltene PMF using two approaches: unconstrained and constraint MD simulations. In the unconstrained simulations, radial distribution function was used to calculate the PMF. In the constrained simulation, the center of mass distance between two molecules were restrained at specific distances from 0.5 to 1.5 nm with an interval of 0.1 nm; the force needed to keep molecules at a specific distance was then calculated and the PMF profile was obtained by integration of the force-distance curve. These two methods yielded similar results: values for free energy of dimer formation were very close in both toluene and heptane. This was a surprising result, as asphaltenes do exhibit very different solubility behaviors in these two solvents. A completely different result was reported by Sedghi et al. [216], where the umbrella sampling method was applied to calculate the dimerization free energy of 8 island-type asphaltenes. Their results suggested that due to the $\pi - \pi$ interaction between toluene and asphaltene, the association free energy (absolute value) of asphaltene was lower in toluene than in heptane. In addition, $\pi - \pi$ interaction was shown to be the driving force for asphaltene association, which was dependent on both the number of aromatic rings and the presence of heteroatoms in the aromatic core.

The time and length scales of the systems studied in MM and MD simulations are restricted by available computational capacity. The application of mesoscopic simulation techniques, as reviewed below, allows the study of larger and more complex systems.

1.4.3 Studies Using Mesoscopic Simulation Techniques

Compared with MM and MD simulations, mesoscopic simulation techniques adopt coarse-grained models to describe molecular structures, instead of considering all the atomic details inside each molecule. These techniques include coarse-grained Monte Carlo (CGMC), coarse-grained molecule dynamics (CGMD), and dissipative particle dynamics (DPD) methods.

Aguilera-Mercado et al. [229] modeled resin molecule as a single sphere and described asphaltene molecule as a central sphere surrounded by a ring of spheres. The intermolecular potential function for each type of spheres was obtained using the method from the work of Ortega-Rodriguez et al [199]. Moreover, the surrounding crude oil medium was modeled as a continuum, characterized by a screening factor. CGMC simulations were performed to investigate the effect of temperatures as well as asphaltene and resin concentrations on aggregation. It was shown that aggregation was affected by temperature and asphaltene concentration but was insensitive to resin concentration.

Based on the method from previous work [199], Ortega-Rodriguez et al. [230] obtained intermolecular potential functions for asphaltene-asphaltene, asphaltene-resin, and resin-resin molecules, which were subsequently used in their CGMD studies. Inside each system, asphaltene and resin molecules were modeled as dots and the solvent was taking into account using a mean-field approximation. Their results suggested that among the different interactions, asphaltene-asphaltene affinity was the strongest followed by asphaltene-resin and then resin-resin affinities. Therefore it was proposed that resins could stabilize asphaltene aggregates. Regarding the effect of solvent, heptane was shown to have a poor screening effect on the long-range interactions of asphaltenes and thus promote aggregation; pyridine was observed to have the largest screening effect and prevent aggregation.

Zhang et al. [231] implemented a rotational algorithm into the DPD program to evolve the motion of rigid bodies and investigated the aggregation of asphaltenes in model heavy crude oil consisting of SARA components and additional light fractions. For the island-type models studied, three typical aggregated structures were observed: face-to-face stacking, offset π-stacking and T-shaped stacking (see Fig. 1.5). In their subsequent work [232], three rotational algorithms were compared, and the results confirmed the superiority of the quaternion method. Therefore the aggregation of island-type asphaltenes in heavy oil and very diluted toluene was investigated by performing DPD simulation integrated with Quaternion method. It was shown that in the system of model heavy oil, although both offset π-stacking and T-shaped stacking were observed, as many as five molecules could concurrently form a well-ordered face-to-face stacking geometry. For asphaltenes in very diluted toluene (model asphaltene concentration: 0.47 g/L–1.7 g/L), the coexistence of monomers and aggregates were confirmed and aggregates were suggested to undergo the dynamics of forming, breaking up, and reforming. In the most recent work [233] of the same group, the above method was implemented on graphics processing units, which speeded up the simulations by more than tenfolds. It was shown that the determined diffusion coefficients of island-type asphaltenes in toluene were similar to those obtained in experimental studies.

1.5 Summary and Future Perspectives

In this chapter, we provided a detailed review on the theoretical and computational works performed on asphaltene aggregation and precipitation, together with the key results from relevant experimental studies. As can be seen from Sect. 1.2, great discrepancies exist regarding the molecular structure of asphaltenes found from experimental studies. Several reasons that may contribute to the existence of the discrepancies are: the usage of diverse samples from different sources, the employment of various techniques, as well as the different conditions under which the experiments were performed. These discrepancies have impeded the clear understanding of asphaltene aggregation and precipitation, and prevented the development of unified theoretical models for predicting asphaltene behaviors. As reviewed in Sect. 1.3, calibrated by different experimental data, almost all the reviewed theoretical works have been shown to agree well with the experimental observations on the specific asphaltene sample used. However, none of these models has been able to reproduce the experimental results for all the available samples in literature. Compared with theoretical works, computational studies, as reviewed in Sect. 1.4, require explicit molecular models as surrogates for real asphaltenes, thus leading to model-dependent conclusions (shown in Table 1.1). For instance, for the aggregated structures, parallel stacking and T-shaped stacking of PA cores as well as tail-tail association were all reported to be the favorite configurations. Moreover, computational results are restricted by the time and length dimensions of the systems studied. Hence caution must be taken in drawing generalized conclusions from specific studies.

Despite the discrepancies, theoretical efforts have clearly demonstrated the possibility of developing predictive models for asphaltene aggregation and precipitation. These models can be very helpful in the cases where experimental operations are extremely difficult and costly. On the other hand, with the development of computational capacities, complex and large systems that correspond to real crude oil mixtures may be simulated to generate a comprehensive picture for asphaltene aggregation from microscopic level. All these studies, although may be sample related, can help with probing the underlying mechanisms and proposing suitable ways to prevent, and further utilize, asphaltene aggregation and precipitation.

References

1. Hall, C. A., & Day, J. W., Jr. (2009). Revisiting the limits to growth after peak oil. *American Scientist, 97*, 230–237.
2. David, A. (1973). Asphaltene flocculation during solvent stimulation of heavy oils. *AIChE (American Institute of Chemical Engineers) Symposium Series, 69*, 56–61.
3. Lichaa, P. (1977). Asphaltene deposition problem in Venezuela crudes-usage of asphaltenes in emulsion stability. *Oil Sands, 609*.

4. Ali Mansoori, G. (1997). Modeling of asphaltene and other heavy organic depositions. *Journal of Petroleum Science and Engineering, 17*, 101–111.
5. Gaspar, A., Zellermann, E., Lababidi, S., Reece, J., & Schrader, W. (2012). Characterization of saturates, aromatics, resins, and asphaltenes heavy crude oil fractions by atmospheric pressure laser ionization Fourier transform ion cyclotron resonance mass spectrometry. *Energy and Fuels, 26*, 3481–3487.
6. Jewell, D., Weber, J., Bunger, J., Plancher, H., & Latham, D. (1972). Ion-exchange, coordination, and adsorption chromatographic separation of heavy-end petroleum distillates. *Analytical Chemistry, 44*, 1391–1395.
7. Mitchell, D. L., & Speight, J. G. (1973). The solubility of asphaltenes in hydrocarbon solvents. *Fuel, 52*, 149–152.
8. Permsukarome, P., Chang, C., & Fogler, H. S. (1997). Kinetic study of asphaltene dissolution in amphiphile/alkane solutions. *Industrial and Engineering Chemistry Research, 36*, 3960–3967.
9. Speight, J. G., Long, R. B., Trowbridge, T. D., & Linden, N. (1982). On the definition of asphaltenes. *American Chemical Society, Division of Petroleum Chemistry, 27*, 268–275.
10. Kokal, S. L., Sayegh, S. G., & Asphaltenes (1995). The cholesterol of petroleum. In *9th SPE Middle East Oil Show Conference*, Manama, Bahrain, March 11−14, 1995; Society of Petroleum Engineers: Richardson, Texas, SPE 29787.
11. de la Cruz, M., José, L., Argüelles-Vivas, F. J., Matías-Pérez, V., Durán-Valencia, C. de l. A., & López-Ramírez, S. (2009). Asphaltene-induced precipitation and deposition during pressure depletion on a porous medium: an experimental investigation and modeling approach. *Energy and Fuels, 23*, 5611–5625.
12. Vafaie-Sefti, M., & Mousavi-Dehghani, S. (2006). Application of association theory to the prediction of asphaltene deposition: Deposition due to natural depletion and miscible gas injection processes in petroleum reservoirs. *Fluid Phase Equilibria, 247*, 182–189.
13. Cosultchi, A., Rossbach, P., & Hernández-Calderon, I. (2003). XPS analysis of petroleum well tubing adherence. *Surface and Interface Analysis, 35*, 239–245.
14. Bartholomew, C. H. (2001). Mechanisms of catalyst deactivation. *Applied Catalysis A: General, 212*, 17–60.
15. Gawel, I., Bociarska, D., & Biskupski, P. (2005). Effect of asphaltenes on hydroprocessing of heavy oils and residua. *Applied Catalysis A: General, 295*, 89–94.
16. Park, S. J., & Ali Mansoori, G. (1988). Aggregation and deposition of heavy organics in petroleum crudes. *Energy Sources, 10*, 109–125.
17. Andersen, S. I., & Birdi, K. S. (1991). Aggregation of asphaltenes as determined by calorimetry. *Journal of Colloid and Interface Science, 142*, 497–502.
18. Leon, O., Rogel, E., Espidel, J., & Torres, G. (2000). Asphaltenes: structural characterization, self-association, and stability behavior. *Energy and Fuels, 14*, 6–10.
19. Speight, J., Wernick, D., Gould, K., Overfield, R., & Rao, B. (1985). Molecular weight and association of asphaltenes: A critical review. *Oil & Gas Science and Technology, 40*, 51–61.
20. Strausz, O. P., Safarik, I., Lown, E., & Morales-Izquierdo, A. (2008). A critique of asphaltene fluorescence decay and depolarization-based claims about molecular weight and molecular architecture. *Energy and Fuels, 22*, 1156–1166.
21. Strausz, O. P., Mojelsky, T. W., Faraji, F., Lown, E. M., & Peng, P. (1999). Additional structural details on Athabasca asphaltene and their ramifications. *Energy and Fuels, 13*, 207–227.
22. Mullins, O. C., Sabbah, H., Eyssautier, J., Pomerantz, A. E., Barré, L., Andrews, A. B., et al. (2012). Advances in asphaltene science and the Yen–Mullins model. *Energy and Fuels, 26*, 3986–4003.
23. Strausz, O. P., Mojelsky, T. W., & Lown, E. M. (1992). The molecular structure of asphaltene: an unfolding story. *Fuel, 71*, 1355–1363.
24. Dickie, J. P., & Yen, T. F. (1967). Macrostructures of the asphaltic fractions by various instrumental methods. *Analytical Chemistry, 39*, 1847–1852.

25. Mullins, O. C. (2011). The asphaltenes. *Annual Review of Analytical Chemistry, 4*, 393–418.
26. Sabbah, H., Morrow, A. L., Pomerantz, A. E., Mullins, O. C., Tan, X., Gray, M. R., et al. (2010). Comparing laser desorption/laser ionization mass spectra of asphaltenes and model compounds. *Energy and Fuels, 24*, 3589–3594.
27. Sabbah, H., Morrow, A. L., Pomerantz, A. E., & Zare, R. N. (2011). Evidence for island structures as the dominant architecture of asphaltenes. *Energy and Fuels, 25*, 1597–1604.
28. Hortal, A. R., Hurtado, P., Martínez-Haya, B., & Mullins, O. C. (2007). Molecular-weight distributions of coal and petroleum asphaltenes from laser desorption/ionization experiments. *Energy and Fuels, 21*, 2863–2868.
29. Groenzin, H., & Mullins, O. C. (1999). Asphaltene molecular size and structure. *Journal of Physical Chemistry A, 103*, 11237–11245.
30. Groenzin, H., & Mullins, O. C. (2000). Molecular size and structure of asphaltenes from various sources. *Energy and Fuels, 14*, 677–684.
31. Andrews, A. B., Guerra, R. E., Mullins, O. C., & Sen, P. N. (2006). Diffusivity of asphaltene molecules by fluorescence correlation spectroscopy. *Journal of Physical Chemistry A, 110*, 8093–8097.
32. Bergmann, U., Groenzin, H., Mullins, O. C., Glatzel, P., Fetzer, J., & Cramer, S. (2003). Carbon K-edge X-ray Raman spectroscopy supports simple, yet powerful description of aromatic hydrocarbons and asphaltenes. *Chemical Physics Letters, 369*, 184–191.
33. Pinkston, D. S., Duan, P., Gallardo, V. A., Habicht, S. C., Tan, X., Qian, K., et al. (2009). Analysis of asphaltenes and asphaltene model compounds by laser-induced acoustic desorption/Fourier transform ion cyclotron resonance mass spectrometry. *Energy and Fuels, 23*, 5564–5570.
34. Qian, K., Edwards, K. E., Siskin, M., Olmstead, W. N., Mennito, A. S., Dechert, G. J., et al. (2007). Desorption and ionization of heavy petroleum molecules and measurement of molecular weight distributions. *Energy and Fuels, 21*, 1042–1047.
35. Lisitza, N. V., Freed, D. E., Sen, P. N., & Song, Y. (2009). Study of Asphaltene Nanoaggregation by Nuclear Magnetic Resonance (NMR). *Energy and Fuels, 23*, 1189–1193.
36. Bouhadda, Y., Bormann, D., Sheu, E., Bendedouch, D., Krallafa, A., & Daaou, M. (1855–1864). Characterization of Algerian Hassi-Messaoud asphaltene structure using Raman spectrometry and X-ray diffraction. *Fuel, 2007*, 86.
37. Payzant, J., Rubinstein, I., Hogg, A., & Strausz, O. (1979). Field-ionization mass spectrometry: application to geochemical analysis. *Geochimica et Cosmochimica Acta, 43*, 1187–1193.
38. Rubinstein, I., & Strausz, O. (1979). Thermal treatment of the Athabasca oil sand bitumen and its component parts. *Geochimica et Cosmochimica Acta, 43*, 1887–1893.
39. Rubinstein, I., Spyckerelle, C., & Strausz, O. (1979). Pyrolysis of asphaltenes: a source of geochemical information. *Geochimica et Cosmochimica Acta, 43*, 1–6.
40. Calemma, V., Rausa, R., D'Anton, P., & Montanari, L. (1998). Characterization of asphaltenes molecular structure. *Energy and Fuels, 12*, 422–428.
41. Ferris, S., Black, E., & Clelland, J. (1967). Aromatic structure in asphalt fractions. *Industrial and Engineering Chemistry Product Research and Development, 6*, 127–132.
42. Su, Y., Artok, L., Murata, S., & Nomura, M. (1998). Structural analysis of the asphaltene fraction of an arabian mixture by a ruthenium-ion-catalyzed oxidation reaction. *Energy and Fuels, 12*, 1265–1271.
43. Murgich, J., Abanero, J. A., & Strausz, O. P. (1999). Molecular recognition in aggregates formed by asphaltene and resin molecules from the Athabasca oil sand. *Energy and Fuels, 13*, 278–286.
44. Artok, L., Su, Y., Hirose, Y., Hosokawa, M., Murata, S., & Nomura, M. (1999). Structure and reactivity of petroleum-derived asphaltene. *Energy and Fuels, 13*, 287–296.

45. Siskin, M., Kelemen, S., Eppig, C., Brown, L., & Afeworki, M. (2006). Asphaltene molecular structure and chemical influences on the morphology of coke produced in delayed coking. *Energy and Fuels, 20*, 1227–1234.
46. Gray, M. R. (2003). Consistency of asphaltene chemical structures with pyrolysis and coking behavior. *Energy and Fuels, 17*, 1566–1569.
47. Karimi, A., Qian, K., Olmstead, W. N., Freund, H., Yung, C., & Gray, M. R. (2011). Quantitative evidence for bridged structures in asphaltenes by thin film pyrolysis. *Energy and Fuels, 25*, 3581–3589.
48. Ralston, C. Y., Mitra-Kirtley, S., & Mullins, O. C. (1996). Small population of one to three fused-aromatic ring moieties in asphaltenes. *Energy and Fuels, 10*, 623–630.
49. Buenrostro-Gonzalez, E., Groenzin, H., Lira-Galeana, C., & Mullins, O. C. (2001). The overriding chemical principles that define asphaltenes. *Energy and Fuels, 15*, 972–978.
50. Groenzin, H., Mullins, O. C., Eser, S., Mathews, J., Yang, M., & Jones, D. (2003). Molecular size of asphaltene solubility fractions. *Energy and Fuels, 17*, 498–503.
51. Badre, S., Carla Goncalves, C., Norinaga, K., Gustavson, G., & Mullins, O. C. (2006). Molecular size and weight of asphaltene and asphaltene solubility fractions from coals, crude oils and bitumen. *Fuel, 85*, 1–11.
52. Mullins, O. (2009). C. Rebuttal to Strausz et al. regarding time-resolved fluorescence depolarization of asphaltenes. *Energy and Fuels, 23*, 2845–2854.
53. Nalwaya, V., Tantayakom, V., Piumsomboon, P., & Fogler, S. (1999). Studies on asphaltenes through analysis of polar fractions. *Industrial and Engineering Chemistry Research, 38*, 964–972.
54. Kaminski, T. J., Fogler, H. S., Wolf, N., Wattana, P., & Mairal, A. (2000). Classification of asphaltenes via fractionation and the effect of heteroatom content on dissolution kinetics. *Energy and Fuels, 14*, 25–30.
55. Fish, R. H., Komlenic, J. J., & Wines, B. K. (1984). Characterization and comparison of vanadyl and nickel compounds in heavy crude petroleums and asphaltenes by reverse-phase and size-exclusion liquid chromatography/graphite furnace atomic absorption spectrometry. *Analytical Chemistry, 56*, 2452–2460.
56. Ancheyta, J., Centeno, G., Trejo, F., Marroquin, G., Garcia, J., Tenorio, E., et al. (2002). Extraction and characterization of asphaltenes from different crude oils and solvents. *Energy and Fuels, 16*, 1121–1127.
57. George, G. N., & Gorbaty, M. L. (1989). Sulfur K-edge X-ray absorption spectroscopy of petroleum asphaltenes and model compounds. *Journal of the American Chemical Society, 111*, 3182–3186.
58. Waldo, G. S., Mullins, O. C., Penner-Hahn, J. E., & Cramer, S. (1992). Determination of the chemical environment of sulphur in petroleum asphaltenes by X-ray absorption spectroscopy. *Fuel, 71*, 53–57.
59. Kelemen, S., George, G., & Gorbaty, M. (1990). Direct determination and quantification of sulphur forms in heavy petroleum and coals: 1. The X-ray photoelectron spectroscopy (XPS) approach. *Fuel, 69*, 939–944.
60. Mitra-Kirtley, S., Mullins, O. C., Van Elp, J., George, S. J., Chen, J., & Cramer, S. P. (1993). Determination of the nitrogen chemical structures in petroleum asphaltenes using XANES spectroscopy. *Journal of the American Chemical Society, 115*, 252–258.
61. Desando, M. A., & Ripmeester, J. A. (2002). Chemical derivatization of Athabasca oil sand asphaltene for analysis of hydroxyl and carboxyl groups via nuclear magnetic resonance spectroscopy. *Fuel, 81*, 1305–1319.
62. Moschopedis, S. E., & Speight, J. G. (1976). Oxygen functions in asphaltenes. *Fuel, 55*, 334–336.
63. Ignasiak, T., Strausz, O. P., & Montgomery, D. S. (1977). Oxygen distribution and hydrogen bonding in Athabasca asphaltene. *Fuel, 56*, 359–365.
64. Mullins, O. C. (2010). The modified Yen model. *Energy and Fuels, 24*, 2179–2207.

65. Yen, T. (1974). Structure of petroleum asphaltene and its significance. *Energy Sources, 1*, 447–463.
66. Andreatta, G., Goncalves, C. C., Buffin, G., Bostrom, N., Quintella, C. M., Arteaga-Larios, F., et al. (2005). Nanoaggregates and structure-function relations in asphaltenes. *Energy and Fuels, 19*, 1282–1289.
67. Zeng, H., Song, Y., Johnson, D. L., & Mullins, O. C. (2009). Critical nanoaggregate concentration of asphaltenes by direct-current (DC) electrical conductivity. *Energy and Fuels, 23*, 1201–1208.
68. Andersen, S. I., del Rio, J. M., Khvostitchenko, D., Shakir, S., & Lira-Galeana, C. (2001). Interaction and solubilization of water by petroleum asphaltenes in organic solution. *Langmuir, 17*, 307–313.
69. Evdokimov, I., Eliseev, N. Y., & Akhmetov, B. (2003). Assembly of asphaltene molecular aggregates as studied by near-UV/visible spectroscopy: I. Structure of the absorbance spectrum. *Journal of Petroleum Science and Engineering, 37*, 135–143.
70. Evdokimov, I., Eliseev, N. Y., & Akhmetov, B. (2003). Assembly of asphaltene molecular aggregates as studied by near-UV/visible spectroscopy: II. Concentration dependencies of absorptivities. *Journal of Petroleum Science and Engineering, 37*, 145–152.
71. Hoepfner, M. P., & Fogler, H. S. (2013). Multiscale scattering investigations of asphaltene cluster breakup, nanoaggregate dissociation, and molecular ordering. *Langmuir, 29*, 15423–15432.
72. Tanaka, R., Hunt, J. E., Winans, R. E., Thiyagarajan, P., Sato, S., & Takanohashi, T. (2003). Aggregates structure analysis of petroleum asphaltenes with small-angle neutron scattering. *Energy and Fuels, 17*, 127–134.
73. Gawrys, K. L., & Kilpatrick, P. K. (2005). Asphaltenic aggregates are polydisperse oblate cylinders. *Journal of Colloid and Interface Science, 288*, 325–334.
74. Eyssautier, J., Levitz, P., Espinat, D., Jestin, J., Gummel, J., Grillo, I., et al. (2011). Insight into asphaltene nanoaggregate structure inferred by small angle neutron and X-ray scattering. *Journal of Physical Chemistry B, 115*, 6827–6837.
75. Seki, H., & Kumata, F. (2000). Structural change of petroleum asphaltenes and resins by hydrodemetallization. *Energy and Fuels, 14*, 980–985.
76. Soorghali, F., Zolghadr, A., & Ayatollahi, S. (2014). Effect of resins on asphaltene deposition and the changes of surface properties at different pressures: A microstructure study. *Energy and Fuels, 28*, 2415–2421.
77. González, G., Neves, G. B. M., Saraiva, S. M., Lucas, E. F., & Anjos de Sousa, M. d. (2003). Electrokinetic characterization of asphaltenes and the asphaltenes-resins interaction. *Energy and Fuels, 17*, 879–886.
78. Mullins, O. C., Betancourt, S. S., Cribbs, M. E., Dubost, F. X., Creek, J. L., Andrews, A. B., et al. (2007). The colloidal structure of crude oil and the structure of oil reservoirs. *Energy and Fuels, 21*, 2785–2794.
79. Sedghi, M., & Goual, L. (2009). Role of resins on asphaltene stability. *Energy and Fuels, 24*, 2275–2280.
80. Breure, B., Subramanian, D., Leys, J., Peters, C. J., & Anisimov, M. A. (2012). Modeling asphaltene aggregation with a single compound. *Energy and Fuels, 27*, 172–176.
81. Tan, X., Fenniri, H., & Gray, M. R. (2007). Pyrene derivatives of 2, 2′-Bipyridine as models for asphaltenes: synthesis, characterization, and supramolecular organization. *Energy and Fuels, 22*, 715–720.
82. Gray, M. R., Tykwinski, R. R., Stryker, J. M., & Tan, X. (2011). Supramolecular assembly model for aggregation of petroleum asphaltenes. *Energy and Fuels, 25*, 3125–3134.
83. Murgich, J. (2002). Intermolecular forces in aggregates of asphaltenes and resins. *Petroleum Science and Technology, 20*, 983–997.
84. Tan, X., Fenniri, H., & Gray, M. R. (2009). Water enhances the aggregation of model asphaltenes in solution via hydrogen bonding. *Energy and Fuels, 23*, 3687–3693.

85. Stoyanov, S. R., Yin, C., Gray, M. R., Stryker, J. M., Gusarov, S., & Kovalenko, A. (2010). Computational and experimental study of the structure, binding preferences, and spectroscopy of nickel (II) and vanadyl porphyrins in petroleum. *Journal of Physical Chemistry B, 114*, 2180–2188.
86. Rogel, E. (1928–1937). Asphaltene aggregation: a molecular thermodynamic approach. *Langmuir, 2002*, 18.
87. Rogel, E. (2004). Thermodynamic modeling of asphaltene aggregation. *Langmuir, 20*, 1003–1012.
88. Rogel, E. (2008). Molecular thermodynamic approach to the formation of mixed asphaltene—resin aggregates. *Energy and Fuels, 22*, 3922–3929.
89. Agrawala, M., & Yarranton, H. W. (2001). An asphaltene association model analogous to linear polymerization. *Industrial and Engineering Chemistry Research, 40*, 4664–4672.
90. Zhang, L., Shi, Q., Zhao, C., Zhang, N., Chung, K. H., Xu, C., et al. (2013). Hindered stepwise aggregation model for molecular weight determination of heavy petroleum fractions by vapor pressure osmometry (VPO). *Energy and Fuels, 27*, 1331–1336.
91. Zhang, L., Zhao, S., Xu, Z., Chung, K. H., Zhao, C., Zhang, N., et al. (2014). Molecular weight and aggregation of heavy petroleum fractions measured by vapor pressure osmometry and hindered stepwise aggregation model. *Energy and Fuels, 28*, 6179–6187.
92. Acevedo, S., Caetano, M., Ranaudo, M. A., & Jaimes, B. (2011). Simulation of asphaltene aggregation and related properties using an equilibrium-based mathematical model. *Energy and Fuels, 25*, 3544–3551.
93. Andersen, S. I., & Speight, J. G. (1999). Thermodynamic models for asphaltene solubility and precipitation. *Journal of Petroleum Science and Engineering, 22*, 53–66.
94. Hirschberg, A., DeJong, L., Schipper, B., & Meijer, J. (1984). Influence of temperature and pressure on asphaltene flocculation. *Society of Petroleum Engineers Journal, 24*, 283–293.
95. Leontaritis, K., & Mansoori, G. (1987). Asphaltene flocculation during oil production and processing: A thermodynamic colloidal model. *In Proceedings of the SPE International Symposium on Oilfield Chemistry,* San Antonio, Texas, Feb 4—5 1987; Society of Petroleum Engineers: Richardson, Texas. SPE 16258.
96. Flory, P. J. (1942). Thermodynamics of high polymer solutions. *Journal of Chemical Physics, 10*, 51–61.
97. Huggins, M. L. (1941). Solutions of long chain compounds. *Journal of Chemical Physics, 9*, 440–440.
98. Victorov, A. I., & Firoozabadi, A. (1996). Thermodynamic micellizatin model of asphaltene precipitation from petroleum fluids. *AICHE Journal, 42*, 1753–1764.
99. Pan, H., & Firoozabadi, A. (1998). Thermodynamic micellization model for asphaltene aggregation and precipitation in petroleum fluids. *SPE Production & Facilities, 13*, 118–127.
100. Hinze, W. L., & Pramauro, E. (1993). A critical review of surfactant-mediated phase separations (cloud-point extractions): Theory and applications. *Critical Reviews in Analytical Chemistry, 24*, 133–177.
101. Peng, D., & Robinson, D. B. (1976). A new two-constant equation of state. *Industrial and Engineering Chemistry Fundamentals, 15*, 59–64.
102. Ashoori, S., Shahsavani, B., Ahmadi, M. A., & Rezaei, A. (2014). Developing thermodynamic micellization approach to model asphaltene precipitation behavior. *Journal of Dispersion Science and Technology, 35*, 1325–1338.
103. Wu, J., Prausnitz, J. M., & Firoozabadi, A. (1998). Molecular-thermodynamic framework for asphaltene-oil equilibria. *AICHE Journal, 44*, 1188–1199.
104. Victorov, A. I., & Smirnova, N. A. (1998). Thermodynamic model of petroleum fluids containing polydisperse asphaltene aggregates. *Industrial and Engineering Chemistry Research, 37*, 3242–3251.
105. Victorov, A. I., & Smirnova, N. A. (1999). Description of asphaltene polydispersity and precipitation by means of thermodynamic model of self-assembly. *Fluid Phase Equilibria, 158*, 471–480.

106. Scott, R., & Hildebrand, J. (1951). *The solubility of nonelectrolytes*. New York: Reinhold.
107. Correra, S., & Donaggio, F. (2000). Onset-constrained colloidal asphaltene model. In *Proceedings of the International Symposium on Formation Damage*, Lafayette, LA, Feb 23–24 2000; Society of Petroleum Engineers: Richardson, Texas; SPE 58724.
108. Correra, S., & Merino-Garcia, D. (2007). Simplifying the thermodynamic modeling of asphaltenes in upstream operations. *Energy and Fuels, 21*, 1243–1247.
109. Merino-Garcia, D., & Correra, S. (2007). A shortcut application of a Flory-like model to Asphaltene precipitation. *Journal of Dispersion Science and Technology, 28*, 339–347.
110. Burke, N. E., Hobbs, R. E., & Kashou, S. F. (1990). Measurement and Modeling of Asphaltene Precipitation (includes associated paper 23831). *Journal of Petroleum Technology, 42*, 1,440–1,446.
111. Novosad, Z., & Costain, T. (1990). Experimental and modeling studies of asphaltene equilibria for a reservoir under $_{CO2}$ injection. In *Proceedings of the 65th Annual Technical Conference and Exhibition of the Society of Petroleum Engineers*, New Orleans, LA, Sept 23–26 1990; Society of Petroleum Engineers: Richardson, Texas; SPE 20530.
112. Hirschberg, A., & Hermans, L. (1984) Asphaltene phase behavior: A molecular thermodynamic model, Presented at International Symposium on Characterization of Heavy Crude Oils and Petroleum Residues, Lyon, France, June15–17; Editions Technip, 492–497.
113. Kawanaka, S., Park, S., & Mansoori, G. (1991). Organic deposition from reservoir fluids: a thermodynamic predictive technique. *SPE Reservoir Engineering, 6*, 185–192.
114. Scott, R. L., & Magat, M. (1945). The Thermodynamics of High-Polymer Solutions: I. The Free Energy of Mixing of Solvents and Polymers of Heterogeneous Distribution. *Journal of Chemical Physics, 13*, 172–177.
115. Manshad, A. K., & Edalat, M. (2008). Application of continuous polydisperse molecular thermodynamics for modeling asphaltene precipitation in crude oil systems. *Energy and Fuels, 22*, 2678–2686.
116. Browarzik, D., Laux, H., & Rahimian, I. (1999). Asphaltene flocculation in crude oil systems. *Fluid Phase Equilibria, 154*, 285–300.
117. Browarzik, C., & Browarzik, D. (2005). Modeling the onset of asphaltene flocculation at high pressure by an association model. *Petroleum Science and Technology, 23*, 795–810.
118. Browarzik, D., Kabatek, R., Kahl, H., & Laux, H. (2002). Flocculation of asphaltenes at high pressure. II. Calculation of the onset of flocculation. *Petroleum Science and Technology, 20*, 233–249.
119. Yarranton, H. W., & Masliyah, J. H. (1996). Molar mass distribution and solubility modeling of asphaltenes. *AICHE Journal, 42*, 3533–3543.
120. Nor-Azian, N., & Adewumi, M. (1993). Development of asphaltene phase equilibria predictive model. In *Proceedings of the Eastern Regional Conference and Exhibition of the SPE*, Pittsburgh, Pennsylvania, Nov 2–4, 1993; Society of Petroleum Engineers:Richardson, Texas; SPE 26905.
121. Alboudwarej, H., Akbarzadeh, K., Beck, J., Svrcek, W. Y., & Yarranton, H. W. (2003). Regular solution model for asphaltene precipitation from bitumens and solvents. *AICHE Journal, 49*, 2948–2956.
122. Akbarzadeh, K., Dhillon, A., Svrcek, W. Y., & Yarranton, H. W. (2004). Methodology for the characterization and modeling of asphaltene precipitation from heavy oils diluted with n-alkanes. *Energy and Fuels, 18*, 1434–1441.
123. Wiehe, I. A., Yarranton, H. W., Akbarzadeh, K., Rahimi, P. M., & Teclemariam, A. (2005). The paradox of asphaltene precipitation with normal paraffins. *Energy and Fuels, 19*, 1261–1267.
124. Tharanivasan, A. K., Yarranton, H. W., & Taylor, S. D. (2010). Application of a Regular Solution-Based Model to Asphaltene Precipitation from Live Oils. *Energy and Fuels, 25*, 528–538.
125. Cimino, R., Correra, S., Sacomani, P., & Carniani, C. (1995). Thermodynamic modelling for prediction of asphaltene deposition in live oils. In *The SPE International Symposium on*

Oilfield Chemistry, San Antonio, Texas, Feb 14-17, 1995; Society of Petroleum Engineers: Richardson, Texas; SPE 28993.
126. Wang, J., & Buckley, J. (2001). A two-component solubility model of the onset of asphaltene flocculation in crude oils. *Energy and Fuels, 15*, 1004–1012.
127. Buckley, J., Hirasaki, G., Liu, Y., Von Drasek, S., Wang, J., & Gill, B. (1998). Asphaltene precipitation and solvent properties of crude oils. *Petroleum Science and Technology, 16*, 251–285.
128. Mohammadi, A. H., & Richon, D. (2007). A monodisperse thermodynamic model for estimating asphaltene precipitation. *AICHE Journal, 53*, 2940–2947.
129. Mohammadi, A. H., Eslamimanesh, A., & Richon, D. (2012). Monodisperse thermodynamic model based on chemical Flory–Hüggins polymer solution theories for predicting asphaltene precipitation. *Industrial and Engineering Chemistry Research, 51*, 4041–4055.
130. Pazuki, G., & Nikookar, M. (2006). A modified Flory-Huggins model for prediction of asphaltenes precipitation in crude oil. *Fuel, 85*, 1083–1086.
131. Mofidi, A. M., & Edalat, M. (2006). A simplified thermodynamic modeling procedure for predicting asphaltene precipitation. *Fuel, 85*, 2616–2621.
132. Nourbakhsh, H., Yazdizadeh, M., & Esmaeilzadeh, F. (2011). Prediction of asphaltene precipitation by the extended Flory–Huggins model using the modified Esmaeilzadeh–Roshanfekr equation of state. *Journal of Petroleum Science and Engineering, 80*, 61–68.
133. Orangi, H. S., Modarress, H., Fazlali, A., & Namazi, M. (2006). Phase behavior of binary mixture of asphaltene solvent and ternary mixture of asphaltene solvent precipitant. *Fluid Phase Equilibria, 245*, 117–124.
134. Mousavi-Dehghani, S., Mirzayi, B., Mousavi, S., & Fasih, M. (2010). An Applied and Efficient Model for Asphaltene Precipitation In Production and Miscible Gas Injection Processes. *Petroleum Science and Technology, 28*, 113–124.
135. Miller, A. R. (1943). The Vapor-Pressure Equations Of Solutions And The Osmotic Pressure Of Rubber. *Proceedings of the Cambridge Philosophical Society, 39*, 131–131.
136. Yang, Z., Ma, C., Lin, X., Yang, J., & Guo, T. (1999). Experimental and modeling studies on the asphaltene precipitation in degassed and gas-injected reservoir oils. *Fluid Phase Equilibria, 157*, 143–158.
137. Akbarzadeh, K., Ayatollahi, S., Moshfeghian, M., Alboudwarej, H., & Yarranton, H. (2004). Estimation of SARA fraction properties with the SRK EOS. *Journal of Canadian Petroleum Technology, 43*, 31–39.
138. Du, J. L., & Zhang, D. (2004). A thermodynamic model for the prediction of asphaltene precipitation. *Petroleum Science and Technology, 22*, 1023–1033.
139. Anderko, A. (1989). Extension of the AEOS model to systems containing any number of associating and inert components. *Fluid Phase Equilibria, 50*, 21–52.
140. Ikonomou, G., & Donohue, M. (1986). Thermodynamics of hydrogen-bonded molecules: The associated perturbed anisotropic chain theory. *AICHE Journal, 32*, 1716–1725.
141. Ikonomou, G. D., & Donohue, M. D. (1988). Extension of the associated perturbed anisotropic chain theory to mixtures with more than one associating component. *Fluid Phase Equilibria, 39*, 129–159.
142. Vafaie-Sefti, M., Mousavi-Dehghani, S. A., & Mohammad-Zadeh, M. (2003). A simple model for asphaltene deposition in petroleum mixtures. *Fluid Phase Equilibria, 206*, 1–11.
143. Sabbagh, O., Akbarzadeh, K., Badamchi-Zadeh, A., Svrcek, W., & Yarranton, H. (2006). Applying the PR-EoS to asphaltene precipitation from n-alkane diluted heavy oils and bitumens. *Energy and Fuels, 20*, 625–634.
144. Agrawal, P., Schoeggl, F., Satyro, M., Taylor, S., & Yarranton, H. (2012). Measurement and modeling of the phase behavior of solvent diluted bitumens. *Fluid Phase Equilibria, 334*, 51–64.
145. Li, Z., & Firoozabadi, A. (2010). Modeling asphaltene precipitation by n-alkanes from heavy oils and bitumens using cubic-plus-association equation of state. *Energy and Fuels, 24*, 1106–1113.

146. Kontogeorgis, G. M., Voutsas, E. C., Yakoumis, I. V., & Tassios, D. P. (1996). An equation of state for associating fluids. *Industrial and Engineering Chemistry Research, 35*, 4310–4318.
147. Wertheim, M. (1984). Fluids with highly directional attractive forces. I. Statistical thermodynamics. *Journal of Statistical Physics, 35*, 19–34.
148. Wertheim, M. (1984). Fluids with highly directional attractive forces. II. Thermodynamic perturbation theory and integral equations. *Journal of Statistical Physics, 35*, 35–47.
149. Wertheim, M. (1986). Fluids of dimerizing hard spheres, and fluid mixtures of hard spheres and dispheres. *Journal of Chemical Physics, 85*, 2929–2936.
150. Wertheim, M. (1986). Fluids with highly directional attractive forces. III. Multiple attraction sites. *Journal of Statistical Physics, 42*, 459–476.
151. Wertheim, M. (1986). Fluids with highly directional attractive forces. IV. Equilibrium polymerization. *Journal of Statistical Physics, 42*, 477–492.
152. Wertheim, M. (1987). Thermodynamic perturbation theory of polymerization. *Journal of Chemical Physics, 87*, 7323–7331.
153. Li, Z., & Firoozabadi, A. (2010). Cubic-plus-association equation of state for asphaltene precipitation in live oils. *Energy and Fuels, 24*, 2956–2963.
154. Saeedi Dehaghani, A., Sefti, M. V., & Amerighasrodashti, A. (2012). The application of a new association equation of state (AEOS) for prediction of asphaltenes and resins deposition during CO_2 gas injection. *Petroleum Science and Technology, 30*, 1548–1561.
155. Chapman, W. G., Jackson, G., & Gubbins, K. E. (1988). Phase equilibria of associating fluids: chain molecules with multiple bonding sites. *Molecular Physics, 65*, 1057–1079.
156. Chapman, W. G., Gubbins, K. E., Jackson, G., & Radosz, M. (1990). New reference equation of state for associating liquids. *Industrial and Engineering Chemistry Research, 29*, 1709–1721.
157. David Ting, P., Hirasaki, G. J., & Chapman, W. G. (2003). Modeling of asphaltene phase behavior with the SAFT equation of state. *Petroleum Science and Technology, 21*, 647–661.
158. Gross, J., & Sadowski, G. (2001). Perturbed-chain SAFT: An equation of state based on a perturbation theory for chain molecules. *Industrial and Engineering Chemistry Research, 40*, 1244–1260.
159. Barker, J. A., & Henderson, D. (1967). Perturbation theory and equation of state for fluids: The square-well potential. *Journal of Chemical Physics, 47*, 2856–2861.
160. Barker, J. (1967). A.; Henderson, D. Perturbation theory and equation of state for fluids. II. A successful theory of liquids. *Journal of Chemical Physics, 47*, 4714–4721.
161. Vargas, F. M., Gonzalez, D. L., Hirasaki, G. J., & Chapman, W. G. (2009). Modeling asphaltene phase behavior in crude oil systems using the perturbed chain form of the statistical associating fluid theory (PC-SAFT) equation of state. *Energy and Fuels, 23*, 1140–1146.
162. Panuganti, S. R., Tavakkoli, M., Vargas, F. M., Gonzalez, D. L., & Chapman, W. G. (2013). SAFT model for upstream asphaltene applications. *Fluid Phase Equilibria, 359*, 2–16.
163. Panuganti, S. R., Vargas, F. M., Gonzalez, D. L., Kurup, A. S., & Chapman, W. G. (2012). PC-SAFT characterization of crude oils and modeling of asphaltene phase behavior. *Fuel, 93*, 658–669.
164. Gonzalez, D. L., Hirasaki, G. J., Creek, J., & Chapman, W. G. (2007). Modeling of asphaltene precipitation due to changes in composition using the perturbed chain statistical associating fluid theory equation of state. *Energy and Fuels, 21*, 1231–1242.
165. Tavakkoli, M., Panuganti, S. R., Taghikhani, V., Pishvaie, M. R., & Chapman, W. G. (2014). Understanding the polydisperse behavior of asphaltenes during precipitation. *Fuel, 117*, 206–217.
166. Zúñiga-Hinojosa, M. A., Justo-García, D. N., Aquino-Olivos, M. A., Román-Ramírez, L. A., & García-Sánchez, F. (2014). Modeling of asphaltene precipitation from -alkane diluted heavy oils and bitumens using the PC-SAFT equation of state. *Fluid Phase Equilibria, 376*, 210–224.

167. Gil-Villegas, A., Galindo, A., Whitehead, P. J., Mills, S. J., Jackson, G., & Burgess, A. N. (1997). Statistical associating fluid theory for chain molecules with attractive potentials of variable range. *Journal of Chemical Physics, 106*, 4168–4186.
168. Artola, P., Pereira, F. E., Adjiman, C. S., Galindo, A., Müller, E. A., Jackson, G., et al. (2011). Understanding the fluid phase behaviour of crude oil: Asphaltene precipitation. *Fluid Phase Equilibria, 306*, 129–136.
169. Wu, J., Prausnitz, J. M., & Firoozabadi, A. (2000). Molecular thermodynamics of asphaltene precipitation in reservoir fluids. *AICHE Journal, 46*, 197–209.
170. Buenrostro-Gonzalez, E., Lira-Galeana, C., Gil-Villegas, A., & Wu, J. (2004). Asphaltene precipitation in crude oils: Theory and experiments. *AICHE Journal, 50*, 2552–2570.
171. Aquino-Olivos, M. A., Grolier, J. E., Randzio, S. L., Aguirre-Gutiérrez, A. J., & García-Sánchez, F. (2013). Determination of the asphaltene precipitation envelope and bubble point pressure for a Mexican crude oil by scanning transitiometry. *Energy and Fuels, 27*, 1212–1222.
172. Rassamdana, H., Dabir, B., Nematy, M., Farhani, M., & Sahimi, M. (1996). Asphalt flocculation and deposition: I. The onset of precipitation. *AICHE Journal, 42*, 10–22.
173. Rassamdana, H., & Sahimi, M. (1996). Asphalt flocculation and deposition: II. Formation and growth of fractal aggregates. *AICHE Journal, 42*, 3318–3332.
174. Hu, Y., Chen, G., Yang, J., & Guo, T. (2000). A study on the application of scaling equation for asphaltene precipitation. *Fluid Phase Equilibria, 171*, 181–195.
175. Duda, Y., & Lira-Galeana, C. (2006). Thermodynamics of asphaltene structure and aggregation. *Fluid Phase Equilibria, 241*, 257–267.
176. Zahedi, G., Fazlali, A., Hosseini, S., Pazuki, G., & Sheikhattar, L. (2009). Prediction of asphaltene precipitation in crude oil. *Journal of Petroleum Science and Engineering, 68*, 218–222.
177. Na'imi, S. R., Gholami, A., & Asoodeh, M. (2014). Prediction of crude oil asphaltene precipitation using support vector regression. *Journal of Dispersion Science and Technology, 35*, 518–523.
178. Alvarez-Ramirez, F., Ramirez-Jaramillo, E., & Ruiz-Morales, Y. (2006). Calculation of the interaction potential curve between asphaltene-asphaltene, asphaltene-resin, and resin-resin systems using density functional theory. *Energy and Fuels, 20*, 195–204.
179. Harris, J. (1985). Simplified method for calculating the energy of weakly interacting fragments. *Physical Review B, 31*, 1770.
180. Vosko, S., Wilk, L., & Nusair, M. (1980). Accurate spin-dependent electron liquid correlation energies for local spin density calculations: a critical analysis. *Canadian Journal of Physics, 58*, 1200–1211.
181. Perdew, J. P., & Wang, Y. (1992). Accurate and simple analytic representation of the electron-gas correlation energy. *Physical Review B, 45*, 13244–13249.
182. Delley, B. (1990). An all-electron numerical method for solving the local density functional for polyatomic molecules. *Journal of Chemical Physics, 92*, 508–517.
183. Moreira da Costa, L., Stoyanov, S. R., Gusarov, S., Seidl, P. R., Carneiro, W. de M., & José, K., A. (2014). Computational study of the effect of dispersion interactions on the thermochemistry of aggregation of fused polycyclic aromatic hydrocarbons as model asphaltene compounds in solution. *Journal of Physical Chemistry A, 118*, 896–908.
184. Grimme, S. (2003). Improved second-order Møller–Plesset perturbation theory by separate scaling of parallel-and antiparallel-spin pair correlation energies. *Journal of Chemical Physics, 118*, 9095–9102.
185. Takatani, T., Hohenstein, E. G., & Sherrill, C. D. (2008). Improvement of the coupled-cluster singles and doubles method via scaling same-and opposite-spin components of the double excitation correlation energy. *Journal of Chemical Physics, 128*, 124111.
186. Frisch, M. J., Head-Gordon, M., & Pople, J. A. (1990). A direct MP2 gradient method. *Chemical Physics Letters, 166*, 275–280.

187. Head-Gordon, M., & Head-Gordon, T. (1994). Analytic MP2 frequencies without fifth-order storage. Theory and application to bifurcated hydrogen bonds in the water hexamer. *Chemical Physics Letters, 220*, 122–128.
188. Frisch, M. J., Pople, J. A., & Binkley, J. S. (1984). Self-consistent molecular orbital methods 25. Supplementary functions for Gaussian basis sets. *Journal of Chemical Physics, 80*, 3265–3269.
189. Rassolov, V. A., Ratner, M. A., Pople, J. A., Redfern, P. C., & Curtiss, L. A. (2001). 6-31G* basis set for third-row atoms. *Journal of Computational Chemistry, 22*, 976–984.
190. Barone, V., & Cossi, M. (1998). Quantum calculation of molecular energies and energy gradients in solution by a conductor solvent model. *Journal of Physical Chemistry A, 102*, 1995–2001.
191. Cossi, M., Rega, N., Scalmani, G., & Barone, V. (2003). Energies, structures, and electronic properties of molecules in solution with the C-PCM solvation model. *Journal of Computational Chemistry, 24*, 669–681.
192. da Costa, L. M., Stoyanov, S. R., Gusarov, S., Tan, X., Gray, M. R., & Stryker, J. M., et al. (2012). Density functional theory investigation of the contributions of p–p stacking and hydrogen-bonding interactions to the aggregation of model asphaltene compounds. *Energy Fuels 26*, 2727–2735.
193. Chai, J., & Head-Gordon, M. (2008). Long-range corrected hybrid density functionals with damped atom–atom dispersion corrections. *Physical Chemistry Chemical Physics, 10*, 6615–6620.
194. Tomasi, J., Mennucci, B., & Cammi, R. (2005). Quantum mechanical continuum solvation models. *Chemical Reviews, 105*, 2999–3094.
195. da Costa, L. M., Hayaki, S., Stoyanov, S. R., Gusarov, S., Tan, X., Gray, M. R., et al. (2012). 3D-RISM-KH molecular theory of solvation and density functional theory investigation of the role of water in the aggregation of model asphaltenes. *Physical Chemistry Chemical Physics, 14*, 3922–3934.
196. Carauta, A. N., Correia, J. C., Seidl, P. R., & Silva, D. M. (2005). Conformational search and dimerization study of average structures of asphaltenes. *Journal of Molecular Structure (Theochem), 755*, 1–8.
197. Castellano, O., Gimon, R., & Soscun, H. (2011). Theoretical study of the σ–π and π–π interactions in heteroaromatic monocyclic molecular complexes of benzene, pyridine, and thiophene dimers: Implications on the resin–asphaltene stability in crude oil. *Energy and Fuels, 25*, 2526–2541.
198. Brandt, H., Hendriks, E., Michels, M., & Visser, F. (1995). Thermodynamic modeling of asphaltene stacking. *The Journal of Physical Chemistry, 99*, 10430–10432.
199. Ortega-Rodriguez, A., Lira-Galeana, C., Ruiz-Morales, Y., & Cruz, S. (2001). Interaction energy in Maya-oil asphaltenes: A molecular mechanics study. *Petroleum Science and Technology, 19*, 245–256.
200. Pacheco-Sánchez, J., Alvarez-Ramírez, F., & Martínez-Magadán, J. (2003). Aggregate asphaltene structural models. *American Chemical Society, Division of Petroleum Chemistry, 48*, 71–73.
201. Pacheco-Sánchez, J., Alvarez-Ramirez, F., & Martínez-Magadán, J. (1676–1686). Morphology of aggregated asphaltene structural models. *Energy and Fuels, 2004*, 18.
202. Murgich, J., Merino-Garcia, D., Andersen, S. I., Manuel del Río, J., & Galeana, C. L. (2002). Molecular mechanics and microcalorimetric investigations of the effects of molecular water on the aggregation of asphaltenes in solutions. *Langmuir, 18*, 9080–9086.
203. Murgich, J., Rodríguez, J., & Aray, Y. (1996). Molecular recognition and molecular mechanics of micelles of some model asphaltenes and resins. *Energy and Fuels, 10*, 68–76.
204. Murgich, J., & Strausz, O. P. (2001). Molecular mechanics of aggregates of asphaltenes and resins of the Athabasca oil. *Petroleum Science and Technology, 19*, 231–243.
205. Murgich, J. (2003). Molecular simulation and the aggregation of the heavy fractions in crude oils. *Molecular Simulation, 29*, 451–461.

206. Rogel, E. (2000). Simulation of interactions in asphaltene aggregates. *Energy and Fuels, 14*, 566–574.
207. Pacheco-Sánchez, J., Zaragoza, I., & Martinez-Magadan, J. (2003). Asphaltene aggregation under vacuum at different temperatures by molecular dynamics. *Energy and Fuels, 17*, 1346–1355.
208. Takanohashi, T., Sato, S., Saito, I., & Tanaka, R. (2003). Molecular dynamics simulation of the heat-induced relaxation of asphaltene aggregates. *Energy and Fuels, 17*, 135–139.
209. Boek, E. S., Yakovlev, D. S., & Headen, T. F. (2009). Quantitative molecular representation of asphaltenes and molecular dynamics simulation of their aggregation. *Energy and Fuels, 23*, 1209–1219.
210. Rogel, E. (1995). Studies on asphaltene aggregation via computational chemistry. *Colloids and Surfaces, A: Physicochemical and Engineering Aspects, 104*, 85–93.
211. Headen, T. F., Boek, E. S., & Skipper, N. T. (2009). Evidence for asphaltene nanoaggregation in toluene and heptane from molecular dynamics simulations. *Energy and Fuels, 23*, 1220–1229.
212. Ungerer, P., Rigby, D., Leblanc, B., & Yiannourakou, M. (2014). Sensitivity of the aggregation behaviour of asphaltenes to molecular weight and structure using molecular dynamics. *Molecular Simulation, 40*, 115–122.
213. Kuznicki, T., Masliyah, J. H., & Bhattacharjee, S. (2008). Molecular dynamics study of model molecules resembling asphaltene-like structures in aqueous organic solvent systems. *Energy and Fuels, 22*, 2379–2389.
214. Jian, C., Tang, T., & Bhattacharjee, S. (2014). Molecular dynamics investigation on the aggregation of Violanthrone78-based model asphaltenes in toluene. *Energy and Fuels, 28*, 3604–3613.
215. Jian, C., & Tang, T. (2014). One-dimensional self-assembly of poly-aromatic compounds revealed by molecular dynamics simulations. *Journal of Physical Chemistry B, 118*, 12772–12780.
216. Sedghi, M., Goual, L., Welch, W., & Kubelka, J. (2013). Effect of asphaltene structure on association and aggregation using molecular dynamics. *Journal of Physical Chemistry B, 117*, 5765–5776.
217. Teklebrhan, R. B., Ge, L., Bhattacharjee, S., Xu, Z., & Sjöblom, J. (2012). Probing structure–nanoaggregation relations of polyaromatic surfactants: a molecular dynamics simulation and dynamic light scattering study. *Journal of Physical Chemistry B, 116*, 5907–5918.
218. Carauta, A. N., Seidl, P. R., Chrisman, E. C., Correia, J. C., Menechini, P. D. O., Silva, D. M., et al. (2005). Modeling solvent effects on asphaltene dimers. *Energy and Fuels, 19*, 1245–1251.
219. Frigerio, F., & Molinari, D. (2011). A multiscale approach to the simulation of asphaltenes. *Computational and Theoretical Chemistry, 975*, 76–82.
220. Takanohashi, T., Sato, S., & Tanaka, R. (2003). Molecular dynamics simulation of structural relaxation of asphaltene aggregates. *Petroleum Science and Technology, 21*, 491–505.
221. Pacheco-Sánchez, J., Zaragoza, I., & Martínez-Magadán, J. (2004). Preliminary study of the effect of pressure on asphaltene disassociation by molecular dynamics. *Petroleum Science and Technology, 22*, 927–942.
222. Headen, T. F., & Boek, E. S. (2010). Molecular dynamics simulations of asphaltene aggregation in supercritical carbon dioxide with and without limonene. *Energy and Fuels, 25*, 503–508.
223. Hu, M., Shao, C., Dong, L., & Zhu, J. (2011). Molecular dynamics simulation of asphaltene deposition during CO_2 miscible flooding. *Petroleum Science and Technology, 29*, 1274–1284.
224. Jian, C., Tang, T., & Bhattacharjee, S. (2013). Probing the effect of side-chain length on the aggregation of a model asphaltene using molecular dynamics simulations. *Energy and Fuels, 2057–2067*, 27.

225. Zhang, L., & Greenfield, M. L. (2007). Molecular orientation in model asphalts using molecular simulation. *Energy and Fuels, 21*, 1102–1111.
226. Diallo, M. S., Strachan, A., Faulon, J., & Goddard, W. A., III. (2004). Thermodynamic properties of asphaltenes through computer assisted structure elucidation and atomistic simulations 1. Bulk Arabian Light asphaltenes. *Petroleum Science and Technology, 22*, 877–899.
227. Aray, Y., Hernández-Bravo, R., Parra, J. G., Rodríguez, J., & Coll, D. S. (2011). Exploring the Structure–solubility relationship of asphaltene models in toluene, heptane, and amphiphiles using a molecular dynamic atomistic methodology. *Journal of Physical Chemistry A, 115*, 11495–11507.
228. Hansen, C. M. (2007). *Hansen solubility parameters: a user's handbook* (2nd ed.). Hoboken: CRC Press.
229. Aguilera-Mercado, B., Herdes, C., Murgich, J., & Müller, E. (2006). Mesoscopic simulation of aggregation of asphaltene and resin molecules in crude oils. *Energy and Fuels, 20*, 327–338.
230. Ortega-Rodríguez, A., Cruz, S., Gil-Villegas, A., Guevara-Rodriguez, F., & Lira-Galeana, C. (2003). Molecular view of the asphaltene aggregation behavior in asphaltene-resin mixtures. *Energy and Fuels, 17*, 1100–1108.
231. Zhang, S., Sun, L. L., Xu, J., Wu, H., & Wen, H. (2010). Aggregate structure in heavy crude oil: using a dissipative particle dynamics based mesoscale platform. *Energy and Fuels, 24*, 4312–4326.
232. Zhang, S., Xu, J., Wen, H., & Bhattacharjee, S. (2011). Integration of rotational algorithms into dissipative particle dynamics: modeling polyaromatic hydrocarbons on the meso-scale. *Molecular Physics, 109*, 1873–1888.
233. Wang, S., Xu, J., & Wen, H. (2014). The aggregation and diffusion of asphaltenes studied by GPU-accelerated dissipative particle dynamics. *Computer Physics Communications, 185*, 3069–3078.

Chapter 2
Advancement in Numerical Simulations of Gas Hydrate Dissociation in Porous Media

Zhen Liu and Xiong Yu

Abstract The amount of research on gas hydrates has been rising dramatically due to the significant role gas hydrates play as a persistent trouble for gas industry, a promising energy source, and a potential threat to environment. In the energy exploration perspective, numerical simulations play a major role in improving our understanding of the fundamentals gas hydrate dissociation as well as hydrate reservoir behaviors. This chapter presents an integrative review on the computer simulation models of gas hydrate dissociation, which have boomed since their first appearance in 1980s. Necessary background knowledge for gas hydrates and the existing investigations on this topic are firstly summarized. A unified framework is then developed for the purpose of integrating and classifying the existing models. The major mechanisms involved in the phase change process are illustrated and explained on the level of governing equations. The similarities and discrepancies among the models are demonstrated and discussed using this framework. Discussions continue on the auxiliary relationships for describing the material properties based on their categories. The various auxiliary relationships employed in the existing computational models are summarized and compared. Finally, the results obtained by previous simulations as well as other laboratory or field data are discussed. Noteworthy trends in the numerical simulations of gas hydrates behaviors are also unveiled. Recommendations are provided for future research. By providing an overview of the topic area, this chapter intends to provide scientific basis to understand the existing gas hydrate simulation models as well as serve as a guide for future research on advanced gas hydrate simulations.

Z. Liu
Department of Civil and Environmental Engineering, Michigan Technological University, 1400 Townsend Drive, Houghton, MI 49931, USA

X. Yu (✉)
Department of Civil Engineering, Case Western Reserve University, 2104 Adelbert Road, Bingham 206, Cleveland, OH 44106-7201, USA
e-mail: xxy21@case.edu

C. Jin, G. Cusatis (eds.), *New Frontiers in Oil and Gas Exploration*,
DOI 10.1007/978-3-319-40124-9_2

2.1 Introduction

This chapter intends to provide a systematic review on the theoretical basis of numerical simulation models for gas hydrates in porous media. The progresses in gas hydrate exploration have advanced to the stage that requires holistic computational simulations of reservoir behaviors. Examination of the existing literature sees a clear trend in developing advanced computational simulation models. The primary goal of this review chapter is to lay down a unified framework for mathematical models for the numerical simulations of gas hydrate dissociation as well as to integrate existing studies. Only models based on continuum mechanics are reviewed due to the fact that this type of model is more popular than alternative models such as the discrete element method and molecular dynamics. The chapter firstly reviews the history of gas hydrates research to lay down the corner stone for the subsequent introduction to numerical simulations. Existing gas hydrate simulation models are reviewed, categorized, compared, and discussed. Both analytical and numerical models are reviewed because analytical methods share a similar theoretical background with numerical ones. A new framework is proposed where the discrepancies among existing models are reconciled. Besides, auxiliary relationships needed for implementing computational simulations are summarized, compared and evaluated. The solution techniques are not included in this review due to the scope of this review as well as length limitation. Furthermore, the emphasis is placed on the dissociation process rather than the formation process considering the relevance to gas recoveries.

This chapter is composed of six sections, i.e., introduction, background, governing equation system, auxiliary relationships, discussion, and conclusion. The ***Introduction*** provides a brief technical and historical overview on gas hydrate related activities. The ***Background*** section firstly reviews the basic knowledge about gas hydrate including the definitions, physics, practical values, reserves and explorations as well as different recovery methods. This is followed by the introduction to previous research based on application types in chronological order. Then, common simulation methods for gas hydrates are described. The ***Governing Equation*** section reviews the various aspects about the governing equation system. It first discusses the basic mechanisms that might be involved in a recovery process of gas hydrates. This is followed by the establishment of a unified framework for the computer simulation models of gas hydrates. The existing models are then reviewed, categorized, and compared within this unified framework. Finally, discussions are made with respect to the chemical models and the recovery schemes that these models are associated with. In the ***Auxiliary Relationship*** section, different auxiliary relationships are discussed and categorized in the order of physical fields that they are related to. The differences and relationships among functions in the existing models are demonstrated. In the ***Discussion*** section, the performance of the existing models is evaluated. Comments are made on current simulations and disputes. Several trends which may indicate a direction for future research are revealed. The ***Conclusion*** section briefly summarizes the results of this review.

2.2 Background

2.2.1 Introduction to Gas Hydrates

Gas hydrates, or clathrate hydrates, exist in a solid, ice-like form that consists of a host lattice of water molecules enclosing cavities occupied by molecules of guest gases [1]. Common guest gases in gas hydrates include CH_4, C_2H_6, C_3H_8, i-C_4H_{10}, CO_2, and H_2S; they also include other gases such as Ne, Ar, Kr, Xe, N_2, O_2, and hydrocarbons such as cyclopropane [2, 3]. Gas hydrates can be categorized into Structure I, II, and H according to the type of cavity in the host lattice. Each cubic meter of gas hydrates can hold approximately 160 m^3 of natural gas at standard temperature and pressure [4]. Gas hydrates stay stable under certain thermodynamic conditions, i.e., low temperatures and high pressures [5, 6]. Such a condition can be provided by geologic formations such as permafrost and suboceanic sediments [7–11]. Most marine gas hydrates are formed of microbially generated gas [12]. In general, gas hydrates can contain different guest molecules in different cages, depending on their sizes and the availability of guest molecules under given thermodynamics conditions. But methane is the prevalent gas in natural gas hydrates [13]. Therefore, many studies under the name of gas hydrates are actually directed to methane hydrate.

Gas hydrates are important energy sources mainly due to the huge amount of hydrocarbons in concentrated forms they contain [12, 14]. Of primary interest are the hydrates that contain combustible low molecular weight hydrocarbons such as methane, ethane, and propane [1, 15, 16]. According to Makogon [11], there are tremendous amounts of natural gas trapped in hydrates in the permafrost and the continental shell in the ocean around the globe [7, 11]. Worldwide, gas hydrate was estimated to hold about 10^{16} kg of organic carbon in the form of methane [17]. The surveys by the US Geological Survey (USGS) have estimated that reserves of methane in hydrate form exceed all the other fossil fuel forms of organic carbon [11, 17–20]. Therefore, naturally occurring gas hydrates on the earth, containing mostly methane, have the potential to become a major source of energy in the second half of the 21st century [21]. Gas hydrates have aroused great interest in disciplines such as chemical engineering, chemistry, earth sciences, and environmental sciences [3, 22]. But in fact, gas hydrates were initially regarded as a source of problems in the energy industry because the conditions under which oil and gas are produced, transported, and processed are frequently conducive to gas hydrates formation [2, 5, 14, 23–25]. Recently, considerable concern over the potential threat of gas hydrates to the global environment has been raised because of the great green house effect of methane. It was argued that release of the large volumes of greenhouse gas stored in hydrates into the ocean and atmosphere may have played a role in the past climate change [26, 27]. Besides, rapid hydrate dissociation may lead to landslides along continental margins as well as other geohazards [3, 28, 29].

Natural hydrate deposits can be divided into four classes [30–32]. Class 1 deposits are composed of two layers: an underlying two-phase fluid zone with mobile (free) gas, and an overlying hydrate-bearing layer (HBL) containing water and hydrate (Class 1 W) or gas and hydrate (Class 1G). In Class 1 deposits, the bottom of the hydrate stability zone (HSZ) coincides with the bottom of the HBL. Class 2 deposits comprise two zones: the HBL overlying a mobile water zone. Class 3 deposits are composed of a single zone, the HBL. And this type of deposit is characterized by the absence of an underlying zone of mobile fluids. In Class 2 and 3 deposits, the entire hydrate interval may be at or within the HSZ. Class 4 deposits involve disperse, low-saturation accumulations in marine geologic media that are not bounded by confining strata and can extend over large areas. Within these four classes, Class 1 reservoirs are thought to be the easiest and probably the first type of hydrate reservoir to be produced [33]. Although most of the seafloor lies within the low-temperature and high-pressure conditions necessary for hydrate formation, hydrates are generally found in sediments along continental margins, where adequate supplies of biogenic gases are available [34, 35].

Technologies for recovering methane (CH_4) from gas hydrates reservoirs are very challenging and still under development [36]. The general concept of producing natural gas from geologic deposits of gas hydrates is to alter the reservoir environment (i.e., temperature or pressure) so that the gas hydrates transit from being thermodynamically stable to unstable [37]. The thermodynamic stability of gas hydrates is dependent upon the temperature and pressure of guest molecules as well as aqueous solute concentrations [35, 37]. Accordingly, the three most practical methods for gas hydrates harvesting are: (1) depressurization, in which the pressure of an adjacent gas phase or water phase is reduced to trigger gas hydrates decomposition; (2) thermal stimulation, in which an external source of energy is provided to increase the temperature; and (3) inhibitor injection, in which methanol or a combination of inhibitors is used to de-equilibrate the system by raising the aqueous solute concentration [8, 15, 20, 38, 39]. An alternative approach is to reduce the partial pressure of the guest molecule by introducing a substitute guest molecule, such as carbon dioxide [40]. For example, introducing carbon dioxide into geologic media filled with methane hydrate results in the displacement of methane with carbon dioxide as a guest molecule without dissociating the hydrate [41, 42]. Among these methods, thermal stimulation models are produced from laboratories [5, 39, 43, 44]. Depressurization, on the other hand, has been the method used in the field production such as in the Messoyakha field, USSR [45].

2.2.2 Existing Research on Gas Hydrates

The research on gas hydrates can be categorized based on the research subjects of the efforts and application types. In terms of the types of research subjects, the categories include: (1) basic research, which focuses on the physical properties, phase equilibrium and kinetics of gas hydrate and so on; (2) geologic explorations of natural gas hydrate resources, which focus on the geological setting, bottom simulating reflectors, geochemical anomalies around gas hydrate reservoirs and so on; (3) simulation techniques for the formation and dissociation processes, which include analytical, numerical and experimental methods; (4) hydrate-based new technologies, such as thermal energy storage, separation of gas mixtures, storage of natural gas [46]. Alternatively, previous studies on gas hydrates can be grouped based on their application types where the advances in this topic can be followed in chronological order, i.e., (1) flow assurance; (2) energy recovery; (3) hydrate-based new technologies (gas storage/transportation); and (4) environmental applications (safety and climate change) [47]. The following context is based on the second classification method.

Interest in gas hydrates was firstly sparked in the early 19th century by chemists when making hydrates of different gases, mostly as a curiosity in the laboratory [48]. In 1778, gas hydrates were first obtained by Priestley by means of bubbling SO_2 through 2 °C water at atmosphere pressure. In 1810, Sir Humphry Davy observed that a solid could be formed when an aqueous solution of chlorine was cooled to a temperature below 9 °C [49]. Faraday [50] confirmed the existence of this solid compound, the composition of which was believed to be 1 part of chlorine and 10 parts of water. It is now recognized that more than 100 species of gases can be combined with water to form nonstoichiometric solid compounds, to which the term "gas hydrates" has been applied [51]. During the first 100 years after the discovery of gas hydrates, most interest in these compounds was pure academic [3]. And the primary efforts were to understand (1) what species can form hydrates and (2) what are the thermodynamic conditions for the gas hydrates formation.

Industrial interest in gas hydrates began in the 1930s, accompanying the booming of gas and oil industry, due to the discovery that hydrate formation could plug natural gas pipelines. In 1934, Hammerschmidt noted that blockage observed in some gas-transmission pipelines was gas hydrates rather than ice [48]. Thereafter much research including extensive thermodynamic studies was carried out to understand the conditions of hydrate formation [2]. As commented by Phale et al. [14], until recently, the natural gas industry considered methane hydrates only as a nuisance, which occasionally plugs up pipelines or causes wellbores to collapse. However, this promoted intense research efforts on natural gas hydrates by industry, government, and academia. More information about these research efforts can be found in the monograph of Deaton and Frost [52].

The third period in the history of gas hydrates studies was initiated by the discovery of naturally occurring gas hydrates under different geological formations [11, 24]. In the 1960s, it was realized that clathrate hydrates of natural gas exist in vast quantities in the earth's crust [9, 53, 54]. For example, gas production from naturally occurring hydrate deposits was reported in the Messoyakha field in western Siberia where an interval saturated with gas hydrates overlaid a gas-saturated formation [48]. The discovery of naturally occurring gas hydrates coincided with the peak of a global energy crisis, which pushed forward the study of gas hydrates in the 1970s [24]. Starting in the 1970s, the search for oil and gas extended into regions which were more difficult to explore, but where geological temperatures and pressures were suitable for the formation of natural gas hydrates [6]. Natural gas hydrates, once considered merely as a nuisance in gas pipelines, were then examined as a long-term energy resource [5]. A bottom simulating reflector was commonly employed to mark the base of gas hydrates in marine sediments [55, 56]. The large amounts of gas in hydrate form justified efforts to find economic recovery schemes [57]. The need to understand gas hydrates together with other technological considerations motivated most of these research efforts [3].

Along with the investigations in gas hydrates as an energy resource, more concern regarding the influence of gas hydrates recovery or naturally dissociation on the climate change, ecosystems, and stability of gas hydrate reservoirs including wellbore safety has been expressed [26–29, 35, 48, 58–65]. For example, studies on the processes of formation and dissociation of gas hydrates in many recent studies are focused on the problems of hydrate formation in the face zones of boreholes, inside boreholes, and in pipelines [25]. Despite the concern with the potentially detrimental effects, there have also been efforts in developing innovative technologies based on the characteristics of gas hydrates for various purposes, such as natural gas storage [66–69], hydrogen storage [70, 71], thermal energy storage ([72]; McCormack, 1990; [73]), and separation of gas mixtures [74].

By reviewing the research on gas hydrates, the state of the art and the key challenges for the future advances in gas hydrates research can be summarized as follows. (1) In flow assurance research, a new approach, known as risk management is being developed to take the place of conventional methods such as thermodynamic chemical inhibitor injection due to economic and ecological concerns [61]. (2) Various aspects for gas recovery from hydrate-bearing formation are being investigated for commercial recovery due to the huge potentials. (3) The environmental impact of naturally occurring hydrates is still mostly unknown, but has been arousing increasing interest. (4) The use of hydrates to store fuel is an exciting prospect that has potential advantages over other storage materials. It should be noted that recovering gas hydrates from naturally occurring reservoirs serves as the major stimulus for various research on gas hydrates. And substantial computer simulations and laboratory experiments have been conducted while in situ explorations have just started. However, the behaviors of gas hydrate in porous media have not been completely understood. Permafrost hydrates are being considered for production tests in the USA, while efforts for marine gas hydrate are also being explored due to the huge amount of marine hydrates (by several orders of magnitude larger than that in the permafrost areas) [61].

2.2.3 Numerical Simulations of Gas Hydrates

Despite the vast reserves, gas hydrates have not been significantly exploited for energy. One of the reasons for this paradox is the limitation in our understanding of the hydrate reservoirs behaviors [75]. A large amount of literature exists on the fundamental behaviors of pure gas hydrates. However, the behaviors of hydrates in porous media, especially those during dissociation processes, are not yet completely understood. Progress is needed to understand the behaviors of gas hydrate dissociation inside pore space, especially at the reservoir scale. This will help develop efficient and economic recovering methods. Theoretical studies including analytical and numerical models need to be coordinated with laboratory studies to address the knowledge gaps that are critical to the prediction of gas production [32].

Numerous analytical methods were proposed for predicting the dissociation rate of gas hydrates to understand the feasibility of recovering gas hydrates, especially in early times. As little was known about naturally occurring hydrates at that time, relatively simple models were used to yield good "ball park" numbers in many studies [76]. Examples of these models include those proposed by Selim et al. [44], Makogon [11], Goel et al. [77], Ji et al. [78], Hong and Pooladi-Darvish [8], Hong et al. [8], and Vasil'ev et al. [79]. In these models, hydrate dissociation was generally viewed as a moving boundary ablation process similar to solid melting [80, 81]. Accordingly, these analytical models usually assumed that decomposition happens at a sharp interface that divides the medium into two regions: the hydrate zone and the dissociated zone. The analytical solutions to these models were obtained based on the use of the governing equation linearization method [79, 82]. Geramiand and Pooladi-Darvish [83] provided a good review for analytical models for the dissociation of gas hydrates.

More detailed studies employ numerical simulation studies. Numerical studies give useful information for predicting the potential and technical viability of a gas recovery process. Modeling gas production from hydrate reservoirs involves solving the coupled equations of mass and energy balances [84]. Numerical models are developed based on similar mathematical equations to those used in analytical models. However, numerical models could be much more rigorous yet more computing resource demanding as they eliminate the simplifications which are necessary for analytical solutions. The theoretical bases of these numerical models are the mathematical model consists of a governing equation system and auxiliary relationships. Governing equations describe the basic mechanisms involved in the physical process of gas dissociation or formation, such as heat transfer (thermal field), mass transfer (extended hydraulic field), conservation of momentum (extended mechanical field), and the kinetics of chemical reactions (chemical field). One governing equation is used to formulate one basic mechanism for the whole system or for one phase/component within the system. Based on the types of basic mechanisms considered, the existing numerical models for gas hydrate recovery can be categorized as Thermo-Chemical models (TC), Hydro-Chemical models (HC), Thermo-Hydro-Chemical models (THC), and Thermo-Hydro-Mechano-Chemical models (THMC). Therefore, these models are in general of a multiphysics nature. On the other hand, various auxiliary relationships are necessary to mathematically close the equation system. These auxiliary relationships represent different material properties of gas hydrates and are affiliated with different physical fields. A mathematical equation system is usually solved with numerical methods such as the finite difference method or the finite element method.

Contribution of this Review: There have been numerous reviews conducted on the various aspects of gas hydrates research. However, a direct comprehensive review on the existing computational models is rare. Mandelcorn [85] reviewed the structure, molecular and thermodynamic properties, and uses of clathrates. The review of van der Waals and Platteeuw [86] included details regarding hydrate structures, a summary of the field and an application of thermodynamics. Byk and Fomina [87] summarized the known structures of hydrates, the thermodynamics of hydrate formation, and methods of determining the composition of hydrates and so on. Hand et al. [88] presented a general view of gas hydrates. Davidson [51] offered the most comprehensive review on the properties and structures. Jeffrey and McMullan [89] and Jeffrey [90] also reviewed the structures of gas hydrate. Holder et al. [91] offered a review on the gas hydrate equilibrium predictive methods. Makogon [92] summarized the work in the Soviet Union on the thermodynamics and kinetics of hydrate formation and the formation of hydrates in porous media. Berecz and Balla-Achs [93] reviewed the literature up to 1980 with emphasis on properties which are important to engineering applications of gas hydrates. The volume edited by Cox [94] contained eight peer-reviewed papers on the properties, phase equilibriums, kinetics, occurrence, and recovery of gas hydrates. Sloan [95] presented a comprehensive overview on the clathrate hydrates of natural gas. More recent advances could also be found in reviews conducted by Englezos [3], Makogon [11], Sloan [96], Buffett [97], Koh [98], Guo [46], Waite et al. [99]. Despite the numerous numerical models that have been developed, a critical review on the theoretical basis for numerical simulations of gas hydrate recovery is unfortunately absent at this time. Such a review paper could be extremely useful for researchers by offering a bird's view and laying down directions of future research. Research of Hong et al. and Sung et al. (i.e., [8, 100]) only very briefly summaries the previous simulation models in the introduction part of their research papers. This chapter **contributes** by providing an integrative review, which is conducted within a unified theoretical framework, to stress the similarities and differences among existing models and to highlight the underlying mechanisms. This will address an urgent need in advancing computational simulations for gas hydrate utilization.

Structure of Review: A mathematical model is essential in numerical simulations because such simulations involve the solutions of the mathematical model using numerical techniques. However, the reality is in spite of the large number of investigations on hydrate dissociation in reservoirs, there still has no complete, even qualitative picture of the physical processes accompanying hydrate dissociation [79]. One major reason is that gas hydrate dissociation is a multiphysical process. This process occurs in porous media system of multiphases and multicomponents. The properties of these porous materials are not completely understood, partly due to the fact that these materials were formed in remote environments under high pressures and low temperatures. Therefore, it is needed to thoroughly understand the individual process in order to construct a complete picture of the dissociation process [8, 79].

With this in mind, we firstly categorize the various mathematical descriptions involved in the dissociation process of gas hydrates into two components: (1) equations for the basic mechanisms, which determine the form of governing equation system (i.e., the number of equations needed, the dependent variables, and the terms in each equation). (2). equations describing the auxiliary relationships, which determine the form of individual terms in the governing equations and the relationships between different terms and dependent variables (i.e., how to formulate each term, the relationships between two terms, between a term and a dependent variables, or between two dependent variables). Section 2.3 focuses on the first component, i.e., equations for the basic mechanisms; while Sect. 2.4 focuses on the second component, i.e., equations describing the auxiliary relationships. The first part of Sect. 2.3 is a review on the primary mechanisms. Based on this, a unified framework for these models is proposed. The integration of existing models within the proposed framework is then discussed.

2.3 Basic Mechanisms in Hydrate Disassociation: Governing Equation System

2.3.1 Basic Mechanisms Involved in Gas Hydrate Dissociation

The term, ***basic mechanism***, is used here to describe the fundamental processes involved in gas hydrate dissociation, which typically involves chemical, thermal, hydraulic, and mechanical fields.

The chemical field describes the dissociation reactions in gas hydrate. Chemical reactions, which describe the dissociation reaction of gas hydrates into gas and water, is the major driving force for the whole multiphysical process. These dissociation reactions can be triggered by breaking the thermodynamic equilibriums by means of lowering the pressure, raising the temperature or changing the concentrations. The chemical field is the dominant mechanism in most simulation models for gas hydrate dissociation in porous media. The mass and/or energy produced by the dissociation reactions can be established based on either an equilibrium model or a kinetic model. For equilibrium models, the computational domain is divided into an un-dissociated hydrate zone and a dissociated zone, which are separated by a boundary. Thermodynamic equilibrium is ensured on the boundary to provide one more equation for the equation system. The two zones are related by the use of conservation laws along the boundary. For kinetic models, chemical kinetics is used to predict the rate of dissociation, that is, the rate of dissociation reactions is determined by the relative positions of current state (a point) with regard to the equilibrium state (a line or a surface) in the phase diaphragm.

The thermal field (i.e., heat transfer) and hydraulic field (i.e., extended mass transfer) are two other important mechanisms frequently involved in modeling gas hydrate dissociations in the existing simulation models. For example, Sun et al. [12] proposed and implemented two different simulation schemes, i.e., the dissociation-controlled scheme (which corresponds to chemical field domination) and the flow-controlled scheme (which corresponds to hydraulic field domination). It was found that dissociation produced in the laboratory-scale experiments was often dissociation-controlled while that produced in the field-scale processes was typically flow-controlled. A few other studies claimed that the chemical field and thermal field dominated the dissociation process [1], but models established based on this viewpoint are rare.

Other than treating dissociation as a coupled chemo-thermal (TC) process or a coupled chemo-hydraulic (HC) process, researchers have also described the dissociation process of gas hydrates as a coupled chemo-thermo-hydraulic (THC) process. Maksimov [25] claimed that the process of dissociation (formation) of gas hydrates was linked to the change in the governing thermodynamic parameters (pressure and temperature) and was accompanied by transport of heat, gas, and water. Similarly, Hong et al. [8] reported that the heat transfer to the dissociation zone, intrinsic kinetics of hydrate decomposition, and gas-water two phase flow are the three primary mechanisms involved in the hydrate dissociation processes in porous media. Tonnet and Herri [101] commented that mass and heat transfers can be coupled in a complex way, firstly because of the permeability changes, and secondly due to material conduction changes.

All of the studies cited in the previous paragraphs treated the gas-bearing sediments as rigid porous media, i.e., a soil in which a hydrate is found is assumed to be non-deformable at all time [60]. While fluid flow, energy flow, and chemical kinetics were believed to be the major mechanisms governing the dissociation process of gas hydrates, the geomechanical effects could also be significant in some hydrate-bearing formations where the hydrates form parts of the solid matrix and contribute to the stiffness [48, 62, 63, 65, 102]. Kim et al.[59] reported that coupled flow and geomechanics processes play an important role in gas hydrate reservoirs because the stiffness of the rock skeleton, porosity and permeability are directly influenced by changes in fluid (water and gas) and solid (hydrate and ice) phase saturations, and in the deformation of the reservoir. The coexistence of fluid and solid phases yields a high nonlinearity for flow and mechanics, and consequently complicated the coupled problems for hydrates. The geomechanical field also needs to be involved for solving practical issues. For example, the stability of gas hydrate reservoir is related to coupled flow and geomechanics such as stability of borehole and surface facilities, hydraulic fracturing, reservoir compaction, heavy oil and oil sand production, CO_2 sequestration, and gas hydrate production [59, 103–109]. Besides, experimental observations also suggested to include the mechanical field. For example, in a series of experiments conducted by Kneafsey [4], the computerized tomography work showed significant shifting of mineral grains in response to hydrate formation and dissociation.

Based on the previous discussions, it can be concluded that energy transfer, mass transfer, geomechanical response, and dissociation reactions are the four major mechanisms which are needed to holistically describe the dissociation process of gas hydrates in porous media. Accordingly, a comprehensive Thermo-Hydro-Mechano-Chemical (THMC) model is needed in order to simulate the phenomena accompanying this type of dissociation process.

But it is also worthwhile to mention that most existing models only accounted for some of these four mechanisms, as the result of simplifications. Certain mechanisms are ignored due to focus of these research as well as their limited effects under given conditions. Besides, some mechanisms are unnecessary for a particular purpose. For example, the state-of-art simulators, such as TOUGH-Hydrate and STOMP, were developed for the simulations of gas recovery from gas hydrate reservoirs. Due to this reason, geomechanical responses were not a primary concern and thus assumed to have negligible effects on the recovery processes. Another important comment is that each individual mechanism mentioned is associated with several affiliated sub-mechanisms. For example, heat transfer typically involves conduction, convection, and radiation; mass transfer typically involves advection, diffusion, and dispersion; momentum transfer may or may not involve a convective part. For each of them, a source/sink term due to dissociation reactions may need to be included. These sub-mechanisms will be further discussed based on the mathematical equations.

2.3.2 A Unified Mathematical Framework for Different Mechanisms

In this part, we propose a unified mathematical framework. That is used to organize all the existing simulation models. All of the mechanisms mentioned in Sect. 2.3.1 can be expressed by mathematical equations. Mathematic equations for the heat transfer, mass transfer, and geomechanical responses boil down to the conservation of energy, mass, and momentum. They are all typical transport phenomena and can be formulated within the framework of continuum mechanics. The role of dissociation reactions is to determine the rate of energy/mass release or consumption in the chemical reactions, while the influence of these reactions on momentum balance is usually neglected. Therefore, the chemical field can be mathematically formulated as a source/sink term in the transport equations of heat and mass. A unified framework is introduced in this subsection to integrate the various mathematical descriptions of existing simulation models for gas hydrate dissociations. The derivation is started from global or integral balance laws as (Eq. 2.1).

$$\frac{d}{dt}\int_{\Omega} u d\mathbf{x} + \oint_{\partial\Omega} \mathbf{J} \cdot \mathbf{n} d\mathbf{x} = \int_{\Omega} Q d\mathbf{x} \tag{2.1}$$

where Ω corresponds to the Representative Element Volume (REV) in continuum mechanics, u is the generic conserved quantity defined by unit mass, $\mathbf{x}$ is the coordinate vector, $\mathbf{J}$ is the flux of this conserved quantity. Material points occupying $\Omega(0)$ may move under the influence of inertia and applied loads, so that at any time t they may occupy a volume $\Omega(t)$ which is different from $\Omega(0)$. The function $\Omega(t)$ is a material volume which is defined by a set of material points rather than by any explicit spatial coordinates.

After applying the Gauss theorem, Reynolds transport theorem, and du Bois-Reymond Lemma theorem [110], we can obtain the local balance equations as (Eq. 2.2).

$$\frac{\partial u}{\partial t} + \nabla \cdot (u\mathbf{v}) + \nabla \cdot \mathbf{J} = Q \tag{2.2}$$

where the flux, $\mathbf{J}$, can be induced by different mechanisms such as diffusion, dispersion, electromagnetic field and so on. This flux term can be formulated as (Eq. 2.3).

$$\mathbf{J} = f(t, \mathbf{x}, u, \nabla u) = -K\nabla u - D\nabla u + \ldots \tag{2.3}$$

By substituting this flux term into the local balance equation, we then obtain the general conservation equation for transport phenomena (Eq. 2.4). More mechanisms responsible for flux across the boundary can be added by extending the flux term.

$$\underbrace{\frac{\partial u}{\partial t}}_{\text{Accumulation}} \underbrace{+\nabla \cdot (u\mathbf{v})}_{\text{Advection}} \underbrace{-\nabla \cdot (K\nabla u)}_{\text{Diffusion (Conduction)}} \underbrace{-\nabla \cdot (D\nabla u)}_{\text{Dispersion}} = \underbrace{Q}_{\text{Source}} \tag{2.4}$$

The conservation equation states that the accumulation of conserved quantity is balanced with the conserved quantity advected into the control volume via mobile phases, diffused into the control volume via the aqueous and gas phases, plus the component mass associated with sources [14, 22]. For the conservation of mass, momentum and energy, the conserved quantities are ρ ($\rho = m/V$, mass per unit volume (density)), $\rho\mathbf{v}$ ($\rho\mathbf{v} = m\mathbf{v}/V$, momentum per unit volume) and e (internal energy per unit volume), respectively. The corresponding conservation equations are formulated as (Eqs. 2.5–2.7).

$$\frac{\partial \rho}{\partial t} + \nabla \cdot (\rho \mathbf{v}) + \nabla \cdot \mathbf{J}_{\mathrm{H}} = Q_{\mathrm{H}} \tag{2.5}$$

$$\frac{\partial (\rho \mathbf{v})}{\partial t} + \nabla \cdot (\rho \mathbf{v} \mathbf{v}_i) + \nabla \cdot \mathbf{J}_{\mathrm{M}} = Q_{\mathrm{M}} \tag{2.6}$$

$$\frac{\partial e}{\partial t} + \nabla \cdot (e \mathbf{v}) + \nabla \cdot \mathbf{J}_{\mathrm{T}} = Q_{\mathrm{T}} \tag{2.7}$$

where the subscripts, H, M, and T, denote the conservation of mass, momentum and energy, which are extended definitions of hydraulic, mechanical and thermal fields, respectively. The conservation of quantities and the corresponding physical fields are used interchangeably throughout this chapter. It is seen that the dependent variable for momentum conservation equation, $\rho \mathbf{v}$, is a vector. Therefore, the flux term, $\mathbf{J}_{\mathrm{M}}$, is a second order tensor and $\mathbf{J}_{\mathrm{M}} = -\boldsymbol{\sigma}$, in which $\boldsymbol{\sigma}$ is the stress tensor.

A composite such as methane hydrate, can be conceptualized as a mixture of several phases, such as gas (g), water (w), hydrate (h) and solid matrix (m). Under such conditions, the conservation equations for each phase are ($i =$ g, w, h, and m):

$$\frac{\partial (\rho_i \theta_i)}{\partial t} + \nabla \cdot (\rho_i \theta_i \mathbf{v}_i) + \nabla \cdot \mathbf{J}_{\mathrm{H},i} = Q_{\mathrm{H},i} \tag{2.8}$$

$$\frac{\partial (\rho_i \theta_i \mathbf{v}_i)}{\partial t} + \nabla \cdot (\rho_i \theta_i \mathbf{v}_i \mathbf{v}_i) + \nabla \cdot \mathbf{J}_{\mathrm{M},i} = Q_{\mathrm{M},i} \tag{2.9}$$

$$\frac{\partial (e_i \theta_i)}{\partial t} + \nabla \cdot (e_i \theta_i \mathbf{v}_i) + \nabla \cdot \mathbf{J}_{\mathrm{T},i} = Q_{\mathrm{T},i} \tag{2.10}$$

where $\theta_i \mathbf{v}_i$ is written as $\overline{\mathbf{v}}_i$ in some cases, in which $\overline{\mathbf{v}}_i$ represents the superficial velocity or Darcy's velocity. One thing needs to be pointed out is that the multiphase nature is usually solely considered for mass balance. For momentum balance, most of the existing THMC models only account for the momentum balance of the solid phase(s). If a solid phase is believed to be immobile, then the governing equation, which is in fact a Navier–Stokes equation, degenerates into a Navier's equation. Energy balance is mostly considered for the whole system instead of individual phases. This treatment is based on the assumption that thermodynamic equilibrium is maintained between different phases. In some petroleum engineering studies, the consideration of mass balance can be even more complicated. For example, the transformation between different phases is possibly considered. In such cases, a composite is a mixture of several components existing in different phases. For each component, a phase change may happen between any two phases that a component is possibly in, depending on the underlying assumptions. Accordingly, mass balance equations are usually established with respect to individual components as the following equation.

$$\frac{\partial\left(\sum_i \rho_i\theta_i\chi_i^\alpha\right)}{\partial t} + \nabla\cdot\left(\sum_i \rho_i\theta_i\chi_i^\alpha \mathbf{v}_i\right) + \nabla\cdot\left(\sum_i \chi_i^\alpha \mathbf{J}_{\mathrm{H},i}\right)$$
$$= \sum_i \chi_i^\alpha Q_{\mathrm{H},i} \tag{2.11}$$

where χ_i^α is the mass fraction of component α in phase i. It is possible that in the future the coexistence of components in different phases and concomitant phase changes will also be considered in the thermal and mechanical fields as a result of increasing computational capacities and emerging application needs.

2.3.3 Classification of Existing Methods

The total number of numerical models for simulating gas hydrate dissociation in porous media or related phenomena is unclear but this number has been increasing in a nonlinear way since 1970s. Hundreds of models can be found by a quick internet search. It can be rather confusing when one is exposed to such a great pool of models with seemingly large discrepancies. However, great similarities can be identified between different models with a careful comparison of their theoretical bases. ***Therefore, it is of great value to summary, classify and compare the existing models to elucidate how they represent the basic mechanisms and what assumptions they make***. In such a way the similarities and differences would be identified. Furthermore, uncovering the fundamental mechanisms underlying those similarities and differences will also help deepen our understanding of gas hydrate dissociation. This chapter reviews representative types of simulation models for gas hydrate dissociation currently available. These selected models are classified and compared with reference to the unified framework introduced in the previous subsection.

In existing studies, the simulation models for gas hydrate dissociation have been classified according to the solution method (analytical or numerical), application (production by depressurization, or thermal stimulation, etc.), and chemical reaction description (equilibrium or kinetic). Classifications based on these criteria helped researchers to categorize and differentiate different models. However, they are far from enough in clarifying the different mechanisms of hydrate disassociation described by the mathematical equation systems. This study proposes new criteria to classify computer simulation models for gas hydrate simulations. These criteria are established based on the unified framework described in the previous section. The key idea for establishing the criteria is to allow classify the selected models based on the simplifications employed in the models. This is based on the fact that different models are derived from the general framework using different simplifications. The four classification criteria we propose in this study are listed as follows.

1. ***Types of the governing equations***. This criterion is based on the number of independent variables used in the governing equations, or equivalently, the types of physical phenomena simulated by the model. By reviewing existing literature, it is found that most models fall into categories such as Hydro-Chemical (HC), Thermo-Chemical (TC), Thermo-Hydro-Chemical (THC), and Hydro-Thermo-Mechano-Chemical (THMC).
2. ***Number of phases and components***. Generally speaking, more phases or components require more governing equations for each field. Taking mass transfer for example, the hydraulic field can be described as single phase flow, two-phase flow, and multiple (more than two) phase flow. If phase exchange happens, the components (species) in different phases may need to be considered. Hydrate, methane, and water are usually the phases formulated in the hydraulic field. When there are more gases than just methane, or when there are other components such as inhibitors, carbon dioxide, and other hydrocarbons, or when mass exchanges between phases occur, the compositional model frequently employed in petroleum engineering has to be used for mass transfer.
3. ***Sub-mechanisms considered.*** Each term in a governing equation has a specific physical meaning. The accumulation term is necessary for any transient process while the source term is reserved for chemical reactions. All other terms represent the contribution of affiliated sub-mechanisms (such as convection, diffusion (conduction), and dispersion) in different physical fields. The inclusion of these sub-mechanisms is dependent on the relative contribution of individual item and the purpose of the simulation model. Therefore the existing models can also be differentiated by the number and type of sub-mechanisms considered.
4. ***Method for the source term***. The treatment of the source/sink generated by dissociation reactions determines whether the model is an equilibrium model or a kinetic model. This criterion was frequently adopted in existing studies. However, its physical meaning with respect to the mathematical equations has seldom been clarified.

The existing methods were also frequently classified based on their applications. This criterion is also somewhat meaningful as the recovery schemes that a model was designed for may imply certain features of that model. This is because there is, to some extent, a relationship between the recovery method and the thermodynamic conditions for the applications of that model. Simplifications may thus be made accordingly to obtain the model. For example, some models used for depressurization neglected the heat transfer process as it was assumed to be insignificant. However, this does not mean that heat transfer is not included in every simulator for depressurization. In fact, many new simulation models for gas recovery from gas hydrate dissociation are all-purpose. Hence this criterion is not as physics-based as the four criteria introduced above. In the following classification process, the recovery scheme will be marked only if it was clearly stated in the original paper.

The selected model will be classified using the proposed criteria with the following expression.

$$\mathrm{X}|imjn\ldots\cdot\mathrm{Y}|\ldots\cdot\ldots\cdot\mathrm{C}|\gamma$$

where,

Classification criterion 1: $\mathrm{X,Y = T,H,M}$ (T = Thermal, H = Hydraulic, M = Mechanical, C = Chemical);

Classification criterion 2: $i, j = \mathrm{w,g,h,s,i,c,b,a}$ (w = water, g = gas (or methane); h = hydrate; s = salt; i = ice; c = carbon dioxide; b = inhibitor; a = all components);

Classification criterion 3: $\mathrm{m,n} = 1,2,3,4$ (1 = advection; 2 = diffusion/conduction; 3 = dispersion; 4 = radiation);

Classification criterion 4: $\gamma = \varepsilon, \kappa$ (ε = equilibrium; k = kinetic).

Based on these classification criteria and the notation convention, the selected models between 1980 to 2011 for gas hydrate simulations are summarized and categorized in Table 2.1.

2.3.4 *Comparison and Integration*

The classifications presented in the last subsection grouped and marked the representative models that are currently available. However, the classifications are not to indicate that one group of models are superior to others. The judgments will be saved for readers based on their experience. One fact is that a comprehensive reservoir model without many simplifying assumptions would be extremely complex and difficult to solve numerically especially in the early stage. Sometimes simplifications could be made without significantly compromising the integrity of the model for the purposes intended [15]. Also, the uncertainty about major hydrate variables significantly limited the practice usefulness of sophisticated multidimensional multicomponent reservoir models [76]. But these factors will not put the brakes on the trend that more complicated and powerful simulators are being developed due to the growing needs and the increasing computational capacity.

2.3.4.1 Classifications Based on Criteria 1, 2, and 3

The classification based on criterion 1 divides the existing models into different categories from a viewpoint of multiphysics. Thus, the classification determines what types of physical phenomena are taken into account. As shown in Table 2.1, different types of multiphysics models have been proposed for the dissociation of gas hydrates. The major reason is that although a dissociation/formation process is a THMC or an even more complicated process, the influences of different physical

Table 2.1 Summary of the selected models between 1980 to 2011 for gas hydrate simulations

Model	Year	Classification	Additional information
Verigin et al.	1980	$H\vert g1 \cdot C\vert \varepsilon$	Depressurization, Analytical
Holder and Angert	1982	$T\vert a2 \cdot H\vert g1 \cdot C\vert \varepsilon$	Depressurization, H for gas layer only
Selim and Sloan	1985	$T\vert a2 \cdot C\vert \varepsilon$	Thermal, Analytical, Undissociated zone
Burshears et al.	1986	$T\vert a2 \cdot H\vert g1w1 \cdot C\vert \varepsilon$	Depressurization, 3D, Penetrating well
Selim and Sloan	1990	$T\vert a2 \cdot H\vert g1 \cdot C\vert \varepsilon$	Thermal, Analytical, H for dissociated zone only
Yousif et al.	1991	$H\vert g1w1h \cdot C\vert \kappa$	Depressurization, 1D
Maksimov	1992	$T\vert a2 \cdot H\vert g1w1 \cdot C\vert \varepsilon$	Analytical, Dissociated and undissociated zones
Tsypkin	1993	$T\vert a2 \cdot H\vert g1w1 \cdot C\vert \varepsilon$	Analytical, phase change, three zones
Rempel and Buffett	1997	$T\vert a12 \cdot H\vert g123 \cdot C\vert \varepsilon\kappa$	Analytical with ε; Numerical with κ, Formation and dissociation
Tsypkin	1998	$T\vert a12 \cdot H\vert g1w1 \cdot C\vert \varepsilon$	Analytical, phase change, three zones
Sung et al.	2000	$H\vert g1w1h \cdot C\vert \kappa$	Depressurization, 1D, Gas dissolves in water
Goel et al.	2001	$H\vert g1 \cdot C\vert \kappa$	Analytical, Dissociated and undissociated zones, Axisymmetric, Depressurization
Masuda et al.	2002	$T\vert a12 \cdot H\vert g12w12h \cdot C\vert \kappa$	The three conventional recovery schemes were tested
Jeannin et al.	2002	$T\vert a2 \cdot H\vert g12w12 \cdot C\vert \kappa$	3D, Formation and dissociation
Sung et al.	2002	$H\vert g1w1b1h \cdot C\vert \kappa$	Depressurization and inhibitor injection, Gas and inhibitor dissolve in water
Hong et al.	2003	$T\vert a2, H\vert a1, C\vert \kappa$	Analytical, solutions to 3 regimes
Ahmadi et al.	2004	$T\vert a12 \cdot H\vert g1 \cdot C\vert \varepsilon$	TH for two zones and linked by C
Chen et al.	2005	$T\vert a12 \cdot H\vert g1s1 \cdot C\vert \varepsilon$	Salt is considered
Li et al.	2005	$H\vert g1w1h \cdot C\vert \kappa$	Depressurization, Compared with Yousif et al. [45], parametric study
Klar and Soga	2005	$H\vert g1w1h \cdot M\vert m1 \cdot C\vert \kappa$	Bishop's model used for unsaturated mechanics
Pawar and Zyvoloski	2005	$T\vert a12 \cdot H\vert g1w1h \cdot C\vert \kappa$	Hot water injection and depressurization
Hong and Pooladi-Darvish	2005	$T\vert a12 \cdot H\vert g1w1h \cdot C\vert \kappa$	2D cylindrical simulator, penetrating well though two layers
Sun et al.	2005	$T\vert a12 \cdot H\vert g1w1h \cdot C\vert \kappa$	Depressurization, 1D

(continued)

Table 2.1 (continued)

Model	Year	Classification	Additional information
Phale et al.	2006	T|a123 · H|g123w123c123s123 · C|?	Injection of CO_2 Microemulsion, STOMP-HYD, 6 phases 4 components, Phase change
Scott et al.	2006	T|a12 · H|g1w1 · C|ε	Hot water injection, FEM
Sun and Mohanty	2006	T|a12 · H|g12w12s12h · C|κ	PVSM for ice melting, 4 component 5 phases
Vasil'ev et al.	2006	T|a12 · H|g1w1 · C|ε	Gas-hydrate (g) and gas-water regions (g, w), analytical and numerical
Sean et al.	2007	T|w12h2 · H|g12w · M|w1 · C|κ	Dissociation under water flow, for testing a proposed kinetic model
Ahmadi et al.	2007	T|a12 · H|g1w1 · C|ε	Depressurization, axisymmetric, T|a1 for gas zone, T|a12 for hydrate zone
Du et al.	2007	T|a12 · H|g1w1h · C|κ	Thermal stimulation
Kimoto et al.	2007	T|a2 · H|g1w1hm · M|g1w1m1 · C|κ	Viscoplastic
Liu and Flemings	2007	T|a12 · H|g12w12s12 · C|ε	Switch primary variables (ε)
Nazridoust and Ahmadi	2007	T|a12 · H|g1w1h · C|κ	Depressurization, FLUENT, axisymmetric
Tsypkin	2007	T|a2 · H|g1w1 · C|ε	Isothermal and nonisothermal, developed for reservoir with high pressure gradients
Tsimpanogiannis and Lichtner	2007	T|a2 · H|g1 · C|ε	H is only for dissociated zone, parametric study
Bai et al.	2008	T|a12 · H|g1w1h · C|κ	Ice was considered as a part of water
Esmaeilzadeh et al.	2008	T|a12 · H|g1w1h · C|κ	Depressurization and thermal stimulation
Garg et al.	2008	T|a12 · H|g1w1s1m · M|(hm)1 · C|ε	Organic carbon is considered in H, kinetics for organic carbon dissociation, formation and dissociation
Liu et al.	2008	T|a12 · H|g1w1 · C|ε	Depressurization, No water in hydrate zone, Class III hydrate accumulation
Ng and Klar	2008	T|a12 · H|g1w1h · M|m1 · C|κ	Use FLAC, based on Klar and Soga [60]
Uddin et al.	2008	T|a12 · H|g1w1c1h · C|κ	CO_2 sequestration, Formation and dissociation, Phase change, h includes CH_4 hydrate and CO_2 hydrate

White and McGrail	2008	T|a123 · H|g123w123c123i123 · C|εκ	STOMP, Equilibrium or kinetic
Bai et al.	2009	T|a12 · H|g1w1h · C|κ	ice formation (switching primary variables), depressurization
Rutqvist et al.	2009	T|a123 · H|g123w123c123i123 · M|g1w1m1 · C|εκ	depressurization , TOUGH-Hydrate + FLAC3D
Salehabadi et al.	2008	T|a12 · H|w1 · M|m1 · C|ε	T is coupled to HM, but the coupling is weak
Tonnet and Herri	2009	T|a12 · H|g1w1h · C|κ	Depressurization, Gas dissolves in water
White et al.	2009	T|a123 · H|g123w123c123i123 · C|εκ	CO_2 sequestration, Class 1 gas hydrate
Bai and Li	2010	T|a12 · H|g1w1h · C|κ	Ice was considered as a part of water, Combination of warm water flooding and depressurization
Gamwo and Liu	2010	T|a124 · H|g12w12h · C|εκ	Three phases three components, Consider phase change, HydrateResSim
Konno et al.	2010	T|a12 · H|g1w1s1i1n1 · C|κ	MH21-HRATES, 4 phases 5 components
Kwon et al.	2010	T|a2 · H|w1 · M|m1 · C|ε	FLAC2D, one-way coupling
Peszynska et al.	2010	T|a12 · H|g12w12s12 · C|ε	Adaptive modeling
Phirani and Mohanty	2010	T|a12 · H|g1w1c1h · C|κ	CO_2 sequestration, UH-hydrate, 5 phases 6 components, h include CO_2 and CH_4 hydrates, Phase change
Janicki	2011	T|g12w12h12m12 · H|g1w1c1h1 · C|κ	UMSICHT HyReS, h includes CO_2 and CH_4 hydrates, consider mass balance of 4 phases 2 components
Kim et al.	2011	T|a123 · H|g123w123c123i123 · M|m1 · C|εκ	TOUGH-Hydrate + FLAC3D, Coussy's theory for M, "two way" but not fully coupled, depressurization, thermal stimulation, and plasticity
Kimoto et al.	2011	T|a2 · H|g1w1hm · M|g1w1m1 · C|κ	Very close to Kimoto et al. [111]
Li et al.	2011	T|a12 · H|g1w1h · M|m1 · C|κ	m is fully coupled into H, effective stress and Drucker–Prager elasto-plastic
Schnurle and Liu	2011	T|a12 · H|g1w1h · C|κ	Simulate gas hydrate and free gas accumulation in marine sediments
Chin et al.	2011	T|a123 · H|g123w123c123i123 · M|m1 · C|εκ	Depressurization, Developed M one way coupled with TH (TOUGH-Hydrate), effective stress theory for M
Ruan et al.	2012	T|a12 · H|g1w1h · C|κ	2D axisymmetric

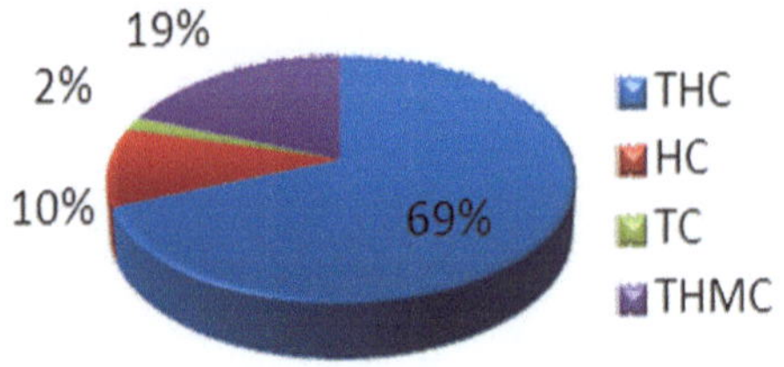

Fig. 2.1 Numbers of different types of multiphysics models

mechanisms differ under various conditions. As a result, the effects of some physical fields on the multiphysical process as well as the strength of their coupling to other physical fields are insignificant. Under such conditions, these physical fields can be excluded without significantly affecting the analysis results. Such models within which less physical mechanisms were considered might have certain advantages for specific problems.

Figure 2.1 shows the percentage of each common types of model based on the multiphysics fields involved. As can be seen from this figure, the majority of existing models belong to THC, while THMC and HC models also take a considerable share. Models at least include two physical fields. The chemical field is an essential component for all the models.

The classification based on Criterion 2 classifies the existing models based on the number of phases or components simulated. Because multiple phases or components are to some extent unavoidable due to the composite nature of gas hydrates, especially for mass balance, a rigorous analysis requires considering at least gas and water. It is impossible to consider the composite as an ensemble because the mass transfers of different phases and components are not coordinated. For heat transfer, under the assumption of the thermodynamic equilibrium between different phases and components, analyses can be conducted with respect to the whole system [112]. For momentum balance, the inclusion of geomechanical responses, in which only the momentums of solids are considered, is still in a preliminary stage. Thus the treatment of gas hydrates as a multiphase multicomponent composite has not been discussed extensively in momentum balance analyses of gas hydrate dissociation.

Due to the aforementioned reasons, the classification based on criterion 2 is only summarized for the hydraulic field in this study. As shown in Fig. 2.2, the majority of the existing models (around 60 %) are two-phase (gas and water) flow model. Immobile phases such as solid matrix and hydrates might or might not be included in these models. Around 18 % of existing models are single phase models. Around 22 % of existing models are "Multi" models, which represents models with more complicated considerations for mass transfer. Examples of these comprehensive models are those simulators (TOUGH-Hydrate, from LBNL; STOMP-Hyd, from PNNL; MH-21, from Japan Oil Engineering Company; CMG-STARS, from University Calgary and University Alaska at Fairbanks; UH-hydrate, from University of Houston) that have been developed for methane hydrate reservoirs [113]. These simulators are the state of the art within the framework of THC.

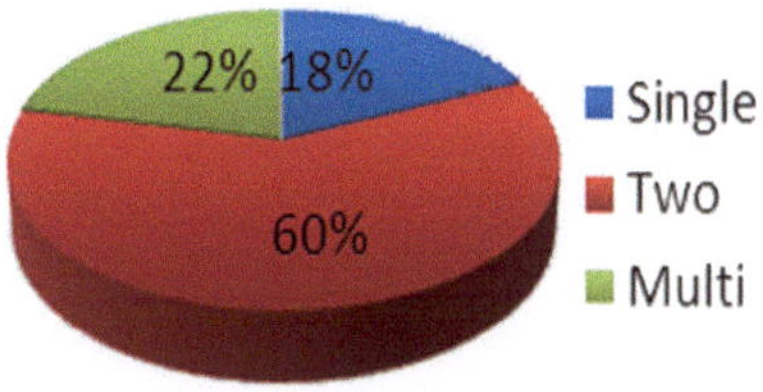

Fig. 2.2 Numbers of models classified by consideration of phase and component

The classification based on Criterion 3 categorizes the existing models based on the number of sub-mechanisms (i.e., convection, diffusion (conduction), and dispersion) considered. The governing equations of the thermal, hydraulic and mechanical fields can be derived from the same form of local balance equation. This implies that these three types of transport phenomena share certain similarities. All of them might include terms representing advection, diffusion, dispersion, and so on. These terms might be called differently when describing different physical phenomena. For example, diffusion term in the governing equation of the thermal field is conventionally called conduction. In addition, some terms are only associated with specific phenomena, for example, radiation occurs in the thermal field only. Inclusion of a certain term in a physical field is dependent upon the nature of this field (as can be seen in Table 2.1). Specially, the conduction term is included in the thermal field and the advection term is included in the hydraulic field for almost every simulation model. Besides, the advection term is always included for the momentum balance of solid phases. Taking the advection term in the hydraulic field as an example, it is usually implemented with Darcy's law which relates the superficial velocity of a phase to the phase pressure. In fact, momentum balance is satisfied implicitly because Darcy's law can be derived from the momentum balance of the fluid at low velocity. That is, Darcy's law is just a special case of the momentum balance equations of fluids. However, in this chapter we treat Darcy's law as an auxiliary relationship for momentum balance of fluids rather than a governing equation describing the mechanical field.

2.3.4.2 Classifications Based on Criteria 4

The classification with Criteria 4, which is based on the chemical model, divides the simulation models into equilibrium models and kinetic models. However, some recent simulators include both modules. The statistical review of published models is plotted in Fig. 2.3, which shows that more than one half of the models are kinetic models. The equilibrium model is used by more than one third of the existing models. By comparison with Table 2.1, it is interesting to find that all analytical models reviewed used equilibrium models [114], except the study conducted by Goel et al. [77]. It is worthwhile to mention that only one type of chemical model, either equilibrium or kinetic, can be used for a single simulation though a simulator can include both.

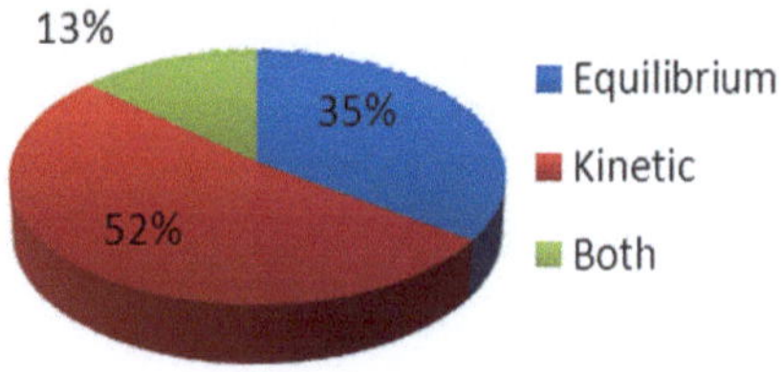

Fig. 2.3 Numbers of models with different chemical modules

In the equilibrium model, the relationships between thermodynamic variables (pressure, temperature, concentration) follow the equilibrium curve (phase) in the reaction sites. These reaction sites are usually distributed on a moving boundary (thin layer) between dissociated and undissociated zones. Therefore, the system stays at equilibrium all the time. The production of the various phases and the amount of phase transitions are determined by the availability and relative distribution of heat and reactants [35]. In the kinetic model, thermodynamic equilibrium does not necessarily coincide with any boundary. Instead, dissociation reactions occur at any locations where the local thermodynamic state falls in the unstable region of the phase diagram. And the hydrate is not treated as a thermodynamic state of methane and water but as a third distinct compound. Rates of phase changes are decided by the kinetic rate of dissociation or formation, which depends on the relative position between the current thermodynamic status and the equilibrium curve.

Hong et al. [8] suggested to differentiate the two approaches, i.e., equilibrium and nonequilibrium (kinetic), with respect to conditions at the decomposition zone. The underlying assumption of the models using the equilibrium approach is that the intrinsic rate of hydrate dissociation is fast enough so that the overall rate of hydrate dissociation is controlled by other mechanisms, i.e., fluid flow or heat transfer [84]. In kinetic models, however, the condition at the hydrate-gas-water interface was kinetically approaching equilibrium. Kowalsky and Moridis [35] reported that for large-scale systems undergoing thermal stimulation and depressurization, the calculated responses for both reaction models were remarkably similar, though some differences were observed at early times. However, for modeling short-term processes, such as the rapid recovery of a hydrate-bearing core, the kinetic characteristic could be important, and neglecting them may lead to significant underprediction of recoverable hydrate. The systematic parametric study of the kinetic reaction constants conducted by Liu and Gamwo [114] argued that the equilibrium model was only a limiting case of the kinetic model for both constant temperature and adiabatic thermal boundary conditions. Their study clearly showed that results obtained the kinetic model approached that by the equilibrium model when the intrinsic mole dissociation constant far exceeded the common range found in the literature. Additionally, the equilibrium model exhibited a moving front pattern for hydrate dissociation while the kinetic model showed a moving zone pattern under adiabatic boundary conditions. For the constant temperature boundary condition, the hydrate dissociated by shrinking in all dimensions for the equilibrium model; while for the kinetic model, it dissociated with no specific pattern throughout the entire reservoir.

The use of equilibrium model often appears to be justified and preferred for simulating the behavior of gas hydrates, given that the computational demand for the kinetic reaction model far exceeds that for the equilibrium reaction model [35]. Evidence in favor of the use of equilibrium model has also been presented for the marine environment. This is because hydrate formation and dissociation in the marine environment occur over an extended period (thousands to millions of years). It is thus permissible to assume local thermodynamic equilibrium and to neglect any chemical kinetic effects [112]. On the other hand, Dendy and Sloan [47] used the Raman spectra to show that the hydrate formation was a rate-dependent process to transform methane into methane hydrate. And the rate-dependent hydrate dissociation process predicted by the kinetic model is similar to the experimental observations. This indicated that kinetic model more realistically describe the natural gas hydrate dissociation process. As commented by Koh and Sloan [61], the paradigm has shifted from thermodynamics (time-independent properties) to hydrate formation and dissociation kinetics in all the gas hydrate technological applications.

2.4 Materials Properties for Gas Hydrate Modeling: Auxiliary Relationships

This section discusses the auxiliary relationships that describe material properties required for gas hydrate simulation. These relationships are needed to mathematically close the equation system for simulating the dissociation process of gas hydrates in porous media. For example, an accurate simulation of hydrate depressurization requires accurate petro-physical and thermophysical data [12]. Typical information needed for reliably predicting the feasibility of natural gas production from hydrates includes but is not limited to: the abundance of the hydrates in the selected reservoir; lithologic and geologic structure of the reservoir; presence or absence of a free gas zone; arrangement of hydrate within the porous medium; permeability, relative permeability-saturation relationships; capillary pressure–saturation relationships; thermal conductivity of the hydrate-bearing and hydrate free medium; energy required to dissociate the hydrate (how close the hydrate to equilibrium); and kinetics of dissociation [4]. Some information is reservoir specific, while the others such as dissociation kinetics, thermal conductivity, and relative permeability can be evaluated in the laboratory [4]. For both types of information, a mathematical representation is required to incorporate the information into a computer simulator. Additionally, natural hydrate samples are not widely available and can be compromised by collection, recovery, transport, and handling [4]. Also, improvements in sampling and remote sensing of hydrate deposits as well as laboratory technologies to study both natural and artificial gas hydrate samples are in crucial demand to determine these relationships [12]. Therefore, more reliable and accurate mathematical expressions for formulating these relationships are especially helpful. Generally, these relationships can be expressed by constants, functions (for continuous relationship), or matrices (for discrete relationship). In the subsections of this section, the various auxiliary relationships used by different simulation models are categorized and discussed based on their roles in the different governing equations (fields). A few mathematical equations for the auxiliary relationships that have not been adopted by the existing models are also introduced.

2.4.1 Material Properties Related to Heat Transfer

The governing equation for heat transfer in Sect. 2.3.2 can be further transformed to obtain the solution. Generally, terms such as energy density, heat flux, and source can be formulated as functions of the dependent variables and material properties. In such a way, only the dependent variables, materials properties depending on the dependent variables, and constants will appear in the equation. One governing equation may contain multiple dependent variables, but the total numbers of dependent variables and that of the governing equations should be equal in the equation system to guarantee a unique solution. Auxiliary relationships necessary for the internal energy and heat flux mainly refer to the heat capacity and thermal conductivity, which are discussed in this subsection. The auxiliary relationships for the source term will be considered in the subsection for chemical reactions. Let us start from the formation of the energy density term as (Eq. 2.12).

$$e = \underbrace{\left(\int_{T_0}^{T} \rho C_v T + U_0 \right)}_{\text{Internal Energy}} + \underbrace{\left(\frac{1}{2} \rho \mathbf{v} \cdot \mathbf{v} \right)}_{\text{Kinetic Energy}} + \underbrace{(\rho g z)}_{\text{Potential Energy}} \tag{2.12}$$

where C_v is the gravimetric heat capacity at constant volume, U_0 is the internal energy per unit volume at the reference temperature T_0, C_p is the heat capacity at a constant pressure.

The internal energy is a part of the accumulation term in addition to the kinetic energy and potential energy. But for gas hydrate dissociation, in general, the internal energy is dominant over the other energy types. As a result, the energy density is frequently represented by the internal energy density only as (Eq. 2.13) [115], in which the pressure term will vanish for solids [112, 116].

$$e = \int_{T_0}^{T} \rho C_{\mathrm{v}} T + U_0 = \int_{T_0}^{T} \rho C_{\mathrm{p}} T - \int p dV + U_0 \tag{2.13}$$

Secondly, the flux term is linked to temperature by Fourier's law of heat conduction as

$$\mathbf{J}_{\mathrm{T}} = -\lambda \nabla T \tag{2.14}$$

where λ is the thermal conductivity. Substituting the above equations for energy density and thermal conductivity into the energy balance equation, we can obtain the heat equation including advection and source terms as (Eq. 2.15).

$$\frac{\partial(\rho C_{\mathrm{v}} T)}{\partial t} + \nabla \cdot (\rho C_{\mathrm{v}} T \mathbf{v}) - \nabla \cdot (\lambda \nabla T) = Q_{\mathrm{T}} \tag{2.15}$$

As introduced before, most existing models of gas hydrate dissociation employed one governing equation for gas transfer for the whole system. Governing equations are established for individual phases/components only in a few cases [13]. The general form of the governing equation for heat transfer of the whole system is as the following.

$$\frac{\partial\left(\sum \rho_i \theta_i C_{\mathrm{v},i} T\right)}{\partial t} + \sum \nabla \cdot (\rho_i \theta_i C_{\mathrm{v},i} T \mathbf{v}_i) - \nabla \cdot (\lambda \nabla T) = Q_{\mathrm{T}} \tag{2.16}$$

2.4.1.1 Heat Capacity

Heat capacities of individual components are mostly used to directly construct the accumulation term in (Eq. 2.16) based on the formulation of the internal energy (Eq. 2.13) [21, 22, 33, 63, 84, 101, 112, 117–121]. This formulation, which is the weighted average of the thermal conductivities of the components with respect to mass, has long been used for soils [122, 123]. This mass weighted average method was also used in many simulation methods [16, 25, 79–81, 111, 124–127].

$$C_{\mathrm{v}} = \sum (\rho_i \theta_i C_{\mathrm{v},i}) / \sum (\rho_i \theta_i) \tag{2.17}$$

This physics-based treatment is based on the definition of the thermal energy and can be derived from thermodynamics.

A constant value was also used for the average heat capacity of a gas hydrate composite for simplicity [8, 128]. There are also more complex relationships that allow more effects on the heat capacity to be considered. For example, the capacity of gas can be described as a function of temperature [44].

$$C_g = C_0 + C_1 \cdot T + C_2 \cdot T^2 + C_3 \cdot T^3 \quad (2.18)$$

where C_0 is 1238.79, C_1 is 3.1303, C_2 is 7.905×10^{-4}, C_3 is -6.858×10^{-7} and C_i is employed to denote fitting constants throughout this section. Similar considerations were also taken by Schnurle and Liu [120]. When the phase changes of water occur, especially for freezing and thawing, the latent heat is typically dealt with the concept of apparent heat capacity [121]. This term was introduced by Williams [129] and later used by Anderson and Morgenstern [130]) to ensure the computational stability. In this treatment, the released or absorbed energy was incorporated into the heat capacity term using (Eq. 2.19).

$$C_a = C + L_f \frac{d\theta_i}{dT} \quad (2.19)$$

where C_a is the apparent gravimetric heat capacity, L_f is the latent heat of phase transition.

2.4.1.2 Thermal Conductivity

The formulation of thermal conductivity is not as straightforward as that of the heat capacity. The average thermal conductivity of a composite is usually used. In several simulation models, this average thermal conductivity was assumed to be a constant [8, 44, 75, 119, 128]. This treatment was adopted to reduce the nonlinearities of the models to allow the solution of the analytical models. However, a volume weighted value of thermal conductivities was more frequently employed in the existing models [13, 21, 22, 25, 33, 63, 79–81, 84, 111, 115, 117, 127, 131]. The mathematical formulation of this volume weighted average value is as follows:

$$\lambda = \sum \theta_i \lambda_i / \sum \theta_i \quad (2.20)$$

The formulation in (Eq. 2.20), which has been extensively used, is essentially an empirical equation. There are still other ways for the calculation of the average thermal conductivity, such as the one used by Tonnet and Herri [101], which was originally used by Henninges et al. [132]. This method used a geometric average as the following.

$$\lambda = \left(\prod \lambda_i^{\theta_i}\right)^{1/\sum \theta_i} \quad (2.21)$$

Russell's equation [133] has also been applied to the formulation of thermal conductivity by Scott et al. [121].

$$\lambda = \lambda_{\mathrm{m}} \sum \frac{[\theta_i/(1-\theta_{\mathrm{m}})]\left[(1-\theta_{\mathrm{m}})^{2/3}(\lambda_i/\lambda_{\mathrm{m}}-1)+1\right]}{\left[(1-\theta_{\mathrm{m}})^{2/3}-(1-\theta_{\mathrm{m}})\right](\lambda_i/\lambda_{\mathrm{m}}-1)+1} \tag{2.22}$$

where m indicates solid matrix, and i denotes the phases other than the solid matrix.

Thermal conductivity can also be calculated using a nested Revil-type mixing rule according to Waite et al. [134] in the following form [120]

$$\lambda = \lambda_{\mathrm{m}} + \frac{(\lambda_{\mathrm{f}}-\lambda_{\mathrm{m}})(1-\theta_{\mathrm{m}})^2}{2\left\{1-\lambda_{\mathrm{m}}/\lambda_{\mathrm{f}}+\left[(1-\lambda_{\mathrm{m}}/\lambda_{\mathrm{f}})^2+4\lambda_{\mathrm{m}}/\lambda_{\mathrm{f}}/(1-\theta_{\mathrm{m}})^2\right]^{0.5}\right\}} \tag{2.23}$$

$$\lambda_{\mathrm{f}} = \lambda_{\mathrm{h}} + \frac{(\lambda_{\mathrm{wg}}-\lambda_{\mathrm{h}})[1-\theta_{\mathrm{h}}/(1-\theta_m)]^2}{2\left\{1-\lambda_{\mathrm{h}}/\lambda_{\mathrm{wg}}+\left[(1-\lambda_{\mathrm{h}}/\lambda_{\mathrm{wg}})^2+4\lambda_{\mathrm{h}}/\lambda_{\mathrm{wg}}/[1-\theta_{\mathrm{h}}/(1-\theta_{\mathrm{m}})]^2\right]^{0.5}\right\}} \tag{2.24}$$

where λ_{f} is the fluid thermal conductivity, and $\lambda_{\mathrm{wg}} = \lambda_{\mathrm{w}}\theta_{\mathrm{w}} + \lambda_{\mathrm{g}}\theta_{\mathrm{g}}$.

There are also other approaches that are popular yet have not been used in the existing models for gas hydrate simulations. These models also deserve attention. One example is the physics-based models proposed by de Vries [123] [135]. Another example is the empirical relationship proposed by Johansen [136], which was later modified by Cote and Konrad [137] and Lu et al. [138]. The key concept in the latter one is the unique relationship between the normalized thermal conductivity and normalized saturation. The basic relations in these models are expressed by (Eqs. 2.25–2.26).

$$\lambda_{\mathrm{r}} = \frac{\lambda - \lambda_{\mathrm{dry}}}{\lambda_{\mathrm{sat}} - \lambda_{\mathrm{dry}}} \tag{2.25}$$

$$\lambda_{\mathrm{r}} = f(\Theta) \tag{2.26}$$

where λ_{r} is the normalized thermal conductivity; and λ, λ_{dry} and λ_{sat} are the actual thermal conductivity and the thermal conductivity of dry and saturated soils, respectively. Θ is the normalized saturation. The relationship between normalized thermal conductivity and the normalized saturation (function f) can be different for different materials. A simplified version of this relationship was once used in the USGS/NETL code comparison project with f assumed as a linear function [139].

The thermal conductivities of individual phases are mostly assumed to be a constant. But in some cases, the thermal conductivities of some phases are considered to be a function of temperature. For example, Schnurle and Liu [120] used the following relations to represent the thermal conductivities of water and hydrate

[134], with a table of 201 precomputed values for the Web Book of Chemistry at NIST (http://webbook.nist.gov).

$$\lambda_i = C_{1,i} + C_{2,i} \cdot (T - 273) \tag{2.27}$$

where $C_{1,w}$ is 0.562, $C_{2,w}$ is 1.75×10^3, $C_{1,h}$ is 0.624, $C_{2,h}$ is -2.78×10^3.

Another example is the equation suggested by Sean et al. [140] for the thermal conductivity of aqueous phase.

$$\lambda_w = C_1 \ln T - C_2 \tag{2.28}$$

where C_1 is 487.85 and C_2 is 2173.8.

2.4.1.3 Thermal Diffusivity

Thermal diffusivity has also been used in a few early simulation models [1, 15, 44]. As thermal diffusivity is the combination of heat capacity and thermal conductivity, the purpose for using this term is to simplify the governing equation and solution process. So in these methods, a constant value of thermal diffusivity was usually employed. Later researchers such as Rempel and Buffett [16] claimed that the effective thermal diffusivity should include dispersive effects.

2.4.2 Material Properties Related to Mass Transfer

Relationships between phase pressures and phase saturations as well as the relationship between the phase saturations and relative phase permeabilities are generally necessary to compute the transport properties of the system [37]. Relations that are additionally required include the relationship between absolute permeability and porosity, and that between permeability and hydrate saturation. Also, coefficients for diffusion and dispersion have to be determined if these two transport mechanisms need to be taken into account. In this subsection, permeability is discussed first because convection is the primary mass transport mechanism in the dissociation processes of gas hydrates. Methods for formulating the influences of porosity, hydrate saturation, and fluids saturations are summarized. Then, auxiliary relationships related to diffusion, dispersion, and mass transfer between phases (phase change) as well as other correlations are reviewed successively.

2.4.2.1 Absolute Permeability and Permeability Considering Hydrate Saturation

Absolute permeability is the permeability when the solid matrix serves as the only component (no fluids and no hydrates). A constant value is frequently assigned for this parameter. However, it can also be calculated with porosity by assuming a unique relationship between absolute permeability and porosity, such as the function used by Garg et al. [112].

$$\kappa_0 = \kappa_1 \cdot (\phi/\phi_1)^{C_1} \cdot [(1-\phi_1)/(1-\phi)]^{C_2} \cdot \exp[C_3(\phi - \phi_1)] \tag{2.29}$$

where κ_0 is the absolute permeability corresponding to any porosity, ϕ; κ_1 is the permeability at the original porosity, ϕ_1. Garg et al. [112] commented that this equation was a rather general expression. With a suitable choice of material constants C_1, C_2, and C_3, the equation could reproduce virtually any measured variation of absolute permeability with porosity.

In addition to porosity, the presence of hydrates in pores can also alter the permeability significantly. There are at least seven methods, in which the influence of porosity is also included, for quantifying this influence of hydrates. The first method (Eq. 2.30) was firstly used by Masuda et al. [119] and later adopted by other simulation models [21, 75, 116, 125, 141, 142].

$$\kappa = \kappa_0 \cdot [1 - \theta_h/\phi]^{C_1} \tag{2.30}$$

Kleinberg et al. obtained two functions for the variation of the permeability with respect to hydrate saturation according to the way in which hydrates form in pores. If the porous medium is approximated as a bundle of capillaries and hydrates form in the center of each capillary (pore-filling), the (Eq. 2.31) can be obtained.

$$\kappa = \kappa_0 \cdot \left[1 - (\theta_h/\phi)^2 + \frac{2(1-\theta_h/\phi)^2}{\log(\theta_h/\phi)}\right] \tag{2.31}$$

This method was used by Liu and Flemings [117], Garg et al. [112], and Schnurle and Liu [120]. However, if the formation of hydrate starts by coating the walls of pores (pore-coating), a function similar to Masuda's method is obtained.

$$\kappa = \kappa_0 \cdot [1 - \theta_h/\phi]^2 \tag{2.32}$$

Garg et al. [112] believed that a large value for the exponent in (Eq. 2.32) corresponds to a faster reduction in permeability with the hydrate saturation, which in fact makes the model the same as Masuda's model. It is noted that the first hydrate formation behavior was supported by experimental evidence with Nuclear Magnetic Resonance.

The fourth method is the power-law model proposed by Civan [143],

$$\kappa = \kappa_0 \cdot (\phi_e/\phi_0) \cdot [(\phi_e/\phi_0) \cdot (1 - \phi_0)/(1 - \phi_e)]^{C_1} \tag{2.33}$$

where $\phi_e = (\phi - \theta_h)$.

This method was used in the simulation models of Tonnet and Herri [101], Bai et al. [80, 81], and Bai et al. [124]. Jeannin et al. [144] linked the permeability to the porosity and hydrate saturation through a logarithmic law.

$$\log(\kappa) = C_1\phi_e - C_2 \tag{2.34}$$

Similar relationships were used by Uddin et al. [36], Janicki et al. [13], and STARS. For instance, the Carmen–Kozeny formula used by Uddin et al. [36] is as following.

$$\kappa = \kappa_0 \cdot (\phi_e/\phi_0)^{C_1} \cdot [(1 - \phi_0)/(1 - \phi_e)]^2 \tag{2.35}$$

2.4.2.2 Relative Permeability

Relative permeability describes the reduction in permeability because of the presence of fluids. Except for a few simulation models which assumed constant permeability [8, 16, 25], relative permeability is generally considered as long as fluids are involved. Some simulation models used the concepts of relative permeability and matric suction without specifying their relationships [15, 115, 127]. In other cases, the following methods have been proposed to formulate this parameter. The first method is to directly import the relationship between relative permeabilities and fluid saturations from experimental data [45, 100, 112, 128, 145]. The second method is to use an adaptation of the equation suggested by van Genuchten [146] and later extended to multiphase flow by Parker et al. [147].

$$\kappa_{rw} = \kappa_{rw0}\overline{S}_w^{1/2}\left[1 - \left(1 - \overline{S}_w^{1/C_1}\right)^{C_1}\right]^2 \tag{2.36}$$

$$\kappa_{rg} = \kappa_{rg0}\overline{S}_g^{1/2}\left(1 - \overline{S}_w^{1/C_1}\right)^{2C_1} \tag{2.37}$$

where $\overline{S}_w = \frac{S'_w - S_{wr}}{1 - S_{wr} - S_{gr}}$, $\overline{S}_g = \frac{S'_g - S_{gr}}{1 - S_{wr} - S_{gr}}$, and the effective water and gas saturations [117] are $S'_w = \frac{S_w}{S_w + S_g}$, $S'_g = \frac{S_g}{S_w + S_g}$.

Klar and Soga [60] used the original form in which κ_{rw0} and κ_{rg0} are equal to 1. Hong and Pooladi-Darvish [84], [36, 125] assumed that κ_{rw0} equals 0.5, and κ_{rg0} equals 1.

The third method includes those variants of the Brooks and Corey model [148–150]. The general form of this model for a system containing two or three fluid phases is

$$\kappa_{\mathrm{rw}} = \kappa_{\mathrm{rw0}} \overline{S}_{\mathrm{w}}^{\mathrm{C_1}} \tag{2.38}$$

$$\kappa_{\mathrm{rg}} = \kappa_{\mathrm{rg0}} \overline{S}_{\mathrm{g}}^{\mathrm{C_2}} \tag{2.39}$$

This type of method seems to be the most popular one in existing simulation models. Different values of constants have been used. For example, C_1 and C_2 were assumed to be equal to 4 and 2, and κ_{rw0} and κ_{rg0} to be 1 in some models [21, 33, 101, 116, 117, 120]. Sun and Mohanty [22] assumed C_1 equals 4, C_2 equals 2.5, and κ_{rw0} equals 0.2, κ_{rg0} equals 1. Bai et al. [80, 81], [124]) assigned C_1 to be 4 and C_2 to be 1.5. In the model proposed by Jackini et al. [13] C_1 was 4 and C_2 was 1.75. Both C_1 and C_2 were assumed to be 3 in the simulation models of White et al. [37] and Gamwo and Liu [151]. There are still some other methods such as the empirical equations used by Verigin et al. [152] and the power law correlations [153] used in the model proposed by Scott et al. [121].

2.4.2.3 Capillary Pressure–Saturation Relationship

The relationship between capillary pressures and phase saturations in porous media (multiphase) is an extension of the soil water characteristic curve in soils (two phase). The soil water characteristic curve (water retention curve or soil moisture characteristic curve) is the relationship between water content (volumetric or gravimetric, or saturation) and soil water potential (or suction, [154]). In the past decades, numerous empirical equations have been proposed for SWCCs [146, 149, 155–157]. In terms of thermodynamics, the SWCC is attributable to the chemical thermodynamics of interfacial phenomena [158–160].

The equation proposed by van Genuchten [146] and later extended to multiphase flow by Parker et al. [147] and the one suggested by Brooks and Corey [149] have gained popularity in simulation models for gas hydrate dissociation in porous media. The van Genuchten equation (Eq. 2.40) has been applied in several simulations [13, 33, 36, 60, 84, 111, 120].

$$\psi = \psi_0 \left(\overline{S}_w^{1/\mathrm{C_1}} - 1 \right)^{1\text{-}\mathrm{C_1}} \tag{2.40}$$

This general form of the model of Brooks and Corey [149] is expressed by the following equation.

$$\psi = \psi_0 \left(\overline{S} \right)^{\mathrm{C_1}} \tag{2.41}$$

Typical applications of this method include the simulations of Sun and Mohanty [22], Bai et al. [124], and Janicki et al. [13]. It is worthwhile to point out that the dependence of relative permeability on fluids saturations is capable of being derived from the relationship between capillary pressure and fluid saturations due to their common basis on internal structure morphology and interface physical chemistry [156, 161, 162]. So it is not surprised that the models used for these two types of auxiliary relationships are to some extent related.

2.4.2.4 Diffusion Coefficients

Diffusion is usually incorporated by those state-of-the-art simulators for THC modeling of gas hydrate dissociation in porous media. The reason may be that its effects on the mass transport could be considerable in a composite with multiphases and multicomponents, especially for the species transported in the fluids. Dispersion could also be considerable due to the same reason, though mechanical dispersion is not as significant as advection. These two phenomena of different natures: dispersion is caused by nonideal flow patterns (i.e., deviations from plug flow) and is a macroscopic phenomenon; while diffusion is caused by random molecular motions (i.e., Brownian motion) and is a microscopic phenomenon. But both of them describe the spread of particles from regions of higher concentration to regions of lower concentration. As the result, both of them can be described using Fick's first law. For example, in the reservoir simulator, STOMP, a combined diffusion–dispersion coefficient replaces the classical Fickian diffusion coefficient.

$$\mathbf{J}^{\alpha} = -D_i^{\alpha}\nabla\chi_i^{\alpha} \tag{2.42}$$

where D are is the coefficient for diffusion or dispersion depending on the transport mechanisms considered. Taking the model of Sun and Mohanty [22] for instance, the advective-dispersive model (ADM) used by Webb [163] was employed to consider the diffusion in gas phase. Based on Fick's law, the molecular diffusions of methane, salt, and water are calculated by (Eqs. 2.43–2.45), respectively.

$$\mathbf{J}^{\mathrm{m}} = -D_{\mathrm{g}}^{\mathrm{m}}\chi_{\mathrm{g}}^{\mathrm{m}} - D_{\mathrm{w}}^{\mathrm{m}}\chi_{\mathrm{w}}^{\mathrm{m}} \tag{2.43}$$

$$\mathbf{J}^{\mathrm{s}} = -D_{\mathrm{w}}^{\mathrm{s}}\chi_{\mathrm{w}}^{\mathrm{s}} \tag{2.44}$$

$$\mathbf{J}^{\mathrm{w}} = -\mathbf{J}^{\mathrm{m}} - \mathbf{J}^{\mathrm{s}} \tag{2.45}$$

where the effective diffusion coefficient of component α in phase i, D_i^{α}, is determined with the model proposed by Pruess and Moridis [164]:

$$D_i^{\alpha} = \rho_i\phi^{4/3}S_i^{10/3}D_{i,\mathrm{B}}^{\alpha} \tag{2.46}$$

where $D_{\mathrm{j,B}}^{\mathrm{i}}$ is the diffusion coefficient of component α in phase i in bulk phase system. For the gas phase composed of methane and water vapor, the diffusion is a binary diffusion. Then the binary diffusion coefficient, $D_{\mathrm{g,B}}^{m}$, can be calculated as a function of pressure and temperature as

$$D^m_{g,B} = 1.89 \times 10^{-5}\left(\frac{101325}{p_G}\right)\left(\frac{T}{273.2}\right)^{3/2} \tag{2.47}$$

For aqueous-phase composed of methane, salt, and water, $D^m_{w,B}$ can be evaluated with the following equation [165]. The diffusivity of salt in aqueous phase which is dependent on salt type was set to be a constant.

$$D^m_{w,B} = 10^{-1.6865-920.576/T} \times 10^{-4} \tag{2.48}$$

The diffusion coefficient of methane in the aqueous phase can be calculated using the second-order polynomial function [140]

$$D^m_{w,B} = C_0 + +C_1 \cdot T + C_2 \cdot T^2 \tag{2.49}$$

In the simulation conducted by Janicki et al. [13], diffusion coefficients were calculated from the IFM-GEOMAR data functions [166].

2.4.2.5 Hydraulic Diffusivity

Similar to thermal diffusivity, hydraulic diffusivity is also used in a few models [7, 77, 167]. The employment of this concept makes the mass balance equation appear similar to the heat equation. Hydraulic diffusivity could be linked to permeability via the following correlation.

$$\alpha = \frac{\kappa}{\mu}\left|\frac{dp}{d\theta}\right| \tag{2.50}$$

2.4.2.6 Mass Transfer Between Phases

Mass transfer between phases could be significant in some cases, e.g., in a system containing highly volatile soils and thus was considered in complex simulation models [168]. For example, Henry's law was used to quantify the solubility of gases in solvents while partial pressure rule could be used to define the molecular fraction of water in the gas phase [151]. Mass transfer of gaseous components to liquids and vice versa were also modeled by a two-film theory linear approach [13]. Other simple functions are also available, such as the one used in the model proposed by Sean et al. [140] for methane solubility in water.

$$\chi = C_1[C_1 p + C_2 \cdot \exp(C_3 T)] \tag{2.51}$$

2.4.3 Material Properties Related to Chemical Reactions

2.4.3.1 Thermodynamic State

A thermodynamic state is a set of values of properties of a thermodynamic system that must be specified to reproduce the system. Thermodynamic state variables include entropy, pressure, temperature, density, and so on. Some variables such as pressure and temperature are the dependent variables of the hydraulic and thermal fields, respectively. Therefore no auxiliary relationships are necessary for their calculations when no phase transition occurs. However, an auxiliary relationship is required to specify the relationship between these variables to close the system when dissociation reactions occur. This relationship is called phase diagram and is discussed later. Other variables such as the densities of phase components are also important parts in the governing equations for thermal, hydraulic, and mechanical fields. Similar observations apply to the enthalpy of gas hydrate dissociation. In the following paragraphs the auxiliary relationships on density and enthalpy of gas hydrate dissociation will be summarized successively.

While the density of solid matrix typically does not need complicated treatments, the densities of other phases such as water, gas, and hydrates are frequently considered as functions of other state variables. For example, water is regarded to be slightly compressive, and its density can be given by (Eq. 2.52) [124].

$$\rho_w = \rho_{w0}[1 - C_1(\rho_w - \rho_{w0})] \tag{2.52}$$

In the model of Schnurle and Liu [120], the density of pore-water is related to the temperature as follows [134, 169],

$$\rho_w = 999.9 + 5.33 \times 10^{-2}(T - 273) - 6.834 \times 10^{-3}(T - 273)^2 \tag{2.53}$$

The variation of density of hydrate with temperature is described in Waite et al. [134] after Shpakov et al. [170]):

$$\rho_h = 926.45 - 0.239(T - 273) - 3.73 \times 10^{-4}(T - 273)^2 \tag{2.54}$$

The ideal gas law was commonly used for the densities of gas components [115, 171]. In more accurate considerations, property data of gas phase components were taken from NIST Standard Reference Database and are molar-weighted; for the gas phase density, the Peng–Robinson equation was used. Data for the density of common liquid phase were typically taken from NIST and other available sources; special corrections were applied on the density of saline seawater (UNESCO Standard Reference Equation) [13].

Although the enthalpy of gas hydrate dissociation could be approximated by a constant [144], its variation with temperature has been considered in many simulation models. According to Sloan [172], the enthalpy of hydrate dissociation can be computed using the Clapeyron equation as follows,

$$\Delta H = T\Delta V\frac{dp}{dT} \tag{2.55}$$

where $\frac{dp}{dT}$ is the slope of the phase line, ΔV is the volume change during the phase transition. This equation can be further transformed into the Clausius–Clapeyron equation if the volume of hydrate approximates that of water trapped in hydrates.

$$\Delta H = n_h ZRT^2\frac{d\ln p}{dT} \tag{2.56}$$

where n_h is the number of moles of hydrate per kilogram.

Moridis [173] obtained a graph for ΔH based on (Eqs. 2.55–2.56) and found that ΔH increased by a factor of 3 to 4 over the temperature range 0 °C to 40 °C [112]. The following function was usually adopted for ΔH as a function of T in simulation models for gas hydrate dissociation [7, 15, 33, 63, 80, 81, 116, 119, 124, 140, 167].

$$\Delta H = \mathrm{C}_1 \cdot T + \mathrm{C}_2 \tag{2.57}$$

A piecewise function for ranges on the two sides of the freezing point of bulk water was also suggested [44]

$$\Delta H = \mathrm{C}_1 \cdot T + \mathrm{C}_2, T < 273\ \mathrm{K};\ \Delta H = \mathrm{C}_3 \cdot T + \mathrm{C}_4, T > 273\ \mathrm{K} \tag{2.58}$$

When the hydrates contain different species of gases, the influence of composition on the dissociation of enthalpy may be considerable. Under such a condition, ΔH could be calculated as the following equation [15],

$$\ln\left(\frac{\Delta H}{\Delta H_0}\right) = \sum \left(\mathrm{C}_{1,i} \cdot \chi_i + \mathrm{C}_{2,i} \cdot \chi_i^2\right) \tag{2.59}$$

where χ_i is the mole fractions of the guest gas, and $\mathrm{C}_{1,i}$ and $\mathrm{C}_{2,i}$ are the binary coefficients which depend on the gas type and temperature range.

2.4.3.2 Equilibrium: Phase Diagram

A phase diagram is usually used to show conditions at which thermodynamically distinct phases can occur at equilibrium, which could possibly involve pressure, temperature, specific volume, specific enthalpy, or specific entropy. But for gas hydrate dissociation simulations in porous media, a 2D diagram with pressure and temperature as the ordinate and abscissa, respectively, is usually used. Common components of a phase diagram are lines of equilibrium or phase boundaries, which refer to the lines that mark the conditions under which multiple phases can coexist at equilibrium. Among these boundaries, the one indicating the coexistence of methane hydrate, methane, and water is the one most frequently adopted in the simulations. Several empirical equations have been used to represent this boundary. The first type of equation is as follows

$$p = C_1 \cdot \exp(C_2 + C_3/T) \tag{2.60}$$

where C_1 was assumed to be 1 in some studies [15, 33, 171] as well as other values in other simulation models [111, 115, 116]. Values assigned to the other two constants were close in these models, in which C_3 was usually a negative value. A similar equation to the above one was used by Maksimov [25] ($p = C_1 \cdot \exp(C_2 + C_3 \cdot T)$).

The second type of equation for the line of phase boundary was in the polynomial form, i.e.,

$$\log p = C_0 + C_1 \cdot (T - T_0) + C_2 \cdot (T - T_0)^2 \tag{2.61}$$

This type of phase equilibrium condition was used by Verigin et al. [152], Admadi et al. [167], Ahmadi et al. [7], Nazridoust and Ahmadi [21], and Liu et al. [118].

A similar empirical fit for the hydrate melting pressure as a function of temperature was proposed by Bakker [174] and later included in the simulation model of Garg et al. [112].

$$\log p = C_0 + C_1 \cdot T + C_2 \cdot T^2 + C_3 \cdot T^3 \tag{2.62}$$

The third type of mathematical expression for the phase equilibrium was developed from the work of Adisasmito et al. [175] [36]. The experimental data summarized in the form of average curves were expressed by a three-parameter K value correlation as (Eq. 2.63),

$$K = \frac{C_1}{p} \exp\left(\frac{C_2}{T - C_3}\right) \tag{2.63}$$

where C_1 is in kPa and C_2 and C_3 are in °C.

In addition, a detailed piecewise function obtained by Moridis [176] based on experimental data [172] was also used in several models (TOUGH-Hydrate/TX; [113, 124]).

The above equations are designated for single substance under simple conditions. Tishchenko et al. [177] proposed empirical algorithms for hydrate dissociation pressures and methane concentrations, in seawater and pore water equilibrated with methane hydrate, as functions of chlorinity Cl (in ppt), temperature, and hydrostatic pressure. The algorithms described pore-water dissociation pressure for a temperature range of 273–293 K and chlorinity range of 0–70 ppt [120]. More advanced methods consider the influence of composition. For example, the one proposed by Burshears et al. [15] considered this influence by using the equation proposed by Holder and John [178]:

$$\ln(p/p_0) = \sum \left(C_{1,i} \cdot \chi_i + C_{2,i} \cdot \chi_i^2\right) \tag{2.64}$$

where χ_i is the mole fractions of the guest gases, and $C_{1,i}$ and $C_{2,i}$ are the binary coefficients which depend on the gas type and temperature range.

2.4.3.3 Kinetic: Dissociation Kinetics

Different from phase boundaries, which are required for every simulation model, the chemical kinetics is only necessary for models using chemical kinetics for reactions (dissociation and formation). However, this term is critical due to the high nonlinearity it could result in. And the fact is that, in contrast to the advances made in the thermodynamics of hydrates, the kinetics are less understood [2]. Most of the kinetic models used in simulation models for gas hydrate dissociation are related to or in a similar form to the Kim and Bishnoi model [179]. This model assumed that the hydrate decomposition rate is proportional to a driving force defined by the difference between the fugacities of gas molecules at the three-phase equilibrium pressure and that in the bulk state [36]. The general form of the model could be expressed as

$$\frac{dn_g}{dt} = k \cdot A \cdot (f_e - f) \tag{2.65}$$

$$k = k_0 \exp\left(\frac{\Delta E}{RT}\right) \tag{2.66}$$

where n_g is the moles of gas, A is the total surface area of hydrates per unit volume, ΔE is the activation energy, f and f_e are the gas fugacity and equilibrium gas fugacity, respectively. It is noticed that the assumption of ideal gas was not adopted in the studies of Masuda et al. [119] and Esmaeilzadeh et al. [33]; and accordingly, the gas fugacities were calculated using the Peng–Robinson equation of state [180] or modified equations [33].

To relate the rate of generated gas to the source term, the following equation should be used, which considers the fact that only a fraction of total volume is occupied by gas hydrates.

$$\frac{dm_g}{dt} = M_g \phi \frac{dn_g}{dt} = M_g \cdot \phi \cdot k \cdot \Gamma \cdot A \cdot (f_e - f) \tag{2.67}$$

where M_g is the molecular weight of gas, Γ is the effective ratio of particle area or active fraction coefficient. A simple approach to implement the above model was presented by Yousif [45] and later used by Sung et al. [145], Jeannin et al. [144], Du et al. [125], and Bai et al. [124]. In this approach, $M_g \cdot \phi \cdot k$ was simplified to be a constant k_d (kinetic constant), whose value was assigned to be 4.4×10^{-13} (mol/(m$^2 \cdot$ Pa $\cdot$ s))) in the above-mentioned studies (2×10^{-9} was used by Du et al. [125]).

A was calculated using the parallel-cylinder model proposed by Amyx et al. [181] as follows:

$$A = \left[\frac{(\theta_w + \theta_g)}{2\kappa_a}\right]^{1/2} \quad (2.68)$$

Γ was usually not considered and the fugacities were replaced by the gas pressures under the assumption of ideal gas [8]. In the simulation studies of Sun and Mohanty [22] and Phirani and Mohanty [113], the kinetic constant was assumed to be 0.5875×10^{-11} $(\text{mol}/(\text{m}^2 \cdot \text{Pa} \cdot \text{s}))$, which is the average of the results measured by Englezos et al. [182]; or 0.35×10^{-10} $(\text{mol}/(\text{m}^2 \cdot \text{Pa} \cdot \text{s}))$ for carbon dioxide hydrate in the latter study. The method adopted by Sun and Mohanty [22] and Phirani and Mohanty [113] is a little complicated than Yousif's. In this method, $M_g \cdot \phi \cdot k$ was calculated with $k_d = k_{d0}\exp(\Delta E/RT)$; and an active fraction coefficient, Γ, equal to S_h, was used to allow for the fact that only a portion of hydrates get involved in the reaction.

The methods used by some other researchers [21, 33, 63, 75, 116, 119] took the form of the general formulation, but the value of intrinsic kinetic constant varies from model to model. In the models of Masuda et al. [119], Ruan et al. [116], Nazridoust and Ahmadi [21], and Esmaeilzadeh et al. [33], A was calculated using a different equation as follows.

$$A = \theta_h \cdot A_s \quad (2.69)$$

where A_s is the surface area of the spherical particles.

Other forms of the Kim–Bishnoi model also exist. For example, Kimoto et al. [111] used an equation which was suggested in the original paper of Kim et al. [179] as following:

$$\frac{dn_g}{dt} = 0.585 \times 10^7 \cdot \exp(-9400/T) \cdot (f_e - f) \cdot n_{h0}^{1/3} \cdot n_h^{2/3} \quad (2.70)$$

where $\frac{dn_g}{dt}$ is in mol/s, n_h is the moles of hydrates, and gas fugacities are represented by the average fluid pressure. Uddin et al. [36] extended the Kim–Bishnoi model used by Hong and Pooladi-Darvish [8] to allow for both formation and dissociation reactions and also the influence of water saturation on the reaction area. Other kinetic models have also been developed, which were in a similar form to that of the Kim–Bishnoi model. For example, the model proposed by Rempel and Buffett [16] assumed the driving force for formation/dissociation reactions was proportional to the difference between mass fraction and equilibrium mass fraction; while Sean et al. [140] believed the driving force for dissociation was proportional to the difference in volumetric molar concentrations of methane under current and equilibrium conditions. In general, the Kim–Bishnoi model for chemical kinetics is used as a chemical kinetic model. However, Goel et al. [77] extended the Kim–Bishnoi model and incorporated it into an equilibrium model.

2.4.4 Material Parameters for Momentum Balance

The incorporation of momentum balance module into the simulation of gas hydrate dissociation in porous media is still in a very preliminary stage. This is mainly due to the limited understanding of the mechanical properties of solid and fluid phases during the dissociation process and of the interactions between phases. However, problems relevant to the momentum balance, such as the geomechanical behaviors, have started to attract attention recently.

2.4.4.1 Solid: Geomechanical Properties, Solid–Fluid Coupling, Constitutive Relations

Gas hydrates are typically found in unconsolidated or weakly consolidated sands in shallow marine floors and permafrost areas [183, 184]. In such sediments, gas hydrate crystals work as cementing materials for bonding solid grains and increasing the stiffness of sediments.

The microscopic distribution of hydrates in sediments was described using three modes, i.e., pore-filling, load-bearing, and cementation [37, 185, 186]. Macroscopic gas hydrate structures inside geologic formation were categorized as massive hydrate blocks on the seabed, massive or thin streak, pore-filling in sand, and vein or fracture filling in clay and so on. In the pore space scale, gas hydrates have different forms such as floating, frame building, coating, and cementation [65]. Under a specific condition, i.e., high pressure and low temperature, methane hydrate is stable, and is a relatively strong and ductile material [187]. However, once the hydrate dissociates, due to the decrease in pressure or increase in temperature, the bonded structure disappears and the soils may behave as unconsolidated materials, leading to some geotechnical engineering problems. Of particular interest are submarine geo-hazards, such as initiation of marine landslides due to hydrate dissociation [28, 188, 189] and wellbore instability during methane gas production from the hydrates [60, 64, 126]. Besides, understanding of the physical and mechanical properties of hydrate-bearing sediments is important for interpreting geophysical data, borehole and slope stability analyses, and reservoir simulations and for developing production models [190].

Current knowledge on the accurate constitutive behaviors of the soil-hydrate structure, e.g., knowledge about how hydrate pattern influences the intermediate and large strain behavior and the failure condition, are limited. The current knowledge of geophysical and geotechnical properties of hydrate-bearing sediments has been largely derived from laboratory experiments conducted on disparate soils at different confining pressures, water saturations, and hydrate concentrations

[37, 186, 190]. This is because the mechanical properties of methane hydrate bearing sediments are highly dependent on factors such as the saturation of methane hydrate, pressure, temperature, and hydrate structure in pore spaces [65].

The mechanical properties influence various scales of geological formation stabilities. Taking Young's modulus for example, Ng et al. [126] used the following relationship for the Young's modulus of gas hydrates:

$$E = 125 + 1000S_{\mathrm{h}} \tag{2.71}$$

where E is Young's modulus in MPa, S_{h} is the degree of gas hydrate saturation.

The mechanical stability of methane hydrate is a function of temperature and pressure. Besides, its mechanical properties are coupled to fluid dynamics and heat transfer phenomena [65]. To evaluate the mechanical properties of sediment samples, Ran et al. [191] simulated loading of a disordered pack of spherical grains by incremental displacements of its boundaries. A series of possible scenarios was simulated, showing the degradation of sediment strength as a reduction in the macroscopic elastic moduli. This trend agreed qualitatively with the published results of experimental and numerical studies. It was concluded that dissociation might lead to a loss of solid support of the skeleton, causing seafloor instabilities such as landslides and subsidence.

The underlying mechanisms for geomechanical responses of gas hydrate dissociation have been investigated. Soga et al. [186] reviewed the mechanical properties of methane hydrate bearing soils and highlighted the following major characteristics. (1) The strength of the gas hydrate bearing soil is dependent on the hydrate saturation; and the contribution to the strength by hydrate is of a cohesive nature rather than frictional nature. (2) The dilation angle is a function of hydrate saturation. (3) The Young's modulus of soil-hydrate formation is higher than that of the soil without hydrates, but the Poisson's ratio was found to be independent of hydrate saturation. Yamamoto [65] summarized the following major effects of the hydrate dissociation on the mechanical behavior of the geological formation [65]: (1) reduction of the bonding between grains; (2) the change of the water content and gas saturation, and hence change in the consistency and capillary suction force; (3) change in the capillary force due to production of water and gas; (4) reduction in the effective stress and shear strength due to increase in the pore pressure in low permeability soils; (5) increase in the permeability and advection fluid transfer due to increases in the effective porosity; (6) decrease in the pore space compressibility, decrease in shear wave velocity, and increase in Poisson's ratio; (7) thermal stress due to decrease in the temperature; (8) change of the dilatancy character from positive to negative because of the change in sand character from dense to loose. The shear strength of artificial hydrate-bearing sediments was found to be similar to that of natural hydrate-bearing sediments [192, 193]. Their strength varies with the content of hydrate, sediment properties and experimental conditions [194–196]. Furthermore, during hydrate dissociation without axial loading, the volumetric strain has dilative tendency regardless whether there is a change in the effective confining pressure [197].

2.4.4.2 Liquid: Darcy's Law, Viscosity

The primary auxiliary relationship for the fluid phases in the mechanical field is Darcy's law. The employment of Darcy's law predominates in the simulations of gas hydrate dissociation in porous media. Thus, the non-Darcy's effects [198] for non-Newtonian fluids, e.g., very low pressure (slip phenomenon) or high velocity range (turbulence, inertia, and other high velocity effects) behaviors, are not discussed in this chapter. The Darcy's law is formulated as the following:

$$\overline{\mathbf{v}}_i = \theta_i \mathbf{v}_i = \frac{k}{\mu}(\nabla p - \rho g \mathbf{i}) \tag{2.72}$$

where $\overline{\mathbf{v}}_i$ is the superficial (or Darcy) velocity, and $\mathbf{i}$ is the unit vector along the gravitational direction.

Darcy's law is extended to multiphase flow by postulating that the same phase pressures are involved in causing each fluid to flow [168]. However, each phase interferes with the flow of the other due to the simultaneous flow of the multiple phases. That is why relative permeability is introduced in most of the existing models.

$$k_i = k_0 k_{\mathrm{r},i} \tag{2.73}$$

where k_i and $k_{\mathrm{r},i}$ is the permeability and relative permeability of phase i, k_0 is the intrinsic permeability of the porous material.

Since all the terms in the Darcy's law except viscosity have been discussed in previous parts of the text, only typical mathematical formulations for viscosity are summarized in the current subsection. Viscosity is usually formulated as a function of some state variables, such as temperature and density. For gas phase, one typical function is that proposed by Selim and Sloan [44], which was later used by Schnurle and Liu [120].

$$\mu = C_0 + C_1 T + C_2 T^2 + C_3 T^3 + C_4 T^4 + C_5 \rho + C_6 \rho^2 + C_7 \rho^3 + C_8 \rho \tag{2.74}$$

where μ is the viscosity. For aqueous phase, Sean et al. [140] used the following equation in their simulation model.

$$\mu = C_0 + C_1 T + C_2 T^2 \tag{2.75}$$

For the overall viscosity of one phase consisting of several components, Janicki et al. [13] suggested to use the following molar-weighted functions:

$$\mu = \sum_{\alpha} \chi_i^{\alpha} \mu_i \tag{2.76}$$

where μ_i is the viscosity of phase i.

2.4.4.3 Solid–Liquid Interaction: Stress Formulation

The solid–liquid interaction, which describes the couplings between the solid matrix and pore liquids, has received very limited attention in gas hydrate simulations. This is because the momentum balance of fluids has hardly been considered except in very few works [111, 141], although the geotechnical responses, i.e., the momentum balance of solid phases, have started to attract increasing attention. But the topic is not new to relevant disciplines, i.e., Geomechanics (e.g., consolidation), Hydrogeology (e.g., groundwater storage), and Petroleum Engineering (subsidence, stress around boreholes). For single phase flow, Tezarghi's consolidation theory lent us a good practical approach. For this two-way coupling problem, the coupling from the pore liquid (water) to solid (soil grains) was dealt with the stress formulation (effective stress) while the coupling from solid to pore liquid was tackled with experimental relations. The later consolidation theory presented by Biot [199] provided a fully theoretical explanation to the single phase (saturated with compressible water) solid–fluid coupling problem. Biot's theory, as the departure point of poroelasticity, could be formulated by the following equations [200]:

$$\varepsilon_{\mathrm{v}} = \frac{1}{K}\sigma + \frac{1}{H}p \tag{2.77}$$

$$\Delta\phi = \frac{1}{H}\sigma + \frac{1}{R}p \tag{2.78}$$

where ε_{v} is the volumetric strain (positive in expansion), σ is the isotropic applied stress (positive if tensile), p is the fluid pressure (greater than atmosphere is positive), $\Delta\phi$ is the increment of water content (positive as fluid is added to the volume), K is the drained bulk modulus, $1/H$ is the poroelastic expansion coefficient, $1/R$ is the specific storage coefficient.

It is seen that the first equation describes the liquid-to-solid coupling: a change in fluid pressure or fluid mass produces a change in volume of the porous skeleton (solid matrix), while the second equation represents the solid-to-fluid coupling: a change in applied stress produces a change in fluid pressure or mass. The first equation could be reformulated as:

$$\sigma = K\varepsilon_{\mathrm{v}} - Bp \tag{2.79}$$

where $B = K/H$ is the Biot–Willis coefficient, which can be calculated with the classical micro–macro relations of poroelasticity as

$$B = 1 - \frac{K_{\mathrm{hom}}}{K_{\mathrm{s}}} \tag{2.80}$$

where K_{hom} is the homogenized bulk modulus and K_{s} is the bulk modulus of solid grains.

By employing the assumption used in Terzaghi's theory, that is, solid grain is incompressible, the above equation degenerates into the formulation of effective stress. On the other hand, Biot's theory for one liquid was extended to two liquid condition (water and air) by Bishop [201], Fredlund and Morgenstern [202], and Lu and Likos [203] in the framework of unsaturated soil mechanics as well as Coussy [204] in the framework of poromechanics. When more liquids get involved, the mixture theory needs to be employed [205]. It is noted that these solid–fluid coupling theories for multiple phases have not been explicitly validated by experiments.

For geomechanical responses, Li et al. [63] developed the stress equilibrium equation based on the effective stress law. However, the definition of pore fluid pressure is unclear for the two-phase flow. In the study of Rutqvist et al. [64], the basic couplings between hydraulic and mechanical processes in the deformable porous media were considered through constitutive laws that define how changes in pressure, temperature, and hydrate saturation affect deformation and stresses, and how changes in stress and strain affect fluid flow. And the numerical simulation involved linking the TOUGH + Hydrate simulator to the FLAC3D commercial geomechanical code. Kim et al. [59] computed the geomechanical responses by connecting a continuum theory in coupled flow to geomechanics. The governing equations of fluid and heat flow were based on Moridis [176] (TOUGH + Hydrate). The constitutive relationships were extended from Coussy's model [204]. A similar study was conducted by Chin et al. [206]. In the work of Klar and Soga [60] and Ng et al. [126], the mechanical behaviors of hydrate soils which depend on effective stresses formulated with the pore pressure (including gas and water) were formulated by Bishop's theory [201] for unsaturated soils. Of great importance is the work of Kimoto et al. [111], which considered the momentum balance of both solids and fluids. And the liquid-to-solid couplings were considered with the mixture theory [205]. In addition, empirical models for solid–fluid interactions have also been used. For example, pore fluid pressure generation was postulated to be proportional to the initial hydrate fraction and the sediment bulk stiffness [62]. Garg et al. [112] calculated the intrinsic sediment stress with the equation proposed by Morland et al. [207], in which a relationship between the sediment porosity and effective stress was assumed.

2.5 Discussions

This section presents discussions on several critical aspects of gas hydrate dissociation simulations including validations, applications, recovery schemes, critical factors in recovery, governing mechanisms, research trends, and needs for future research. All the discussions are made on basis of published simulation results, or direct analyses of existing data, or other solid facts. In such a way, this review study lays down a solid foundation for future research.

2.5.1 Validation of the Performance of Existing Models

As shown in Sect. 2.3, a number of simulation models have been developed for predicting gas hydrate dissociation in porous media. In addition to their common features and differences, the functions and potential applications also need to be evaluated. Strategies for the performance validation include: validation by experiments, validation by comparisons between models, and validation by field production applications. The ultimate success of a given model, i.e., the usefulness and applicability of the model, can only be more effectively demonstrated based on their applications to real-world problems.

2.5.1.1 Validations by Experiments

Commercial attempts in gas recovery from gas hydrate reservoirs are limited at the current stage. Besides, the natural samples are not readily available because of the remote locations of the accumulations and the difficulties in collecting and transporting samples without compromising them. Consequently, only a few models have been validated against results from laboratory experiments. Listed below are a few of these studies, for which details can be found in the cited references. The predictions yielded by the model of Yousif et al. [45] closely matched their experimental data of gas and water productions, the progress of the dissociation front, and the pressure and saturation profiles. The simulation results of Bai et al. [80, 81] were well validated by their laboratory experiments. Laboratory scale experiments were conducted at the Pacific Northwest National Laboratory (PNNL) to analyze the feasibility of the simulation method of Phale et al. [14]. Mass and heat transfers have been studied both experimentally and numerically by Tonnet and Herri [101]. And their numerical model was validated by the experiments. The model-predicted performance by Goel et al. [77], i.e., production rate, compared well with the published experimental studies on the hydrate dissociation in porous media [45, 208, 209]. The predicted vertical distribution of hydrate in the study of Rempel and Buffett [16] was consistent with geophysical inferences from observed hydrate occurrences along the Cascadia margin. The simulation results of Garg et al. [112] applied at the Blake Ridge (site 997) and Hydrate Ridge (site 1249) were compared with the chlorinity, sulfate, and hydrate distribution data. However, field validations of gas hydrate simulation models are still rare at this moment. With global wise interest in gas hydrate pilot production, it is anticipated that there will be more opportunities for direct field validations of simulation models.

2.5.1.2 Mutual Validation Between Models

In addition to direct validations of simulation models with in situ or laboratory data, mutual validations between different models have also been carried out frequently. Li et al. [210] validated the model proposed by Yousif et al. [45]. A close agreement was achieved between the results reported by Hong and Pooladi-Darvish [8] and that of Uddin et al. [36]. The numerical solution obtained by Gamwo and Liu [151] was verified against the state-of-the-art simulator TOUGH-Hydrate. For comparisons between analytical and numerical models, Admadi et al. [7, 167] compared their numerical simulation results with that of the linearization approach. Similarly, Hong [8] compared the analytical solution with the numerical solution to the original equation system. The code comparison project conducted by USGS/NETL is of great importance for evaluating the existing large-scale simulators for THC modeling of gas hydrate dissociation.

2.5.1.3 Applications

A number of simulation models have been designed for applications. For example, results from the solution of Selim and Sloan [1] were presented in graphs to estimate the amount of hydrate dissociated as a function of time. The model of Garg et al. [112] was applied to studying hydrate distributions and predicting chlorinity, sulfate, and hydrate distribution data at the Blake Ridge (site 997) and Hydrate Ridge (site 1249). The simulation study of Konno et al. [142] confirmed the advantage of employing depressurization as a gas production method with the hydrate sample in sandy turbidite sediments at the Eastern Nankai Trough. Their numerical analysis was proven to be effective in analyzing the dissociation behavior of hydrate-bearing cores obtained at natural hydrate reservoirs; and it enabled evaluation of gas productivity in those reservoirs. The code presented by Klar and Soga [60] was used for investigating the stability of a methane extraction well by depressurizing the well. The simulation by Janicki et al. [13] considered two scenarios: the depressurization of an area 1000 m in diameter and a one/two-well scenario with CO_2 injection. Realistic rates for injection and production were estimated. Their study also discussed the limitations of these processes. The production modeling conducted by Phale et al. [14] would help in deciding the critical operating factors such as pressure and temperature conditions of the injecting CO_2-slurry, the injecting flow rate the slurry, and the concentration of the slurry. The work of Rutqvist et al. [64] analyzed the geomechanical responses during the depressurization production from two known hydrate-bearing permafrost deposits: the Mallik (Northwest Territories, Canada) deposit and Mount Elbert (Alaska, USA) deposit. The simulation results showed that general thermodynamic and geomechanical responses for these two sites were similar.

2.5.2 Suggestion on Practice Production by Model Simulations

2.5.2.1 Recovery Schemes

For the conventional recovery schemes, depressurization received particular attention in the early stage of the development of simulation models for gas hydrate dissociation in porous media. This is possibly attributed to the application of this method in the Messoyakha field, USSR [45]. Special interest was repeatedly expressed for a type of reservoir with a gas layer overlaid by a hydrate layer, which was later classified as Class 1 accumulation. Numerous simulations proved the feasibility and effectiveness of applying depressurization to this type of gas hydrate reservoir [8, 84] and indicated that the hydrate layer have a large impact on improving the productivity of the underlying gas reservoir [84]. For example, the simulation results of Burshears et al. [15] inferred that massive hydrate can be dissociated without external heat energy source, which would not be significantly affected by the water from gas hydrate dissociation. In addition, gas produced by hydrate dissociation contributed significantly to the total production [124], i.e., 20 % to 30 % of the total gas production [6]. Also, the computer simulation by Bai et al. [124] showed that overlying hydrate zone could evidently prolong the lifespan of the gas reservoir. The study of Konno et al. [142] confirmed the advantage of employing depressurization as a gas production method using the hydrate in sandy turbidite sediments at the Eastern Nankai Trough. With depletion of reservoir energy, methods such as thermal stimulation and inhibitor injection were suggested to replace depressurization (Bai et al. [80, 81]). For example, the simulation of Yang et al. [211] suggested to use the depressurization method in the initial stage and the thermal stimulation method in the later stage. On the contrary, the simulation of Masuda et al. [119] indicated that the production schemes of depressurization and heating well were not effective to economically obtain feasible gas production rates.

Computer models also provided important insights into other recovery methods. For example, for thermal stimulation, an energy efficiency value about 9 % was found in the study of Selim and Sloam [44], which appeared encouraging for natural gas production from hydrates. The thermal stimulation using constant temperature at the reservoir with a single well was found to have a limited effect on gas production compared to gas production due to depressurization [118]. Within various techniques of thermal stimulation, hot water injection was proven to be able to remarkably improve gas productivity and a maximum production rate of 860 m^3/day was obtained per unit length of well [119]. And the energy balance for this scheme was proven to be high enough to be a promising gas recovery method. Bai and Li [212] demonstrated that, under certain conditions, the combination of warm water flooding and depressurization has the advantage of longer stable period of high gas rate than single production method. The theoretical analysis presented by Kamath and Godbole [39] indicated that there was good potential in the brine injection technique for thermally efficient production of gas from hydrate reservoirs if the hydrates exist in a pure state and the economics is favorable. From the results of numerical studies, Sung et al. [100] noticed that the effect of methanol injection only appeared near the well, rather than whole reservoir, possibly due to the extremely low permeability of hydrate reservoirs before dissociation occurred. From their parametric study, methanol content was identified as the most sensitive parameter in a sense of the time that a well block to be completely dissociated. Phirani and Mohanty [113] used computer model to study the gas hydrate production from CO_2 injection. It was found that to dissociate methane hydrate by CO_2 injection, the CO_2 mole fraction needs to be kept very high in the fluid phase or to be operated at a relative lower pressure. The results obtained by Phale et al. [14] indicated that the injection of CO_2-microemulsion produced considerably larger amount of methane than warm water injection alone.

2.5.2.2 Critical Factors in Recovery

The recovery process is influenced by various factors ranging from material properties, environmental conditions to operating factors. Results obtained by Bayles et al. [5] indicated that increasing the hydrate-filled porosity or increasing the zone thickness contributed to better gas production and energy efficiency ratios for steam stimulation. And deeper reservoirs appeared to yield more favorable gas production and energy efficiency ratios, a result that was dependent on the natural geothermal gradient. Also for steam and hot water stimulation, the parametric study conducted by Selim and Sloan [44] showed that the dissociation rate was a strong function of the thermal properties of the system and the porosity of the porous medium for thermal stimulation. Pawar and Zyvoloski [75] reported that the gas production predictions were sensitive to operational parameters such as injection temperature, injection rate, and pumping well pressure. For depressurization, it was observed that numerical grid size has a significant impact on simulation results. Admadi et al. [7, 167] and Esmaeilzadeh et al. [33] showed that the gas production rate was a sensitive function of well pressure. The simulation results from the study of Ruan et al. [116] showed that the depressurizing range has a significant influence on the final gas production the in the depressurization process. In contrast, the depressurizing rate only affects the production lifetime. A greater amount of cumulative gas can be produced with a larger depressurization range or a lower depressurizing rate for a certain depressurizing range. For CO_2 sequestration, the simulation results of White et al. [37] inferred that low injection pressures can be used to reduce secondary hydrate formation, and that direct contact of injected CO_2 with the methane hydrate in the formation was limited due to bypass through the higher permeability gas zone. The simulation of Phale et al. [14] indicated that a higher hydraulic conductivity resulted in much faster methane production for CO_2-slurry injection. This lent supports to the use of fracturing the formation to improve injectivity. It was also concluded that moderate concentrations and higher temperatures might help in optimizing methane recovery. However, in general, Sun and

Mohanty [22] found that an increase in initial temperature, an introduction of salt component into system, a decrease in outlet pressure, or an increase in boundary heat transfer coefficient can lead to faster hydrate dissociation. Nazridoust and Ahmadi [21] illustrated that the rate of hydrate dissociation in a core was a sensitive function of the surrounding environment temperature, outlet pressure condition, and permeability. But in the study of Sun et al. [12], gas production was not very sensitive to the temperature boundary conditions of the well. It was also found that an increase in the reservoir and well pressures and a decrease in the permeability suppressed the formation of an extended hydrate dissociation region [79].

Several factors relevant to chemical models have been identified to be able to affect simulation results. The simulation conducted by Li et al. [210] reported that higher permeability led to a lower initial saturation, a lower production pressure, a higher production rate, and a faster moving dissociation front. Liu et al. [118] also concluded that the speed of the moving front and the gas production rate were strong functions of the well pressure and the absolute permeability of porous media. In addition, their simulations showed that the assumption of stationary water phase underpredicted gas production and overpredicted the speed of the moving front. The combustion of methane yielded 44 % less CO_2 than coal, per unit energy release, and 29 % less CO_2 than oil. Similarly, Kwon et al. [62] demonstrated the speed of the moving front and the gas production rate were strong functions of the well pressure and the absolute permeability of the porous media. Gamwo and Liu [151] found that the hydrate dissociation patterns differed significantly when the thermal boundary condition was shifted from adiabatic to constant temperature. In their study, the surface area factor was found to have an important effect on the rate of hydrate dissociation for the kinetic model. The deviation between the kinetic and equilibrium models was found to increase with a decreasing surface area.

Various factors critical to geomechanical responses have also been identified. Results obtained by Ng et al. [126] showed that heat flows from the clay layer to the sand layer led to a faster hydrate dissociation rate in the hydrate region near the clay–sand boundary than that at the center of the hydrate-sand layer, which influenced the stress distribution around the wellbore. As the soil relaxed toward the wellbore, arching effect in the vertical plane can be seen in the sand layer in addition to the usual increase in the circumferential stress. This was due to the force transferred from the casing to the clay layer, which deformed greater than the sand layer during depressurization. The modeling results of Chin et al. [206] predicted the possible range of peak surface subsidence and the maximum downward displacement within the modeled formations. Li et al. [63] showed that mechanical properties of the formation became worse sharply with the thermal decomposition of hydrates, which easily resulted in the instability of the formation. It was demonstrated by Tsimpanogiannis and Lichtner [171] that, at low permeabilities of oceanic sediments, extremely high dissociation pressures developed at the hydrate dissociation front and would result in fracturing of the hydrate-sediments and led to break down of the model. In the study of Rutqvist et al. [64], depressurization resulted in an increasing shear stress within the body of the receding hydrate and caused a vertical compaction of the reservoir. The increasing shear stress might lead to shear failures in the hydrate-free zone bounded by the hydrate layer overburden and the downward-receding upper dissociation interface. The Poisson's ratio of the hydrate-bearing formation was a particularly important parameter that determined whether the evolution of the reservoir stresses will increase or decrease the likelihood of shear failure. Kim et al. [59] demonstrated that noticeable differences between one- and two-way couplings for several cases.

2.5.2.3 Governing Mechanisms for Hydrate Dissociation

The computer modeling also provided insights into the governing mechanisms of gas hydrate dissociation. There have been debates over questions such as if gas hydrate dissociation is localized or globalized or if the process is thermal conduction controlled, convection controlled, or diffusion controlled. Yang et al. [211] found that gas hydrate dissociation occurred throughout the hydrate zone, which was controlled by both mass transfer and heat transfer throughout the stages. But the sharp-interface was hardly found during the hydrate dissociation. And the ice arisen from hydrate dissociation slowed the hydrate dissociation rate below the ice point, which would affect gas production rate. The simulations and experiments of Tonnet and Herri [101] demonstrated that the dissociation limiting step switched from thermal transfer to mass transfer depending on the initial permeability and conductivity of the porous medium. In the study of Hong et al. [8], the analytical solutions for the rates of decomposition in a semi-infinite zone controlled by separate mechanisms were obtained by ignoring the other two mechanisms. For the case studied, the effect of two-phase flow was shown to be significantly smaller than heat transfer and intrinsic kinetics of hydrate decomposition. In the study of Sun et al. [12], the results showed that laboratory-scale experiments were often dissociation-controlled, but the field-scale processes were typically flow-controlled. Vasil'ev et al. [79] reported that for high-permeability rocks the convective heat transfer in the near-well space of the reservoir predominated over the conductive transfer. This indicated that the use of intra-well heaters was ineffective. Admadi et al. [7, 167] and Esmaeilzadeh et al. [33] showed that both heat conduction and convection in the hydrate zone were important during gas recovery from gas hydrates by depressurization method. Moreover, the study of Tonnet and Herri [101] established criteria for distinguishing dissociation governed by heat transfer and that governed by mass transfer. For high permeability sediments, the heat conductivity of the sediment was identified as a key parameter that controlled the dissociation kinetics; for low permeability sediment ($<5e^{-12}$ m^2), the dissociation kinetics became dependent on the parameters that influence mass transfer, i.e., the initial hydrate saturation and the absolute permeability of the sediment.

2.5.3 Research Trends and Future Needs

2.5.3.1 Physical Fields

Shown in Fig. 2.4 is a summary on the development of simulation models with respect to physical field over the last three decades. It is seen that the importance of mass transfer has been more and more emphasized. This explains the fact that there were few efforts for TC modeling, while THC models achieved a sustained growth in numbers. On the other hand, THMC models which emerged just a few years ago have been increasing dramatically. It is thus estimated that, with the maturity of THC models representing by those state-of-the-art simulators, more and more attention will be directed to the incorporation of the momentum balance module, ranging from geotechnical responses to fluid momentum.

2.5.3.2 Phases and Components

Trends in the development of simulation models regarding the number of phases are summarized in Fig. 2.5, from which a few trends can be clearly observed. It is seen that the development of two-phase models, which dominate in the current models, is slowing down in recent years. The possible reason is that two phase models have became mature and possessed the ability to deal with the problems that they are designed for. On the other hand, the amount of models designed for more than two phases and those complicated behaviors between phases are increasing rapidly. This trend may imply that, with the settlement of simple issues with single phase or two phase models, the demand for more complicated multiphase multicomponent simulation models for gas hydrate dissociation is increasing. The increase in computer capacity also helps to promote and to support the births of these more powerful simulators.

2.5.3.3 Equilibrium Versus Kinetic Models

As discussed in Sect. 2.3, there are still some disputes over the difference between equilibrium and kinetic models. Figure 2.6 summarizes a quantitative evaluation of the numbers of different types of models developed at different times. As can be seen, equilibrium models received major attention in the early stage. This is possibly due to that chemical kinetics for dissociation reactions was not well known at that time. Equilibrium models are attractive because they are able to yield qualitatively reasonable results with limited computational resources. However, after 2000, the number of kinetic models started to increase in a greater and more stable rate. At the same time, the idea to include both chemical methods (or modules) into one simulator has also attracted considerable attention.

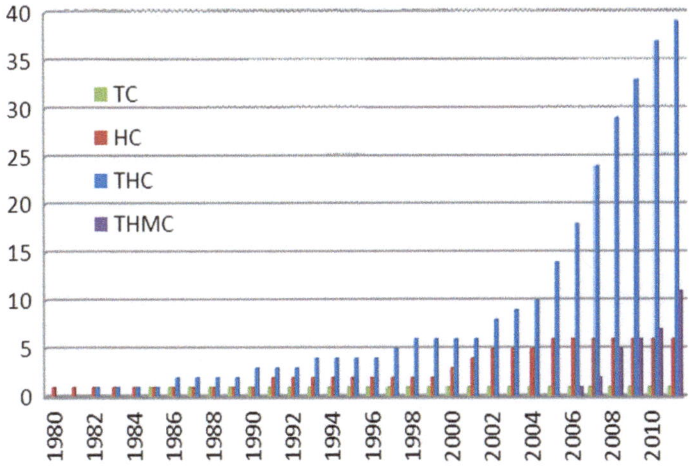

Fig. 2.4 Development of models with respect to physical fields

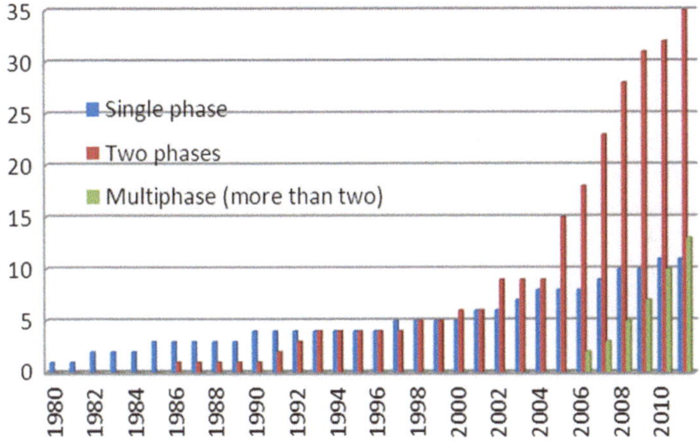

Fig. 2.5 Development of models with respect to phases in H

2.5.3.4 Environmental Effects

The detrimental effects of the naturally gas hydrate dissociation and the gas recovery from gas hydrates have been debated. As commented by Beauchamp [58], many researchers believed that, if released in the environment, the methane from hydrates would be a significant hazard to marine ecosystems, coastal populations and infrastructures, or worse, would dangerously contribute to global warming. But evidence indicated that the greatest threat to gas hydrate stability in oceanic settings does not come from minor environmental fluctuations, but rather from the buildup of free gas beneath the gas hydrate stability zone, which leads to overpressurizing and catastrophic release of gas through pockmarks expulsion,

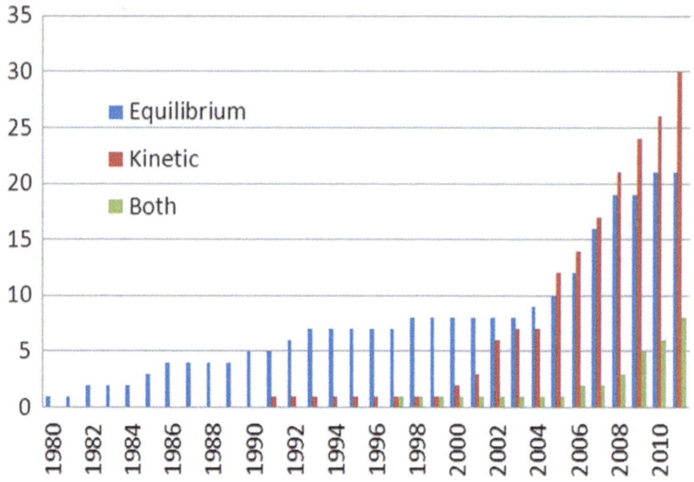

Fig. 2.6 Development of models with respect to chemical model

volcanoes, or surface seepage [213]. A truth conflicting with the assumed threats of released gas during gas recovery processes is that methane has to bypass normal fermentation processes to be a warming agent and thus needs to be released very quickly and massively. However, that type of releasing process is impossible within a conventional recovery process. Therefore, while more and more research has been launched due to the possible detrimental effects of gas hydrate dissociation, attention may also be needed to identify solid evidence to support the original judgment.

2.6 Conclusion

Understanding and modeling the gas hydrate dissociation (formation) process have been stimulating a tremendous amount of research due to the potential role of gas hydrates as a major energy resource in the future. Various computational simulation models have been developed over the past three decades. But a systematical organization of the rich collection of the literature is currently lacking. Such an effort, however, will help researchers to build a broad picture on the research in this area. This chapter conducts an integrative review on the topic of computer simulations of gas hydrate simulation with emphasis on the theoretical basis for the simulation models. A unified framework is developed to classify and integrate the existing models. The intention is to provide not only an overview but also the specific angles for understanding the literature. The major mechanisms involved in the process are illustrated and explained on the level of governing equation. The discrepancies among models are demonstrated and discussed with reference to the governing equation system. To facilitate implementation, the auxiliary relationships, which are intended to describe material properties of gas hydrates, are discussed according to their categories. The auxiliary relationships used by existing models together with other possible useful information are summarized and compared. Finally, discussions are made based on the results obtained by existing models as well as other solid data. Specific trends are revealed and beneficial conclusions are reached to provide reference for future research.

Acknowledgement This work is partially supported by the US National Science Foundation via project CMMI-0856407.

References

1. Selim M. S. & Sloan, E. D. (1985). Modeling of the dissociation of in-situ hydrate. SPE 1985 California Regional Meeting, Bakersfield, CA, March 27–29.
2. Bishnoi, P. R., & Natarajan, V. (1996). Formation and decomposition of gas hydrates. *Fluid Phase Equilibria, 117*, 168–177.
3. Englezos, P. (1993). Clathrate hydrates. *Industrial and Engineering Chemistry Research, 32*, 1251–1274.
4. Kneafsey, T. J., Tomutsa, L., Moridis, G. J., Seol, Y., Freifeld, B. M., Taylor, C. E., et al. (2007). Methane hydrate formation and dissociation in a partially saturated core-scale sand sample. *Journal of Petroleum Science and Engineering, 56*, 108–126.
5. Bayles, G. A., Sawyer, W. K., & Malone, R. D. (1986). A steam cycling model for gas production from a hydrate reservoir. *Chemical Engineering Communication, 47*, 225–245.
6. Holder, G. D., & Angert, P. F. (1982). *Simulation of gas production from a reservoir containing both gas hydrates and free natural gas*. SPE Annual Technical Conference and Exhibition, 26–29 September 1982, New Orleans, Louisiana.
7. Ahmadi, G., Ji, C., & Smith, D. H. (2007). Natural gas production from hydrate dissociation: an axisymmetric model. *Journal of Petroleum Science and Engineering, 58*, 245–258.
8. Hong, H., Pooladi-Darvish, M., & Bishnoi, P. R. (2003). Analytical modeling of gas production from hydrates in porous media. *Journal of Canadian Petroleum Technology, 42* (11), 45–56.
9. Katz, D. L. (1971). Depths to which frozen gas fields (gas hydrates) may be expected. *Journal of Petroleum Technology, 23*(4), 419–423.
10. Makogon, Y. F. (1965). Hydrate formation in the gas-bearing beds under permafrost conditions. *Gazovaia Promyshlennost, 5*, 14–15.
11. Makogon, Y. F. (1997). *Hydrates of natural gas*. Tulsa, Oklahoma: Penn Well Books.
12. Sun, X., Nanchary, N., & Mohanty, K. K. (2005). 1-D modeling of hydrate depressurization in porous media. *Transport in Porous Media, 58*, 315–338.
13. Janicki, G., Schluter, S., Hennig, T., Lyko, H., & Deergerg, G. (2011). Simulation of methane recovery from gas hydrates combined with storing carbon dioxide as hydrates. *Journal of Geological Research, 2011*, 1–15.
14. Phale, H. A., Zhu, T., White, M. D., & McGrail B. P. (2006). Simulation study on injection of CO_2-microemulsion for methane recovery from gas hydrate reservoirs. SPE Gas Technology Symposium, Calgary, Alberta, Canada, 15–17 May 2006.
15. Burshears, M., O'Brien, T. J., & Malone, R. D. (1986). *A multi-phase, multi-dimensional, variable composition simulation of gas production from a conventional gas reservoir in contact with hydrates*. Unconventional Gas Technology Symposium of the Society of Petroleum Engineers, Louisville, KY, May 18–21.
16. Rempel, A. W., & Buffett, B. A. (1997). Formation and accumulation of gas hydrate in porous media. *Journal of Geophysical Research, 102*(5), 151–164.

17. Kvenvolden, K. A., Carlson, P. R., & Threlkeld, C. N. (1993). Possible connection between two Alaskan catastrophes occurring 25 years apart (1964 and 1989). *Geology, 21*, 813–816.
18. Booth, J. S., Rowe, M. M., & Fischer, K. M. (1996). *Offshore gas hydrate sample database with an overview and preliminary analysis*. U.S. Geological Survey, Open File Report 96-272, Denver, Colorado.
19. MacDonald, G. J. (1990). The future of methane as an energy resource. *Annual Review of Energy, 15*, 53–83.
20. White, M. D., & McGrail, B. P. (2008). Numerical simulation of methane hydrate production from geologic formations via carbon dioxide injection. 2008 Offshore Technology Conference, Houston, Texas, 5–8 May.
21. Nazridoust, K., & Ahmadi, G. (2007). Computational modeling of methane hydrate dissociation in a sandstone core. *Chemical Engineering Science, 62*, 6155–6177.
22. Sun, X., & Mohanty, K. K. (2006). Kinetic simulation of methane hydrate formation and dissociation in porous media. *Chemical Engineering Science, 61*, 3476–3495.
23. Hammerschmidt, E. G. (1934). Formation of gas hydrates in natural gas transmission lines. *Industrial, 26*(8), 851–855.
24. Makogon, Y. F., Holditch, S. A., & Makogon, T. Y. (2007). Natural gas-hydrates—a potential energy source for the 21st Century. *Journal of Petroleum Science and Engineering, 56*, 14–31.
25. Maksimov, A. M. (1992). *Mathematical model of the volume dissociation of gas-phase hydrates in a porous medium with water-phase mobility*. Moscow: Institute for Gas and Oil Problems, Academy of Sciences of the USSR and GKNO of the USSR.
26. Dickens, G. R. (2003). Rethinking the global carbon cycle with a large dynamic and microbially mediated gas hydrate capacitor. *Earth and Planetary Science Letters, 213*, 169–183.
27. Kennett, J. P., Cannariato, K. G., Hendy, I. L., & Behl, R. J. (2000). Carbon isotopic evidence for methane hydrate stability during Quaternary Interstadials. *Science, 288*, 128–133.
28. Kayen, R. E., & Lee, H. J. (1991). Pleistocene slope instability of gas hydrate-laden sediment on the Beaufort Sea margin. *Marine Georesources & Geotechnology, 10*, 125–141.
29. Paull, C. K., Buelow, W. J., Ussler, W., & Borowski, W. S. (1996). Increased continental-margin slumping frequency during sea-level lowstands above gas hydrate-bearing sediments. *Geology, 24*, 143–146.
30. Moridis, G. J., & Sloan, E. D. (2006). Gas production potential of disperse low-saturation hydrate accumulations in oceanic sediments. LBNL-52568, Berkeley, CA: Lawrence Berkeley National Laboratory.
31. Moridis, G. J., & Collett, T. S. (2003). *Strategies for gas production from hydrate accumulations under various geologic conditions*. LBNL-52568, Berkeley, CA: Lawrence Berkeley National Laboratory
32. Moridis, G. J., Kneafsey, T. J., Kowalsky, M., & Reagan, M. (2006). *Numerical, laboratory and field studies of gas production from natural hydrate accumulations in geologic media*. Berkeley, CA: Earth Science Division, Lawrence Berkeley National Laboratory.
33. Esmaeilzadeh, F., Zeighami, M. E., & Fathi, J. (2008). 1-D modeling of hydrate decomposition in porous media. *Proceedings of World Academy of Science, Engineering and Technology, 41*, 647–653.
34. Hyndman, R. D., & Davis, E. E. (1992). A mechanism for the formation of methane hydrate and seafloor bottom-simulating reflectors by vertical fluid expulsion. *J. Geophys. Res., 97*, 7025–7041.
35. Kowalsky, M. B., & Moridis, G. J. (2007). *Comparison of kinetic and equilibrium reaction models in simulating gas hydrate behavior in porous media*. Berkeley, CA: Earth Science Division, Lawrence Berkeley National Laboratory.
36. Uddin, M., Coombe, D., Law, D., & Gunter, B. (2008). Numerical studies of gas hydrate formation and decomposition in a geological reservoir. *Journal of Energy Resources Technology, 130*, 032501-1–032501-14.

37. White, M. D., Wurstner, S. K., & McGrail, B. P. (2009). Numerical studies of methane production from Class 1 gas hydrate accumulations enhanced with carbon dioxide injection. *Marine and Petroleum Geology, 28*(2), 546–560.
38. Goel, N. (2006). In situ methane hydrate dissociation with carbon dioxide sequestration: current knowledge and issues. *Journal of Petroleum Science and Engineering, 51*, 169–184.
39. Kamath, V. A., & Godbole, S. P. (1987). Evaluation of hot-brine simulation technique for gas production from natural gas hydrates. *Journal of Petroleum Technology, 39*, 1379–1388.
40. Sloan, E. D., & Koh, C. A. (2007). *Clathrate hydrates of natural gases* (3rd ed.). Boca Raton: CRC Press.
41. Graue, A., Kvamme, B., Baldwin, B., Stevens, J., Howard, J., & Aspenes, E. (2006). Environmentally friendly CO_2 storage in hydrate reservoirs benefits from associated spontaneous methane production. In Proceedings of the Offshore Technology Conference (OTC-18087), Huston, Texas, United States.
42. Stevens, J., Howard, J., Baldwin, B., Ersland, B., Huseb, J., & Graue, A. (2008). Experimental hydrate formation and production scenarios based on CO2 sequestration. Proceedings of the 6th International Conference on Gas Hydrates (ICGH 2008), Vancouver, British Columbia, Canada, July 6–10, 2008.
43. McGruire, P. L. (1982). Methane hydrate gas production by thermal stimulation. Proceedings of the Fourth Canadian Permafrost Conference, H.M. French (ed.), Calgary.
44. Selim, M. S., & Sloan, E. D. (1990). Hydrate dissociation in sediments. *SPE Reservoir Engineering, 5*(2), 245–251.
45. Yousif, M. H., Abass, H. H., Selim, M. S., & Sloan, E. D. (1991). Experimental and theoretical investigation of methane-gas-hydrate dissociation in porous media. *SPE Reservoir Engineering, 6*(1), 69–76.
46. Guo, T., Wu, B., Zhu, Y., Fan, S., & Chen, G. (2004). A review on the gas hydrate research in China. *Journal of Petroleum Science and Engineering, 41*, 11–20.
47. Sloan, E. D. (2003). Clathrate hydrate measurements: microscopic mesoscopic, and macroscopic. *The Journal of Chemical Thermodynamics, 35*, 41–53.
48. Pooladi-Darvish, M. (2004). Gas production from hydrate reservoirs and its modeling. *Society of Petroleum Engineers, 56*(6), 65–71.
49. Davy, H. (1881). *The bakerian lecture on some of the combinations of oxymuriatic gas and oxygen, and on the chemical relations of these principles to inflammable bodies.* Philosophical Transactions of the Royal Society, London 1811,101, (Part I), pp. 1-35.
50. Faraday, M. (1823). On fluid chlorine. *Philosophical Transactions of the Royal Society B: Biological Sciences, London, 113*, 160–165.
51. Davidson, D. W. (1973). Gas hydrates. In F. Frank (Ed.), *Water: A comprehensive treatise* (Vol. 2, pp. 115–234). New York: Plenum Press. Chapter 3.
52. Deaton, W. M., & Frost, E. M. Jr. (1946). US Bureau of Mines Monograph 8, No. 8.
53. Chersky, N. J., & Makogon, Y. F. (1970). Solid gas world reserves are enormous. *Oil Gas International, 10*(8), 82–84.
54. Makogon, Y. F., Trebin, F. A., Trofimuk, A. A., Tsarev, V. P., & Chersky, N. V. (1972). Detection of a pool of natural gas in a solid (hydrate gas) state. *Doklady Akademii Nauk SSSR, 196*, 203–206. originally published in Russian, 1971.
55. Shipley, T. H., Houston, K. J., Buffler, R. T., Shaub, F. J., McMillen, K. J., Ladd, J. W., et al. (1979). Seismic evidence for widespread possible gad hydrate horizons on continental slopes and rises. *American Association of Petroleum Geologists Bulletin, 63*, 2204–2213.
56. Stoll, R. D., & Bryan, G. M. (1979). Physical properties of sediments containing gas hydrates. *Journal of Geophysical Research, 84*, 1629–1634.
57. Finlay, P., & Krason, J. (1990). Evaluation of geological relationships to gas hydrate formation and stability: Summary report., Gas Energy Rev. *Vol., 18*, 12–18.
58. Beauchamp, B. (2004). Natural gas hydrates: myths facts and issues. *Comptes Rendus Geoscience, 226*, 751–765.

59. Kim, J., Yang, D., & Rutqvist, J. (2011). *Numerical studies on two-way coupled fluid flow and geomechanics in hydrate deposits*. SPE Reservoir Simulation Symposium, Woodlands, Texas, 21–23 February.
60. Klar, A., & Soga, K. (2005). *Coupled deformation-flow analysis for methane hydrate production by depressurized wells*. Proceeding of 3rd International Biot Conference on Poromechanics, pp. 653–659.
61. Koh, C. A., & Sloan, E. D. (2007). Natural gas hydrates: recent advances and challenges in energy and environmental applications. *American Institute of Chemical Engineers, 53*(7), 1636–1643.
62. Kwon, T., Song, K., & Cho, G. (2010). Destabilization of marine gas hydrate-bearing sediments induced by a hot wellbore: a numerical approach. *Energy Fuels, 24*, 5493–5507.
63. Li, L., Cheng, Y., Zhang, Y., Cui, Q., & Zhao, F. (2011). A fluid-solid coupling model of wellbore stability for hydrate bearing sediments. *Procedia Engineering, 18*, 363–368.
64. Rutqvist, J., Moridis, G. J., Grover, T., & Collett, T. (2009). Geomechanical response of permafrost-associated hydrate deposits to depressurization-induced gas production. *Journal of Petroleum Science and Engineering, 67*, 1–12.
65. Yamamoto, K. (2008). Methane hydrate bearing sediments: a new subject of geomechanics. The 12th International Conference of International Association for Computer Methods and Advances in Geomechanics (IACMAG), 1–6 October 2008, Goa, India.
66. Gudmundsson, J., & Borrehaug, A. (1996) Frozen hydrate for transport of natural gas. Proc. 2nd Int. Conf.on Natural Gas Hydrates, pp. 415–422.
67. Gudmundsson, J., Andersson, V., Levik, O. I., Mork, M., & Borrehaug, A. (2000). Hydrate technology for capturing stranded gas. *Ann. NY Acad. Science, 912*, 403–410.
68. Stern, L. A., Circone, S., Kirby, S. H., & Durham, W. B. (2001). Anomalous preservation of pure methane hydrate at 1 atm. *The Journal of Physical Chemistry, 105*(9), 1756–1762.
69. Takaoki, T., Hirai, K., Kamei, M., & Kanda, H. (2005). Study of natural gas hydrate (NGH) carriers. Proceedings of the Fifth International Conference on Natural Gas Hydrates, June 13-16, Trondheim, Norway. Paper 4021.
70. Florusse, L. J., Peters, C. J., Schoonman, J., Hester, K. C., Koh, C. A., Dec, S. F., et al. (2004). Stable low-pressure hydrogen clusters stored in a binary clathrate hydrate. *Science., 306* (5695), 469–471.
71. Mao, W. L., Mao, H., Goncharov, A. F., Struzhkin, V. V., Guo, Q., Hu, J., et al. (2002). Hydrogen clusters in clathrate hydrate. *Science, 297*, 2247–2249.
72. Hunt, S. C. (1992). *Gas hydrate thermal energy storage system*. United States Patent No. 5140824.
73. Guo, K. H., Shu, B. F. & Yang, W. J. (1996). *Advances and applications of gas hydrate thermal energy storage technology*. Proceedings of 1st Trabzon Int. Energy and Environment
74. Chen, G. J., Sun, C. Y., Ma, C. F., & Guo, T. M. (2002). *A new technique for separating (Hydrogen + Methane) gas mixtures using hydrate technology*. Proceedings of the 4th International Conference on Gas Hydrates, May 19-23, 2002, Yokohama, Japan, pp. 1016–1020.
75. Pawar, R. J., & Zyvoloski, G. A. (2005). Numerical simulation of gas production from methane hydrate reservoirs. Proceedings of the Fifth International Conference on Gas Hydrates, Trondheim, Norway, pp. 259–267.
76. McGuire, P. L. (1981). Methane hydrate gas production by thermal stimulation. Proceedings of the Fourth Canadian Permafrost Conference, March 2-6, 1981, Calgary, Alberta.
77. Goel, N., Wiggins, M., & Shah, S. (2001). Analytical modeling of gas recovery from in situ hydrates dissociation. *Journal of Canadian Petroleum Technology, 29*, 115–127.
78. Ji, C., Ahmadi, G., & Smith, D. H. (2001). Natural gas production from hydrate decomposition by depressurization. *Chemical Engineering Science, 56*, 5801–5814.
79. Vasil'ev, V. N., Popov, V. V., & Tsypkin, G. G. (2006). Numerical investigation of the decomposition of gas hydrates coexisting with gas in natural reservoirs. *Fluid Dynamics, 41* (4), 599–605.

80. Bai, Y., Li, Q., Li, X., & Du, Y. (2008). The simulation of nature gas production from ocean gas hydrate reservoir by depressurization. *Science in China Series E: Technological Sciences, 51*(8), 1272–1282.
81. Bai, Y., Li, Q., Li, X., & Du, Y. (2008). The simulation of nature gas production from ocean gas hydrate reservoir by depressurization. *Science in China Series E: Technological Sciences, 51*(8), 1272–1282.
82. Tsypkin, G. G. (2007). Analytical solution of the nonlinear problem of gas hydrate dissociation in a formation. *Fluid Dynamics, 42*(5), 798–806.
83. Gerami, S., & Pooladi-Darvish, M. (2007). Predicting gas generation by depressurization of gas hydrates where the sharp-interface assumption is not valid. *Journal of Petroleum Science and Engineering, 56*, 146–164.
84. Hong, H., & Pooladi-Darvish, M. (2005). Simulation of depressurization for gas production from gas hydrate reservoirs. *Journal of Canadian Petroleum Technology, 44*(11), 39–46.
85. Mandelcorn, L. (1959). Clathrates. *Chemical Reviews, 59*, 827–839.
86. van der Waals, J. H., & Platteeuw, J. C. (1959). Clathrate Solutions. *Advances in Chemical Physics, 2*, 1–57.
87. Byk, S. S., & Fomina, V. J. (1968). Gas Hydrates. *Russian Chemical Reviews, 37*(6), 469–491.
88. Hand, J. H., Katz, D. L., & Verma, V. K. (1974). Review of gas hydrates with implication for ocean sediments. In I. R. Kaplan (Ed.), *Natural Gases in Marine Sediments* (pp. 179–194). New York: Plenum.
89. Jeffrey, G. A., & McMullan, R. K. (1967). The clathrate hydrates. *Progress in inorganic chemistry, 8*, 43–108.
90. Jeffrey, G. A. (1984). Hydrate inclusion compounds. *Journal of Inclusion Phenomena and Macrocyclic Chemistry, 1*(3), 211–222.
91. Holder, G. D., Zetts, S. P., & Pradham, N. (1988). Phase behavior in systems containing clathrate hydrates. *Reviews in Chemical Engineering, 5*(l), 1–69.
92. Makogon, Y. F. (1981). *Hydrates of natural gas*. Tulsa, OK: Penn Well Publishing. (translated by W. J. Cieslewicz).
93. Berecz, E., & Balla-Achs, M. (1983). *Studies in Inorganic Chemistry 4: Gas Hydrates* (pp. 184–188). Amsterdam: Elsevier.
94. Cox, J. L. (1983). *Natural gas hydrates: Properties, occurrence and recovery*. Woburn, MA: Butterworth Publiehere.
95. Sloan, E. D. (1990). *Clathrate hydrates of natural gases*. New York: Dekker.
96. Sloan, E. D., Jr. (1998). *Clathrate hydrates of natural gases* (2nd ed.). New York, NY: Marcel Deckker Inc.
97. Buffett, B. A. (2000). Clathrate hydrates. *Annual Review of Earth and Planetary Sciences, 28*, 477–507.
98. Koh, C. A. (2002). Towards a fundamental understanding of natural gas hydrates. *Chemical Society Reviews, 31*, 157–167.
99. Waite, W. F., Santamarina, J. C., Cortes, D. D., Dugan, B., Espinoza, D. N., Germaine, J., et al. (2009). Physical properties of hydrate-bearing soils. *Reviews of Geophysics, 47*, RG4003.
100. Sung, W., Lee, H., Lee, H., & Lee, C. (2002). Numerical study for production performances of a methane hydrate reservoir stimulated by inhibitor injection. *Energy Sources, 24*, 499–512.
101. Tonnet, N., & Herri, J. M. (2009). Methane hydrates bearing synthetic sediments-experimental and numerical approaches of the dissociation. *Chemical Engineering Science, 64*(19), 4089–4100.
102. Yu, F., Song, Y., Liu, W., Li, Y., & Lam, W. (2011). Analyses of stress strain behavior and constitutive model of artificial methane hydrate. *Journal of Petroleum Science and Engineering, 77*, 183–188.

103. Bagheri, M., & Settari, A. (2008). Modeling of geomechanics in naturally fractured reservoirs. *SPE Reservoir Evaluation & Engineering, 11*(1), 108–118.
104. Freeman, T. L, Chalatumyk, R. J., & Bogdanov, I. I. (2009). *Geomechanics of heterogeneous bitumen carbonates*. SPE Reservoir Simulation Symposium, 2-4 February 2009, The Woodlands, Texas.
105. Kosloff, D., Scott, R., & Scranton, J. (1980). Finite element simulation of Wilmington oil field subsidence: I linear modelling. *Tectonophysics, 65*, 339–368.
106. Lewis, R. W., & Schreflei, B. A. (1998). *The finite element method in the deformation and consolidation of porous media*. Wiley, Chichester, Great Britain, 2nd edition.
107. Merle, H. A., Kentie, C. J. P., van Opstal, G. H. C., & Schneider, G. M. G. (1976). The Bachaquero study—a composite analysis of the behavior of a compaction drive/solution gas drive reservoir. *Journal of Petroleum Technology, 28*(9), 1107–1115.
108. Morris, J. P. (2009). Simulations of injection-induced mechanical deformation: A study of the In Salah CO_2 storage project. Society of Exploration Geophysicists 2009 Summer Research Workshop, Banff, Canada, August, 2009.
109. Rutqvist, J., & Moridis, G. J. (2009). Numerical Studies on the Geomechanical Stability of Hydrate-Bearing Sediments. *SPE Journal, 14*(2), 267–282.
110. Allen, M. B. (1954). In: M.B. Allen, G.A. Behie, and J.A. Trangenstein (Eds.), *Multiphase flow in porous media: Mechanics, mathematics, and numerics*. New York, Berlin: Springer-Verlag, 1988.
111. Kimoto, S., Oka, F., Fushita, T., & Fujiwaki, M. (2007). A chemo-thermo-mechanically coupled numerical simulation of the subsurface ground deformations due to methane hydrate dissociation. *Computers and Geotechnics, 34*, 216–228.
112. Garg, S. K., Pritchett, J. W., Katoh, A., Baba, K., & Fujii, T. (2008). A mathematical model for the formation and dissociation of methane hydrates in the marine environment. *Journal of Geophysical Research, 113*, B01201.
113. Phirani, J., & Mohanty, K. L. (2010). *Kinetic simulation of CO_2 flooding of methane hydrates*. SPE Annual Technical Conference and Exhibition, Florence, Italy, 19–22 September.
114. Liu, Y., & Gamwo, I. K. (2012). Comparison between equilibrium and kinetic models for methane hydrate dissociation. *Chemical Engineering Science, 69*, 193–200.
115. Tsypkin, G. G. (1998). Decomposition of gas hydrates in low-temperature reservoirs. *Fluid Dynamics, 33*(1), 82–90.
116. Ruan, X., Song, Y., Zhao, J., Liang, H., Yang, M., & Li, Y. (2012). Numerical simulation of methane production from hydrates induced by different depressurizing approaches. *Energies, 5*, 438–458.
117. Liu, X., & Flemings, P. B. (2007). Dynamics multiphase flow model of hydrate formation in marine sediments. *Journal of Geophysical Research, 112*, B03101.
118. Liu, Y., Strumendo, M., & Arastoopour, H. (2008). Numerical simulation methane production from a methane hydrate formation. *Industrial & Engineering Chemistry Research, 47*, 2817–2828.
119. Masuda, Y., Kurihara, M., Ohuchi, H., & Sato, T. (2002). A field-scale simulation study on gas productivity of formations containing gas hydrates. Proceedings of 4th International Conference on Gas Hydrates, Yokohama, Japan, May 19–23.
120. Schnurle, P., & Liu, C. (2011). Numerical modeling of gas hydrate emplacements in oceanic sediments. *Marine and Petroleum Geology, 28*, 1856–1869.
121. Scott, D. M., Das, D. K., & Subbaihaannadurai, V. (2006). A finite element computational method for gas hydrate. Part I: theory. *Petroleum Science and Technology, 24*, 895–909.
122. Campbell, G. S. (1985). *Soil physics with BASIC: transport models for soil-plant systems* (1st ed.). BV Amsterdam, Netherlands: Elsevier Sci.
123. De Vries, D. A. (1963). Thermal properties of soils. In W. R. van Wijk (Ed.), *Physics of plant environment* (pp. 210–235). Amsterdam: North-Holland Publ. Co.
124. Bai, Y., Li, Q., Li, F., & Du, Y. (2009). Numerical simulation on gas production from a hydrate reservoir underlain by a free gas zone. *Chinese Science Bulletin, 54*, 865–872.

125. Du, Q., Li, Y., Li, S., Sun, J., & Jiang, Q. (2007). Mathematical model for natural gas hydrate production by heat injection. *Petroleum Exploration and Development, 34*(4), 470–487.
126. Ng, M. Y. A., Klar, A., & Soga, K. (2008). Coupled soil deformation-flow-thermal analysis of methane production in layered methane hydrate soils. 2008 Offshore Technology Conference, Houston, Texas, 5–8 May.
127. Tsypkin, G. G. (1993). Mathematical model of the dissociation of gas hydrates coexisting with ice in natural reservoirs. Translated from Izvestiya Rossiiskoi Akademii Nauk, Mekhanika Zhidkosti i Gaza, No. 2, pp. 84-92, March–April.
128. Chen, Z., Bai, W., & Xu, W. (2005). Prediction of stability zones and occurrence zones of multiple composition natural gas hydrate in marine sediment. *Chinese Journal of Geophysics, 48*(4), 939–945.
129. Williams, P. J. (1964). Specific heat and apparent specific heat of frozen soils. pp. 225–229. In 1st International Conference of Permafrost, 1964, National Academy of Sciences, Washington, DC.
130. Anderson, D. M., & Morgenstern, N.R. (1973). *Physics, chemistry and mechanics of frozen ground: A review*. In Proceeding of 2nd International Conference of Permafrost, Yakutsk, Siberia, 13-28 July 1973, National Academy of Sciences, Washington, DC, pp. 257–288.
131. Quintard, M., & Whitaker, S. (1995). Local thermal equilibrium for transient heat conduction: theory and comparison with numerical experiments. *International Journal of Heat and Mass Transfer, 38*(15), 2779–2796.
132. Henninges, J., Schrötter, J., Erbas, K., & Huenges, E. (2002). Temperature field of the Mallik gas hydrate occurrence. Implications on phase changes and thermal properties, GEO Technologien 2002.
133. Perry, R. H., & Chilton, C. H. (1973). *Chemical engineers handbook*. New York, NY: McGraw Hill.
134. Waite, W. F., Stern, L. A., Kirby, S. H., Winters, W. J., & Mason, D. H. (2007). Simultaneous determination of thermal conductivity thermal diffusivity and specific heat in sI methane hydrate. *Geophysical Journal International, 169*, 767–774.
135. Liu, Z., Sun, Y., & Yu, X. (2012). Theoretical basis for Modeling Porous Geomaterials under Frost Actions: A Review. *Soil Science Society of America Journal, 76*(2), 313–330.
136. Johansen, O. (1975). Thermal conductivity of soils. Ph.D. dissertation. Norwegian University of Science and Technology, Trondheim (CRREL draft transl. 637, 1977).
137. Cote, J., & Konrad, J. M. (2005). A generalized thermal conductivity model for soils and construction materials. *Canadian Geotechnical Journal, 42*, 443–458.
138. Lu, S., Ren, T., Gong, Y., & Horton, R. (2007). An improved model for predicting soil thermal conductivity from water content at room temperature. *Soil Science Society of America Journal, 71*, 8–14.
139. Gaddipati, M. (2008). Code comparison of methane hydrate reservoir simulators using CMG STARS, Master Thesis. West Virginia University, Morgantown, West Virginia.
140. Sean, W., Sato, T., Yamasaki, A., & Kiyono, F. (2007). CFD and experimental study on methane hydrate dissociation part No. dissociation under water flow. *American Institute of Chemical Engineers, 53*(1), 262–274.
141. Kimoto, S., Oka, F., & Fushita, T. (2011). A chemo-thermo-mechanically coupled analysis of ground deformation induced by methane hydrate dissociation. *Bifurcations, Instabilities and Degradations in Geomaterials, Springer Series in Geomechanics and Geoengineering, 0*, 145–165.
142. Konno, Y., Oyama, H., Nagao, J., Masuda, Y., & Kurihara, M. (2010). Numerical analysis of the dissociation experimental of naturally occurring gas hydrate in sediment cores obtained at the Eastern Nankai Trough, Japan. *Energy Fuels, 24*(12), 6353–6358.
143. Civan, F. C. (2001). Scale effect on porosity and permeability: Kinetics, model and correlation. *AIChE Journal, 47*(2), 271–287.

144. Jeannin, L., Bayi, A., Renard, G., Bonnefoy, O., & Herri, J. M. (2002). *Formation and dissociation of methane hydrates in sediments part II: numerical modeling*. Proceeding of 4th International Conference on Gas Hydrates, Yokahama, Japan, May 19–23.
145. Sung, W., Huh, D., Ryu, B., & Lee, H. (2000). Development and application of gas hydrate reservoir simulator based on depressurizing mechanism. *Korean Journal of Chemical Engineering, 17*(3), 344–350.
146. Van Genuchten, M. T. (1980). A close-form equation for predicting the hydraulic conductivity of unsaturated soil. *Soil Science Society of America Journal, 44*, 892–898.
147. Parker, J. C., Lenhard, R. J., & Kuppusamy, T. (1987). A parametric model for constitutive properties governing multiphase flow in porous media. *Water Resources Research, 23*, 618–624.
148. Bear, J. (1972). *Dynamics of Fluids in Porous Media*. Mineola, NY: Dover.
149. Brooks, R. H., & Corey, A. T. (1964). *Hydraulic properties of porous media*. Hydrology Papers, No. 3, Colorado State University, Fort Collins.
150. Lake, L. W. (1989). *Enhanced Oil Recovery*. Upper Saddle River, NJ: Prentice-Hall.
151. Gamwo, I. K., & Liu, Y. (2010). Mathematical modeling and numerical simulation of methane production in a hydrate reservoir. *Industrial & Engineering Chemistry Research, 49*, 5231–5245.
152. Verigin, N. N., No, L. K., & Khalikov, G. A. (1980). Linear problem of the dissociation of the hydrates of a gas in a porous medium. *Fluid Dynamics, 15*(1), 144–147.
153. Willhite, P. G. (1986). *Waterflooding. Society of Petroleum Engineers Textbook Series* (Vol. 3). Texas: Society of Petroleum Engineers.
154. Williams, P. J., & Smith, M. W. (1989). *The frozen earth: Fundamentals of geocryology*. Cambridge: Cambridge University Press.
155. Fayer, M. J. (2000). UNSAT-H version 3.0: Unsaturated soil water and heat flow model, theory, user manual, and examples. Rep. 13249.Pac. Northwest Natl. Lab., Richland, WA.
156. Fredlund, D. G., & Xing, A. (1994). Equations for the soil-water characteristic curve. *Canadian Geotechnical Journal, 31*, 521–532.
157. Vogel, T., van Genuchten, M. T., & Cislerova, M. (2001). Effect of the shape of the soil hydraulic functions near saturation on variably saturated flow predictions. *Advances in Water Resources, 24*, 133–144.
158. Grant, S. A., & Salehzadeh, A. (1996). Calculation of temperature effects on wetting coefficients of porous solids and their capillary pressure functions. *Water Resources Research, 32*(2), 261–270.
159. Hassanizadeh, S. M., & Gary, W. G. (1993). Thermodynamic basics of capillary pressure in porous media. *Water Resources Research, 29*, 3389–3405.
160. Morrow, N. R. (1969). *Physics and thermodynamics of capillary. In Symposium on Flow Through Porous Media*. Washington, DC: The Carnegie Inst.
161. Burdine, N. T. (1953). Relative permeability calculations from pore-size distribution data. *Transactions of the American Institute of Mining, Metallurgical and Petroleum Engineers, 198*, 71–78.
162. Childs, E. C., & Collis-George, G. (1950). The permeability of porous materials. *Proceedings of the Royal Society of London. Series A, 201*, 392–405.
163. Webb, S. W. (1998). Gas-phase diffusion in porous media-evaluation of an advective-dispersive formulation and the dusty-gas model for binary mixtures. *Journal of Porous Media, 1*(2), 187–199.
164. Pruess, K., & Moridis, G. (1999). TOUGH2 User's Guide, Version 2.0. LBNL-43134. Lawrence Berkley National Laboratory, University of California, Berkley, CA.
165. Yaws, C. L. (1995). *Handbook of Transport Property Data: Viscosity, Thermal Conductivity, and Diffusion Coefficients of Liquids and Gases*. Houston, TX: Gulf Publishing Company.
166. Haeckel, M., & Wallmann K., et al. (2010). *Main equations for gas hydrate modeling*. SUGAR Internal Communication.

167. Ahmadi, G., Ji, C., & Smith, D. H. (2004). Numerical solution for natural gas production from methane hydrate dissociation. *Journal of Petroleum Science and Engineering, 41*, 269–285.
168. Peaceman, D.W. (1977). *Fundamentals of numerical reservoir simulation*. Amsterdam: Elsevier Scientific Pub. Co.
169. Weast, R. C. (1987). *CRC handbook of chemistry and physics*. Boca Raton: CRC Press, Inc.
170. Shpakov, V. P., Tse, J. S., Tulk, C. A., Kvamme, B., & Belosludov, V. R. (1998). Elastic moduli calculation and instability in structure I methane clathrate hydrate. *Chemical Physics Letters, 282*(2), 107–114.
171. Tsimpanogiannis, I. N., & Lichtner, P. C. (2007). Parametric study of methane hydrate dissociation in oceanic sediments driven by thermal stimulation. *Journal of Petroleum Science and Engineering, 56*, 165–175.
172. Sloan, E. D. (1998). *Clathrate hydrates of natural gases* (2nd ed.). New York, NY: Marcel Dekker.
173. Moridis, G. J. (2002). Numerical studies of gas production from methane hydrates. *SPE Journal, 8*(4), 1–11.
174. Bakker, R. (1998). Improvements in clathrate modeling II: The H_2O-CO_2-CH_4-N_2-C_2H_6 fluid system. In J. P. Henriet & J. Mienert (Eds.), *Gas hydrates: Relevance to world margin stability and climate change* (Vol. 137, pp. 75–105). London: Geological Society Special Publication.
175. Adisasmito, S., Frank, R. J., & Sloan, E. D. (1991). Hydrates of carbon dioxide and methane mixtures. *Journal of Chemical & Engineering Data, 36*, 68–71.
176. Moridis, G. J. (2003). Nummerical Studies of Gas Production from Methane Hydrates. *SPE Journal, 8*(4), 359–370.
177. Tishchenko, P., Hensen, C., Wallmann, K., & Wong, C. S. (2005). Calculation of the stability and solubility of methane hydrate in seawater. *Chemical Geology, 219*, 37–52.
178. Holder, G. D., & John, V. T. (1985). Thermodynamics of multicomponent hydrate forming mixtures. *Fluid Phase Equilibria, 14*, 353–361.
179. Kim, H. C., Bishnoi, P. R., Heidemann, R. A., & Rizvi, S. S. H. (1987). Kinetics of methane hydrate decomposition. *Chemical Engineering Science, 42*(7), 1645–1653.
180. Peng, D., & Robinson, D. B. (1976). A new two-constant equation of state. *Industrial & Engineering Chemistry Fundamentals, 15*(1), 59–64.
181. Amyx, J. W., Bass, D. M., & Whiting, R. L. (1960). *Petroleum reservoir engineering-physical properties*. New York City: McGraw-Hill Book Co.
182. Englezos, P., Kalogerakis, N., Dholabhai, P. D., & Bishnoi, P. R. (1987). Kinetics of formation of methane and ethane gas hydrates. *Chemical Engineering Science, 42*, 2647–2658.
183. Boswell, R., Kleinberg, R., Collett, T., & Frye M. (2007). *Exploration priorities for methane gas hydrate resources*. Fire in the Ice, 1194 Spring/Summer 2007. pp. 11-13. (USDOE National Energy Technology Laboratory, Hydrate Newsletter).
184. Yamamoto, K., Yasuda, M., & Osawa, O. (2005). Geomechanical condition of deep water unconsolidated and hydrate related sediments off the Pacific coast of central Japan. Proceeding of 5th International Conference on Gas Hydrate, Trondheim, Norway, Vol.3, 922 (Paper ref.3031), 13–16 June, 2005.
185. Brugada, J., Cheng, Y. P., Soga, K., & Santamarina, J. C. (2010). Discrete element modelling of geomechanical behaviour of methane hydrate soils with pore-filling hydrate distribution. *Granular Matter, 12*(5), 517–525.
186. Soga, K., Lee, S. L., Ng, M. Y. A., & Klar, A. (2006). Characterization and engineering properties of methane hydrate soils. Proceedings of the Second International Workshop on Characterization and Engineering Properties of Natural Soils, Singapore, 29 November-1 December, Taylor & Francis, London.
187. Durham, W., Kirby, S., & Stern, L. (2003). The strength and rheology of methane hydrate. *Journal of Geophysical Research, A, Space Physics, 108*, 2182–2193.

188. Nixon, M. F., & Grozic, J. L. H. (2007). Submarine slope failure due to hydrate dissociation: A preliminary quantification. *Canadian Geotechnical Journal, 44*, 314–325.
189. Xu, W., & Germanovich, L. N. (2006). Excess pore pressure resulting from methane hydrate dissociation in marine sediments: A theoretical approach. *Journal of Geophysical Research, 111*, B01104.
190. Santamarina, J. C., & Ruppel, C. (2008). The impact of hydrate saturation on the mechanical , electrical, and thermal properties of hydrate-bearing sand, silts, and clay. Proceedings of the 6th International Conference on Gas Hydrates (ICGH 2008).
191. Ran, H., Silin, D. B., & Patzek, T. W. (2008). Micromechanics of hydrate dissociation in marine sediments by grain-scale simulations. 2008 SPE Western Regional and Pacific Section AAPG Joint Meeting, Bakersfield, CA, 31 March-2 April.
192. Masui, A., Miyazaki, K., Haneda, H., Ogata, Y., & Aoki, K. (2008). *Mechanical characteristics of natural and artificial gas hydrate bearing sediments*. Proceedings of the 6th International Conference on Gas Hydrates, ICGH.
193. Winters, W. J., Pecher, I. A., Waite, W. F., & Mason, D. H. (2004). Physical properties and rock physics models of sediment containing natural and laboratory-formed methane gas hydrate. *American Mineralogist, 89*, 1221–1227.
194. Hyodo, M., Nakata, Y., Yoshimoto, N., & Ebinuma, T. (2005). Basic research on the mechanical behavior of methane hydrate-sediments mixture. *Soils & Foundations, 45*(1), 75–85.
195. Masui, A., Haneda, H., Ogata, Y., & Aoki, K. (2005). *The effect of saturation degree of methane hydrate on the shear strength of synthetic methane hydrate sediments*. Proceedings of the 5th International Conference on Gas Hydrates (ICGH 2005), pp. 2657–2663.
196. Winters, W. J., Waite, W. F., Mason, D. H., Gilbert, L. Y., & Pecher, I. A. (2007). Methane gas hydrate effect on sediment acoustic and strength properties. *Journal of Petroleum Science and Engineering, 56*, 127–135.
197. Hyodo, M., Nakata, Y., Yoshimoto, N., & Orense, R. (2007). Shear behavior of methane hydrate-bearing sand. Proceedings of the 17th International Offshore and Polar Engineering Conference, ISOPE, pp. 1326–1333.
198. Aziz, K., & Settari, A. (1979). *Petroleum reservoir simulation*. Imprint London: Applied Science Publishers.
199. Biot, M. A. (1941). General theory of three-dimensional consolidation. *Journal of Applied Physics, 12*(2), 155–164.
200. Wang, H. F. (2000). *Theory of linear poroelasticity: With applications to geomechanics and hydrogeology*. Princeton, NJ: Princeton University Press.
201. Bishop, A. W. (1959). The principle of effective stress.□Teknisk. *Ukeblad, 106*(39), 859–863.
202. Fredlund, D. G., & Morgenstern, N. R. (1977). Stress state variables for unsaturated soils. *Journal of Geotechnical Engineering, ASCE, 103*(5), 447–466.
203. Lu, N., & Likos, W. J. (2006). Suction stress characteristic curve for unsaturated soil. *Journal of Geotechnical and Geoenvironmental Engineering, 132*(2), 131–142.
204. Coussy, O. (Ed.). (2004). *Poromechanics*. Chichester, England: John Wiley & Sons, ltd.
205. Schrefler, B. A., & Gawin, D. (1996). The effective stress principle: incremental or finite form. *International Journal for Numerical and Analytical Methods in Geomechanics, 20*(11), 785–814.
206. Chin, L. Y., Silpngarmlert, S., & Schoderbek, D. A. (2011). Subsidence prediction by coupled modeling of geomechanics and reservoir simulation for methane hydrate reservoirs. 45th U.S. Rock Mechanics/Geomechanics Symposium, June 26-29, 2011, San Francisco, CA.
207. Morland, L. W., Foulser, R., & Garg, S. K. (2004). Mixture theory for a fluid-saturated isotropic elastic matrix. *International Journal of Geomechanics, 4*(3), 207–215.
208. Dominic, K., & Hilton, D. (1987). Gas production from depressurization of bench-scale methane hydrate reservoirs. US Department of Energy, DOE/METC-87/4073, pp. 1–9.

209. Kamath, V. A., Mutalik, P. N., Sira, J. H., & Patil, S. L. (1991). Experimental study of brine injection and depressurization methods for dissociation of gas hydrates. SPE Formation. *Evaluation, 6*(4), 477–484.
210. Li, S., Chen, Y., & Du, Q. (2005). Sensitivity analysis in numerical simulation of natural gas hydrate production. *Geoscience, 19*(1), 108–112.
211. Yang, X., Sun, C., Su, K., Yuan, Q., Li, Q., & Chen, G. (2012). A three-dimensional study on the formation and dissociation of methane in porous sediment by depressurization. *Energy Conversion and Management, 56*, 1–7.
212. Bai, Y., & Li, Q. (2010). Simulation of gas production from hydrate reservoir by the combination of warm water flooding and depressurization. *Science China, 53*, 2469–2475.
213. Hovland, M., Judd, A. (Eds.). (1988). *Seabed pockmarks and seepages*. Impact on Geology, Biology and the Marine Environment, Graham and Trotman, London, 1988, pp. 293.

Chapter 3
Discrete Element Modeling of the Role of In Situ Stress on the Interactions Between Hydraulic and Natural Fractures

Riccardo Rorato, Frédéric-Victor Donzé, Alexandra Tsopela, Hamid Pourpak, and Atef Onaisi

Abstract The interaction between HF (hydrofractures) and NF (natural fractures) is a complex-coupled process which involves several physical parameters. Despite numerous previous works, the respective role of in situ stress, natural fracture properties, and orientations is still difficult to assess. In this chapter, a fully hydromechanical coupled numerical model has been used to simulate different three-dimensional configurations. These configurations provide insight into how a natural fracture is mechanically or hydraulically activated depending on well-defined parameters. It has been shown that the natural fracture can be either activated hydraulically without any shear displacement or mechanically activated while not loaded hydraulically. These configurations are controlled at a first-order level by the combination of the in situ differential stress state and the natural fracture orientation.

3.1 Introduction

Hydraulic fracturing is routinely performed on oil and gas wells in low-permeability layers. During the process of hydraulic fracturing, the pumping rate is maintained at a higher rate than the fluid leak-off rate (loss of fluid inside the formation), and the activated fracture network will continue to propagate and grow in the formation. Special fracturing fluids are pumped at high pressure into the low permeability reservoir to be operated, causing fractures to open [1]. Ideally, the tips of the hydraulic fractures extend away from the well bore according to the regional stresses within the formation. However, the presence of natural fractures can strongly affect the fluid flow pathway [2]. Depending on their orientation, degree

R. Rorato • F.-V. Donzé (✉) • A. Tsopela
Université de Grenoble Alpes, Laboratoire 3SR, Grenoble, France
e-mail: frederic.donze@3sr-grenoble.fr

H. Pourpak • A. Onaisi
TOTAL S.A., Pau, France

C. Jin, G. Cusatis (eds.), *New Frontiers in Oil and Gas Exploration*,
DOI 10.1007/978-3-319-40124-9_3

of connectivity, and level of aperture [3], the fluid can propagate in an uncontrolled way.

Besides the potential presence of these heterogeneities in the rock mass and as the treatment fluid propagates, proppant, such as sands of a particular size, is mixed and pumped together with it, to keep the fractures open after the injection is completed. The objective is to create high-conductivity channels within a large area of formation. The fractured zone needs to be adequately designed to ease the propagation of the proppant [4], i.e., with a low radius of curvature of hydrofractures or a stress stability of reactivated natural fractures for example [5].

Today the cost related to hydraulic fracturing requires that the operator knows how the formation will respond to a stimulation process, and whether its design—the selection of pump rates, fluid properties, pumping schedule, and fracture propagation model—will create the intended fracture network. Therefore, it is necessary to understand how fractures propagate not only in the intact rock matrix but also how they interact hydromechanically with the existing fractures: this will provide an assessment of the possible stimulated volume before the real operation [6].

Design of hydraulic fracturing operations involves fluid and solid mechanics, fluid flow in fracture and diffusion processes (fluid and thermal), as well as fracture propagation. All responses are coupled: the process of hydraulic fracturing lies at the interface of fluid mechanics and rock mechanics [7]. Numerical simulations of the hydraulic fracturing process are now crucial to understand its high degree of complexity.

In this chapter, a fully hydromechanical coupled numerical model is used to study the interaction between hydraulic and natural fractures for a given in-situ stress state in a low permeability rock environment. This approach has the advantage of providing accurate results [8–10]; however the calculation time can be very long and optimization needs to be performed. Based on the discrete element method, case studies have been performed with 3DEC 5.0 (a software developed by ITASCA Consulting Group Inc.). The results of the simulations of idealistic cases were first compared to an analytical solution of fracture mechanics (Perkins-Kern-Nordgren model: [11, 12]) to show the ability of the model to reproduce analytical solutions. Then, the interaction between hydraulic and natural fractures is presented. A parametric study has been carried out to identify the key parameters controlling the interaction between hydraulic and natural fractures in the presence of different in situ stress fields.

3.2 Discrete Element Method

In numerical modeling of engineering problems, some problems can be represented by an adequate model using a finite number of well-defined components. The behavior of such components is either well known, or can be independently treated mathematically. The global behavior of the system can be determined through well-defined inter-relations between the individual components (elements). Such problems are termed *discrete* and the discrete representation and solution of such system by numerical methods are usually straightforward. For geomechanical problems, discontinuous methods are mostly the discrete element method (DEM) [13, 14] or the discrete deformation analysis (DDA) [15].

The key concept of the DEM is that the domain of interest is treated as an assemblage of rigid or deformable tetrahedral blocks (3DEC, [16], for example), or spherical elements (YADE for example, [17, 18]). Bodies and the contacts among them need to be identified and continuously updated during the entire deformation/motion process, and represented by proper constitutive models.

3DEC is a three-dimensional numerical software based on distinct element method based on discontinuum modeling. The response of discontinuous media (such as a jointed rock mass) subjected to either static or dynamic loadings can be simulated with this type of numerical approach. The discontinuous medium is represented by a set of discrete blocks. The discontinuities are represented by boundary conditions between blocks and large displacements along discontinuities and rotations of blocks are allowed. Individual blocks behave as either rigid or deformable material. In this study, deformable blocks are subdivided into a mesh to be used as finite difference elements, and each element can respond according to a prescribed linear stress–strain law. The relative motion of the discontinuities is also governed by linear force-displacement relations for movement in both the normal and shear directions.

Regarding hydraulic fracturing, an essential role is represented by persistent joints or finite size fractures, features which are explicitly represented in 3DEC. Different representations of joint material behavior can be set up and the basic model is the Coulomb slip criterion, which assigns an elastic stiffness, friction angle, cohesive and tensile strengths, and dilation characteristics to a joint has also been used in this study.

3.3 Representing Discrete Fracture

In 3DEC, the fractures have a specific definition. They are circular and of finite extent. Moreover, they correspond to a subdomain of a joint where different joint hydromechanical properties are assigned. This concept is illustrated in Fig. 3.1.

Realistic values of the rock's cohesion and tension strength can be assigned to the joint material relative to the zone outside the circular fractures, since it represents the intact rock matrix joints, considered as fictitious joint. Lower (or zero) values of cohesion and tensile strength can be given to the joint material related to the actual circular fracture in order to simulate an opened purely frictional fracture. A limitation of this formulation is that the rock mass is discretized into many impermeable blocks. Blocks represent intact and deformable rock material, while interfaces represent pre-existing natural fractures. Fluid flow is simulated in a network along the interfaces, which acts as flow planes.

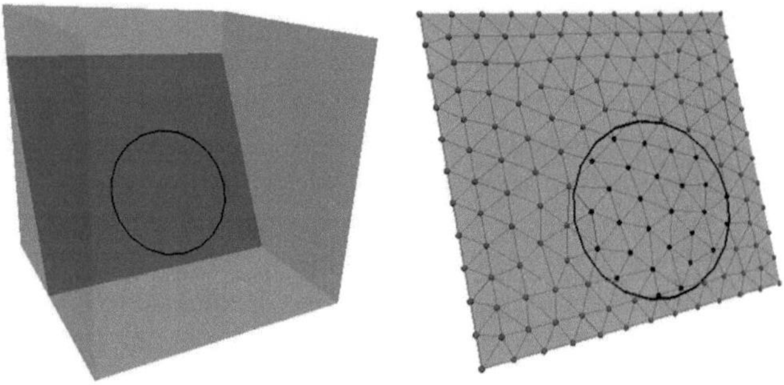

Fig. 3.1 Illustration of how sub-contacts on a joint plane can be assigned different properties inside (*black*) and outside (*gray*) of a circular fracture

The mechanical response of a fracture network stimulated by a fluid injection, as well as the interaction of a created hydraulic fracture with the existing natural fractures, can be modeled explicitly within these limitations. Moreover, in 3DEC, cracks can only propagate along the pre-existing fracture plane, which means that for example wing or twisted cracks due to mode II and III, respectively, cannot propagate in the intact part of the medium. However, such numerical tool can be valuable since it can simulate at high degree of complexity the hydromechanical response in a fully coupled way, as shown in the next sections.

3.4 Hydromechanical Coupling

Before the fluid injection is carried out, it is necessary to reach the initial equilibrium of the model, since it is unbalanced by the initial kinematic and static boundary conditions. To solve this initial state, the mechanical mode is activated, the flow is turned off and the fluid bulk modulus is set to zero. These steps ensure that the stress solution is obtained without changing the fluid pressure. After equilibrium is obtained, the mechanical mode is turned off, flow is turned on and fluid bulk modulus is set back to its real value. The calculation is then stepped to equilibrium again to obtain the initial steady-state pressure. During initialization, the goal is to obtain the equilibrium state and the initial steady-state pressure within the model.

Hydraulic fracturing is a complex process, since it involves several different mechanisms. For example, injecting a viscous fluid into a fractured rock causes elastic deformation, in particular dilatation of existing fractures affecting their hydraulic properties and generally leading to an increase of the effective permeability of the rock, thus influencing the initial fluid flow. To run a hydro-mechanical solution, both mechanical and flow mode must be on. Internally, 3DEC runs a series of mechanical steps followed by a series of fluid steps. The number of mechanical steps per each fluid step and vice versa can be chosen. The time associated with mechanical deformation is significantly less than that associated with fluid flow. In fact, the assumption of the coupled hydromechanical model in 3DEC is that it is quasi-static. Both flow and mechanical mode are explicitly integrated in time, only flow time is associated to the physical time.

In fully-coupled approaches, the number of mechanical step for each fluid step, it is automatically done by 3DEC. The determination of the number of hydraulic cycles is the maximum unbalanced force. For each injection process, the mechanical calculation is carried out right after the hydraulic loading and the corresponding unbalanced force is compared to a fixed value of maximum unbalanced force. If the unbalanced force is higher than this value, then the number of hydraulic cycles remains low (around 10, for the physical properties values assigned in the present model); on the other hand, if it is below the maximum unbalanced force decided by the user, the number of hydraulic cycles can increase up to 100, since the unbalanced force values remain low.

In this chapter, a validation of the 3DEC simulations is proposed. In particular, the solution of an analytical method for hydraulic fracture mechanics [12] is compared to the results of the simulation.

3.5 PKN Model Simulation

The longitudinal propagation of a vertical fracture, which is consistent with the geometrical assumption of the PKN model (all vertical sections can be thought as planes having null orthogonal strains) can be simulated in a straightforward way.

Remind that the PKN model is applicable only to fully confined fractures; it converts a three-dimensional problem into a two-dimensional one, under the hypothesis of plane strain conditions. A plane strain deformation is one in which planes that were parallel before the deformation remains parallel afterward. This is generally a good assumption for fractures in which one dimension is much greater than the other. Perkins and Kern assumed that the pressure at any section is dominated by the height of the section. This is true if the length is much greater than the height.

In the PKN model, fracture mechanics and the effect of the fracture tip are not considered; they focused on the effect of fluid flow in the fracture and the corresponding pressure gradient.

The 3DEC model is represented by an elastic deformable block containing a vertical plane (which will "host" the fracture, represented in yellow in Fig. 3.2) delimitated by two joints (in red) characterized by very large (fictitious values) cohesion and tensile resistance, in order to force the fracture to propagate along the pre-defined direction. Moreover, these joints represent the strong contrast border between the pay layer and the neighboring formation, assumption which makes appropriate the comparison with PKN model. The size of the block is 80 m × 40 m × 20 m and the fracture height is 12 m. It is indeed necessary to guarantee a large ratio between the fracture length and height but also to optimize

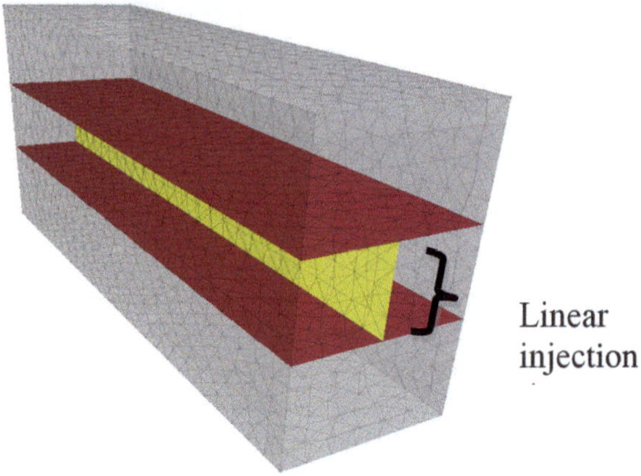

Fig. 3.2 3DEC model for PKN problem

the quality of the results: this shape allows using a fine mesh saving calculation time remaining inside the domain of applicability of the PKN model.

The boundary conditions which are applied are both, mechanical and hydraulic. The block is compressed by a lithostatic stress state. Moreover, the edges of the block cannot move. Regarding the hydraulic boundary conditions, the fluid is injected at the specified range along the vertical edge of the fracture surface as indicated in Fig. 3.2. The injection is performed imposing a flow rate in order to ease the comparison with the analytical results of the PKN model. The fluid will propagate along the yellow joint, because of the large resistance of the contrast joints.

All the parameters concerning rock, joints, and injection are listed in Table 3.1.

After 236 s of injection, the fracture propagates up to 75 m, close to the end of the model. This length is considered sufficient to satisfy the assumption of large ratio between the fracture length (75 m) and height (12 m), in order to assure the plane-strain deformation in the vertical cross section. The contours of the fracture aperture and pressure are shown in Fig. 3.3.

The results of the PKN analytical solution, making use of Eq. 3.1, are compared to the results of the 3DEC simulation, in terms of fracture aperture (Fig. 3.4):

$$w(x) = 3\left[\frac{\mu Q(1-\nu^2)(L-x)}{E}\right]^{\frac{1}{4}} \tag{3.1}$$

The match presents an error less than 1 % along the first 20 m of the fracture, near the injection point. Then, the gap increases towards the fracture tip. A possible reason for this discrepancy could be that in 3DEC a minimum residual aperture needs to be defined, in order to prevent instability of the model.

Table 3.1 List of all the parameters used in 3DEC PKN problem simulation

Size of the block (L × l × H)	80 m × 40 m × 20 m
Boundary conditions	Zero velocity imposed on the edges Lithostatic stress imposed: • $S_H = 1$ MPa • $S_h = 0.5$ MPa • $S_V = 1$ MPa
Parameters of the block	Density: $\rho = 2600$ kg/m^3 Bulk modulus: $K = 5.59 \times 10^9$ Pa Young modulus: $E = 10$ GPa Poisson ratio: $\nu = 0.2$
Parameters of the contrast joints (red)	Normal stiffness: $K_n = 1 \times 10^{10}$ Pa/m Shear stiffness: $K_s = 1 \times 10^{10}$ Pa/m Cohesion: $c = 1 \times 10^{30}$ Pa Friction angle: $\varphi = 30°$ Tensile strength: jtens $= 1 \times 10^{30}$ Pa Initial aperture: $a_0 = 0.5 \times 10^{-4}$ m Residual aperture: $a_{res} = 0.5 \times 10^{-4}$ m Maximum aperture: $a_{max} = 1 \times 10^{-2}$ m
Parameters of the fracture joint (yellow)	Normal stiffness: $K_n = 1 \times 10^{10}$ Pa/m Shear stiffness: $K_s = 1 \times 10^{10}$ Pa/m Cohesion: $c = 1 \times 10^5$ Pa Friction angle: $\varphi = 30°$ Tensile strength: jtens $= 1 \times 10^5$ Pa Initial aperture: $a_0 = 1 \times 10^{-4}$ m Residual aperture: $a_{res} = 1 \times 10^{-4}$ m Maximum aperture: $a_{max} = 1 \times 10^{-2}$ m
Fluid parameters	Bulk modulus: $K = 2 \times 10^8$ Pa Density: $\rho = 1000$ kg/m^3 Viscosity: $\mu = 1$ cP
Injection	Flow rate: $Q = 0.002$ m^3/s Injection time $= 236$ s

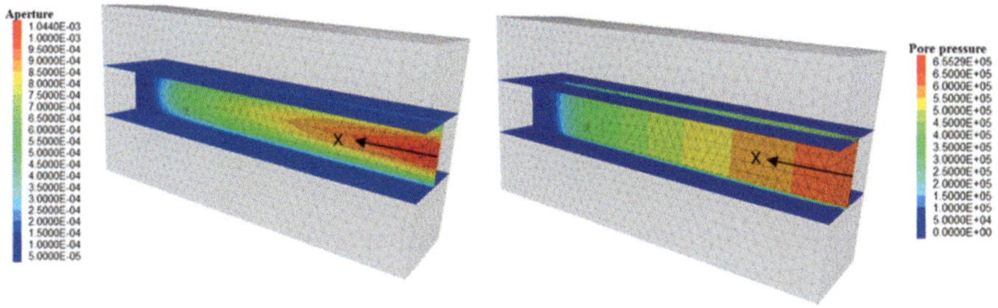

Fig. 3.3 *Left*: Hydraulic fracture aperture contour (*m*). *Right*: Fluid pressure contour inside the hydraulic fracture (Pa)

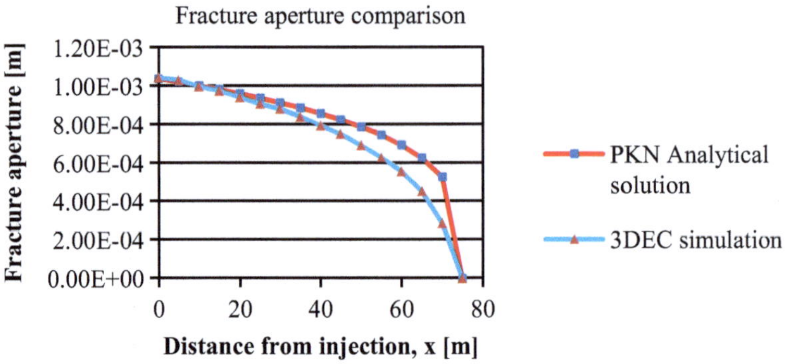

Fig. 3.4 Fracture aperture comparison between the PKN analytical solution and 3DEC simulation

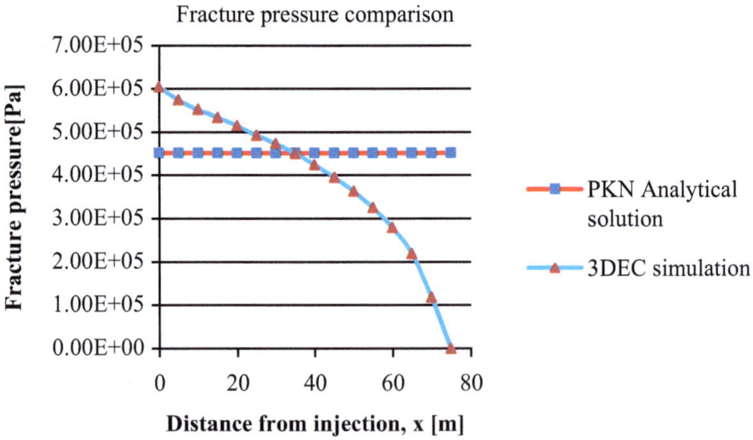

Fig. 3.5 Fracture pressure comparison between the PKN analytical solution and 3DEC simulation

Regarding the pore pressure:

$$p = \left[\frac{16\mu Q E^3}{\left(1 - \upsilon^2\right)^3 \pi H^4} L \right]^{1/4} \tag{3.2}$$

The PKN model gives a constant value for the pore pressure and it is one of the strong hypotheses of the analytical solution. On the other hand, 3DEC provides a decreasing tendency from the injection point to the fracture tip, due to the transient regime for a viscous flow regime (Fig. 3.5). Therefore the trend is different but it is interesting to point out that the two lines intersect near the half-length of the fracture. Another possible comparison is the evolution in time of the crack aperture at the injection boundary (Fig. 3.6). The match is also good, except for the first seconds of injection; this is due to the strong unbalancing of the 3DEC simulation, since it did not have time enough to converge to the right solution. But after 30 s of

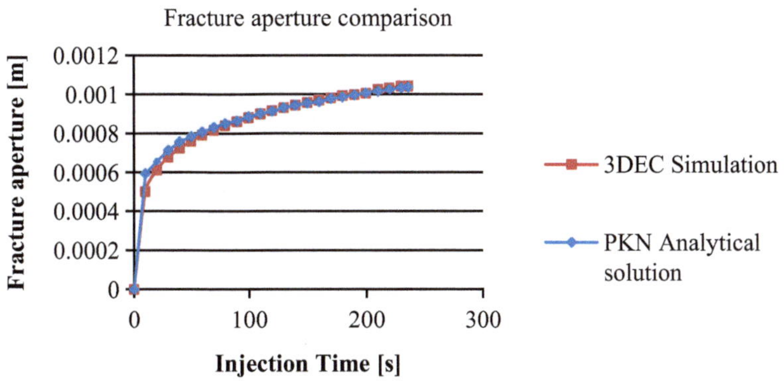

Fig. 3.6 Fracture aperture comparison in time between the PKN analytical solution and 3DEC simulation

injection the difference between the two models is lower than 5 %, and after the first minute it oscillates around the zero.

Concerning the predicted fracture length proposed by Nordgren:

$$L(t) = 0.68\left[\frac{EQ^3}{\mu(1-\nu^2){h_f}^4}\right]^{\frac{1}{5}} t^{4/5} = 71.1\,\text{m} \tag{3.3}$$

After 236 s of injection, the fracture length, 71.1 m, is very close to the length obtained in 3DEC, 75 m, with a difference of 5 %.

All these results are referred to the case of viscous dominated regime. Thus, it is interesting to find the threshold between the viscous and the toughness dominated regime, since the mechanism of propagation is very different (see [19] for example). In the following simulation, all the parameters are exactly the same except the rock tensile strength, which is increased from 0.1 MPa to 10 MPa, i.e., two orders of magnitude. Therefore, the fracture toughness defined by

$$K_{\text{IC}} = T \cdot \sqrt{\pi \cdot a_c} \tag{3.4}$$

where K_{IC} is the fracture toughness, T is the rock tensile strength, and a_c is the length scale (e.g., flaws or grain scale) characteristic of the rock under consideration is increased with a no more negligible value.

The result of the simulation in terms of fracture aperture and pressure is shown in Fig. 3.7.

At 236 s of injection, the fracture geometry is completely different from the previous viscous dominated regime case. In particular, the fracture aperture (2.58 mm at the injection zone) and pressure (1.59 MPa) are increased while the length is decreased from 75 m to 25 m, as expected. Moreover, in contrast to the previous case, the pressure is constant for all the fracture length. This is due

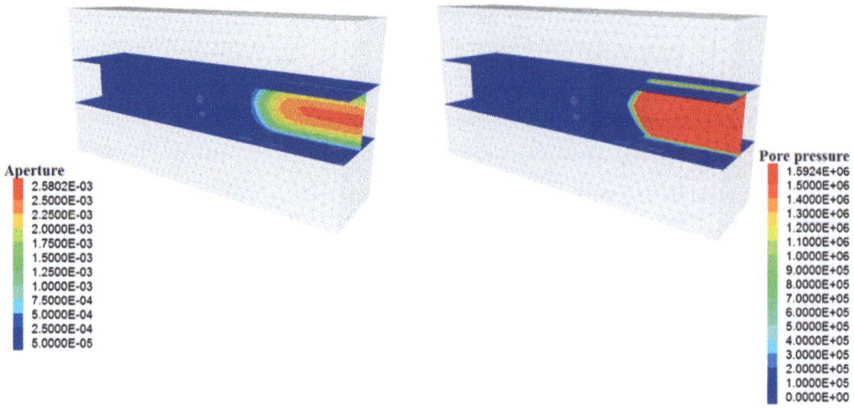

Fig. 3.7 Fracture aperture (*m*) on the left and pressure (Pa) on the right in toughness-dominated regime (same injection time)

to the fact that, since the toughness is large, the fracture propagation velocity is slow and thus the pressure has no need to decrease towards the fracture tip.

3DEC is a very useful tool in this setting, since it provides results for which the analytical solutions merely define the behavior, without giving any values.

3.6 Hydraulic (HF) and Natural (NF) Fracture Interaction

Micro-seismic measurements have shown that complex fracture networks are created during fracture treatment. It has been observed in fracture core-through experiments that the interaction between hydraulic fracture (HF) and the pre-existing natural fracture (NF) in shales could be involved in the formation of this complex fracture network [20]. In an early frac-treatment of vertical wells in Barnett shale, the frac-fluid often unexpectedly connected to several adjacent vertical wells and brought down the well production [21]. This field experience has provided supporting evidence on the creation of a complex fracture network by hydraulic fracturing. In analyzing field cases in Barnett shale, [22] have shown that the predicted fracture length from a planar fracture model analysis far exceeds the fracture length indicated by the microseismic data. A larger fluid volume can be stored in a fracture network for a shorter network length. This fact points to the existence of a network of hydraulically induced fractures.

Different models for simulating the propagation of a fracture network have been developed ([23, 24] for examples). The creation and activation of a fracture network is a complex phenomenon that includes the interaction between HF and NF. It is however important to capture the key elements so that the model represents the realistic process. A hydraulic fracture may cross a natural fracture without change of direction, or it may be arrested and branch off along the NF depending upon the property of the NF and the difference between the minimum horizontal stress and closure stress acting on the fracture. According to Ching [20], there are several possibilities: when the fracture tip reaches the interface, the NF is already under the influence of the stress field generated by the HF. At this point, the fluid has not penetrated into the NF. There are different possible outcomes from this interaction: one is that the NF slips under shear stress and the HF is arrested, and others are a direct crossing, or a crossing with an offset. The current version of 3DEC is not able to represent this latter possibility, since an additional joint should be defined a priori. The HF-NF crossing behavior depends on in situ stress, rock and NF properties, frac-fluid property and its injection rate. Nowadays, extensive theoretical and experimental studies are still carried out for developing a rule governing the HF-NF interaction ([25–28] for examples) but the hypotheses of the models are often strong and therefore new fully HM-coupled simulations are still useful to understand the interaction process.

In the following, some simulations are carried out in order to study the effect of different stress fields in the HF-NF interaction. An initial model, called "reference," is compared to others in which the in situ stresses and natural fracture orientation change.

3.7 Parametric Study: Reference Model

The initial model is represented by a parallelepiped containing a natural fracture and a joint along which the hydraulic load will be performed; both fractures are vertical in order to respect the stress field. A finer mesh is generated along the planes in order to increase the quality of the output.

A linear injection is carried out along the vertical joint and the simulation is stopped after one second of injection, since it has reached the equilibrium. Then, several parameters could be monitored in order to evaluate the cross tendency.

The fluid injection is simulated imposing the pore pressure, instead of flow rate, since it is easier to keep the control on the simulation process and make comparisons with the rock strength parameters. The 3DEC model is represented in Fig. 3.8 and the parameters which have been used for the initial model are listed in Table 3.2.

The contour of the fracture aperture obtained in the 3DEC simulation is represented in Fig. 3.9.

The aperture contours respectively along the HF and NF plane are given in the top-left and top-right of Fig. 3.10, as the bottom-left image shows the contour of the pore pressure along both the fractures.

The bottom-right plot of Fig. 3.10 represents the evolution in time of the amount of water stored in four different flow knots: on the HF plane just before the NF (pink curve), on the HF plane just after the NF (blue curve) and on both sides of the NF, in two points very close to the intersection with the HF (yellow and red curves).

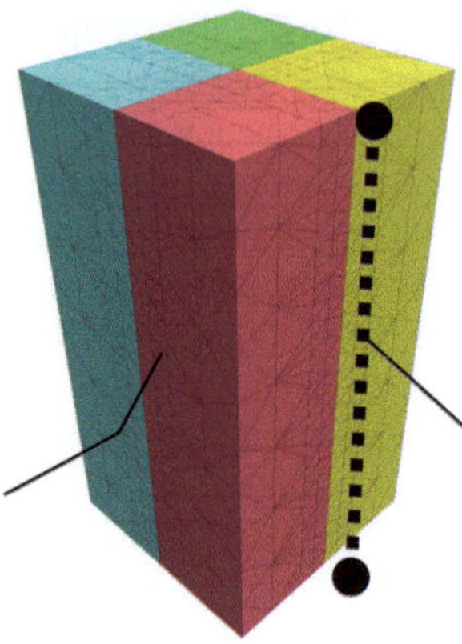

Fig. 3.8 3DEC initial reference model

Table 3.2 Parameters used in the reference case simulation

Size of the block (L × l × H)	10 m × 10 m × 20 m
Boundary conditions	Zero velocity imposed on the edges Lithostatic stress imposed: • $S_H = S_x = 20$ MPa • $S_h = S_y = 20$ MPa • $S_V = S_z = 20$ Mpa • Pore pressure = 20 Mpa
Parameters of the rock matrix	Density: $\rho = 2500$ kg/m^3 Bulk modulus: $K = 5.56 \times 10^9$ Pa Shear modulus: $G = 4.17 \times 10^9$ Pa Young modulus: $E = 10$ GPa Poisson ratio: $v = 0.2$
Parameters of the joints submitted to hydraulic load (parameters of intact rock matrix)	Normal stiffness: $K_n = 1 \times 10^{11}$ Pa/m Shear stiffness: $K_s = 1 \times 10^{11}$ Pa/m Cohesion: $c = 1 \times 10^6$ Pa Friction angle: $\varphi = 30°$ Tensile strength: jtens $= 1 \times 10^6$ Pa Initial aperture: $a_0 = 1 \times 10^{-4}$ m Residual aperture: $a_{res} = 1 \times 10^{-5}$ m Maximum aperture: $a_{max} = 1 \times 10^{-3}$ m
Parameters of the natural fracture	Normal stiffness: $K_n = 1 \times 10^{11}$ Pa/m Shear stiffness: $K_s = 1 \times 10^{11}$ Pa/m Cohesion: $c = 1 \times 10^6$ Pa Friction angle: $\varphi = 30°$ Tensile strength: jtens $= 1 \times 10^6$ Pa Initial aperture: $a_0 = 1 \times 10^{-4}$ m Residual aperture: $a_{res} = 1 \times 10^{-5}$ m Maximum aperture: $a_{max} = 1 \times 10^{-3}$ m
Fluid parameters	Bulk modulus: $K = 2 \times 10^8$ Pa Density: $\rho = 1000$ kg/m^3 Viscosity: $\mu = 1$ cP
Injection	Linear injection: pp = 22 MPa Injection time = 1 s

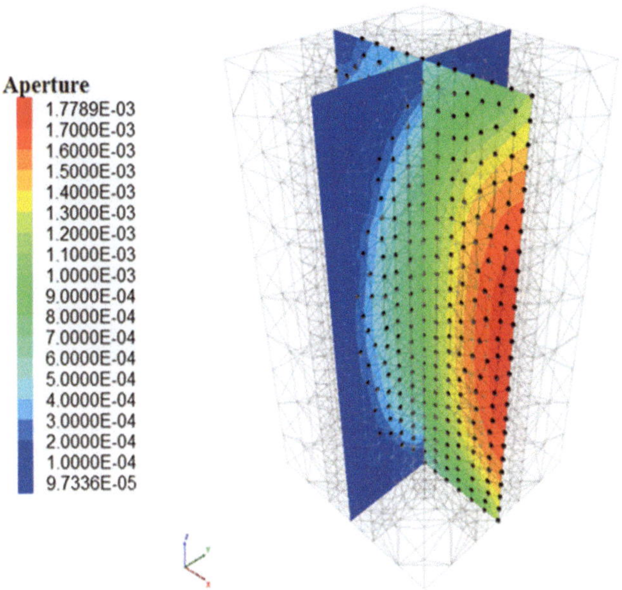

Fig. 3.9 Fracture aperture (*m*) contour after the fluid injection: The *black dots* represent the knots which have failed in tension because of water pressure

As seen from the contours, the hydraulic fracture has propagated crossing the NF but also opening it, in particular in Mode I (tensile failure).

The pore pressure has been able to overcome the normal stress acting on the NF as well as the rock tensile strength.

The four curves show that after 0.2 s of injection the first flow knot on the HF plane (pink curve) starts storing water (the fracture is propagating), then the first tendency is to cross the NF, since the blue curve starts to increase its value. After 0.5 s of injection the water starts entering in the NF, causing its aperture and even more fluid flow through it; therefore the water volume in the HF knots starts to decrease. The opening process along the NF is perfectly symmetric, since the fluid has not any preferential direction in which propagate.

3.7.1 Anisotropic Stress Field

In the presence of an anisotropic stress field: the maximum stress is the vertical one, $S_V = S_z$, and the minimum is the horizontal one perpendicular to the HF plane, $S_h = S_y$ (Table 3.3). This scheme is consistent with fracture propagation conditions, since the fractures are vertical and parallel to the maximum principal stress and the HF propagates in the direction of the maximum horizontal stresses. On the other

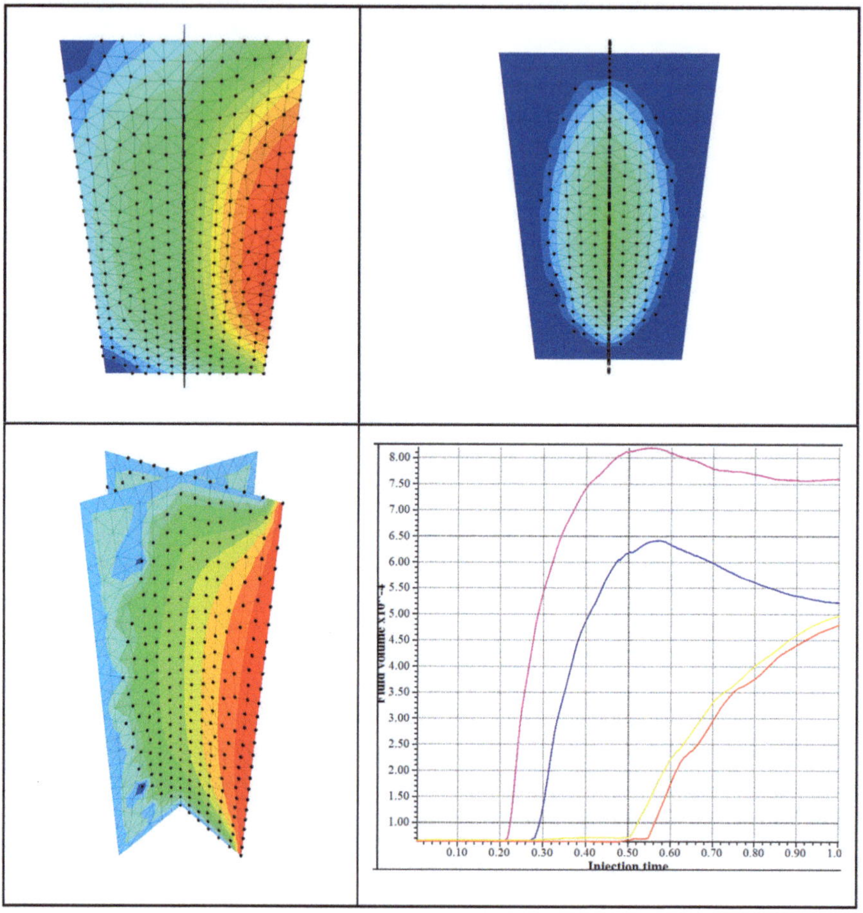

Fig. 3.10 Aperture contours [*m*] at top left: (**a**) HF and right NF (**b**), pore pressure (Pa) contour at bottom left, (**c**) and fluid volume (m^3) stored in four flow knots at bottom right (**d**)

Table 3.3 Differences from "reference model"

Stress field	• $S_H = S_x = 20.5$ MPa • $S_h = S_y = 20$ MPa • $S_V = S_z = 22$ MPa Pore pressure: pp = 20 MPa

hand, the NF tends to remain closed since it is perpendicular to the maximum horizontal stress.

The results of the simulation are represented in Fig. 3.11. As shown in the aperture contours (Fig. 3.11a, b), the fluid pressure is not able to overcome the

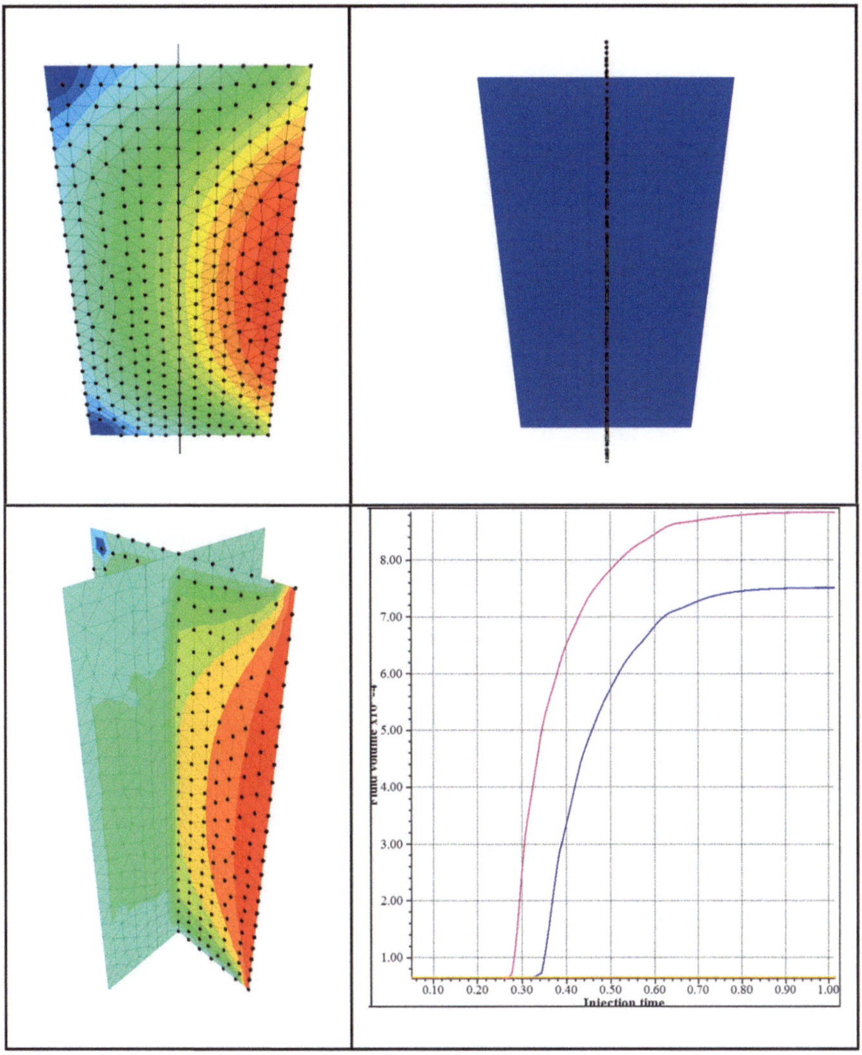

Fig. 3.11 Aperture contours (*m*), HF at top left (**a**) and NF right (**b**), pore pressure (Pa) contour at bottom left, (**c**) and fluid volume (m^3) stored in four flow knots at bottom right (**d**)

compression stress acting on the NF, which stays closed because the natural fracture has not failed in tension. The flow knots positioned on the NF do not store any amount of water; thus the yellow and red curves remain null (Fig. 3.11d).

3.7.2 Effect of Different Orientation of the NF

The orientation of the NF plays an important role in the interaction process. Until now, the NF has been considered perpendicular to the HF plane. The NF rotation is presented in Fig. 3.12.

The inclination of the NF is given respecting the stress field: the NF is still vertical; thus its normal line is always perpendicular to the maximum principal stress. The maximum horizontal stress is not perpendicular to the NF anymore; therefore it is easier to open it. Moreover, since the inclination of the NF is not aligned to one of the principal stresses, it is subjected to shear stresses acting along its plane. Note that now the NF is purely frictional with zero cohesion and tensile strength values. Before the injection process, the stress state acting on the inclined fracture is in accordance with the stress state predicted by the Mohr's circle in terms of normal and shear stresses, once the equilibrium is reached. The contours of the compression and shear stresses on the NF are presented in Fig. 3.13.

The values of the effective stresses are, respectively, $\sigma_n' = 375.00\,\text{KPa}$ and $\tau_{\text{nm}} = 216.51$ KPa. These values can be obtained also making use of the Mohr's circles (represented in Fig. 3.14) and its equations (Eq. 3.5):

$$\begin{aligned} &\sigma n' = \sigma x' + \sigma y' 2 + \sigma x' - \sigma y' 2 \cos 2\alpha + \tau xy \cdot \sin 2a = -375.00 \\ &KPa\tau nm = -\alpha x' - \sigma y' 2s \\ &in2\alpha + \tau xy \cdot \cos 2\alpha = 216.51\text{KPa} \end{aligned} \tag{3.5}$$

The values are identical; therefore the stress state is well computed by 3DEC and it is now possible to carry on with the injection load for different NF angles.

The aperture contour in both HF and NF in function of the NF orientation, ϑ, is presented in Fig. 3.15.

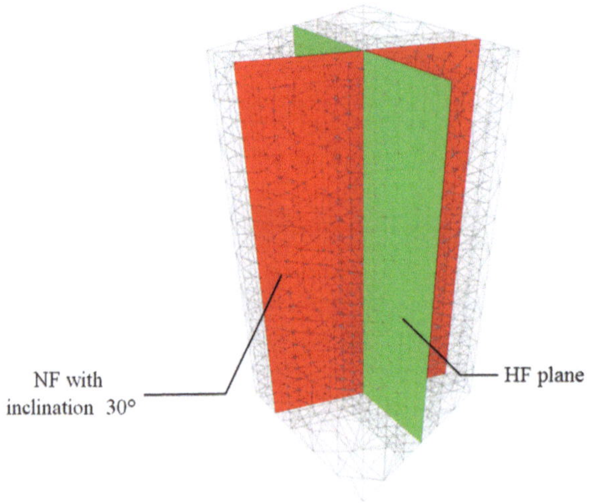

Fig. 3.12 3DEC model with NF (*green plane*) inclined

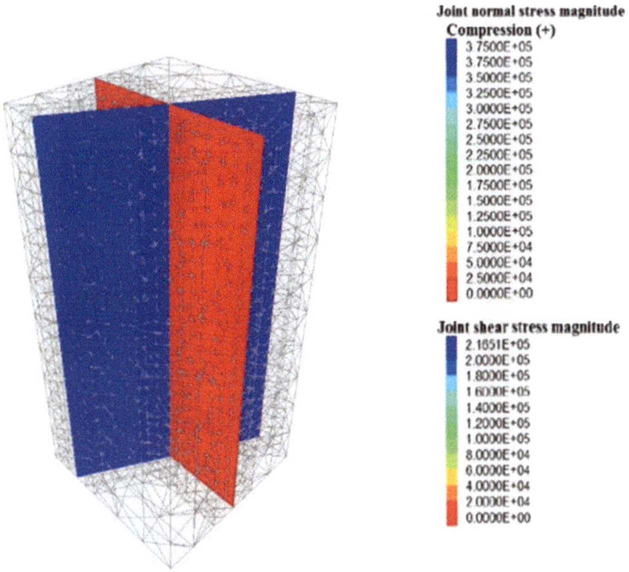

Fig. 3.13 Joint normal and shear stresses contours of the HF (*red plane*) and NF (*blue plane*) in Pa

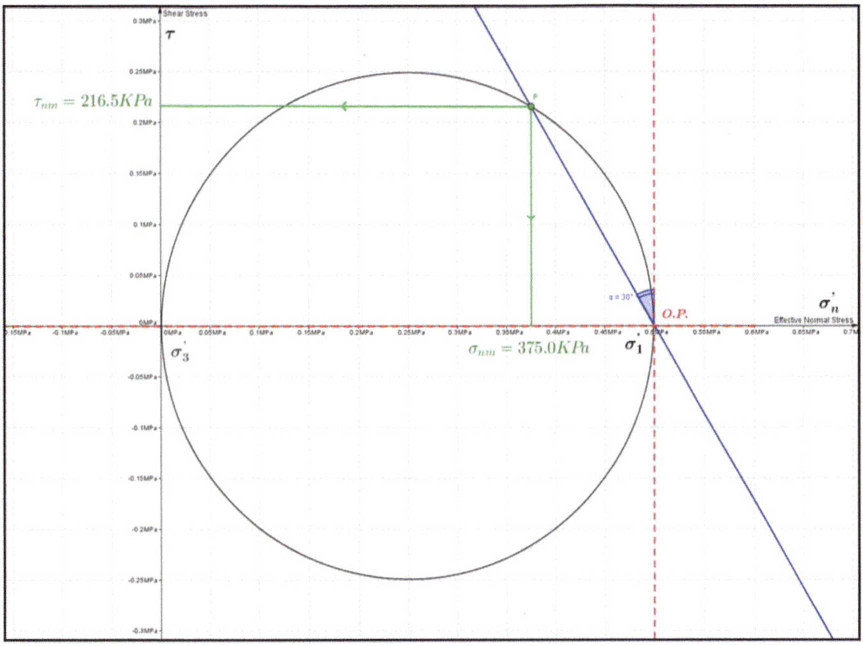

Fig. 3.14 Mohr's circle representing the stress state used in 3DEC simulations

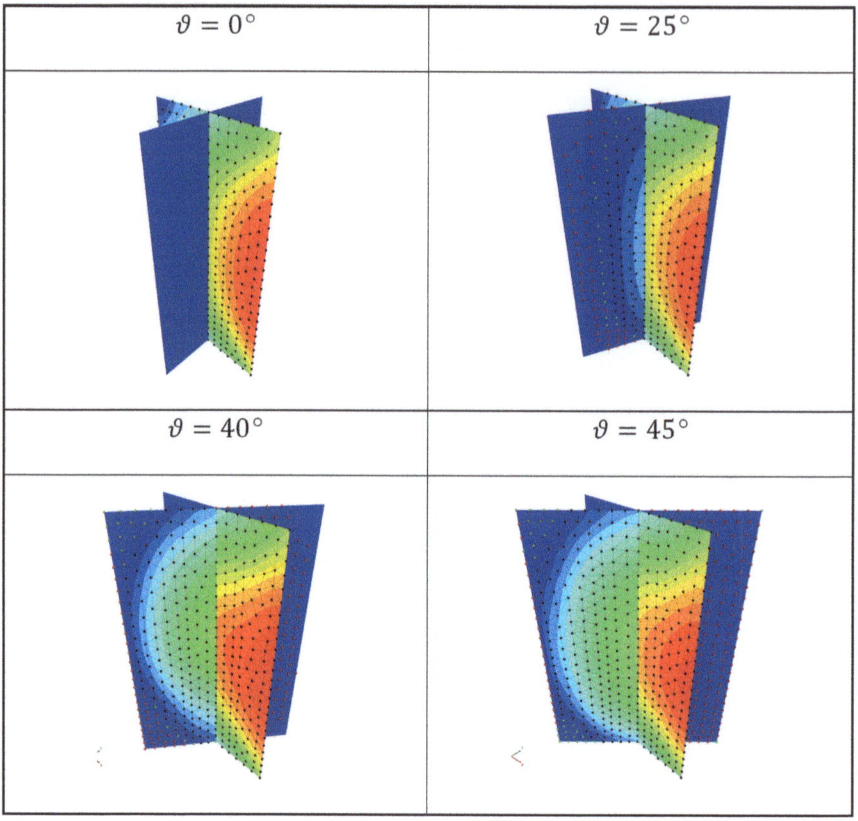

Fig. 3.15 HF and NF aperture contours obtained making use of different NF orientation

When the NF inclination increases, the fluid flow tends even more to open the natural fracture. One side of the NF is preferred than the other one since the aperture on the HF plane due to water pressure, when the fracture is approaching to the NF intersection, causes a modification of the compressive stress on the NF: greater in one side and lower in the other one as shown in the example of Fig. 3.16. Therefore, it is easier to open one side of the NF in Mode I and thus the fluid flow path is not symmetric. Then, it is possible to plot a curve able to describe the tendency to open the NF when its inclination changes, as in Fig. 3.17.

The abscissa is the NF inclination, ϑ. Theta equal to zero means the NF is orthogonal to the HF. The y-axis is the normalized ratio between the amounts of water stored in two different flow knots: the first one is on the HF plane just after the intersection with the NF, and the second one is on the opened side of the NF very close to the HF plane. In other words, the higher this ratio, the higher the tendency of the injected fluid to open the NF. It is expressed in percentage: 100 % means all the fluid is lost into the NF.

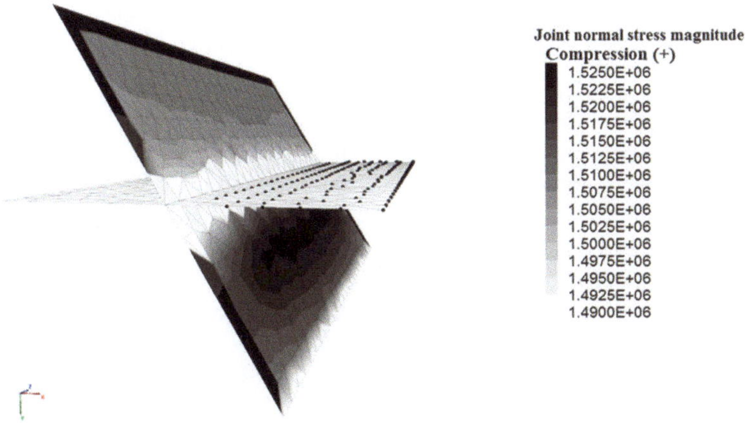

Fig. 3.16 Joint normal stress acting on the NF: The fracture is approaching (*black dots* are *cracked knots*) and the below side of the NF has a higher compression stress

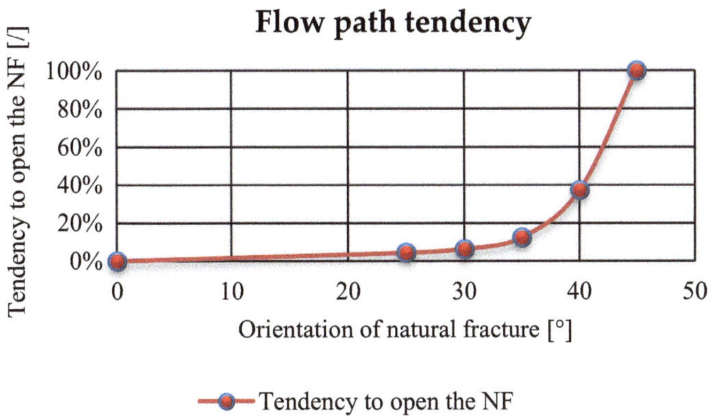

Fig. 3.17 Trend of the tendency to open the NF in function of its orientation: over 45° the tendency to open the NF goes to 100 %

To summarize, an inclination of the natural fracture between 0° and 30° does not lead to a significant change in the HF propagation process, which tends to cross the NF intersection.

On the other hand, between 30° and 40°, there is the critical inclination which induces a completely different tendency: the fluid starts opening the NF that becomes part of the fracture network, rather than cross it.

When the natural fracture intersects the hydraulic one with an angle greater than 45°, it is possible to state that crossing does not occur. Therefore, all the injected fluid is lost into the NF when this "threshold" is reached.

3.7.3 Effect of Different Orientation of a Dilatant NF Combined with Higher Anisotropic Stress Field

In this configuration, the natural fracture has a dilatancy angle of 20° and it is 30° inclined with respect to the HF (i.e., the maximum horizontal stress). Moreover, the stress field is much more anisotropic: the anisotropy ratio is 1.5 (Table 3.4).

Due to the higher horizontal stress acting perpendicularly to the NF plane, it becomes difficult to open it in Mode I (tensile failure): Water pressure is not high enough to counterbalance the normal stress acting on the fracture. On the other hand, because NF is inclined, there is pre-existing shear stress acting on the NF plane, before the injection loading. Then, when injection is performed, the HF opening induces additional shear stress on the NF, due to stress shadow effects, favoring slipping. Therefore, if shear failure occurs, the NF fracture can open because of dilatancy, as shown in Fig. 3.18.

The HF has crossed the NF intersection without opening the NF, as expected. But the upper side of the NF has slipped: the red dots are contacts which have failed in shear. The reason why only one NF side has slipped is the Mohr—Coulomb shear criterion: according to Fig. 3.16, the joint compression stress is higher on the bottom NF side. Therefore, the higher the normal stress the higher the shear resistance, τ_r:

$$\tau_r = \sigma_n \cdot \tan(\varphi) \tag{3.6}$$

Subsequently, it is interesting to introduce a dilatancy angle to the NF plane. The role of dilatancy is not fully assessed in petroleum applications, but it could play a crucial role when NF are reactivated.

The comparison between a first simulation without dilatancy, and a second one assigning a dilatancy angle equal to 20° is presented in Table 3.5.

Table 3.4 Differences from "reference model"

Stress field	• $S_H = S_x = 30$ MPa • $S_h = S_y = 20$ MPa • $S_V = S_z = 30$ MPa Pore pressure: pp = 15 MPa
Parameters of the natural fracture	Normal stiffness: $K_n = 1 \times 10^{11}$ Pa/m Shear stiffness: $K_s = 1 \times 10^{11}$ Pa/m Cohesion: $c = 0$ Pa Friction angle: $\varphi = 30°$ Dilatancy angle: $d = 20°$ Tensile strength: jtens = 0 Pa Initial aperture: $a_0 = 1 \times 10^{-4}$ m Residual aperture: $a_{res} = 1 \times 10^{-5}$ m Maximum aperture: $a_{max} = 1 \times 10^{-3}$ m
Injection	Linear injection: pp = 30 MPa

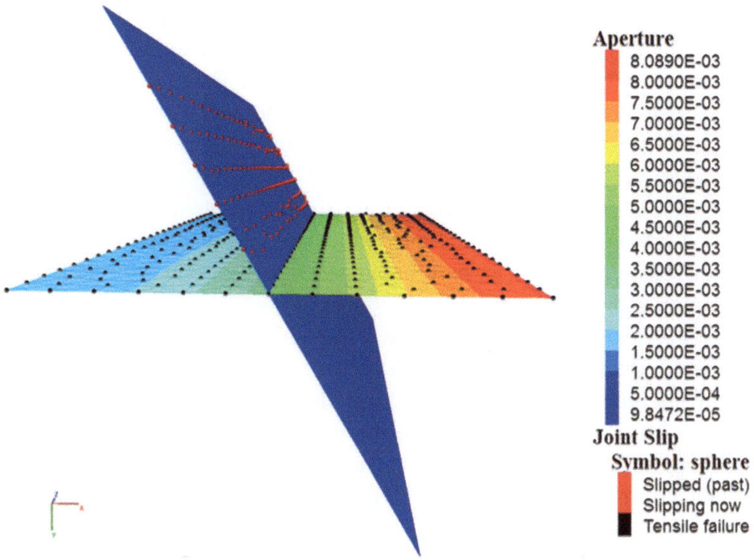

Fig. 3.18 Joint aperture contour (*m*): the NF is closed but it has failed in shear (*red dots*)

Table 3.5 Aperture on the NF obtained assigning different dilatancy angles

NF aperture (mm)	Dilatancy angle = 0°	Dilatancy angle = 20°
Before the injection	0.100 mm	0.100 m
After the injection	0.105 mm	1.290 mm

In the first case, even if the injection induces a shear failure on the NF plane, the aperture does not change from its initial value. On the other hand, assigning a dilatancy angle of 20° to the NF the fluid injection induces a slip on the NF and the fracture aperture increases more the ten times. It could be suspected that dilatancy could play a very important role in NF reactivation process.

3.8 Conclusions

In this chapter, 3DEC software has been used to simulate the interaction between hydraulic and natural fractures in order to highlight the associated mechanisms for given applied stress loadings. Hydraulic fracturing is a complex process to model: it is still difficult to properly take into account the three-dimensionality of the problem, the hydromechanical coupling, and the role of fracture dilatancy.

A large set of values must be set for the rock matrix properties (elastic and plastic parameters for isotropic or anisotropic rocks), joints (initial, residual and maximum aperture, normal and shear stiffness, friction angle, cohesion, tensile strength, dilatancy angle, and others), the fluid (bulk modulus, viscosity, density, thermal conductivity, for Newtonian and non-Newtonian fluids), and stress state (directions, magnitude, gradients, pore pressures, etc.). Therefore, use of numerical approaches like 3DEC is not straightforward: it is necessary to deal with a critical point of view to assess coupled hydraulic and rock mechanics in order to build reliable codes.

Before tackling real hydraulic fracturing, simple problems must first be undertaken to understand the basics of coupled processes. The validation step has shown that discrete element codes can be reliable and represent powerful tools to understand complex geomechanisms.

These numerical methods can involve some critical points. If not well calibrated and optimized, the calculation time can become a major obstacle. The mesh size, the fracture apertures, the fluid bulk modulus, the joint, and block stiffness are some of the key parameters which influence the calculation time. Together with other numerical issues, this might require the use of specific scaling approaches and simplifying assumptions for identified second order parameters, in order to ensure convergence.

Still, discrete element models are interesting tools in treating local problems with a high level of complexity, like HF-NF interaction, in the presence of anisotropic rocks or plastic deformation, and they haven't reached their limit yet. For example, it is possible to set up one or more DFN sets in order to represent the level of connectivity of the pre-existing natural fracture network as well as assigning aperture values depending on the fracture diameter. Thus, each discrete fracture set could be implemented assigning a given spatial and length distributions, connectivity, percolation, orientation, etc. and the injection can be performed in a single point, in a cluster or in multi-stage configurations. Research is currently undergoing in this direction.

Acknowledgment This work is supported by TOTAL SA.

References

1. Britt, L. (2012). Fracture stimulation fundamentals. *Journal of Natural Gas Science and Engineering, 8*, 34–51.
2. Gale, J. F., Reed, R. M., & Holder, J. (2007). Natural fractures in the Barnett Shale and their importance for hydraulic fracture treatments. *AAPG Bulletin, 91*(4), 603–622.
3. Rogers, S., Elmo, D., Dunphy, R., & Bearinger, D. (2010, January). *Understanding hydraulic fracture geometry and interactions in the Horn River Basin through DFN and numerical modeling*. In Canadian Unconventional Resources and International Petroleum Conference. Society of Petroleum Engineers.
4. Liang, F., Sayed, M., Ghaithan, A.-M., Chang, F. F., & Li, L. (2016). *A comprehensive review on proppant technologies*. Petroleum, ISSN 2405-6561.
5. Jeffrey, R. G., Chen, Z. R., Zhang, X., Bunger, A. P., & Mills, K. W. (2015). Measurement and Analysis of Full-Scale Hydraulic Fracture Initiation and Reorientation. *Rock Mechanics and Rock Engineering, 48*(6), 2497–2512.

6. Olson, J. E., & Taleghani, A. D. (2009, January). *Modeling simultaneous growth of multiple hydraulic fractures and their interaction with natural fractures*. In SPE Hydraulic Fracturing Technology Conference. Society of Petroleum Engineers.
7. Neuzil, C. E. (2003). Hydromechanical coupling in geologic processes. *Hydrogeology Journal, 11*(1), 41–83.
8. Nagel, N. B., & Zhang, F. (2013). Coupled Numerical Evaluation of the Geomechanical Interactions between a Hydraulic Fracture Stimulation and a Natural Fracture System in Shale Formation. Rock Mech Rock Eng, Springer.
9. Papachristos, E., Scholtès L., Donzé, F. V., Chareyre, B., & Pourpak, H. (2015). Hydraulic fracturation simulated by a 3D coupled HM-DEM model, 13th International Symposium on Rock Mechanics, ISRM Congress.
10. Riahi, A., & Damjanac, B. (2013, May). *Numerical study of interaction between hydraulic fracture and discrete fracture network*. In ISRM International Conference for Effective and Sustainable Hydraulic Fracturing. International Society for Rock Mechanics.
11. Kovalyshen, Y., & Detournay, E. (2010). A reexamination of the classical PKN model of hydraulic fracture. *Transport in Porous Media, 81*(2), 317–339.
12. Perkins, T. K., & Kern, L. R. (1961). *Width of hydraulic fractures*. Texas: Journal of Petroleum Technology.
13. Cundall, P. A., & Strack, O. D. L. (1979). Geotechnique 29, No. 1, 47–65.
14. Donzé, F. V., Richefeu, V., & Magnier, S. A. (2009). Advances in discrete element method applied to soil, rock and concrete mechanics. State of the art of geotechnical engineering. *Electronic Journal of Geotechnical Engineering, 44*, 31.
15. Shi, G. H. (1992). Discontinuous deformation analysis: a new numerical model for the statics and dynamics of deformable block structures. *Engineering Computations, 9*(2), 157–168.
16. Itasca. (2013). 3DEC, Three Dimensional Distinct Element Code. Version 5.0, Minneapolis.
17. Kozicki, J., & Donzé, F. V. (2009). Yade-open dem: An open-source software using a discrete element method to simulate granular material. *Engineering Computations, 26*(7), 786–805.
18. Kozicki, J., & Donzé, F. V. (2008). A new open-source software developed for numerical simulations using discrete modeling methods. *Computer Methods in Applied Mechanics and Engineering, 197*(49), 4429–4443.
19. Abbas, S., & Lecampion, B. (2013, May). *Initiation and breakdown of an axisymmetric hydraulicfracture transverse to a horizontal wellbore*. In ISRM International Conference for Effective and Sustainable Hydraulic Fracturing. International Society for Rock Mechanics.
20. Yew, C. H. (1997). *Mechanics of hydraulic fracturing*. Amsterdam: Elsevier Science Ltd.
21. Fisher, N. I. (1996). *Statistical analysis of circular data*. Cambridge: Cambridge University Press.
22. Cipolla, C. L., Lolon, E. P., Erdle, J. C., & Rubin, B. (2010). Reservoir modeling in shale-gas reservoirs. *SPE Reservoir Evaluation & Engineering, 13*(04), 638–653.
23. Damjanac, B., & Cundall, P. (2016). Application of distinct element methods to simulation of hydraulic fracturing in naturally fractured reservoirs. *Computers and Geotechnics, 71*, 283–294.
24. Yaghoubi A., & Zoback M., (2012). *Hydraulic fracturing modeling using a discrete fracture network in the Barnett Shale*. Stanford Stress and Geomechanics Group. American Geophysical Union, Fall Meeting 2012.
25. Fu, P., Johnson, S. M., & Carrigan, C. R. (2013). An explicitly coupled hydro-geomechanical model for simulating hydraulic fracturing in arbitrary discrete fracture networks. *International Journal for Numerical and Analytical Methods in Geomechanics, 37*(14), 2278–2300.
26. Grasselli, G., Lisjak, A., Mahabadi, O. K., & Tatone, B. S. (2015). Influence of pre-existing discontinuities and bedding planes on hydraulic fracturing initiation. *European Journal of Environmental and Civil Engineering, 19*(5), 580–597.
27. Khoei, A. R., Vahab, M., & Hirmand, M. (2015). Modeling the interaction between fluid-driven fracture and natural fault using an enriched-FEM technique. *International Journal of Fracture, 197*, 1–24.

28. Rahman, M. M. (2009, January). *A fully coupled numerical poroelastic model to investigate interaction between induced hydraulic fracture and pre existing natural fracture in a naturally fractured reservoir: potential application in tight gas and geothermal reservoirs.* In SPE Annual Technical Conference and Exhibition. Society of Petroleum Engineers.
29. Shah, R.K., and D.P. Sekulic. 2003. *Fundamentals of heat exchanger design.* New York: John Wiley & Son Publisher.
30 Hesselgreaves, J.E. 2007. *Compact heat exchangers: Selection, design and operation.* 3rd edition. Pergamon.

Chapter 4
Rock Physics Modeling in Conventional Reservoirs

Dario Grana

Abstract Seismic reservoir characterization focuses on the interpretation of elastic attributes, such as seismic velocities and impedances, estimated from geophysical data such as surface seismic, crosswell seismic, and well log data. Elastic attributes depend on rock and fluid properties. The discipline of rock physics investigates the physical relations between petrophysical properties of porous rocks and their elastic response. In this chapter, we review the most common rock physics models for conventional hydrocarbon reservoirs. Rock physics models are commonly used to study the effect of variations in porosity, lithology, fluid saturation, and other petrophysical properties in reservoir rocks and the changes in the corresponding elastic and seismic response. These models can then be used to quantitatively interpret geophysical data and build reservoir models conditioned by well log and seismic data.

4.1 Review of Geophysical Concepts

Reservoir models of rock and fluid properties are generally built by integrating several types of measurements: well logs and geophysical data (seismic, gravity, and electromagnetic data). Most of these datasets, however, do not contain direct measurements of the properties of interest. Seismic data for example only provide information about the elastic contrast at layer interfaces in the subsurface. These contrasts depend on the elastic properties (velocity and density) of the upper and lower layers, and the elastic attributes depend on the rock and fluid properties. Similarly, well log data, such as gamma ray, neutron porosity, density, and resistivity, mostly contain indirect measurements of rock and fluid properties. Therefore, in order to interpret and process these measurements, we must introduce physical–mathematical relations to link the measurements to the properties of interest.

Rock physics includes a variety of models: porosity-velocity, porosity-permeability, and saturation-resistivity. In this chapter, we focus on the elastic response of

D. Grana (✉)
Department of Geology and Geophysics, University of Wyoming, Laramie, WY, USA
e-mail: dgrana@uwyo.edu

C. Jin, G. Cusatis (eds.), *New Frontiers in Oil and Gas Exploration*,
DOI 10.1007/978-3-319-40124-9_4

the principal rock and fluid properties. Seismic attributes can be measured at the well location or estimated from seismic data. In particular, we can measure (from well logs) or estimate (from seismic data) P-wave and S-wave velocity. For a complete review of reflection seismology concepts, we refer the reader to Aki and Richards [1] and Yilmaz [2]. In this section, we define elastic properties in terms of elastic moduli and density [3]. Indeed, P-wave and S-wave velocity, V_{P} and V_{S}, can be defined as

$$\begin{aligned} V_{\mathrm{P}} &= \sqrt{\frac{K + \frac{4}{3}G}{\rho}} \\ V_{\mathrm{S}} &= \sqrt{\frac{G}{\rho}}, \end{aligned} \tag{4.1}$$

where K is the bulk modulus, G is the shear modulus, and ρ is the density of the rock. The bulk modulus is the reciprocal of the compressibility and it is defined as the ratio of the hydrostatic stress to the volumetric strain. The shear modulus is defined as the ratio of the shear stress to the shear strain. The quantity $M = K + 4/3G$ is called compressional modulus. If the elastic moduli are measured in *GPa*, and density is measured in g/cm^3, then the resulting velocity is in *km/s*. From Eq. 4.1, we can derive the inverse formula to compute the elastic moduli from velocities:

$$\begin{aligned} G &= \rho {V_{\mathrm{S}}}^2 \\ K &= \rho {V_{\mathrm{P}}}^2 - \frac{4}{3}G. \end{aligned} \tag{4.2}$$

In many practical applications, geophysicists commonly use P- and S-impedance, I_{P} and I_{S}, instead of velocity, because density estimation, especially at the seismic scale, can often be challenging. Impedance is the product of velocity and density; therefore

$$\begin{aligned} I_{\mathrm{P}} &= \rho V_{\mathrm{P}}, \\ I_{\mathrm{S}} &= \rho V_{\mathrm{S}}. \end{aligned} \tag{4.3}$$

All the other elastic constants commonly used in reservoir geophysics, such as the Lame's constant λ, the Young's modulus E, and the Poisson's ratio ν, can be derived from the elastic properties above:

$$E = \frac{9KG}{3K + G},$$
$$\nu = \frac{3K - 2G}{2(3K + G)}, \tag{4.4}$$
$$\lambda = K - \frac{2}{3}G.$$

Elastic moduli depend on the rock and fluid properties of the porous rock. These underlying properties include the properties of the solid phase (indicated in the following with the subscript *mat* as matrix), the properties of the fluid phase (indicated with the subscript fl), and the properties of the dry and saturated rock (indicated with the subscript *dry* and *sat*). These attributes are generally functions of porosity, mineral content, fluid saturations, rock texture, temperature, and pressure.

Laboratory measurements presented in Wyllie et al. [4], Raymer et al. [5], Han [6], Bourbié et al. [7], Zimmerman [8], Nur et al. [9], Avseth et al. [10], Mavko et al. [3], and Dvorkin et al. [11] show common trends in rock physics measurements. For example, if porosity increases, generally P-wave, S-wave velocity, and density decrease. In carbonate and tight sandstone, the relation between porosity and elastic properties is approximately linear; however, in unconsolidated sandstone, the relation is generally nonlinear. For the saturation effect, generally, P-wave velocity in water-filled rocks is higher than hydrocarbon rocks as well as the density and this relation can be described by physical models such as Gassmann's equations [3]. In many studies, we can observe that many changes in rock properties are coupled with other petrophysical variations: for example if clay content increases, porosity generally decreases. Furthermore, because porous rocks are deformable, rock properties such as porosity, permeability, as well as elastic moduli and seismic velocities, are sensitive to applied stresses and pore pressure.

The physical model that links all these properties together is the so-called rock physics model and generally depends on the geological scenario under study. For example, different rock physics models should be applied in different lithologies. In the following, we analyze the most common rock physics models in conventional hydrocarbon reservoirs.

4.2 Empirical Relations

Well log data and core measurements collected in rock physics show a number of general trends. In this section, we use the dataset measured by Han [6]. This dataset includes more than 50 samples of consolidated sands with clay content between 0 and 50 %, measured in dry and brine-saturated conditions. The dataset includes measurements of porosity, clay content, density, P-wave, and S-wave velocity at

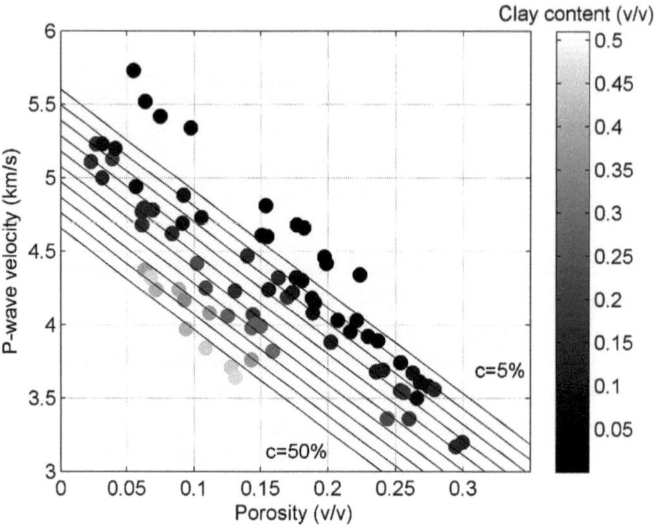

Fig. 4.1 P-wave velocity versus porosity measured for a set of sand and shaly-sand rocks [6]. Data are color coded by clay content. Each *solid line* represents a linear relation between P-wave velocity and porosity (clay content varies from 0.05 to 0.5 from *top* to *bottom*)

different effective pressures. In this section, we illustrate some of the general trends between rock and elastic properties.

As shown in Fig. 4.1, in general, porosity and velocity are anti-correlated, meaning that if porosity increases, P-wave velocity decreases. Therefore, the relation between porosity and P-wave velocity can be expressed by a linear regression. However, the slope of the curve and the intercept of the curve depend on other properties, such as saturation and clay content. In the example in Fig. 4.1, an increase in clay content causes a decrease in P-wave velocity. Han [6] proposed an empirical relation to link P-wave and S-wave velocity to rock properties:

$$\begin{aligned} V_{\mathrm{P}} &= (5.6 - 2.1C) - 6.9\phi \\ V_{\mathrm{S}} &= (3.5 - 1.9C) - 4.9\phi, \end{aligned} \tag{4.5}$$

where C is the clay content and ϕ is the porosity of the rock. The expressions in Eq. 4.5 are multilinear regressions, and the coefficients depend on the fluid conditions, mineral content, rock texture, and pressure and they can vary from one dataset to another one.

Other empirical relations are available in literature. Wyllie et al. [4] proposed the time-average model for P-wave velocity:

$$V_{\mathrm{P}} = \left(\frac{1-\phi}{V_{\mathrm{P_{mat}}}} + \frac{\phi}{V_{\mathrm{P_{fl}}}}\right)^{-1}, \tag{4.6}$$

where $V_{P_{mat}}$ and $V_{P_{fl}}$ are the P-wave velocity of the solid and fluid phase, respectively. Similarly, Raymer et al. [5] proposed a quadratic model for P-wave velocity for porosities lower than 0.37:

$$V_P = (1-\phi)^2 V_{P_{mat}} + \phi V_{P_{fl}}. \tag{4.7}$$

In general, $V_{P_{mat}}$ is a function of the mineralogical volumes of the solid phase of the rock and $V_{P_{fl}}$ is a function of the fluid saturations. Raymer's model was extended to S-wave velocity by Dvorkin [11]:

$$V_S = (1-\phi)^2 V_{S_{mat}} \sqrt{\frac{(1-\phi)\rho_{mat}}{(1-\phi)\rho_{mat} + \phi\rho_{fl}}}, \tag{4.8}$$

where $V_{S_{mat}}$ is the S-wave velocity of the solid phase, ρ_{mat} is the density of the solid phase, and ρ_{fl} is the density of the fluid.

In Fig. 4.2, we imposed Wyllie and Raymer's equations to Han's dataset. The model predictions have been computed for two different rock types: 100 % quartz and a mixture of 40 % clay and 60 % quartz.

Most of these relations are valid for P-wave velocity. Castagna [12] and Greenberg and Castagna [13] proposed empirical relations to estimate V_S from V_P in multi-mineralic, brine-saturated rocks based on empirical polynomial relations in different lithologies. The shear wave velocity in brine-saturated composite

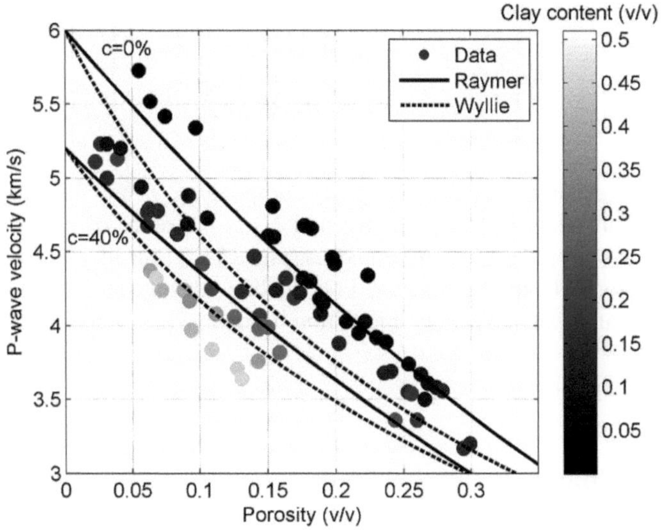

Fig. 4.2 P-wave velocity versus porosity measured for a set of sand and shaley-sand rocks [6]. Data are color coded by clay content. *Dashed lines* represent Wyllie's model for two values of clay content (0 and 0.4 for *top* and *bottom line*, respectively); *solid lines* represent Raymer's model for two values of clay content (0 and 0.4 for *top* and *bottom line*, respectively)

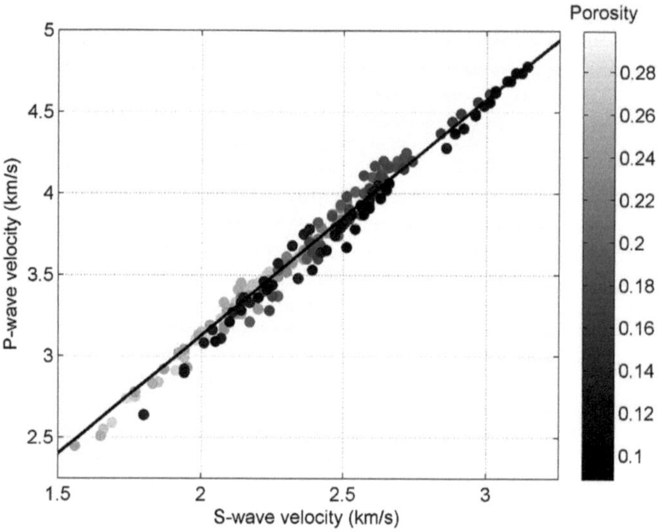

Fig. 4.3 P-wave velocity versus S-wave velocity, measured for a set of sand and shaley-sand rocks [6]. Data are color coded by porosity. The *solid line* represents Greenberg-Castagna relation

lithologies is approximated by the average of the arithmetic and harmonic means of the constituent lithology velocities:

$$V_S = \frac{1}{2}\left[\left(\sum_{i=1}^{N} f_i \sum_{j=1}^{N_d} a_{ij} V_p^j\right) + \left(\sum_{i=1}^{N} \frac{f_i}{\sum_{j=1}^{N_d} a_{ij} V_p^j}\right)^{-1}\right], \tag{4.9}$$

where N is the number of mineralogical components; f_i are the mineralogical fractions; a_{ij} are empirical coefficients; and N_d is the order of polynomial. The coefficients are given in Greenberg and Castagna [13], but in general should be calibrated using real data. Vernik et al. [14] modified this model for soft sediments. An example of S-wave velocity prediction from P-wave velocity is shown in Fig. 4.3.

4.3 Solid Phase

A porous rock is generally a mixture of a number of mineralogical components. For example, a porous sandstone can contain quartz, feldspar, and clay. Clay can be made by various minerals such as illite, kaolinite, or smectite. From core samples and well log, we can generally estimate the mineralogical fractions of the various

Table 4.1 Elastic moduli and densities of minerals used in this book

Mineral	Bulk modulus (GPa)	Shear modulus (GPa)	Density (g/cm^3)
Quartz	36.6	45.0	2.65
Clay (illite)	21.0	7.0	2.55
Calcite	76.8	32.0	2.71
Dolomite	94.9	45.0	2.87

minerals within the solid phase. The elastic moduli and densities of minerals range from low values for water-rich clays to high values for limestone and dolomite. Mavko et al. [3] and Dvorkin et al. [11] provide a set of tables with these properties measured in the laboratory. We report the values of the most common rock types in Table 4.1.

In many applications, however, rather than the elastic properties of each single mineral we want to estimate the elastic properties of the solid phase (the matrix) as an effective medium.

The density of the solid phase can be computed as a linear average of the densities of the mineralogical fractions. If we assume that the solid phase is made by N mineralogical constituents, then the density of the matrix is

$$\rho_{\mathrm{mat}} = \sum_{i=1}^{N} f_i \rho_i, \tag{4.10}$$

where f_i is the mineral fraction of the ith constituent, and ρ_i is the corresponding density. For example, if the rock is a mixture of 70 % quartz $\left(\rho_{\mathrm{quartz}} = 2.65\,\mathrm{g/cm^3}\right)$ and 30 % illite $\left(\rho_{\mathrm{illite}} = 2.55\,\mathrm{g/cm^3}\right)$, then the density of the matrix is

$$\rho_{\mathrm{mat}} = 0.7 \times 2.65 + 0.3 \times 2.55 = 2.62\,\mathrm{g/cm^3}. \tag{4.11}$$

The computation of the elastic moduli is generally more complex. Indeed the exact estimation of the bulk and shear moduli of the solid phase would require the knowledge of the mineralogical fractions, the elastic moduli of each constituent, and the geometrical description of the rock texture. For example, in a mixture of quartz and clay, the effective elastic moduli would depend on the proportions of the two minerals, their elastic moduli and the geometry of the mixture. For instance, clay could be mixed to quartz in a dispersed model, where clay grains fill the pore space of the sandstone, or in a laminar model where clay is organized into thin layers within the sand. The derivation of a mathematical model to describe these geometrical features is generally complex; therefore, the exact estimation of the elastic moduli of the solid material is not possible. However, rock physics provides a set of physical bounds to approximate the elastic behavior of the solid phase.

The upper bound is the so-called Voigt average, and it is a linear average of the elastic moduli of the constituents:

$$K_{\text{Voigt}} = \sum_{i=1}^{N} f_i K_i$$
$$G_{\text{Voigt}} = \sum_{i=1}^{N} f_i G_i, \tag{4.12}$$

where K_i and G_i are the bulk and shear moduli of the ith constituent, respectively. The Voigt bound corresponds to a mixing law that provides the ratio of the average stress to the average strain when all constituents are assumed to have the same strain. The lower bound is the so-called Reuss average, and it is a harmonic average of the elastic moduli of the constituents:

$$K_{\text{Reuss}} = \frac{1}{\sum_{i=1}^{N} \frac{f_i}{K_i}} \qquad G_{\text{Reuss}} = \frac{1}{\sum_{i=1}^{N} \frac{f_i}{G_i}}. \tag{4.13}$$

The Reuss bound corresponds to a mixing law that provides the ratio of the average stress to the average strain when all constituents are assumed to have the same stress. The Voigt and Reuss averages are the most general bounds in rock physics. In a mixture of two minerals, if the two end-members are very different, the elastic responses obtained by applying Voigt and Reuss averages can be significantly different. In order to approximate the elastic behavior of the solid medium, it is common to average these two bounds. This average is often called Hill average, or Reuss-Voigt-Hill average:

$$K_{\text{Hill}} = \frac{K_{\text{Voigt}} + K_{\text{Reuss}}}{2}$$
$$G_{\text{Hill}} = \frac{G_{\text{Voigt}} + G_{\text{Reuss}}}{2}. \tag{4.14}$$

In our previous example of the mixture of quartz-clay, we can compute the effective bulk and shear moduli of the solid phase as follows:

$$K_{\text{Hill}} = \frac{K_{\text{Voigt}} + K_{\text{Reuss}}}{2} = \frac{1}{2}\left(0.7 \times 36 + 0.3 \times 21 + \frac{1}{0.7/36 + 0.3/21}\right) = 30.5\,\text{GPa}$$
$$G_{\text{Hill}} = \frac{G_{\text{Voigt}} + G_{\text{Reuss}}}{2} = \frac{1}{2}\left(0.7 \times 45 + 0.3 \times 7 + \frac{1}{0.7/45 + 0.3/7}\right) = 25.3\,\text{GPa}. \tag{4.15}$$

More accurate approximations are given by Hashin-Shtrikman upper and lower bounds. The upper bound is computed as follows:

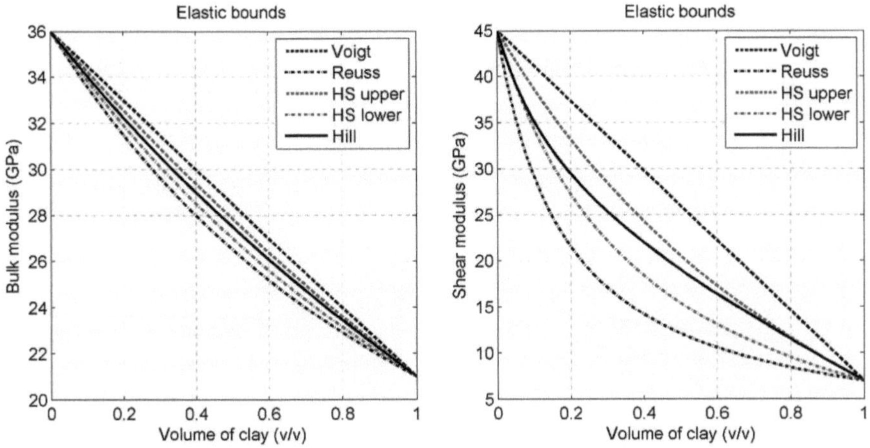

Fig. 4.4 Elastic bounds for a mixture of quartz and clay as a function of clay volume: bulk modulus (*left*) and shear modulus (*right*) of solid rock versus clay content. *Solid lines* represent Voigt-Reuss-Hill averages; *black dotted lines* represent Voigt averages; *grey dotted lines* represent Hashin–Shtrikman (HS) upper bounds; *black dashed lines* represent Reuss averages; *grey dashed lines* represent Hashin–Shtrikman (HS) lower bounds

$$\begin{aligned} K_{\mathrm{HS}+} &= \left(\sum_{i=1}^{N}\frac{f_i}{K_i+\frac{4}{3}G_{\max}}\right)^{-1} - \frac{4}{3}G_{\max} \\ G_{\mathrm{HS}+} &= \left(\sum_{i=1}^{N}\frac{f_i}{G_i+\frac{G_{\max}}{6}\frac{9K_{\max}+8G_{\max}}{K_{\max}+2G_{\max}}}\right)^{-1} - \frac{G_{\max}}{6}\frac{9K_{\max}+8G_{\max}}{K_{\max}+2G_{\max}}, \end{aligned} \tag{4.16}$$

where $K_{\max}$ and $G_{\max}$ represent the maximum of the bulk and shear moduli of the constituents, respectively. The lower bound is

$$\begin{aligned} K_{\mathrm{HS}-} &= \left(\sum_{i=1}^{N}\frac{f_i}{K_i+\frac{4}{3}G_{\min}}\right)^{-1} - \frac{4}{3}G_{\min} \\ G_{\mathrm{HS}-} &= \left(\sum_{i=1}^{N}\frac{f_i}{G_i+\frac{G_{\min}}{6}\frac{9K_{\min}+8G_{\min}}{K_{\min}+2G_{\min}}}\right)^{-1} - \frac{G_{\min}}{6}\frac{9K_{\min}+8G_{\min}}{K_{\min}+2G_{\min}}. \end{aligned} \tag{4.17}$$

In Fig. 4.4, we compare the elastic bounds for a mixture of quartz and clay, as a function of the clay content.

4.4 Fluid Phase

Similarly to the solid phase, the fluid phase can contain different constituents: typically water, oil, and gas. However, the elastic properties of water, oil, and gas depend on a number of factors: temperature, pressure, water salinity, gas-oil ratio,

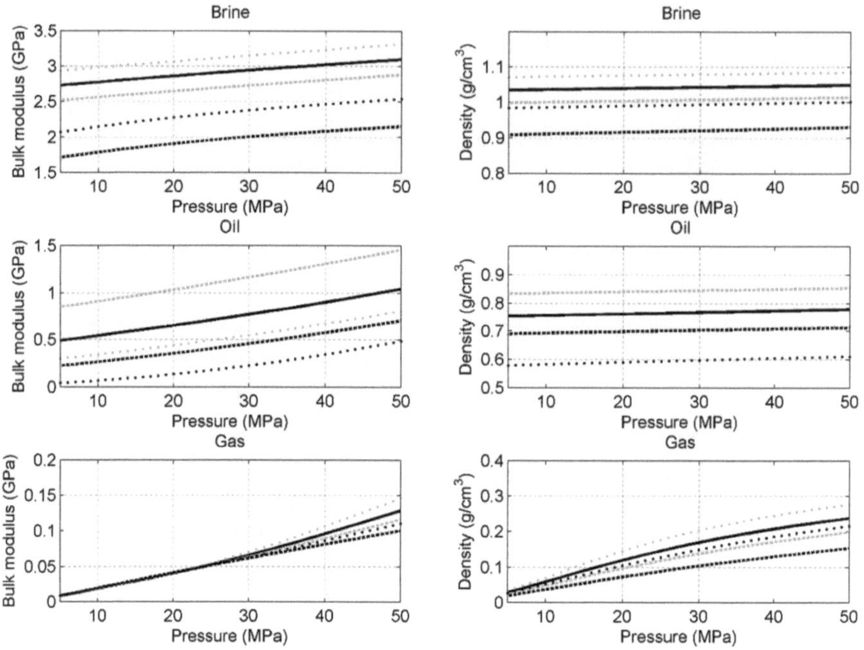

Fig. 4.5 Elastic properties of fluids: left plots show bulk modulus versus fluid pressure; right plots show density versus fluid pressure; top row shows brine properties; mid row shows oil properties; and bottom row shows gas properties. Brine (top plots): *solid lines* represent salinity of 100,000 ppm and temperature of 100 °C; *grey dashed lines* represent salinity of 50,000 ppm and temperature of 100 °C; *grey dotted lines* represent salinity of 150,000 ppm and temperature of 100 °C; *black dashed lines* represent salinity of 50,000 ppm and temperature of 200 °C; *black dotted lines* represent salinity of 150,000 ppm and temperature of 200 °C. Oil (mid plots): *solid lines* represent gas-oil ratio of 200 l/l and temperature of 100 °C; *grey dashed lines* represent gas-oil ratio of 100 l/l and temperature of 100 °C; *grey dotted lines* represent gas-oil ratio of 300 l/l and temperature of 100 °C; *black dashed lines* represent gas-oil ratio of 100 l/l and temperature of 200 °C; *black dotted lines* represent gas-oil ratio of 300 l/l and temperature of 200 °C. Gas (bottom plots): *solid lines* represent gas gravity of 0.6 and temperature of 100 °C; *grey dashed lines* represent gas gravity of 0.5 and temperature of 100 °C; *grey dotted lines* represent gas gravity of 0.7 and temperature of 100 °C; *black dashed lines* represent gas gravity of 0.5 and temperature of 200 °C; *black dotted lines* represent gas gravity of 0.7 and temperature of 200 °C

gas gravity, and oil gravity. Batzle and Wang [15] derived a set of empirical models to link these variables to bulk moduli and density for different fluid types. For a complete mathematical formulation of these equations, we refer the reader to the original paper [15]. In Fig. 4.5, we show a sensitivity analysis to show the effects of the reservoir and fluid conditions on the bulk moduli and densities of the three main fluid types.

The properties of the effective fluid mixture can be computed using mixing laws. The density of a fluid mixture is computed as a linear average of the densities of the fluid components:

$$\rho_{\text{fl}} = s_{\text{w}}\rho_{\text{w}} + s_{\text{o}}\rho_{\text{o}} + s_{\text{g}}\rho_{\text{g}}, \tag{4.18}$$

where s_{w}, s_{o}, and s_{g} are, respectively, the saturation of water, oil, and gas, and ρ_{w}, ρ_{o}, and ρ_{g} are respectively the density of water, oil, and gas. For example, for a fluid mixture of 80 % oil ($\rho_o = 0.7\,\text{g/cm}^3$) and 20 % water ($\rho_{\text{w}} = 1.05\,\text{g/cm}^3$), the density of the fluid phase is

$$\rho_{\text{fl}} = 0.2 \times 1.05 + 0.8 \times 0.7 = 0.77\,\text{g/cm}^3. \tag{4.19}$$

Generally, the bulk modulus of the fluid is computed using Reuss average:

$$K_{\text{fl}} = \frac{1}{\frac{s_{\text{w}}}{K_{\text{w}}} + \frac{s_{\text{o}}}{K_{\text{o}}} + \frac{s_{\text{g}}}{K_{\text{g}}}}. \tag{4.20}$$

Reuss average is often used for fluid in which the fluid constituents are uniformly mixed. In the previous example, a fluid mixture of 80 % oil ($K_{\text{o}} = 0.8\,\text{GPa}$) and 20 % water ($K_{\text{w}} = 2.25\,\text{GPa}$), the effective bulk modulus is

$$K_{\text{fl}} = \frac{1}{\frac{0.2}{2.25} + \frac{0.8}{0.8}} = 0.91\ \text{GPa}. \tag{4.21}$$

However, especially in a fluid mixture with gas, the two fluid components can be distributed in isolated patches. For this scenario, often called patchy saturation, the Voigt average provides a more accurate estimation. The bulk modulus of a patchy-saturated fluid mixture is then

$$K_{\text{fl}} = s_{\text{w}}K_{\text{w}} + s_{\text{o}}K_{\text{o}} + s_{\text{g}}K_{\text{g}}. \tag{4.22}$$

Another common law for fluid mixtures is given by Brie's equation [16]:

$$K_{\text{fl}} = \left(K_{\text{w}} - K_{\text{g}}\right)\left(1 - s_{\text{g}}\right)^e + K_{\text{g}}, \tag{4.23}$$

where e is an empirical constant, typically equal to about 3. A comparison of these three mixing law is given in Fig. 4.6. The shear modulus of water, oil and gas, as well as the shear modulus of the fluid mixture is 0 by definition.

4.5 Dry Rock Properties

In order to compute the elastic properties, bulk and shear moduli, of the dry rock, i.e., a porous rock with no fluid in the pore space, several equations have been proposed for a number of lithologies [3].

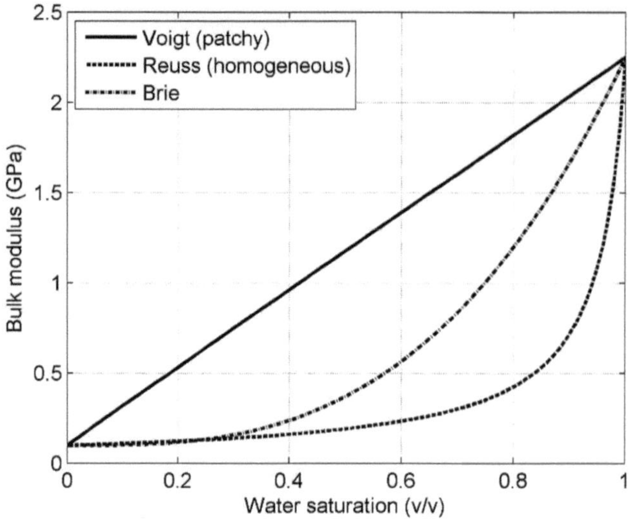

Fig. 4.6 Mixing laws for the bulk modulus of a fluid mixture as a function of water saturation: the solid line represents Voigt average (patchy saturation); the *dashed line* represents Reuss average (homogeneous saturation); the *dashed-dotted line* represents Brie's equation (with exponent e equal to 3)

Nur et al. [9, 17] showed that the elastic properties (velocities and impedances) of dry rocks decrease as a function of porosity and that this relation is approximately linear. According to Nur et al. [9, 17], for most porous materials, there is a critical porosity value that separates their elastic behavior into two domains. In particular, the critical porosity separates the phase in which the rock grains are load-bearing, from the phase in which the fluid phase is load-bearing and the porous rock becomes a suspension. In the suspension domain, the bulk and shear moduli can be estimated using the Reuss average (Sects. 4.3 and 4.4). In the porous rock domain, in well-consolidated rocks, the bulk and shear moduli decrease, approximately linearly, from the mineral values (zero porosity) to the suspension values (critical porosity). Nur et al. [9, 17] approximated this behavior with linear relations in the elastic moduli of the dry rock K_{dry} and G_{dry} domain:

$$\begin{aligned} K_{\mathrm{dry}} &= K_{\mathrm{mat}}\left(1-\frac{\phi}{\phi_{\mathrm{c}}}\right) \\ G_{\mathrm{dry}} &= G_{\mathrm{mat}}\left(1-\frac{\phi}{\phi_{\mathrm{c}}}\right), \end{aligned} \tag{4.24}$$

where ϕ_{c} is the critical porosity of the rock. More sophisticated models have been later developed by introducing a geometrical description of the pore space. In the following, we divide these equations into two main categories: granular media models and inclusion models.

4.5.1 Granular Media Models

To account for the porosity effect in the estimation of the elastic properties of a porous rock, we must account for the volume of the porous space as well as its geometry. The geometrical description of the pore space must be idealized in order to be included in a mathematical model. In rock physics, a common assumption for sandstones is a spherical shape for the rock grains. Spheres provide an analytical description of the pore space geometry and simplify the mathematical-physical models for the description of the grain interactions. Generally, a porous sandstone is assumed to be made by a random pack of spherical grains. The average number of contacts between grains is named the coordination number. From laboratory measurements, we can observe that the coordination number varies throughout different sample, from 4 to 12, and it increases with decreasing porosity due to tighter packing.

Hertz-Mindlin grain-contact theory provides estimations for the effective bulk K_{HM} and shear G_{HM} moduli of a dry rock, assuming that the sand frame is a dense random pack of identical spherical grains subject to an effective pressure P (in GPa) with a given porosity ϕ_{c} (critical porosity) and an average number of contacts per grain n (coordination number):

$$K_{\mathrm{HM}} = \sqrt[3]{\frac{n^2(1-\phi_{\mathrm{c}})^2 G_{\mathrm{mat}}^2}{18\pi^2(1-\nu)^2}P} \quad G_{\mathrm{HM}} = \frac{5-4\nu}{5(2-\nu)}\sqrt[3]{\frac{3n^2(1-\phi_{\mathrm{c}})^2 G_{\mathrm{mat}}^2}{2\pi^2(1-\nu)^2}P}, \tag{4.25}$$

where ν is the Poisson ratio. This model assumes frictionless spheres; for an extension of the model to include the friction coefficient in the shear modulus estimation we refer to Mavko et al. [3]. A number of rock physics models has been built based on Hertz-Mindlin grain-contact theory. Dvorkin et al. [18], Dvorkin and Nur [19], and Gal et al. [20] proposed a set of models for unconsolidated (soft) sand, consolidated (stiff) sand, and cemented sand.

Soft and stiff sand models are based on the modified Hashin–Shtrikman bounds in which the end members are represented by a mineral mixed with a fluid–solid suspension. The soft sand model [11] includes mixing laws for the solid phase and Hertz-Mindlin grain-contact theory at the critical porosity ϕ_{c}. For effective porosity values between zero and the critical porosity, the soft sand model connects the elastic moduli K_{mat} and G_{mat} of the solid phase to the elastic moduli K_{HM} and G_{HM} of the dry rock at porosity ϕ_{c}, by interpolating these two end members at the intermediate porosity values by means of the modified Hashin-Shtrikman lower bound:

$$K_{\mathrm{dry}} = \left(\frac{\frac{\phi}{\phi_c}}{K_{\mathrm{HM}} + \frac{4}{3}G_{\mathrm{HM}}} + \frac{1-\frac{\phi}{\phi_c}}{K_{\mathrm{mat}} + \frac{4}{3}G_{\mathrm{HM}}} \right)^{-1} - \frac{4}{3}G_{\mathrm{HM}}$$
$$G_{\mathrm{dry}} = \left(\frac{\frac{\phi}{\phi_c}}{G_{\mathrm{HM}} + \frac{1}{6}\xi G_{\mathrm{HM}}} + \frac{1-\frac{\phi}{\phi_c}}{G_{\mathrm{mat}} + \frac{1}{6}\xi G_{\mathrm{HM}}} \right)^{-1} - \frac{1}{6}\xi G_{\mathrm{HM}}, \qquad \xi = \frac{9K_{\mathrm{HM}} + 8G_{\mathrm{HM}}}{K_{\mathrm{HM}} + 2G_{\mathrm{HM}}}. \tag{4.26}$$

Similarly, the stiff sand model includes mixing laws for the solid phase and Hertz-Mindlin grain contact theory at the critical porosity, but the interpolation between zero and the critical porosity ϕ_c is done using the modified Hashin-Shtrikman upper bound:

$$K_{\mathrm{dry}} = \left(\frac{\frac{\phi}{\phi_c}}{K_{\mathrm{HM}} + \frac{4}{3}G_{\mathrm{mat}}} + \frac{1-\frac{\phi}{\phi_c}}{K_{\mathrm{mat}} + \frac{4}{3}G_{\mathrm{mat}}} \right)^{-1} - \frac{4}{3}G_{\mathrm{mat}}$$
$$G_{\mathrm{dry}} = \left(\frac{\frac{\phi}{\phi_c}}{G_{\mathrm{HM}} + \frac{1}{6}\xi G_{\mathrm{mat}}} + \frac{1-\frac{\phi}{\phi_c}}{G_{\mathrm{mat}} + \frac{1}{6}\xi G_{\mathrm{mat}}} \right)^{-1} - \frac{1}{6}\xi G_{\mathrm{mat}}, \qquad \xi = \frac{9K_{\mathrm{mat}} + 8G_{\mathrm{mat}}}{K_{\mathrm{mat}} + 2G_{\mathrm{mat}}}. \tag{4.27}$$

Elastic properties computed using the soft and stiff sand models (Eqs. 4.26 and 4.27, respectively) can be fitted to real data by varying the critical porosity and the coordination number.

The cemented sand model [19] estimates the dry rock bulk and shear moduli of cemented sandstones with mid-high-porosity values. The sand framework is assumed to be made by a random pack of spherical grains and the cement is deposited at grain contacts. For the model equations, we refer to Dvorkin and Nur [19] and Dvorkin et al. [11]. A comparison of granular media models is shown in Fig. 4.7. Other models based on granular media theory include Walton, Digby, Jenkins, Brandt, and Johnson relations; see [3].

4.5.2 Inclusion Models

In rock physics, the estimation of elastic properties of a porous rock can be expressed in terms of the elastic behavior of cavities and inclusions. A detailed description of cavity deformation and pore compressibility is given in Zimmerman [8].

In inclusion models, the dry rock is built by placing inclusions into the solid matrix [3]. Inclusions can be in the solid phase (for example clay in sand) or in the fluid phase (fluid in solid structures). These models are suitable, for instance, for carbonate rocks where the pores appear as inclusions in calcite or dolomite matrix. Examples of these models are the differential effective medium model, Kuster-Toksöz, Xu-White, and Berryman [3]. These models are generally more complex than granular media and can include systems of ordinary differential equations. In

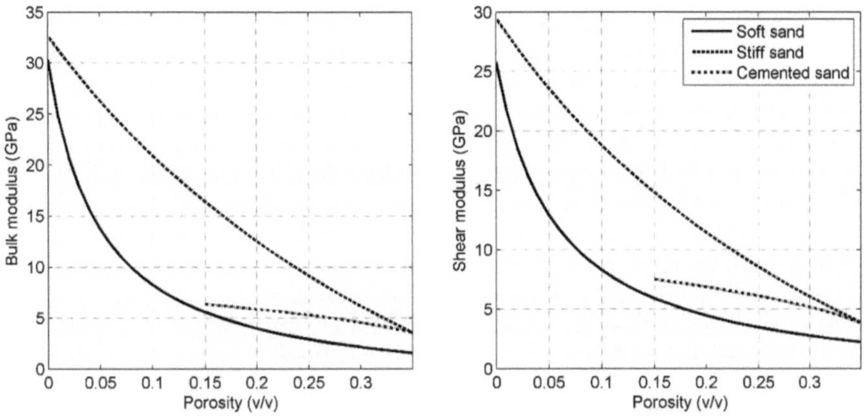

Fig. 4.7 Comparison of granular media models for a mixture of 80 % quartz and 20 % clay: left plot shows bulk modulus versus porosity; right plot shows shear modulus versus porosity. *Solid lines* represent the soft sand model predictions; the *dashed lines* represent the stiff sand model predictions; the *dashed-dotted lines* represent the cemented sand model predictions for porosity higher than 0.15 (assuming that the cement is quartz). The coordination number is 6; the critical porosity is 0.4

general, the input parameters of these models include the elastic moduli of the matrix; the elastic moduli of the inclusions; the volumetric fractions of the inclusions (porosity, volume of minerals, and/or fluid saturations); and the aspect ratio of the inclusions. The aspect ratio is the quotient of the minimum and maximum axis of the inclusion shape. Inclusion models allow modeling different inclusion shapes, such as spheres, penny-shaped cracks, and needle-shaped pores; however, the estimation and calibration of the aspect ratio of these shapes generally require a large number of data. If the inclusions are filled by fluids, the inclusion model might overestimate the actual elastic property values as it provides results comparable to ultrasonic velocity measurements. For a detailed description of inclusion models, we refer the reader to Mavko et al. [3]. A common strategy to mitigate this issue is to assume that the inclusion is not saturated, use the inclusion model to estimate the dry rock properties and combine the model with Gassmann's equations (see Sect. 4.6).

Kuster and Toksöz [21] proposed a model to estimate the elastic properties of dry and saturated rocks by using a long-wavelength first-order scattering theory. The effective moduli K_{KT} and G_{KT} for a number of inclusion shapes can be written as

$$
\begin{aligned}
&(K_{\mathrm{KT}} - K_{\mathrm{mat}})\frac{K_{\mathrm{mat}} + \frac{4}{3}G_{\mathrm{mat}}}{K_{\mathrm{KT}} + \frac{4}{3}G_{\mathrm{mat}}} = \sum_{i=1}^{N_i} f_i(K_i - K_{\mathrm{mat}})P_i \\
&(G_{\mathrm{KT}} - G_{\mathrm{mat}})\frac{G_{\mathrm{mat}} + \frac{G_{\mathrm{mat}}}{6}\xi}{G_{\mathrm{KT}} + \frac{G_{\mathrm{mat}}}{6}\xi} = \sum_{i=1}^{N_i} f_i(G_i - G_{\mathrm{mat}})Q_i, \qquad \xi = \frac{9K_{\mathrm{mat}} + 8G_{\mathrm{mat}}}{K_{\mathrm{mat}} + 2G_{\mathrm{mat}}}.
\end{aligned}
\tag{4.28}
$$

The coefficients P_i and Q_i describe the geometry of N_i inclusions in the background matrix. For example, for a solid material with a pore space filled by hydrocarbon, the summation in Eq. 4.28 consists of a single term where the volumetric fraction f_i is the porosity, and the elastic moduli K_i and G_i are the elastic moduli of the fluid phase (with the inclusion shear modulus equal to zero). For spherical inclusions, for example, the coefficients are

$$
\begin{aligned}
P_i &= \frac{K_{\mathrm{mat}} + \frac{4}{3}G_{\mathrm{mat}}}{K_i + \frac{4}{3}G_{\mathrm{mat}}} \\
Q_i &= \frac{G_{\mathrm{mat}} + \frac{G_{\mathrm{mat}}}{6}\xi}{G_i + \frac{G_{\mathrm{mat}}}{6}\xi}.
\end{aligned}
\tag{4.29}
$$

Other common inclusion shapes are needles, disks, and penny cracks. Non-spherical shapes contain an additional parameter, the aspect ratio. For the analytical expressions of the geometrical coefficients, we refer the reader to Mavko et al. [3]. Inclusions with different material properties or different shapes require separate terms in the summation. Dry inclusions can be modeled by setting the inclusion moduli K_i and G_i to zero.

Wu's self-consistent approximation allows computing the elastic moduli K_{SC} and G_{SC} for a single-inclusion model as

$$
\begin{aligned}
K_{\mathrm{SC}} &= K_{\mathrm{mat}} + f_i(K_i - K_{\mathrm{mat}})P_i \\
G_{\mathrm{SC}} &= G_{\mathrm{mat}} + f_i(G_i - G_{\mathrm{mat}})Q_i.
\end{aligned}
\tag{4.30}
$$

Berryman [22] provided a generalization of the model in Eq. 4.30 for a composite of N_i inclusions:

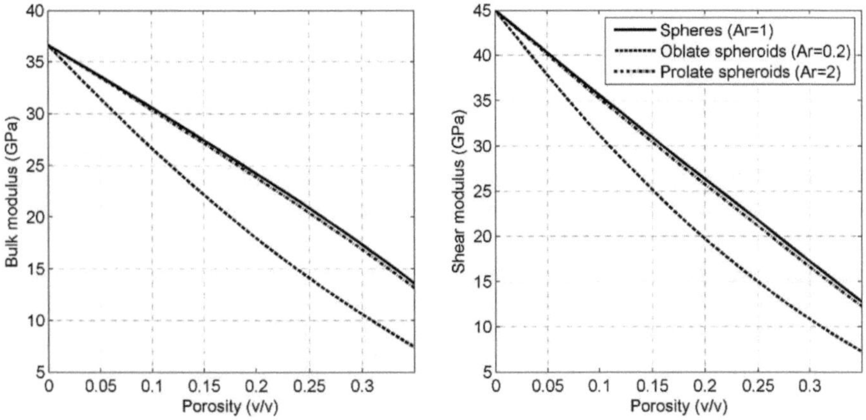

Fig. 4.8 Comparison of inclusion models for dry sandstone: left plot shows bulk modulus versus porosity; right plot shows shear modulus versus porosity. *Solid lines* represent Berryman's model predictions with aspect ratio equal to 1 (spherical pores); the *dashed lines* represent Berryman's model predictions with aspect ratio equal to 0.2 (oblate spheroids); the *dashed-dotted lines* represent Berryman's model predictions with aspect ratio equal to 2 (prolate spheroids)

$$\begin{aligned} &\sum_{i=1}^{N_i} f_i (K_i - K_{\mathrm{SC}}) P_i = 0 \\ &\sum_{i=1}^{N_i} f_i (G_i - G_{\mathrm{SC}}) Q_i = 0. \end{aligned} \tag{4.31}$$

A comparison of inclusion models with different values of aspect ratio is shown in Fig. 4.8. In the differential effective medium models, infinitesimal increments of inclusion volumes are added to the background matrix until the total volume fraction of the inclusion is reached. This model calls for ordinary differential equations to explicitly solve for the elastic moduli of the porous rock [3].

4.6 Saturated Rock Properties

In Sects. 4.3 and 4.4, we reviewed a set of mixing laws for the solid and fluid phases respectively; Sect. 4.5 contains an overview of the most common rock physics models to describe the porosity effect. The final component of the rock physics model is the description of the fluid effect. This task is achieved by combining dry rock properties with the fluid properties through Gassmann's equations:

$$K_{\mathrm{sat}} = K_{\mathrm{dry}} + \frac{\left(1 - \frac{K_{\mathrm{dry}}}{K_{\mathrm{mat}}}\right)^2}{\frac{\phi}{K_{\mathrm{fl}}} + \frac{(1-\phi)}{K_{\mathrm{mat}}} - \frac{K_{\mathrm{dry}}}{K_{\mathrm{mat}}^2}} \qquad G_{\mathrm{sat}} = G_{\mathrm{dry}}. \tag{4.32}$$

These equations were derived to link the dry rock moduli to the saturated rock moduli. Gassmann's model for the saturated bulk modulus is based on three main assumptions: the rock is isotropic; the mineral moduli are homogeneous; and the pore pressure changes induced in the fluid by seismic waves can rapidly equilibrate throughout the pore space. The last assumption is generally verified when the measurement frequency is low enough. The shear modulus equation is based on the assumption that fluids do not affect the shear moduli; therefore, the saturated rock shear modulus equals the dry rock shear modulus. Gassmann's equations are generally valid at the seismic and well log scale. Dry rock properties can be estimated from saturated rock properties by using the inverse equations:

$$\begin{aligned} K_{\mathrm{dry}} &= \frac{K_{\mathrm{sat}}\left(\frac{\phi K_{\mathrm{mat}}}{K_{\mathrm{fl}}} + 1 - \phi\right) - K_{\mathrm{mat}}}{\frac{\phi K_{\mathrm{mat}}}{K_{\mathrm{fl}}} + \frac{K_{\mathrm{sat}}}{K_{\mathrm{mat}}} - 1 - \phi} \\ G_{\mathrm{dry}} &= G_{\mathrm{sat}}. \end{aligned} \tag{4.33}$$

In many practical applications, the computation of the dry rock bulk modulus is still challenging for the lack of calibration data or the complexity of the geological scenario; however, a sensitivity analysis on the fluid effect is possible even if there is no explicit model for the dry rock bulk modulus. This process is called fluid substitution [3]. In the fluid substitution method, we first extract the saturated rock elastic moduli from P-wave and S-wave velocities, with the initial fluid (fluid 1); we compute the dry rock properties using Eq. 4.33; we apply Gassmann's transformation using Eq. 4.32 to compute the saturated rock elastic moduli, with the new fluid (fluid 2); we compute the densities with the new fluid (fluid 2); and finally compute the new P-wave and S-wave velocities with the new fluid. In Fig. 4.9, we show an example of fluid substitution applied to a subset of samples (with low clay content) extracted from Han's dataset. Three fluid conditions are compared: 100 % brine (measured data), 100 % oil and 100 % gas simulated data based on Gassmann's equations. Bulk moduli in gas-saturated rocks are lower than moduli in oil-saturated rocks and bulk moduli in oil-saturated rocks are lower than moduli in brine-saturated rocks. The difference in the bulk moduli increases for high-porosity values (Fig. 4.9, right plot). We point out that Gassmann's fluid substitution does not require the knowledge of S-wave velocity. Indeed, an approximation of fluid substitution can be formulated in terms of the compressional moduli M_{sat} and M_{dry}, as presented in [23].

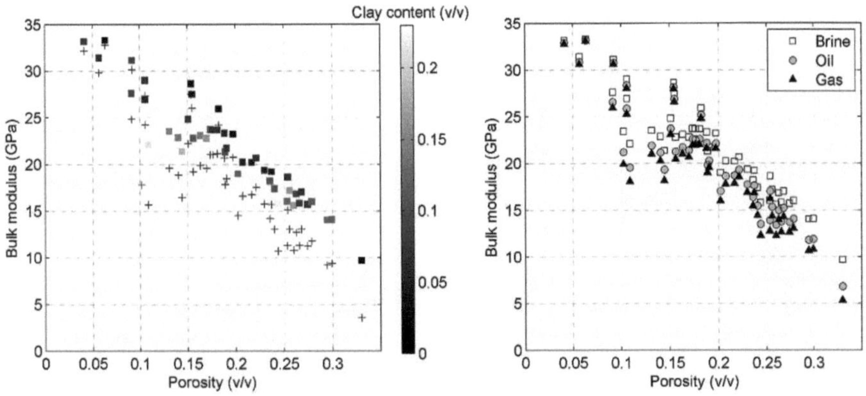

Fig. 4.9 Application of Gassmann's fluid substitution to a subset of Han's dataset. Left plot shows Han's measurements in brine conditions (*square symbols*, color coded by clay content) and dry measurements (*black crosses*). Right plot shows data in brine conditions (*white squares*), oil conditions (*grey circles*), and gas conditions (*black triangles*). Bulk moduli of oil- and gas-saturated samples have been estimated using Gassmann's fluid substitution

4.7 Example

In this section, we show an example of application of rock physics modeling to a well log dataset extracted from a real case study. A rock physics model and the fluid substitution method are applied to an oil-saturated clastic reservoir in the North Sea. Input data for the rock physics model are well log data (density, P-wave, and S-wave velocities) and petrophysical curves obtained in formation evaluation analysis (effective porosity, clay content, and water saturation). The complete dataset is shown in Fig. 4.10. The interval of interest is between 2320 and 2370 m; the main reservoir has a maximum porosity of 0.32 and oil saturation reaches a maximum value of 0.9.

The adopted rock-physics model is the stiff-sand model that combines Hertz-Mindlin contact theory and the modified Hashin–Shtrikman upper bound (Sect. 4.5.1). The critical porosity used in the model application is 0.4 and the coordination number is 7 (Fig. 4.11, left plot). Effective pressure in the reservoir is 62 MPa. For the solid phase, we used a matrix model made by two components: quartz and illite. Matrix parameters have been selected to match well log measurements. A fluid substitution study has been performed to study the effect of water injection during production. In Fig. 4.11 (right plot), we show rock physics model predictions of P-wave velocity in brine conditions according to Gassmann's model (Sect. 4.6). Rock physics model predictions (in situ conditions) of density, P-wave, and S-wave velocities are shown in Fig. 4.10 (grey dashed lines).

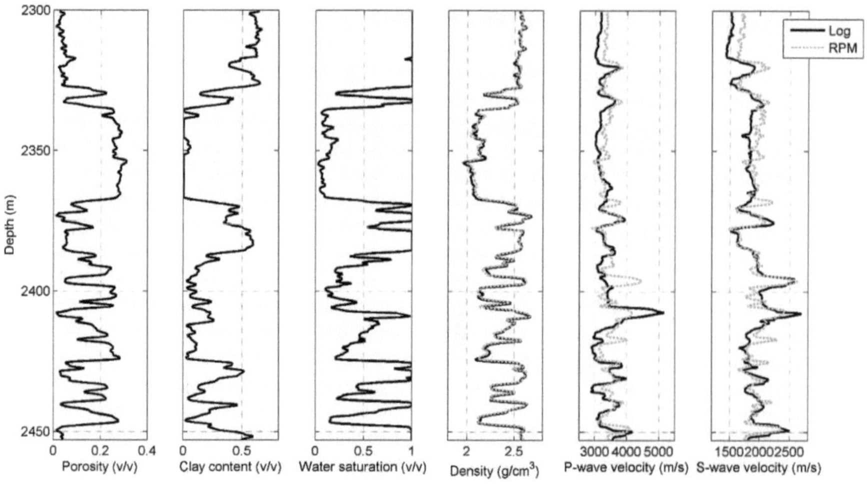

Fig. 4.10 Well log dataset of a well in the North Sea. From *left* to *right*: effective porosity; clay content; water saturation, density, P-wave velocity, and S-wave velocity. *Black solid lines* represent well log measurements and estimated petrophysical curves; *grey dashed lines* represent rock physics model predictions

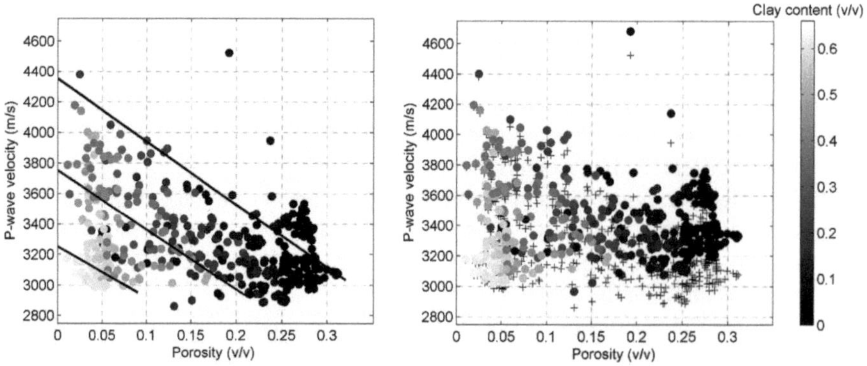

Fig. 4.11 Rock physics model applied to well log data in the North Sea shown in Fig. 4.10. Left plot shows P-wave velocity versus porosity and the calibrated rock physics model (*solid lines* represent the stiff sand model for clay content equal to 0.1, 0.3, and 0.5, from *top* to *bottom*); right plot shows Gassmann's fluid substitution (*black crosses* represent in situ measurements; *color-coded dots* represent rock physics model predictions in brine conditions). Data are color coded by clay content

4.8 Other Rock Physics Models

Rock physics models also include relations between porosity and permeability (for example, Kozeny-Carman relation), velocity and pressure (such as Eberhart-Phillips equations), clay content and porosity (such as Marion's model), and saturation and resistivity (for example, Archie's laws).

Several empirical models have been proposed to include the pressure effect. Experimental data in Han [6], Eberhart-Phillips et al. [44], Prasad and Manghnani [24], and MacBeth [25] show that velocity rapidly increases with effective pressure at low-pressure conditions, and tends to converge to asymptotical values for highly effective pressure values. Eberhart-Phillips et al. [44] presented a multivariate analysis to investigate the combined effects of effective pressure, porosity, and clay content on Han's measurements of velocities in water-saturated shaly sandstones:

$$\begin{aligned} V_{\mathrm{P}} &= 5.77 - 6.94\phi - 1.73\sqrt{C} + 0.446\left(P_e - e^{-16.7P_e}\right) \\ V_{\mathrm{S}} &= 3.70 - 4.94\phi - 1.57\sqrt{C} + 0.361\left(P_e - e^{-16.7P_e}\right). \end{aligned} \tag{4.34}$$

Other rock physics transformations have been proposed to empirically describe the pressure–velocity relation and are generally based on exponential laws [3].

We then focus on saturation-porosity-resistivity relations. These models are mostly common in petrophysics, and are applied in formation evaluation analysis to compute well-log based estimates of saturation from resistivity and porosity logs [26, 27]. Furthermore, these models can be integrated in joint rock physics studies including seismic and electromagnetic data.

The most popular saturation-resistivity relation is Archie's law [3]. This model allows computing the resistivity of a sandstone partially saturated with water when porosity and saturation are known:

$$R = a\frac{R_w}{\phi^m s_w^n} = a\frac{R_w}{\phi^{m-n} v_w^n}, \tag{4.35}$$

where a is a multiplicative constant (often assumed to be 1), R_{w} is the resistivity of water, and m and n are the cementation and saturation exponents, respectively, that must be calibrated with lab measurements. Archie's law can be expressed as a function of porosity and saturation s_{w} or porosity and water volume v_{w}. The cementation exponent m is close to 2 for sandstones and varies between approximately 1.3 and 2.5 for most sedimentary rocks (for example, for unconsolidated spherical grains it is close to 1.3, and for thin disk-like grains it is close to 1.9). However, the cementation exponent in carbonate rocks can reach values close to 5 [3]. The saturation exponent n depends on the fluid type and it is empirically around 2.

Raiga and Clemenceau proposed a relation between permeability and the cementation exponent m:

$$m = 1.28 + \frac{2}{\log(k) + 2}, \tag{4.36}$$

where k is permeability in mD. For mid- to high-permeability sandstones, Eq. 4.36 provides cementation exponents consistent with the previously discussed values. In general, permeability is unknown but can be computed using porosity-permeability relations such as Kozeny-Carman [3]. Experimental data and the corresponding fitted models are shown in Gomez [28].

Archie's law provides accurate results in sand but often fails in shaly sand because it does not account for clay conductivity. A number of formulations have been proposed to model conductivity in shaly sands [29]. Most of the available models modify Archie's relation to introduce a shale conductivity term. These models include Waxman-Smits, the dual-water model, Simandoux, and Poupon-Leveaux (commonly called Indonesia model). Simandoux equation is

$$R = \frac{1}{s_w^n\left(\frac{\phi^m}{R_w} + \frac{C}{R_{\text{clay}}}\right)}, \tag{4.37}$$

where R_{clay} is the resistivity of clay and C is the clay content. Poupon-Leveaux equation (or Indonesia equation) is often used in sand-shale environments in petrophysics:

$$R = \frac{1}{s_w^n\left[\left(\frac{\phi^m}{R_w}\right)^{\frac{1}{2}} + \left(\frac{C^{\alpha-C}}{R_{\text{clay}}}\right)^{\frac{1}{2}}\right]^2}, \tag{4.38}$$

where α is an empirical constant that depends on the clay type.

Typically, resistivity models are mostly used in quantitative log interpretation to estimate saturation from resistivity measurements. In formation evaluation analysis, the deep resistivity log is generally used to avoid the mud filtration effect, and resistivity models are applied to the log measurements after the calibration of the empirical parameters using core samples.

4.9 Rock Physics Inversion

Geophysical datasets (seismic and electromagnetic data) do not provide direct measurements of rock and fluid properties. Seismic data, for example, only provide information about the elastic contrasts at layer interfaces in the subsurface. Electromagnetic data provide information about the resistivity of the reservoir rocks. Elastic attributes (velocity and density) and resistivity depend on the rock and fluid properties, such as porosity, lithology, and saturations. In order to interpret these measurements, we must combine physical relations and mathematical methods to link the measurements to the properties of interest. From a mathematical point of view, the estimation of rock and fluid properties from geophysical measurements is an inverse problem [30]. Indeed, rock physics models (Sects. 4.2–4.6) allow computing the elastic response when the reservoir properties are known. However, the geophysical response is known but the reservoir properties are unknown; hence the modeling problem is an inverse problem.

In reservoir characterization, the input dataset generally consists of seismic data (amplitudes and travel time) and electromagnetic data, and the goal of the modeling study is to estimate rock and fluid properties, such as porosity, saturation, mineralogical content, and rock-type (or facies) classification. In this section, we focus on the rock physics component of the inverse problem. Therefore, we assume that seismic and electromagnetic inversions have been previously performed and the available input dataset consists of elastic properties inverted from seismic data (for example inverted seismic velocities or impedances) and resistivity estimated from electromagnetic data.

Several methods have been proposed to solve the reservoir characterization problem. For a detailed overview of the available methods, we refer to Doyen [31]. These methods can be deterministic, probabilistic, or stochastic. Deterministic methods are generally easier to implement, provide a single solution to the problem but could lead to local minima of the inverse problem, which means that the proposed estimated model is not necessarily the one that best fits the data. Probabilistic methods provide a statistical solution to the problem, therefore rather than obtaining a deterministic model, we obtain a model of probability distributions; from the probabilistic model, we can then extract a deterministic model using statistical estimators, such as the mean, the median or the mode, or sample a set of realizations using geostatistical algorithms. Stochastic methods provide a set of realizations that honor the geophysical dataset; however, the computational cost of stochastic methods is generally large because of the computational time of the optimization algorithm. In this section, we work in a probabilistic setting to assess the uncertainty estimated model of rock and fluid properties.

If $\boldsymbol{d}$ represents the measured geophysical data, and $\boldsymbol{m}$ represents the rock and fluid properties of interest, then we can link the model properties to the measured data, through a set of geophysical equations $\mathbf{F}$:

$$\boldsymbol{d} = \mathbf{F}(\boldsymbol{m}) + \boldsymbol{\varepsilon}, \tag{4.39}$$

where $\boldsymbol{\varepsilon}$ represents the noise in the data. In a general reservoir characterization problem, $\mathbf{F}$ is a set of equations including the rock physics model and seismic (and potentially electromagnetic) modeling. The rock physics relations link the set of rock properties, such as porosity, lithology, and saturations, to elastic and electric attributes, such as P- and S-wave velocities or impedances and resistivity. Seismic modeling allows estimating the seismic response from the set of elastic attributes. Electromagnetic modeling allows computing the electromagnetic field from resistivity. Mathematically, the estimation of the rock properties $\boldsymbol{m}$ is an inverse problem. Because one of the main goals in seismic reservoir characterization is to assess the uncertainty associated to reservoir property predictions, we operate in a probabilistic framework. In this section, we focus only on the rock physics component of the problem. For a review of seismic inversion, we refer to Aki and Richards [1] for the physical model, Buland and Omre [32] for the probabilistic approach to seismic inversion, and Sen and Stoffa [33] for stochastic optimization algorithms. For a review of electromagnetic inversion, we refer to Chen et al. [34], MacGregor [35] and to Buland and Kolbjørnsen [36]. In this section, we assume that only seismic data are available and that seismic amplitudes and travel times have been previously inverted to estimate the corresponding elastic properties.

From a probabilistic point of view, the solution of the inversion problem corresponds to the estimation of the conditional probability $P(\boldsymbol{m}|\boldsymbol{d})$. The assessment of the conditional distribution $\boldsymbol{m}|\boldsymbol{d}$ can be made by Bayesian inversion:

$$P(\boldsymbol{m}|\boldsymbol{d}) = \frac{P(\boldsymbol{d}|\boldsymbol{m})P(\boldsymbol{m})}{P(\boldsymbol{d})} = \frac{P(\boldsymbol{d}|\boldsymbol{m})P(\boldsymbol{m})}{\int P(\boldsymbol{d}|\boldsymbol{m})P(\boldsymbol{m})d\boldsymbol{m}} \tag{4.40}$$

where $P(\boldsymbol{d}|\boldsymbol{m})$ is the likelihood function, $P(\boldsymbol{m})$ is the prior model, and $P(\boldsymbol{d})$ is a normalizing constant ensuring that $P(\boldsymbol{m}|\boldsymbol{d})$ is a valid probability density function. In Bayesian inversion, we often assume that the physical relation $\mathbf{F}$ is linear and that the prior distribution $P(\boldsymbol{m})$ is Gaussian [30]. These two assumptions are not necessarily required to solve the Bayesian inversion problem, but under these assumptions, the inverse solution can be analytically derived [30].

In many applications, the Gaussian assumption does not necessarily provide a realistic description of the model. For multimodal distributions, Grana and Della Rossa [37] proposed a Gaussian mixture model, i.e., a linear combination of Gaussian distribution, whereas Rimstad and Omre [38] proposed skewed distributions. However, the most general statistical assumption is based on non-parametric distributions [31]. Differently from Gaussian distributions, non-parametric distributions cannot be described by a finite set of parameters, therefore the solution of the Bayesian inverse problem must be numerically computed, i.e. the value of the posterior distribution must be evaluated for each value of the model properties. Grana and Della Rossa [37] used the kernel density estimation approach for non-parametric distributions in rock physics inversion and compared it with the Bayesian Gaussian mixture inversion approach.

Rock physics models shown in Sects. 4.2–4.6 combined with mathematical inversion can be used in probabilistic inversion to estimate reservoir properties from geophysical attributes. Mukerji et al. [39] and Eidsvik et al. [40] introduced statistical rock physics to estimate reservoir parameters from seismic attributes and to evaluate the associated uncertainty. Stochastic rock physics models are used by Bachrach [41] for a joint estimation of porosity and saturation and by Sengupta and Bachrach [42] for pay-volume uncertainty evaluation. Spikes et al. [43] developed a probabilistic seismic inversion to constrain reservoir properties estimation with well data and seismic attributes.

Statistical rock physics modeling combines rock physics models with Monte Carlo simulations to simulate the variability of the physical model and the uncertainty of the input data and to mimic different geological scenarios in addition to those sampled by well log data. By using statistical rock physics in a Bayesian inversion setting, we aim to estimate the rock and fluid properties in the reservoir conditioned by the geophysical measurements and to assess the uncertainty associated to the model. For the prior distribution, Grana and Della Rossa [37] proposed to use a multivariate Gaussian mixture model with a fixed number of components corresponding to the different reservoir facies. From the prior distribution, different scenarios can be generated by Monte Carlo simulation: petrophysical variables can be sampled from the prior distribution and the elastic response can be computed through the rock-physics model. The spatial correlation of petrophysical properties can then be modelled by means of a variogram to obtain realistic spatial correlation [31]. This approach allows accounting for uncertainty associated with rock physics model predictions.

In general, more advanced mathematical methods can be applied to solve the rock physics inverse problem and estimate rock and fluid properties from geophysical data [31]. In this section, we apply a Bayesian linearized inversion to a 2D section extracted from a reservoir model in the North Sea (Sect. 4.7). The seismic dataset consists of four angle gathers and contains about 300,000 traces in a time window corresponding to an average depth interval of approximately 250 m. We first applied a Bayesian linearized AVO inversion [32] to estimate P-wave and S-wave velocity and density. A 2D section passing through the calibration well (Sect. 4.7) is shown in Fig. 4.12.

The second step of the seismic reservoir characterization workflow is the Bayesian rock physics inversion. The inversion is based on the Gaussian mixture model assumption, as in Grana and Della Rossa [37]. The input data are the inverted seismic attributes shown in Fig. 4.12. By combining the rock physics model calibrated in Sect. 4.7 (stiff sand model) with Bayesian inversion methods, we can estimate the rock and fluid properties in the reservoir grid. In Fig. 4.13 we show a 2D section of porosity and clay content. The top of the reservoir is clearly visible in the 2D sections, even though the resolution of the dataset is fairly low.

For a complete illustration of the application of rock physics inversion methods to real data, we refer the reader to Avseth et al. [10], Bachrach [41], Doyen [31], Spikes et al. [43], Grana and Della Rossa [37], and Rimstad and Omre [38]. For rock physics inversion methods including electromagnetic data we refer the reader to Chen et al. [34] and MacGregor [35].

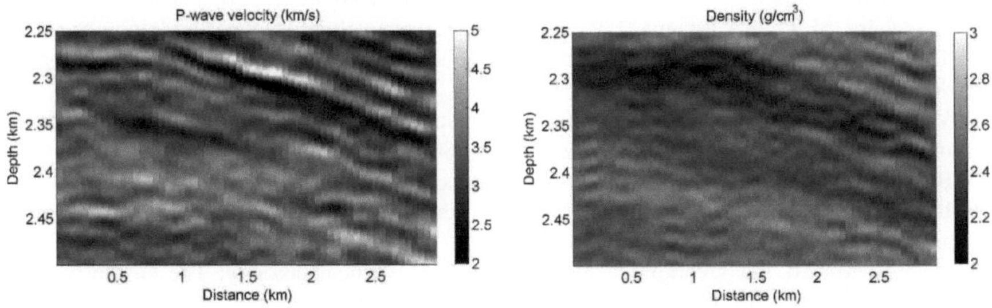

Fig. 4.12 Inverted seismic attributes along a 2D section of a North Sea reservoir: *left plot* shows P-wave velocity; *right plot* shows density

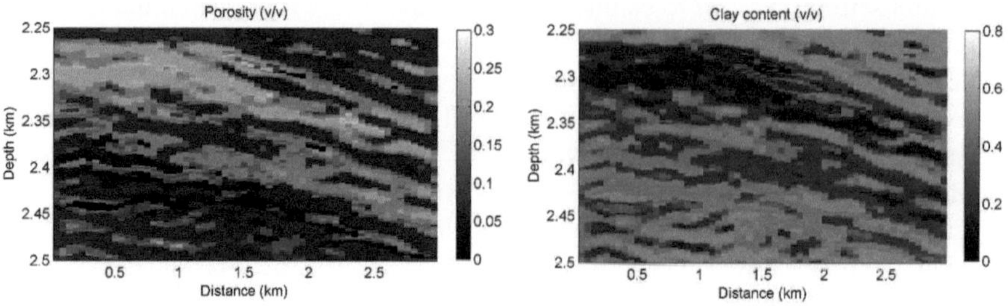

Fig. 4.13 Inverted petrophysical properties along a 2D section of a North Sea reservoir: *left plot* shows porosity; *right plot* shows clay content

References

1. Aki, K., & Richards, P. G. (1980). *Quantitative seismology*. North Bay, ON: W.H. Freeman & Co.
2. Yilmaz, Ö. (2001). *Seismic data analysis, processing, inversion and interpretation of seismic data* (2nd ed., p. 2027). Tulsa, OK: Society of Exploration Geophysicists. ISBN-13: 978–1560800941, ISBN-10: 1560800941.
3. Mavko, G., Mukerji, T., & Dvorkin, J. (2009). *The rock physics handbook*. Cambridge: Cambridge University Press.
4. Wyllie, M. R. J., Gregory, A. R., & Gardner, L. W. (1956). Elastic wave velocities in heterogeneous and porous media. *Geophysics, 21*, 41–70.
5. Raymer, L. L., Hunt, E. R., & Gardner, J. S. (1980). An improved sonic transit time-to-porosity transform: Transactions of SPWLA, Paper P.
6. Han, D.-H. (1986). Effects of porosity and clay content on acoustic properties of sandstones and unconsolidated sediments: Ph.D. dissertation, Stanford University.
7. Bourbié, T., Coussy, O., & Zinszner, B. (1987). *Acoustics of porous media*. Houston: Gulf.
8. Zimmerman, R. W. (1991). *Compressibility of Sandstones*. Amsterdam: Elsevier.
9. Nur, A., Marion, D., & Yin, H. (1991). Wave velocities in sediments. In J. M. Hovem, M. D. Richardson, & R. D. Stoll (Eds.), *Shear waves in marine sediments* (pp. 131–140). Berlin: Kluwer Academic Publishers.
10. Avseth, P., Mukerji, T., & Mavko, G. (2005). *Quantitative seismic interpretation*. Cambridge: Cambridge University Press.
11. Dvorkin, J., Gutierrez, M., & Grana, D. (2014). *Seismic reflections of rock properties*. Cambridge: Cambridge University Press.
12. Castagna, J. P., Batzle, M. L., & Eastwood, R. L. (1985). Relationships between compressional wave and shear-wave velocities in clastic silicate rocks. *Geophysics, 50*, 571–581.
13. Greenberg, M. L., & Castagna, J. P. (1992). Shear-wave velocity estimation in porous rocks: Theoretical formulation, preliminary verification and applications. *Geophysical Prospecting, 40*, 195–209.
14. Vernik, L., Fisher, D., & Bahret, S. (2002). Estimation of net-to-gross from P and S impedance in deepwater turbidites. *The Leading Edge, 21*, 380–387.
15. Batzle, M., & Wang, Z. (1992). Seismic properties of pore fluids. *Geophysics, 57*, 1396–1408.
16. Brie, A., Pampuri, F., Marsala, A. F., & Meazza, O. (1995). Shear sonic interpretation in gas-bearing sands. *SPE, 30595*, 701–710.
17. Nur, A., Mavko, G., Dvorkin, J., & Gal, D. (1995). Critical porosity: The key to relating physical properties to porosity in rocks. SEG Expanded Abstracts, 878.
18. Dvorkin, J., Nur, A., & Yin, H. (1994). Effective properties of cemented granular materials. *Mechanics of Materials, 18*, 351–366.

19. Dvorkin, J., & Nur, A. (1996). Elasticity of high-porosity sandstones: Theory for two North Sea datasets. *Geophysics, 61*, 1363–1370.
20. Gal, D., Dvorkin, J., & Nur, A. (1998). A physical model for porosity reduction in sandstones. *Geophysics, 63*, 454–459.
21. Kuster, G. T., & Toksöz, M. N. (1974). Velocity and attenuation of seismic waves in two-phase media. *Geophysics, 39*, 587–618.
22. Berryman, J. G. (1995). Mixture theories for rock properties. In T. J. Ahrens (Ed.), *Rock physics and phase relations: A handbook of physical constants* (pp. 205–228). Washington: American Geophysical Union.
23. Mavko, G., Chan, C., & Mukerji, T. (1995). Fluid substitution: Estimating changes in VP without knowing VS. *Geophysics, 60*, 1750–1755.
24. Prasad, M., & Manghnani, M. H. (1997). Effects of pore and differential pressure on compressional wave velocity and quality factor in Berea and Michigan sandstones. *Geophysics, 62*, 1163–1176.
25. MacBeth, C. (2004). A classification for the pressure-sensitivity properties of a sandstone rock frame. *Geophysics, 69*, 497–510.
26. Darling, T. (2005). *Well logging and formation evaluation*. Amsterdam: Elsevier Inc.
27. Ellis, D. V., & Singer, J. M. (2007). *Well logging for earth scientists*. New York: Springer.
28. Gomez, C. (2009). Reservoir characterization combining elastic velocities and electrical resistivity measurements: Ph.D. dissertation, Stanford University.
29. Worthington, P. F. (1985). Evolution of shaley sand concepts in reservoir evaluation. *Log Analyst, 26*, 23–40.
30. Tarantola, A. (2005). *Inverse problem theory and methods for model parameter estimation* (p. 352). SIAM. ISBN-13: 978–0898715729, ISBN-10: 0898715725.
31. Doyen, P. (2007). *Seismic reservoir characterization: An earth modelling perspective* (p. 255). EAGE Publications. ISBN 978-90-73781-77-1.
32. Buland, A., & Omre, H. (2003). Bayesian linearized AVO inversion. *Geophysics, 68*, 185–198.
33. Sen, M. K., & Stoffa, P. L. (2013). *Global optimization methods in geophysical inversion*. Cambridge: Cambridge University Press.
34. Chen, J., Hoversten, M., Vasco, D. W., Rubin, Y., & Hou, Z. (2007). Joint inversion of seismic AVO and EM data for gas saturation estimation using a sampling based stochastic model. *Geophysics, 72*, WA85–WA95.
35. MacGregor, L. (2012). Integrating seismic CSEM, and well-log data for reservoir characterization. *The Leading Edge, 31*, 268–277.
36. Buland, A., & Kolbjørnsen, H. (2012). Bayesian inversion of CSEM and magnetotelluric data. *Geophysics, 77*, E33–E42.
37. Grana, D., & Della Rossa, E. (2010). Probabilistic petrophysical-properties estimation integrating statistical rock physics with seismic inversion. *Geophysics, 75*, O21–O37.
38. Rimstad, K., & Omre, H. (2010). Impact of rock-physics depth trends and Markov random fields on hierarchical Bayesian lithology/fluid prediction. *Geophysics, 75*, R93–R108.
39. Mukerji, T., Jørstad, A., Avseth, P., Mavko, G., & Granli, J. R. (2001). Mapping lithofacies and pore-fluid probabilities in a North Sea reservoir: Seismic inversions and statistical rock physics. *Geophysics, 66*, 988–1001.
40. Eidsvik, J., Avseth, P., Omre, H., Mukerji, T., & Mavko, G. (2004). Stochastic reservoir characterization using prestack seismic data. *Geophysics, 69*, 978–993.
41. Bachrach, R. (2006). Joint estimation of porosity and saturation using stochastic rock-physics modeling. *Geophysics, 71*, O53–O63.
42. Sengupta, M., & Bachrach, R. (2007). Uncertainty in seismic-based pay volume estimation: Analysis using rock physics and Bayesian statistics. *The Leading Edge, 24*, 184–189.
43. Spikes, K., Mukerji, T., Dvorkin, J., & Mavko, G. (2008). Probabilistic seismic inversion based on rock-physics models. *Geophysics, 72*, R87–R97.
44. Eberhart-Phillips, D., Han, D. H., & Zoback, M. D. (1989). Empirical relationships among seismic velocity, effective pressure, porosity, and clay content in sandstone. *Geophysics, 54* (1), 82–89.

Chapter 5
Geomechanics and Elastic Anisotropy of Shale Formations

Mehdi Ostadhassan

Abstract Deep shales are the most abundant yet least characterized sedimentary rocks in petroleum industry while they have become significant sources of hydrocarbon unconventional resources. This chapter aims to fulfill an investigation of anisotropy in this rock type in several different facets through integration of field and lab data. I seek to generate key information to better understand elastic anisotropy as well as in situ stresses to better perform drilling, well completion, perforating, and hydraulic fracturing for the purpose of geomechanical modeling.

The first step was to study the anisotropic behavior of shale formations. For such a purpose three necessary independent shear moduli, elastic stiffness coefficients, and principal stresses are calculated and measured. The parameters then are used to generate shear radial profiles and slowness-frequency plots to analyze formation anisotropy, type, and origin.

The next step was to evaluate direction and magnitude of the minimum and maximum anisotropic principal horizontal stresses as the governing element in geomechanical modeling. I also analyzed wellbore behavior and predicted wellbore failure under stress alteration caused by drilling. Elastic anisotropy of the formation is considered in 3D numerical models and calculations, which has improved the results considerably.

5.1 Introduction

Significant activities are underway in the USA to explore and develop America's shale oil plays. The shale oil plays contain "fine grained, organic rich, sedimentary rocks." The shales are both the source of and the reservoir for oil. They are also defined by the extremely small pore sizes which make them relatively impermeable to fluid flow, unless natural or artificial fractures occur.

To gain a better understanding of the potential US domestic shale oil resources, Energy Information Administration (EIA) developed an assessment of onshore

M. Ostadhassan (✉)
Petroleum Engineering Department, University of North Dakota,
Grand Forks, ND 58203, USA
e-mail: mehdi.ostadhassan@engr.und.edu

C. Jin, G. Cusatis (eds.), *New Frontiers in Oil and Gas Exploration*,
DOI 10.1007/978-3-319-40124-9_5

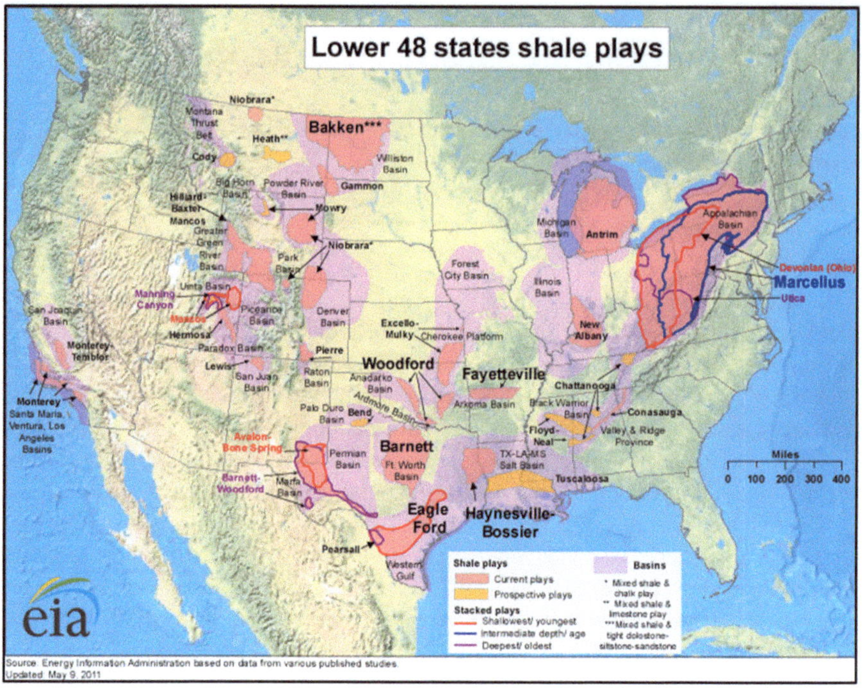

Fig. 5.1 Map of major US shale plays including the Bakken in Williston Basin, ND, [1]

Lower 48 States technically recoverable shale oil resources [1]. This report estimates shale oil resources for the undeveloped portions of 20 shale plays that have been discovered. Eight of those shale plays are subdivided into 2 or 3 areas, resulting in a total of 29 separate resource assessments. The map in Fig. 5.1 shows the location of the shale plays in the Lower 48 States [2].

According to the shale report's assessment there are 23.9 billion barrels of recoverable of shale oil in the onshore Lower 48 States. The largest shale oil formation is the Monterey/Santos play in southern California, which is estimated to hold 15.4 billion barrels or 64 % of the total shale oil resources. The Monterey shale play is the primary source rock for the conventional oil reservoirs found in the Santa Maria and San Joaquin Basins in southern California. The next largest shale oil plays are the Bakken and the Eagle Ford, which are assessed to hold approximately 3.6 billion barrels and 3.4 billion barrels of oil, respectively. Bakken which is comprised of upper and lower shale section and carbonate-clastic middle member with high clay contents was used as the case study for this chapter

Table 5.1 summarizes the amount of revocable resources in Billion Barrels of Oil (BBO) of major US shale oil plays with the area of extension in square miles [2]. The Estimated Ultimate Recovery (EUR) in Thousands Barrels (MBO) per well is denoted in the last column.

Table 5.1 Technically recoverable shale oil recourses summary in the USA

	Technically recoverable resource	Area (sq. miles)	Average EUR
Play	Oil (BBO)		Oil (MBO/well)
Eagle Ford	3.35	3323	300
Total Gulf Coast	3.35	3323	300
Avalon and Bone Springs	1.58	1313	300
Total Southwest	1.58	1313	300
Bakken	3.59	6522	550
Total Rocky Mountain	3.59	6522	550
Monterey/Santos	15.42	1752	550
Total West Cost	15.42	1752	550
Total Lower 48 USA	23.94	12,910	460

5.2 Theory of Anisotropy

5.2.1 Elastic Anisotropy

Variation of an elastic properties in a medium such as: different moduli and wave velocity propagation with respect to the direction of measurement is called anisotropy [3]. A medium that displays this directional dependence is referred to an anisotropic medium.

Sedimentary rocks are often found to be anisotropic. In sedimentary rocks there are many sources of elastic anisotropy; some of them are as follows [4]:

- Aligned crystals,
- Direct stress-induced anisotropy,
- Lithologic anisotropy (i.e., aligned grains),
- Aligned fractures, cracks and pores, subsequently the nature of their infilling material (clays, hydrocarbons, water, etc.)
- Structural anisotropy (i.e., fine layering)

New advancements in sonic logging have made us capable of quantifying anisotropy and the origins azimuthally and radially around the borehole [5].

5.2.2 Classification of Anisotropic Media

Transverse isotropy is defined as having the same properties (e.g., velocity, stiffness, permeability, resistively) in a medium when measured within a plane that is normal to an axis, but having different values when measuring those properties at other angles to that axis normal to the plane of measurement [4, 6]. This axis is

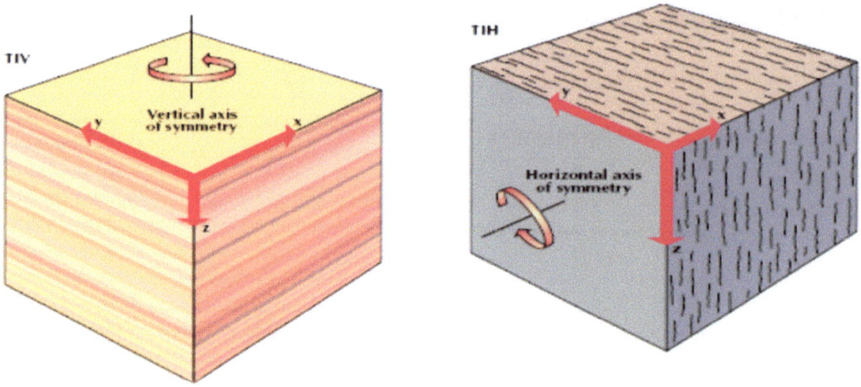

Fig. 5.2 VTI and HTI mediums, [8]

called the symmetry axis and is normal to the alignment direction of different properties.

There are two styles of property alignment in mediums: horizontal alignment of properties with vertical axis of symmetry and vertical alignment of properties with horizontal axis of symmetry. These two types of alignments (symmetries) make two types of anisotropies: transversely isotropic with vertical axis of symmetry (VTI) and transversely isotropic with horizontal axis of symmetry (HTI). These are two oversimplified but convenient models that have been created to describe how elastic properties, such as velocity or elasticity, vary in anisotropic media [4, 7] (Fig. 5.2).

Elastic anisotropy is a key parameter for geomechanical modeling. Most of the rock constituents of the earth's crust exhibit some degree of anisotropy. Shales are a major component of sedimentary basins [9] that exhibit a high degree of intrinsic anisotropy due to their microstructures and platy shape clay minerals [10]. As a matter of fact, the anisotropic behavior of shales could be simplified by making the assumption that they are Vertically Transverse Isotropic (VTI) [11, 12]. This has made transverse isotropy the most common anisotropy model to be acquired in modeling and various analyses.

5.2.3 VTI Medium

Vertical transverse isotropy, also known as polar anisotropy [13], can be quantified in the manner of including transverse isotropic planes with a vertical axis of rotational symmetry. A VTI medium can be characterized by having five independent elastic stiffness coefficients. Considering X_3 as the axis of rotational symmetry in the conventional two index notation [14, 15] and applying general Hook's law (Eq. 5.1), the non-vanishing elastic stiffness coefficients (Eqs. 5.2 and 5.3) of the elasticity matrix (Eq. 5.2) reads as follows:

$$\sigma_{ij} = C_{ijkl}\varepsilon_{kl} - \alpha P_{\mathrm{p}} \tag{5.1}$$

Where σ_{ij}: Stress tensor, C_{ijkl}: Fourth rank stiffness tensor, ε_{kl}: Strain tensor, α: Biot's constant and P_{p}: Pore pressure, and the conventional two index notation [15] of the stiffness tensor will be:

$$C_{ij} = \begin{pmatrix} C_{11} & C_{12} & C_{13} & 0 & 0 & 0 \\ C_{21} & C_{22} & C_{23} & 0 & 0 & 0 \\ C_{31} & C_{32} & C_{33} & 0 & 0 & 0 \\ 0 & 0 & 0 & C_{44} & 0 & 0 \\ 0 & 0 & 0 & 0 & C_{55} & 0 \\ 0 & 0 & 0 & 0 & 0 & C_{66} \end{pmatrix}, \tag{5.2}$$

Hence a VTI medium the five non-vanishing elastic stiffness coefficients along with C_{66} are as follows:

$$\begin{aligned} &C_{11} = C_{22}, C_{33}, C_{12} = C_{21}, C_{13} = C_{23} = C_{32} = C_{31}, \\ &C_{44} = C_{55}, C_{66} = \frac{C_{11} - C_{12}}{2} \end{aligned} \tag{5.3}$$

Elastic anisotropy and stiffness tensor components can be quantified with wave propagation through an elastic medium. Considering the velocity υ_{ij} of an elastic wave traveling along the X_i axis and polarizing along the X_j, the relationship between υ_{ij} and C_{ij} for a TI medium will become:

$$\upsilon_{11} = \upsilon_{22} = \sqrt{\frac{C_{11}}{\rho}, \upsilon_{33} = \sqrt{\frac{C_{33}}{\rho}, \upsilon_{12} = \upsilon_{21} = \sqrt{\frac{C_{66}}{\rho}},}} \tag{5.4}$$

$$\upsilon_{13} = \upsilon_{31} = \upsilon_{23} = \upsilon_{32} = \sqrt{\frac{C_{55}}{\rho}} \tag{5.5}$$

Where ρ is the bulk density, υ_{12} is the velocity of a shear wave propagating along the axis (X_1) and polarized along the axis (X_2), υ_{33} is the velocity of a compressional wave traveling along the axis of symmetry (X_3) and polarized along the same axis.

5.2.4 Shale Anisotropy

Shales are known to be anisotropic [16, 17]. In addition to the mineralogy, the mechanical anisotropy of shales are related to the organic richness of kerogen content. A series of publications [12, 18, 19] discussed shale anisotropy in the lab on a variety of shales, with different clay and kerogen content, clay mineralogy and porosity at different effective pressures. They found that black, kerogen-rich shales are transversely isotropic, and anisotropy of shales increases substantially with compaction and kerogen content. Vernik and Nur [18] pointed out that the anisotropy of shales is enhanced by bedding-parallel microcracks, especially at the high pore pressure. Prasad and Mukerji [20], and Mukerji and Prasad [21, 22] analyzed scanning acoustic microscope (SAM) images of the Bakken shale and found that the textural heterogeneity, P wave impedance and velocity, and density increase with increasing maturity (decreasing kerogen content), while textural anisotropy decreases with maturity.

In order to determine the type of anisotropy in the field to better plan further, field operations such as hydraulic fracturing, dispersion analysis, and sonic logging are widely used [23–25]. Dispersion plots are graphical representations of frequency versus slowness for different wave types gathered through advanced sonic logging. In such graphs, slowness in high frequency represents the near wellbore, while far field slowness is related to the low frequency region. A dispersion plot enables us to determine whether the formation is isotropic or anisotropic, homogeneous or inhomogeneous as well as the cause of the anisotropy. In this regard, four different cases can be expected as follows:

1. Homogeneous–isotropic,
2. Inhomogeneous–isotropic,
3. Homogeneous–anisotropic,
4. Inhomogeneous–anisotropic,

In this regard radial slowness variation profiles (RSVP) are beneficial for formation characterization [26, 27]. Figures 5.3, 5.4, 5.5, 5.6, 5.7, and 5.8 represents data processed from two different wells through three different members of the Bakken Formation [28, 29]. Compressional, shear, and Stoneley wave slowness variations in a deep penetration into the formation provide valuable information of true formation dynamic properties.

5.2.5 Case Study

Figure 5.3 displays the dispersion plot of a depth representative of the Upper Bakken (UB) in well 1, where the fast and slow shear flexural modes overlay each other, but do not match the modeled curves. This validates the assumption that the Bakken Formation is more likely to be homogeneous and vertically transverse isotropic. This elastic anisotropy originates from the platy shape clay particles as the major constituent minerals of the shaley UB. Lower Bakken demonstrates similar behavior, whereas the dispersion plot of middle member (Fig. 5.4) exhibits a perfect match between the flexural modes and the modeled curves, indicating homogeneous isotropic medium.

Figures 5.6 and 5.8 show the dispersion and SRP plots at depths 9720 and 9808 ft in well 2. They represent perfect VTI caused by clay minerals in UB and LB. In Figs. 5.6 and 5.8, flexural modes overlay each other and do not match the modeled curves in dispersion plots. On the SRP plots, flexural modes slowness does not

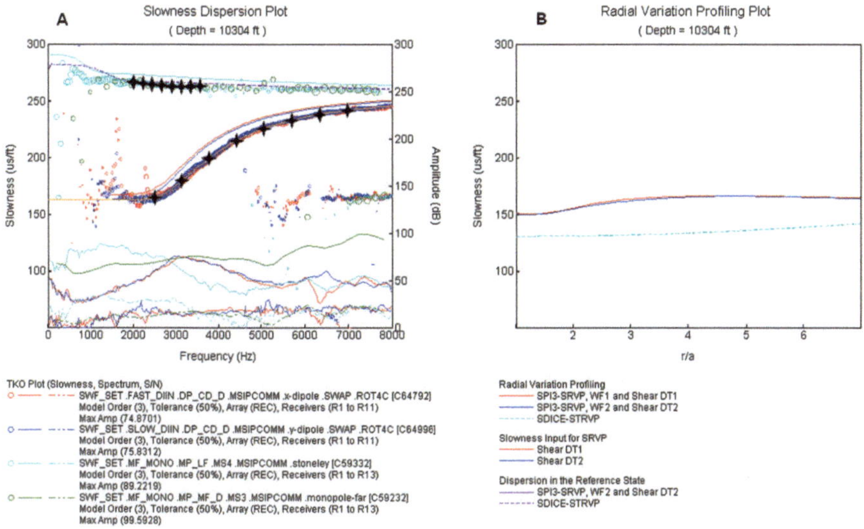

Fig. 5.3 Dispersion plot (**a**) and radial slowness variation profile (**b**) for Upper Bakken, well 1, [28]

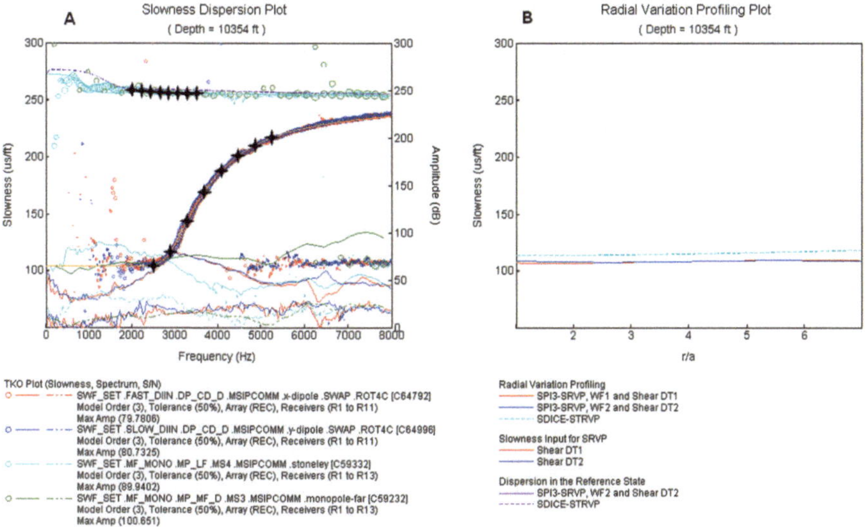

Fig. 5.4 Dispersion plot (**a**) and radial slowness variation profile (**b**) for Middle Bakken, well 1, [28]

separate and are slower than the Stoneley waves. Dispersion and SRP plots of MB shown in Fig. 5.7 have the characteristics of an isotropic medium. Finally, comparing the SRP plots from Middle Bakken (MB) in well 1 and 2, the mismatch of Stoneley slowness with the overlaying flexural modes (slower Stoneley compared to faster shear dipoles) in Well 1, confirms the idea of a quiet permeable Middle Bakken due to the presence of vertical fractures which should be considered in stimulation design.

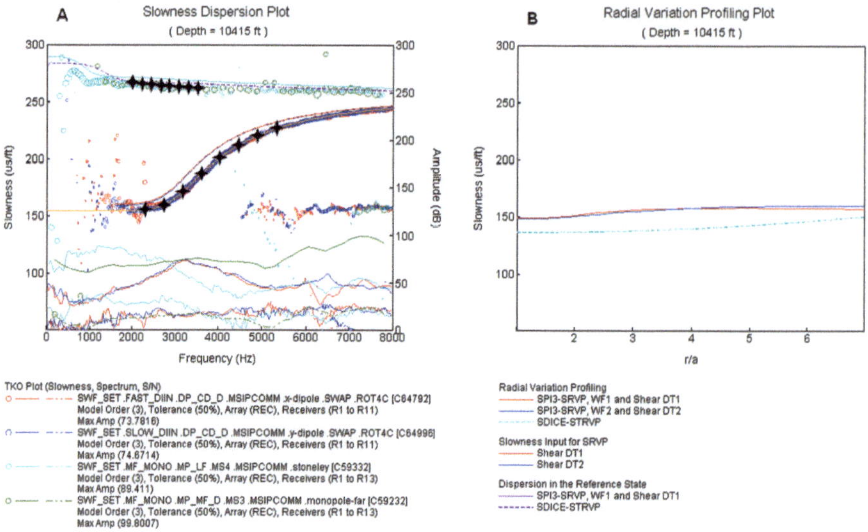

Fig. 5.5 Dispersion plot (**a**) and radial slowness variation profile (**b**) for Lower Bakken, well 1, [28]

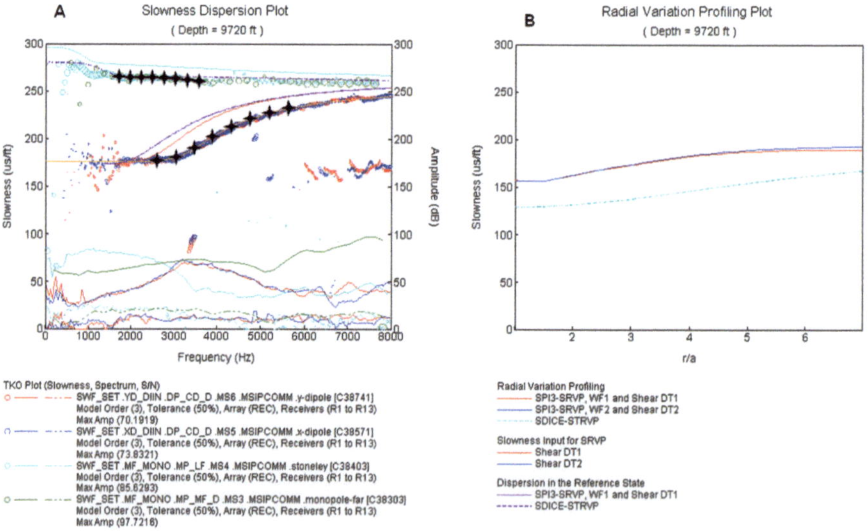

Fig. 5.6 Dispersion plot (**a**) and radial slowness variation profile (**b**) for Upper Bakken in well 2, [28]

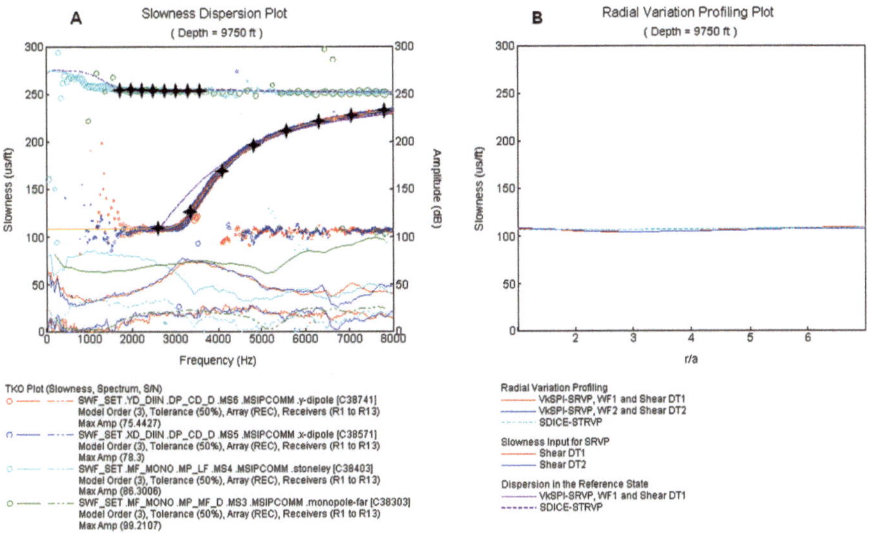

Fig. 5.7 Dispersion plot (**a**) and radial slowness variation profile (**b**) for Middle Bakken in well 2, [28]

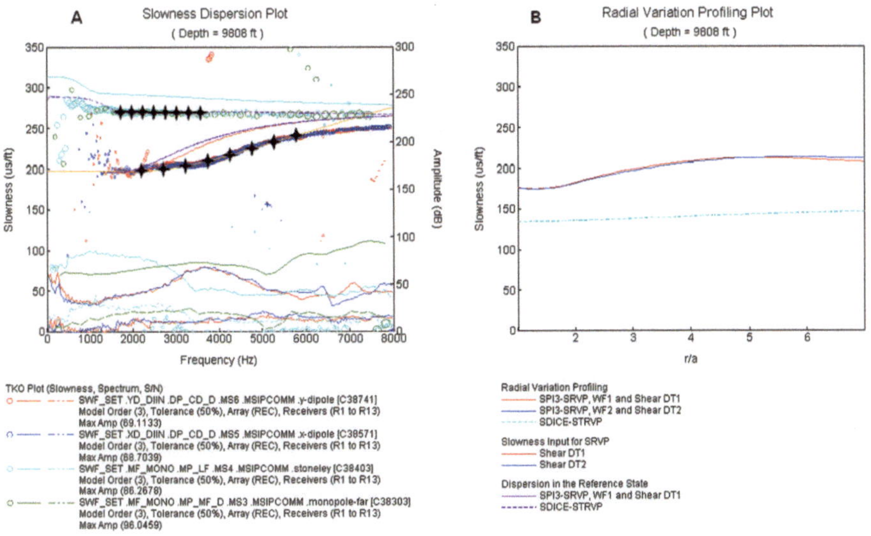

Fig. 5.8 Dispersion plot (**a**) and radial slowness variation profile (**b**) for Lower Bakken in well 2, [28]

5.3 Fundamentals of Geomechanical Modeling for Wellbore Instability

Wellbore instability can be induced chemically or mechanically, or a combination of both, either way, the mechanical anisotropy of the shales if not considered can result in failed operations and financial losses.

5.3.1 Chemically Induced Instability

Chemically induced instability usually takes place in shale bearing zones when shales water absorption and swelling leads to hole enlargement, or in salt layers when the salt is dissolved by the drilling fluid and causes a reduction in the hole size.

5.3.2 Mechanically Induced Instability

Formations at depth are under a state of compressive in situ stresses. When a well is drilled through the formation, a significant amount of rock volume is removed which causes stress alteration. As a result, the surrounding rocks at the borehole must compensate for the eliminated load. Stress concentration around the borehole is the direct result of this process. Thus, in cases of weak formations such as unsolicited sands or shales, the formation may fail. The failures could be categorized under three main classes, as shown in Fig. 5.9:

1. Formation breakdown, or unintended hydraulic fracturing that will result in loss of drilling fluid circulation.
2. Hole enlargement due to brittle rock fracture or rupture.
3. Hole size reduction, which can happen due to ductile yield of the rock.

5.3.3 Factors Influencing Wellbore Stability

To obtain a good prediction of wellbore stability and geomechanical behavior of the reservoir, we should try to recognize the governing parameters and various possible conditions occurring around the borehole. There is a comprehensive list of factors encountered in wellbore stability analysis and geomechanical modeling, but the major ones that are controllable, predictable and measurable are rock strength, stiffness, permeability, temperature, pore pressure and concentration (Fig. 5.10).

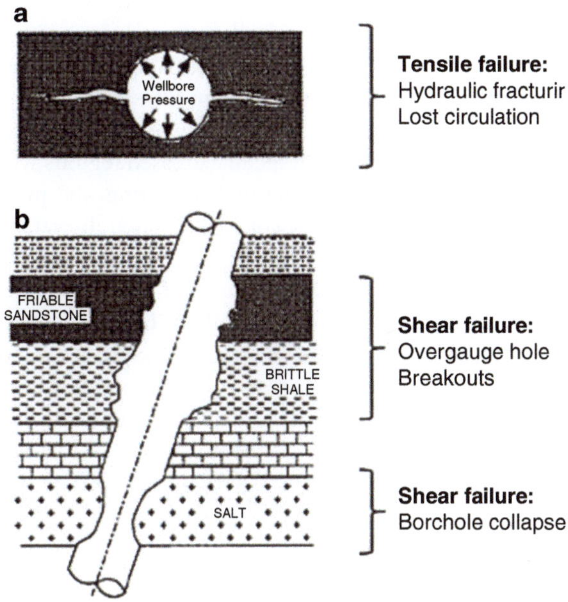

Fig. 5.9 Three main classes of wellbore failure (modified from ref. [30])

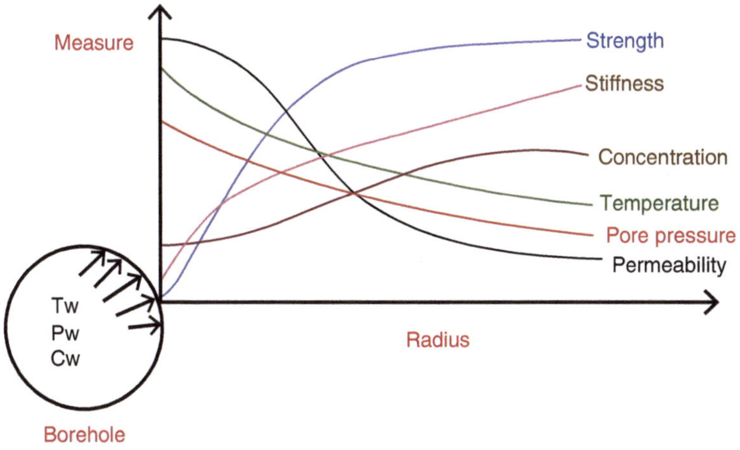

Fig. 5.10 Alteration of different parameters around the borehole (modified from ref. [30])

5.3.4 *In Situ Stress Field*

Formation stresses play an important role in geomechanical modeling and development of oil and gas reservoirs. Both the direction and magnitude of these stresses are required in (a) planning for borehole stability, (b) hydraulic fracturing for enhanced production, and (c) selective perforation for sand control. The formation stress state is characterized by the magnitude and direction of the three principal stresses, one vertical and two horizontal. In this regard formations could be either normally stressed or tectonically stressed. In a normally stressed formation, the maximum principal stress is the vertical effective stress (σ_v), equal to the overburden stress. The other two principal in situ stresses (σ_H, σ_h) are located in a horizontal plane. For well-compacted and cemented formations, the overburden stress varies linearly with depth. In tectonically stressed regions which may contain active faults, salt domes or severe folding and fracturing, the principal in situ stresses are not necessarily oriented in vertical or horizontal directions that have led into structural deformations.

Observations show that changes of in situ stresses can cause major borehole instability problems associated with drilling specifically in tectonically active areas. In addition to in situ stresses, pore pressure plays an important role in wellbore stability. Pore pressure, or reservoir pressure, is the amount of pressure exerted by the in situ pore fluid to the internal pore walls [3]. For normally pressured formations, the pore pressure gradient is constant at approximately 10.4 kPa/m. In geo-pressured (overpressured) formations such as the Bakken, pore pressure gradient can exceed 20.4 kPa/m.

5.3.5 Wellbore Pressure

Drilling wellbore pressure, which is maintained by drilling mud, tries to protect the wellbore from failure. On one hand, excess mud pressure will prevent blow-outs; on the other hand, it can cause pipe sticking problems, washouts or unintended fractures and lead into wellbore failure. It is important to maintain an appropriate and safe mud pressure window to avoid such problems. In general, the safe mud weight window needs to be determined through modeling [31–33].

5.3.6 Fractures and Damages in the Formation

Discontinuities, such as bedding planes, fractures, and damages in the formation which are considered as weak planes, can cause wellbore instability problems as a source of anisotropy. Mud can infiltrate into either natural or drilling-induced open fractures that cross the wellbore wall, increasing tensile stresses and resulting in wellbore instability problems.

Formation damage, which means the invasion of the mud filtrate into the formation and reducing formation permeability, can reduce the wellbore stability. Figure 5.11 explains that the effective radial stress near the wellbore has been reduced due to high pore pressure gradients set up in the damaged zone. In certain conditions, the pore pressure may be high enough to induce tensile effective radial stresses [31, 34].

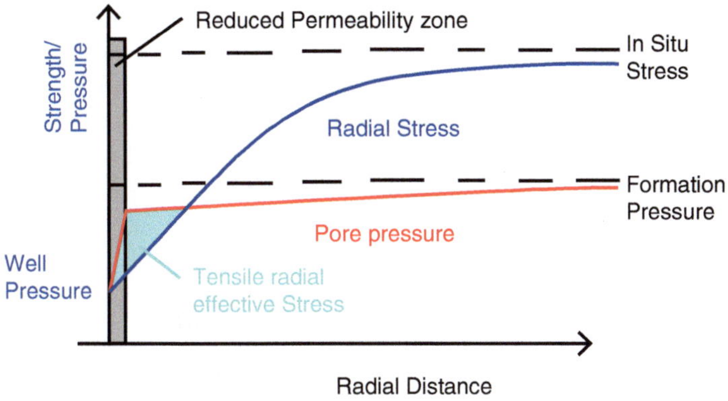

Fig. 5.11 Reduced radial effective stress due to formation damage (modified from ref. [31])

5.3.7 *Thermal Effect*

Temperature can affect borehole stability, and should be considered in stress analysis [35]. Cooling the wellbore wall, which is one of the main roles of the drilling fluid, can cause an increase in mud weight and result in the increase of tangential stresses, causing shear failures [36]. Thermal effects become more important when drilling in naturally fractured reservoirs.

5.3.8 *Fluid Flow into the Wellbore*

Permeability of the formation should also be included for more accurate geomechanical modeling. In high permeable formations, an effective mud cake will form around the wellbore wall, preventing mud losses and wellbore failures. This mud cake is beneficial in borehole stability, and its characteristic is partially dependent on the formation permeability. Mud cakes will only develop on the edge of sufficiently permeable formations. In impermeable (tight) rocks like shales, the mud cake is negligible. Flow of the formation fluid in to the wellbore drops formation pressure, thus the mud pressure remains greater than the pore pressure [32, 37].

5.3.9 *Chemical Effects (in Shales)*

One of the main causes of shale instability is believed to be the unfavorable interactions between the shale and drilling mud [38]. This is primarily due to the shale acting as an osmosis membrane, helping the movement of water/ions into or out of shale. This movement causes alterations in mechanical and physiochemical properties of the shale, and can lead to wellbore instability problems. As previously mentioned water absorption by clay minerals in shale will increase pore pressure, thus decreasing the effective in situ stresses and resulting in wellbore failures. Shale swelling which happens by clay minerals absorbing water will result in clay bond strength loss, thus ending in wellbore failure [34, 39].

5.3.10 Numerical Modeling of Wellbore Stability

There are number of input parameters that influence wellbore stability and impact the accuracy of numerical models. However it is necessary to validate the models with laboratory experiments, but typical laboratory work cannot meet the conditions under which the numerical models were run. Thus, all numerical modeling projects should generally be considered as theoretical until proven under laboratory-controlled conditions. There are an abundance of numerical models developed for analyzing wellbore instability, among which elastic and elastoplastic models are the most common used in the industry [34].

5.3.10.1 Elastic Models

The first elastic solution to predict wellbore stability was developed by Hubbert and Willis [40] in a vertical borehole under non-hydrostatic far field stresses and constant borehole fluid pressure conditions. Fairhurst [41] developed a solution for the stress distribution around an inclined borehole. The equations derived by Fairhurst [41] were used by Bradley [42] to model the stresses around a circular opening in a general case for the inclined well and not parallel to the principal stresses' directions. It should be mentioned that all these solutions were simple, and did not take temperature or pore pressure gradients, mud cake effectiveness, material anisotropy or time dependency into account. Santareli and Brown [43] derived a stress-dependent elastic modulus solution for borehole stability. Aadnoy and Chenevert [44] included the influence of rock anisotropy in inclined boreholes in further stability models to solve for stresses at the borehole wall. This model was later improved by Ong and Roegiers [45], accounting for the stress as a function of borehole radius and also adding an anisotropic shear failure criterion. Roegiers and Detournay [46] developed a model that was able to predict stresses, displacements and fracture initiations at inclined borehole walls. Mody and Hale [47] presented a model taking into account the chemical effects of drilling fluid and the formation. Sherwood and Bailey [48] modified Biot's [49] linear theory of poroelasticity for shale swelling around a cylindrical borehole. Linear elastic solutions are the most widely used models in the industry for mud weight design and borehole stability analyses due to their ease of use and less dependency on input parameters [34].

5.3.10.2 Elastoplastic and Poro-elastoplastic Models

An elastoplastic model suggests that even after the borehole might be stressed beyond its elastic limit to undergo a reversible deformation, the non-elastic region remains intact and load-bearing. A number of elastoplastic and nonlinear analytical and semi-analytical models for borehole stability include the work of Gnirk [50] and Papamichos and Vardoulakis [51]. More recently, poro-elastoplastic models have been developed, which take the effects of the fluid flow and elastoplastic behavior of the rock into consideration [3].

5.3.10.3 Stress Distribution Around the Wellbore

A schematic diagram (Fig. 5.12) and solutions of a vertical borehole in a linear elastic formation with horizontal in situ stresses, and on its internal boundary by a wellbore pressure P_{w} (known as the plane strain problem), is as follows [34, 52]:

$$\sigma_r = \frac{P_0}{2}\left[(1+K)\left(1-\frac{a^2}{r^2}\right) - (1-K)\left(1-\frac{4a^2}{r^2}+\frac{3a^4}{r^4}\right)\cos 2\theta\right] + \frac{P_{\mathrm{w}}a^2}{r^2} \quad (5.6)$$

$$\sigma_\theta = \frac{P_0}{2}\left[(1+K)\left(1+\frac{a^2}{r^2}\right) + (1-K)\left(1+\frac{3a^4}{r^4}\right)\cos 2\theta\right] - \frac{P_{\mathrm{w}}a^2}{r^2} \quad (5.7)$$

$$\sigma_z = \sigma_{\mathrm{v}} - \mu\frac{4a^2}{r^2}\frac{(1-K)P_0}{2} \quad (5.8)$$

$$\tau_{r\theta} = \frac{P_0}{2}\left[(1-K)\left(1+\frac{2a^2}{r^2}-\frac{3a^4}{r^4}\right)\sin 2\theta\right] \quad (5.9)$$

$$\tau_{rz} = \tau_{\theta z} = 0 \quad (5.10)$$

Where μ is the Poisson's ratio, K is at rest earth pressure coefficient, a is borehole radius, and P_0 is the external pressure of the borehole.

A more general solution would be a deviated borehole in an elastic formation (Fig. 5.13) which is solved in a local coordinate system whose z axis is parallel to the wellbore axis and x axis is chosen to be parallel to the lowermost radial direction of the wellbore. The global coordinate (virgin formation stress coordinate, x', y', z') can be then converted into the local coordinate (borehole coordinate, x, y, and z) system [32]:

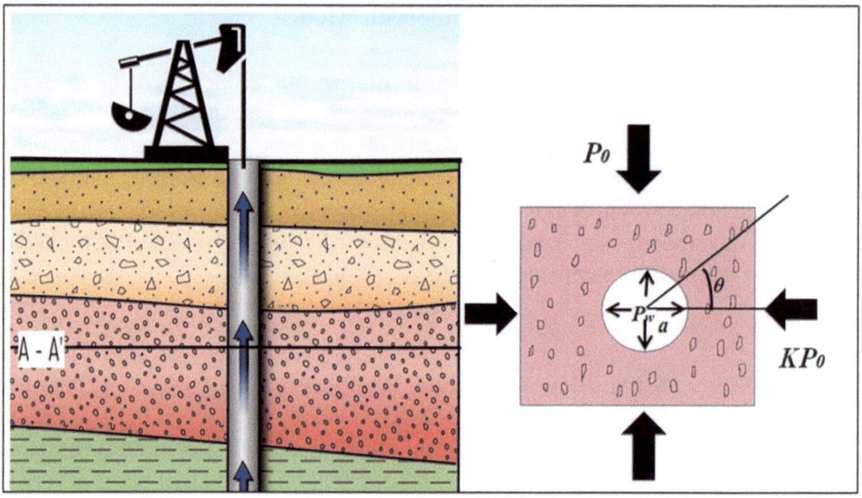

Fig. 5.12 A vertical borehole in an anisotropic in situ stress with internal wellbore pressure (modified from ref. [34])

Fig. 5.13 Schematic diagram for an inclined wellbore subjected to in situ stresses (modified from ref. [34])

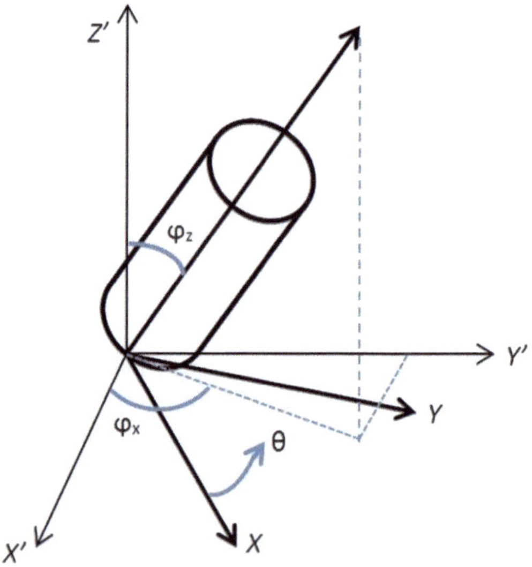

$$\begin{bmatrix} \sigma_x^0 \\ \sigma_y^0 \\ \sigma_z^0 \\ \sigma_{xy}^0 \\ \sigma_{yz}^0 \\ \sigma_{xz}^0 \end{bmatrix} = \begin{bmatrix} l_{xx'}^2 & l_{xy'}^2 & l_{xz'}^2 \\ l_{yz}^2 & l_{yy'}^2 & l_{yz'}^2 \\ l_{zx'}^2 & l_{zy'} & l_{zz'}^2 \\ l_{xx'}l_{yx'} & l_{xy'}l_{yy'} & l_{zz'}l_{yz'} \\ l_{yx'}l_{zx'} & l_{zy'}l_{yy'} & l_{zz'}l_{yz'} \\ l_{yx'}l_{zx'} & l_{zy'}l_{yy'} & l_{zz'}l_{xz'} \end{bmatrix} \begin{bmatrix} \sigma_{\mathrm{H}} \\ \sigma_{\mathrm{h}} \\ \sigma_{\mathrm{v}} \end{bmatrix} \tag{5.11}$$

Where

$$\begin{bmatrix} l_{xx'} & l_{xy'} & l_{xz'} \\ l_{yx'} & l_{yy'} & l_{yz'} \\ l_{zx'} & l_{zy'} & l_{zz'} \end{bmatrix} = \begin{bmatrix} \cos\varphi_x \cos\varphi_z & \sin\varphi_x \cos\varphi_x & -\sin\varphi_z \\ -\sin\varphi_x & \cos\varphi_x & 0 \\ \cos\varphi_x \cos\varphi_z & \sin\varphi_x \cos\varphi_z & \cos\varphi_z \end{bmatrix} \tag{5.12}$$

σ_{H}, σ_{h}, and σ_{v} are the far-field stresses, ${\sigma_x}^0$, ${\sigma_y}^0$, ${\sigma_z}^0$, ${\tau_{xy}}^0$ ${\tau_{yz}}^0$, and ${\tau_{xz}}^0$ are the local wellbore coordinate stresses. ϕ_x is the azimuth angle, and is the counterclockwise angle between the projection of the wellbore axis on the horizontal plane and the direction of the maximum horizontal in situ stress. ϕ_z is the wellbore inclination, the angle between the wellbore axis and the vertical direction. After the conversion, the analysis can be worked out in the local coordinate system. Then the complete stress solutions are:

$$\begin{aligned} \sigma_r = {} & \left(\frac{\sigma_x^0 + \sigma_y^0}{2}\right)\left(1 - \frac{r^2}{a^2}\right) + \left(\frac{\sigma_x^0 - \sigma_y^0}{2}\right)\left(1 + 3\frac{r^4}{a^4} - 4\frac{r^2}{a^2}\right)\cos 2\theta \\ & + \tau_{xy}^0\left(1 + 3\frac{r^4}{a^4} - 4\frac{r^2}{a^2}\right)\sin 2\theta + P_{\mathrm{w}}\frac{r^2}{a^2} \end{aligned} \tag{5.13}$$

$$\begin{aligned} \sigma_\theta = {} & \left(\frac{\sigma_x^0 + \sigma_y^0}{2}\right)\left(1 + \frac{r^2}{a^2}\right) - \left(\frac{\sigma_x^0 - \sigma_y^0}{2}\right)\left(1 + 3\frac{r^4}{a^4}\right)\cos 2\theta \\ & - \tau_{xy}^0\left(1 + 3\frac{r^4}{a^4}\right)\sin 2\theta - P_{\mathrm{w}}\frac{r^2}{a^2} \end{aligned} \tag{5.14}$$

$$\sigma_z = \sigma_z^0 - \mu\left[2\left(\sigma_x^0 - \sigma_y^0\right)\right]\left(\frac{r^2}{a^2}\right)\cos 2\theta + 4\tau_{xy}^0\left(\frac{r^2}{a^2}\right)\sin 2\theta \tag{5.15}$$

$$\tau_{r\theta} = \left(\frac{\sigma_x^0 - \sigma_y^0}{2}\right)\left(1 - 3\frac{r^4}{a^4} + 2\frac{r^2}{a^2}\right)\sin 2\theta + \tau_{xy}^0\left(1 - 3\frac{r^4}{a^4} + 2\frac{r^2}{a^2}\right)\cos 2\theta \tag{5.16}$$

$$\tau_{\theta z} = \left(-\tau_{xz}^0 \sin\theta + \tau_{yz}^0 \cos\theta\right)\left(1 + 2\frac{r^2}{a^2}\right) \tag{5.17}$$

$$\tau_{rz} = \left(\tau_{xz}^0 \cos\theta + \tau_{yz}^0 \sin\theta\right)\left(1 - 2\frac{r^2}{a^2}\right) \tag{5.18}$$

Where σ_θ, σ_r, and σ_z are tangential stress, radial stress, and axial stress, respectively, and $\tau_{r\theta}, \tau_{\theta z}, \tau_{rz}$ are three components of the shear stress. θ is the angle from the maximum principal horizontal stress and a is the borehole radius [34].

5.3.10.4 Mohr–Coulomb Failure Criterion

The analysis begins with a simple case: a vertical borehole in a linear elastic formation with horizontal in situ stresses, and the mud weight P_w on its internal boundary exerted by the fluid pressure. In this case, the largest stress difference occurs at the borehole wall, $r = a$. According to Eqs. (5.19–5.22), the stresses at the borehole wall are:

$$\sigma_r = P_w \tag{5.19}$$

$$\sigma_\theta = P_0[1 + K + 2(1 - K)\cos 2\theta] - P_w \tag{5.20}$$

$$\sigma_z = \sigma_v - 2\mu(1 - K)P_0 \cos 2\theta \tag{5.21}$$

$$\tau_{r\theta} = \tau_{rz} = \tau_{\theta z} = 0 \tag{5.22}$$

Since all shear stresses vanish, σ_θ (tangential stress), σ_r (radial stress), and σ_z (axial stress) are principal stresses, and can be used directly in the failure criterion. The Mohr–Coulomb criterion is given as follows:

$$\sigma_1 - \alpha\sigma_3 = Y \tag{5.23}$$

Where

$$\alpha = \frac{1 + \sin\varphi}{1 - \sin\varphi}, Y = \frac{2C\cos\varphi}{1 - \sin\varphi} \tag{5.24}$$

And φ and C are the friction angle and cohesion respectively.

The Minimum Wellbore Pressure

To prevent wellbore collapse when wellbore pressure is decreasing, the minimum wellbore pressure should be calculated. In this situation, two cases are considered [32, 37]:

Case I: $\sigma_\theta > \sigma_z > \sigma_r$

Therefore, the Mohr–Coulomb criterion can be written as follows according to Eq. (5.25):

$$\sigma_\theta - \alpha\sigma_r = Y \tag{5.25}$$

So the minimum wellbore pressure which can initiate the shear failure can be obtained:

$$P_{\mathrm{w}} = \frac{[(1+K)+2(1-K)\cos 2\theta]P_0 - Y}{1+\alpha} \tag{5.26}$$

Case II: $\sigma_z > \sigma_\theta > \sigma_r$

The minimum wellbore pressure in this case can be obtained:

$$P_{\mathrm{w}} = \frac{[\sigma_{\mathrm{v}} - 2\mu(1-K)P_0 \cos 2\theta] - Y}{\alpha} \lim_{x \to \infty} \tag{5.27}$$

The Maximum Wellbore Pressure

If the wellbore pressure is increased significantly, the tensile failure can occur at the borehole wall when the smallest principal stress becomes tensile and equal to the tensile strength of the rock around the borehole:

$$\sigma_3 + T_r = 0 \tag{5.28}$$

Where T_0 is the tensile strength of the rock around the borehole.

It is likely assumed that the tangential stress is the minor principal stress. Therefore, the maximum wellbore pressure is:

$$P_{\mathrm{w}} = P_0[1 + K + 2(1-K)\cos 2\theta] + T_0 \tag{5.29}$$

Elastoplastic Stress Analysis

After initial yielding takes place with the decrease of the borehole pressure at the borehole wall, a plastic zone within the region $a \leq r \leq R_{\mathrm{p}}$ forms around the inner wall of the borehole. The stresses in the plastic zone must satisfy the equilibrium equation:

$$\frac{\partial \sigma_r}{\partial \sigma} + \frac{\sigma_r - \sigma_\theta}{r} = 0 \tag{5.30}$$

Also, the stresses in the plastic region must satisfy the yield conditions:

$$\sigma_\theta - \alpha\sigma_r = Y \tag{5.31}$$

Substituting Eq. (5.31) into Eq. (5.30) and equating, one obtains:

$$\sigma_r = \frac{C}{\alpha - 1} r^{\alpha-1} - \frac{Y}{\alpha - 1} \tag{5.32}$$

Using the boundary conditions for $r = a$, $\sigma_r = P_{\mathrm{w}}$, a value for C can be obtained:

$$C = \frac{\alpha - 1}{a^{\alpha-1}} P_{\mathrm{w}} + \frac{Y}{a^{\alpha-1}} \tag{5.33}$$

Therefore, the radial stress above the crown in the plastic zone can then be written as:

$$\sigma_r^p = \left(\frac{r}{a}\right)^{\alpha-1} \left[P_{\mathrm{w}} + \frac{Y}{\alpha-1}\right] - \frac{Y}{\alpha-1} \tag{5.34}$$

The radial stress should be continuous across the elastoplastic boundary, and so the following condition must be satisfied:

$$\sigma_r^p = \sigma_r^e atr = R_{\mathrm{p}}, \theta = 90^{\circ} \tag{5.35}$$

Where the superscripts e and p denote the elastic and plastic zones, the borehole pressure P_{w} can then be expressed as:

$$P_{\mathrm{w}} = \left[\frac{3K - 1}{\alpha + 1} P_0 - \frac{2Y}{1 - \alpha^2}\right] \left(\frac{a}{R_{\mathrm{p}}}\right)^{\alpha-1} + \frac{Y}{1 - \alpha} \tag{5.36}$$

An axisymmetric cavity contraction theory is used to represent stresses at point ($r = R_{\mathrm{p}}$, $\theta = 90°$) for the asymmetric problem (where $K \neq 1$), therefore the following assumptions can be made:

- The plastic region, where its radius is the distance between the center of the borehole and the furthest yield point, is axisymmetric.
- The displacement for each point on the elastoplastic zone boundary is the same.
- Displacement depends on the maximum plastic radius and not the position angle θ.

With the above assumptions, the displacement at the interface between the elastic and plastic zone is given by the elasticity solution [52]:

$$u\Big|_{r=R_p} = \frac{P_{1y} - P_0}{2G}\left(\frac{R_p}{r}\right)^2 r = \frac{\left[\left(1 - \frac{3K-1}{\alpha+1}\right)P_0 + \frac{Y}{1+\alpha}\right]}{2G}R_p \tag{5.37}$$

Where G is the shear modulus of the rock around the borehole, $G = E/2(1+\mu)$. For unloading cavity contraction, the following solution can be derived [52]:

$$R_0^{1+\beta} - a^{1+\beta} = \left(R_p - u\right)^{1+\beta} - R_p^{1+\beta} \tag{5.38}$$

Where

$$\beta = \frac{1 + \sin\psi}{1 - \sin\psi} \tag{5.39}$$

Where ψ denotes the dilation angle of the rock.

Since soft rocks have a small dilation angle, it can be assumed that the rock volume does not change within the plastic zone. Therefore $\sin\psi = 0$ can be achieved, which leads to:

$$R_0^2 - a^2 = \left(R_p - u\right)^2 - R_p^2 \tag{5.40}$$

It has been found that the dilation angle has relatively little influence on the solutions.

Substituting Eq. (5.37) into Eq. (5.40) and neglecting the higher-order terms gives:

$$\frac{a}{R_p} = \left(\frac{R_0^2}{R_p^2} - \frac{\left(1 - \frac{3K-1}{1+\alpha}\right)P_0 + \frac{Y}{1+\alpha}}{G}\right)^{0.5} \tag{5.41}$$

And substituting Eq. (5.41) into Eq. (5.36) yields:

$$P_w = \left[\frac{3K-1}{1+\alpha}P_0 - \frac{2Y}{1-\alpha^2}\right]\left(\frac{R_0^2}{R_p^2} - \frac{\left(1 - \frac{3K-1}{1+\alpha}\right)P_0 + \frac{Y}{1+\alpha}}{G}\right)^{0.5} + \frac{Y}{1+\alpha} \tag{5.42}$$

If $K < 1$ then, the initial yield and the furthest points are at the spring-line of the borehole, i.e., $\theta = 0°$.

The critical borehole pressure P_{2y} for initial yield at the spring-line of the borehole is:

$$P_{2y} = \frac{3K - 1}{1 + \alpha} P_0 - \frac{Y}{1 + \alpha} \tag{5.43}$$

A similar approach can be used to obtain the equivalent limiting borehole pressure equation:

$$P_w = \left[\frac{3 - K}{1 + \alpha} P_0 - \frac{2Y}{1 - \alpha^2} \right] \left(\frac{R_0^2}{R_p^2} - \frac{\left(1 - \frac{3-K}{1+\alpha}\right) P_0 + \frac{Y}{1+\alpha}}{G} \right)^{\frac{\alpha-1}{2}} + \frac{Y}{1 - \alpha} \tag{5.44}$$

5.3.10.5 Wellbore Stability in Laminated (VTI) Formations

We discussed that the strength anisotropy, which exists due to the presence of weak planes, is one of the unique properties of laminated shales. Therefore, both the stress state around the wellbore and the variable strength properties of laminated formations should be considered in wellbore stability calculations.

Anisotropic Strength Model

In this model, it is assumed that under triaxial compressive strength the pore pressure within the shale sample is zero (drained-test conditions). A shale sample, as shown in Fig. 5.14, is subjected to a typical compressive strength test. Based on the operation angle (β), which is defined as the angle between the bedding plane and the axial stress, two possibilities for laminated rock failure may exist: either across or along the bedding plane.

- If the sample fails across the bedding plane, then the strength is defined as "normal strength," σ'_{1n}.
- If the sample fails along the bedding plane, then the strength is defined as the "bedding plane strength," σ'_{1b} at that operation angle.

According to Mohr–Coulomb failure criterion, the normal strength can be calculated using the following equation [54]:

$$\sigma'_{1n} = \sigma'_3 + 2\left(C_0 + \mu\sigma'_3\right)\left[\left(\sqrt{1 + \mu^2}\right) + \mu\right] \tag{5.45}$$

Similarly, the bedding plane strength can be calculated using the following equation [54]:

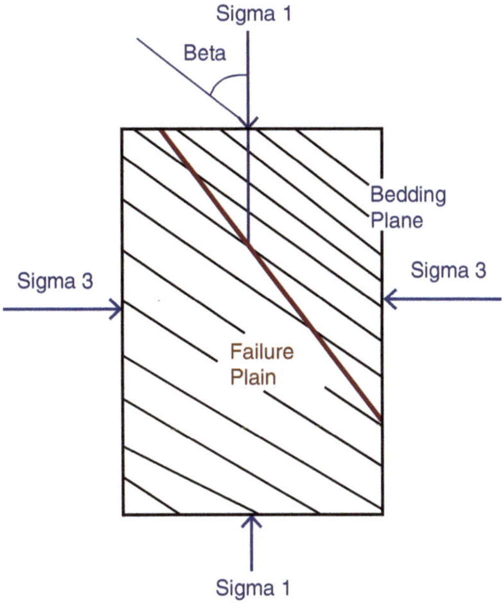

Fig. 5.14 Shale sample under triaxial stress (modified from ref. [53])

$$\sigma'_{1b} = \sigma'_3 + \frac{2\left(C_{0w} + \mu_w \sigma'_3\right)}{(1 + \tan \beta \mu_w) \sin 2\beta} \tag{5.46}$$

Differentiating Eq. (5.46) with respect to β shows that σ'_{bs} has a minimum value when: $\tan 2\beta_w = \frac{1}{\mu_w}$ (β_w is special operation angle at which the bedding strength becomes minimum), this minimal value of $\sigma_{bs,\min}$ is:

$$\sigma'_{1b,\min} = \sigma'_3 + 2\left(C_{0w} + \mu_w \sigma'_3\right)\left[\left(\sqrt{1 + \mu_w{}^2}\right) + \mu_w\right] \tag{5.47}$$

By equating Eqs. (5.46) and (5.47), two values of operation angle, denoted as β_1 and $\beta_2, \beta_1 \prec \beta_2$, can be calculated as:

$$\beta_1, \beta_2 = \tan^{-1}\left(\frac{a \pm \sqrt{a^2 - b^2 - 2ab\mu_w}}{b + 2a\mu_w}\right) \tag{5.48}$$

Where μ_w is the weak plane coefficient of friction and

$a = \left(C_0 + \mu \sigma'_3\right)\left[\left(\sqrt{1 + \mu^2}\right) + \mu\right]$ and $b = C_{0w} + \mu_w \sigma'_3$.

Where μ is the intrinsic coefficient of friction. The cohesion (C_0) can be determined through uniaxial compressive strength (UCS) test, geophysical well logs or from correlations.

5.4 Anisotropic Geomechanical Modeling Case Study-Bakken Formation

To carry out a complete geomechanical study, a mechanical earth model (MEM) should be generated. MEM is a numerical representation of reservoir properties in 1D, 2D, or even 3D style. A MEM contains data related to the rock failure mechanisms, in situ stress, stratigraphy and geologic structure of the reservoir [55–58]. MEM should be made any time before the drilling. It will be updated when new information during drilling and later during the production is obtained (Fig. 5.15).

Incorporating elastic anisotropy in to constructing a more precise MEM for geomechanical modeling is crucial [28, 29, 59] (Fig. 5.16).

5.4.1 Anisotropy in Geomechanical Modeling

Anisotropy has been neglected in constructing MEM and geomechanical (numerical) modeling for decades. In the petroleum industry, isotropy assumptions have been frequently applied to geomechanical modeling not because they are good approximations, but since anisotropic measurements were not available and relevant models were sophisticated. Moreover, isotropy assumptions have led to inaccurate results [14, 60]. Mechanical anisotropy, in particular, means elastic properties of the rock, such as Young's modulus, Poisson's ratio and shear modulus,

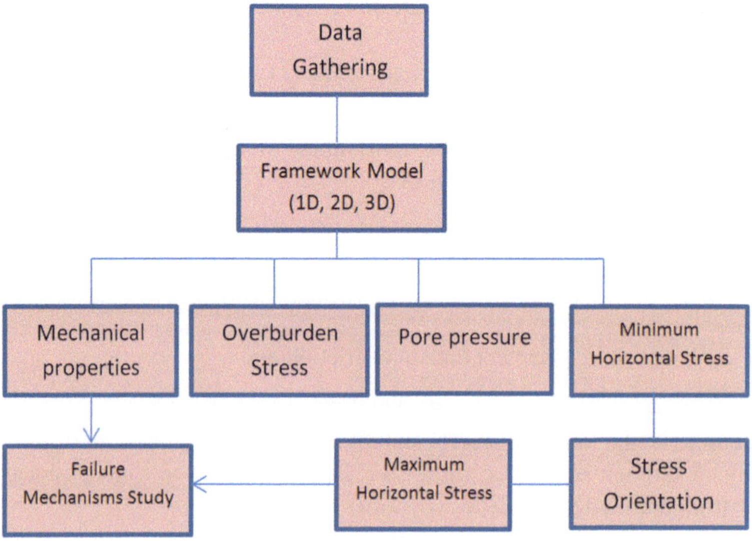

Fig. 5.15 Mechanical earth model (MEM) flowchart, [29]

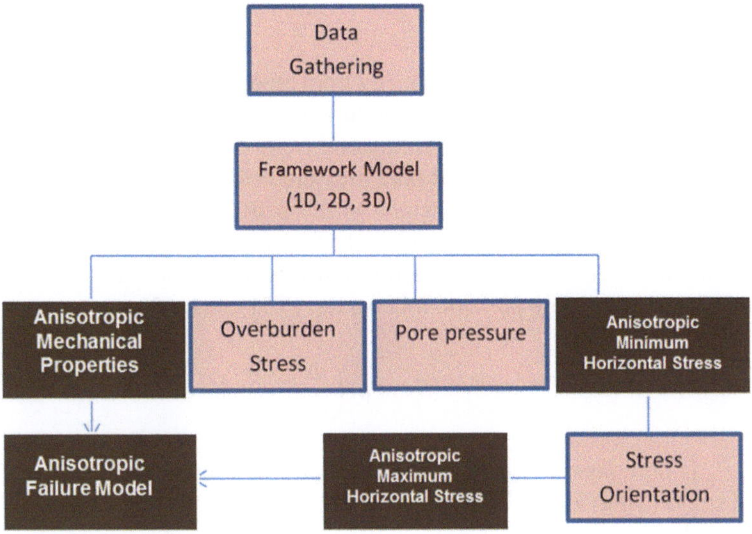

Fig. 5.16 Anisotropic MEM flowchart, [29]

change due to the direction of the measurement. As stated earlier sonic logging technology have made the three dimensional (3D) analysis of mechanical anisotropy possible around the wellbore with high levels of precision [5, 13].

5.4.1.1 Vertical Stress

Total vertical stress is defined as the combination of the weight of the rock matrix which is the vertical effective stress and the pressure exerted by the fluids in the pore spaces overlying the depth of interest, as shown in Eq. (5.49).

$$S(h) = \int_0^h \rho(z) g dz \tag{5.49}$$

Where $S(h)$ is the total vertical stress at depth h, $\rho(z)$ is the density at depth z below the surface, and g is the acceleration due to gravity. Equation (5.49) can be rearranged in the form of Eq. (5.50):

$$\sigma_v = S(h) - P_p \tag{5.50}$$

Where σ_v is the vertical effective stress and P_p is the pore pressure which is caused by the fluid in the pore spaces, $S(h)$ can be computed by integrating bulk densities from the density log data. It should be noted that if water column exists above the surface, $\rho_w g z_w$ should be added to Eq. (5.49); ρ_w is the water density and z_w is water depth.

5.4.1.2 Pore Pressure

Pore pressure (P_p) can be measured directly from repeated formation test (RFT) and drill stem test (DST) with high accuracy during drilling. When these data are not available, qualitative pore pressure estimation from velocity data can be utilized. Several methods for velocity-based pore pressure estimation have been proposed [61, 62]. The most widely used method in the petroleum industry is the Eaton method [57, 63]. This approach is based on the fact that the relationship between the ratio of the observed sonic log value (or slowness) to the normal velocity value (or slowness) and the pore pressure follows Eq. (5.51), [64, 65]:

$$P_p = S - (S - P_p)\left(\frac{\Delta t_{\text{normal}}}{\Delta t_{\text{observed}}}\right)^3 \tag{5.51}$$

Where P_p is pore pressure (MPa), P_h is the hydrostatic pressure (MPa), S is total vertical stress (MPa) and Δt is slowness (μs/ft).

5.4.1.3 Horizontal Stress

Isotropic horizontal stresses for an various types of medium have been extensively used in the industry and was derived from the solution of the linear poroelastic equation [54]. Horizontal stresses for an isotropic poroelastic medium under uniform tectonic horizontal strain can be expressed as follows, Eqs. (5.52) and (5.53), [60]:

$$\sigma_h - \alpha P_p = \frac{\nu}{1-\nu}(\sigma_v - \alpha P_p) + \frac{E}{1-\nu^2}\varepsilon_h + \frac{E\nu}{1-\nu^2}\varepsilon_H \tag{5.52}$$

$$\sigma_H - \alpha P_p = \frac{\nu}{1-\nu}(\sigma_v - \alpha P_p) + \frac{E}{1-\nu^2}\varepsilon_H + \frac{E\nu}{1-\nu^2}\varepsilon_h \tag{5.53}$$

where E is Young's modulus, ν is Poisson's ratio, σ_h is the minimum horizontal principal stress, σ_H is the maximum horizontal principal stress, σ_v is the vertical (overburden) stress, α is Biot's constant, ε_h is the minimum horizontal strain and ε_H is the maximum horizontal strain.

It was made clear that shales are transversely isotropic, thus using isotropic models for horizontal stress calculations is not acceptable. Horizontal stresses for a transversely isotropic medium with a vertical axis of symmetry under uniform tectonic horizontal strain can be developed as [60, 66]

$$\sigma_{\rm h} - \alpha P_{\rm p} = \frac{E_{\rm horz}\nu_{\rm vert}}{E_{\rm vert}(1-\nu_{\rm horz})}\left(\sigma_{\rm v} - \alpha(1-\xi)P_{\rm p}\right) + \frac{E_{\rm horz}}{1-\nu_{\rm horz}{}^2}\varepsilon_{\rm h} + \frac{E_{\rm horz}\nu_{\rm horz}}{1-\nu_{\rm horz}{}^2}\varepsilon_{\rm H} \tag{5.54}$$

$$\sigma_{\rm H} - \alpha P_{\rm p} = \frac{E_{\rm horz}\nu_{\rm vert}}{E_{\rm vert}(1-\nu_{\rm horz})}\left(\sigma_{\rm v} - \alpha(1-\xi)P_{\rm p}\right) + \frac{E_{\rm horz}}{1-\nu_{\rm horz}{}^2}\varepsilon_{\rm h} + \frac{E_{\rm horz}\nu_{\rm horz}}{1-\nu_{\rm horz}{}^2}\varepsilon_{\rm h} \tag{5.55}$$

Where $E_{\rm horz}$ is Young's modulus in the plane of isotropy, $E_{\rm vert}$ is Young's modulus along the axis of symmetry, which is the direction of anisotropy, $\nu_{\rm horz}$ is Poisson's ratio in the plane of isotropy, $\nu_{\rm vert}$ is Poisson's ratio along the axis of symmetry. α is Biot's constant and ξ is the poroelastic constant. It is clear that the components of Eqs. (5.52) and (5.53) are very well improved from isotropic E and ν to anisotropic ones in Eqs. (5.54) and (5.55).

5.4.1.4 Anisotropic Elastic Parameters

For an anisotropic elastic medium assuming vertical transverse isotropy (VTI), by applying linear Hook's law expressed in Eq. (5.56) and taking X_3 as the axis of rotational symmetry, the stiffness matrix becomes Eq. (5.57) in the conventional two index Voigt notation [15]

$$\sigma_{ij} = C_{ijkl}\varepsilon_{kl} \tag{5.56}$$

where σ_{ij} is the stress tensor, C_{ijkl} is the fourth rank stiffness tensor and ε_{kl} is the strain tensor.

$$C_{ij} = \begin{pmatrix} C_{11} & C_{12} & C_{13} & 0 & 0 & 0 \\ C_{21} & C_{22} & C_{23} & 0 & 0 & 0 \\ C_{31} & C_{32} & C_{33} & 0 & 0 & 0 \\ 0 & 0 & 0 & C_{44} & 0 & 0 \\ 0 & 0 & 0 & 0 & C_{55} & 0 \\ 0 & 0 & 0 & 0 & 0 & C_{66} \end{pmatrix} \tag{5.57}$$

The inverse of Eq. (5.57) in which the strain tensor is expressed as a linear function of the stress for a linear elastic transverse isotropic medium can be written as $\varepsilon_{kl} = S_{ijkl}\sigma_{ij}$ where $S = C^{-1}$. Thus the non-vanishing elastic compliance coefficients become $S_{11} = S_{22}$, S_{33}, $S_{12} = S_{21}$, $S_{13} = S_{23} = S_{32} = S_{31}$, $S_{44} = S_{55}$ and $S_{66} = 2(S_{11} - S_{12})$ in the two index notation.

The compliance tensor S_{ij} of a vertical transverse isotropic medium can be expressed in terms of Young's modulus and Poisson's ratio [54, 67]:

$$C_{ij} = \begin{pmatrix} \frac{1}{E_1} & \frac{-\nu_{21}}{E_1} & \frac{-\nu_{31}}{E_1} & 0 & 0 & 0 \\ \frac{-\nu_{12}}{E_1} & \frac{1}{E_1} & \frac{-\nu_{31}}{E_3} & 0 & 0 & 0 \\ \frac{-\nu_{13}}{E_3} & \frac{-\nu_{31}}{E_3} & \frac{1}{E_3} & 0 & 0 & 0 \\ 0 & 0 & 0 & \frac{1}{\mu_{13}} & 0 & 0 \\ 0 & 0 & 0 & 0 & \frac{1}{\mu_{13}} & 0 \\ 0 & 0 & 0 & 0 & 0 & \frac{1}{\mu_{12}} \end{pmatrix} \quad (5.58)$$

If X_3 is taken as the axis of vertical rotational symmetry and $X_1 - X_2$ the plane of symmetry, then $E_1 = E_{\text{horz}}$ is the horizontal Young's modulus in the plane of symmetry, $E_3 = E_{\text{vert}}$ is the vertical Young's modulus along the axis of symmetry, $\nu_{12} = \nu_{21} = \nu_{\text{horz}}$ is the horizontal Poisson's ration in the plane of symmetry (isotropy) and $\nu_{31} = \nu_{13} = \nu_{\text{vert}}$ is the vertical Poisson's ratio along the vertical axis of symmetry. The anisotropic Young's modulus and Poisson's ratio for a vertically transverse isotropic medium in terms of elastic stiffness coefficients (C_{ij}), Eqs. (5.59)–(5.62), will be obtained by mathematical calculations after re-inversing the compliance tensor, Eq. (5.58), $C = S^{-1}$. Thus, the horizontal and vertical E and ν will be [14, 68]:

$$E_{\text{horz}} = \frac{(C_{11} - C_{12})\left(C_{11}C_{33} - 2C_{13}^2 + C_{12}C_{13}\right)}{C_{11}C_{13} - C_{13}^2} \quad (5.59)$$

$$E_{\text{vert}} = C_{33} - \frac{2C_{13}^2}{C_{11} + C_{12}} \quad (5.60)$$

$$\nu_{\text{horz}} = \frac{C_{12}C_{33} - C_{13}^2}{C_{11}C_{33} - C_{13}^2} \quad (5.61)$$

$$\nu_{\text{vert}} = \frac{C_{13}}{C_{11} + C_{33}} \quad (5.62)$$

To evaluate the anisotropic elastic parameters from Eqs. (5.59)–(5.62), elastic stiffness coefficients (C_{ij}) are needed. Advanced sonic logging has made capturing the relevant elastic stiffness coefficients (C_{ij}) azimuthally and radially deep in the intact formation, and in the area around the well [69] possible with high accuracy.

5.4.1.5 Stress Profile

To following figures shows the calculated stress profile and the pore pressure as an input to the MEM (Fig. 5.17).

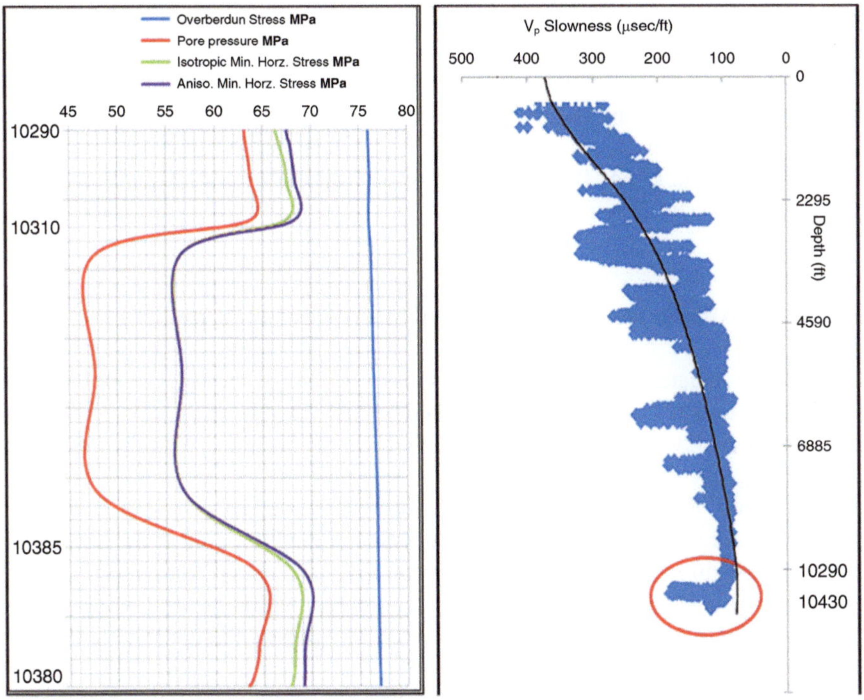

Fig. 5.17 Stress profile through the Bakken formation (*left*), compressional wave slowness trend-deflection from the normal velocity is observed in the Bakken (*red* oval) due to the overpressure nature of this formation (*right*), [29]

The decrease in the effective stress normally takes place due to an increase in the pore pressure in a specific interval [70]. Considering the petroleum system governing the Bakken, this overpressure behavior can be explained by the conversion of kerogen to hydrocarbon in the upper and lower shale members. Additionally, the high clay volume content of the middle member in the vicinity of clastic and carbonate facies has caused a high degree of heterogeneity. Consequently, migration of the generated and expelled hydrocarbon from the upper and lower members into the middle section through the vertical fractures can explain the abnormal overpressure characteristics of the Middle Bakken. Ultimately, the highly compacted shales of UB and LB, along with high clay volume content of MB, are the primary reasons for the high overpressure nature of these intervals.

Comparing the two minimum horizontal stress profiles (anisotropic and isotropic), better results when anisotropy assumptions are included in calculations can be achieved. In order to evaluate the horizontal stress magnitude from Eqs. (5.52)–(5.55), lateral tectonic strains, ε_h, ε_H should be evaluated. It is often assumed that after fluid is withdrawn from a reservoir where the vertical stress is the major acting principal stress, lateral strain is inhibited by the rock adjacent to the reservoir, thus uniaxial strain assumptions maybe acceptable [3]. To approximate the minimum horizontal stress, uniaxial strain assumption was taken into account. This means that the only strain is in the vertical direction thus the lateral tectonic strains, ε_h, ε_H in Eqs. (5.52)–(5.55), were assumed zero [60]. Since Williston Basin is considered an intracratonic basin and is currently under tectonic equilibrium, setting the lateral strains, ε_h, ε_H equal to zero is reasonable. To calculate horizontal stress from Eq. (5.52) and Eq. (5.54) Biot's constant $\alpha \approx 1$ from laboratory experiments, and poroelastic constant $\xi = 0$ were applied.

5.4.1.6 Maximum Horizontal Principal Stress (Second Approach)

In order to calculate horizontal principal stresses, C_{44}, C_{55}, and C_{66} (major stiffness matrix components for a TI medium) should be calculated from the slow shear, fast shear and Stoneley wave velocities or measured in the lab along with the formation density. The calculations are shown in Fig. 5.18.

In addition, a better estimation of anisotropic maximum principal horizontal stress is to take the effect of the elastic properties of the formation in the horizontal and vertical directions (mechanical anisotropy) and three formation moduli (TI medium) into consideration. The results are depicted in the following figure. The black star in the figure shows the measured pore pressure in the Middle Bakken obtained from DST (Drill Stem Test) (Fig. 5.19).

A new approach to calculate maximum principal horizontal stress was developed by Sinha et al. [71] and adapted here. In this method, the minimum principal horizontal and overburden stress's magnitude as a function of depth are used as input to estimate the magnitude of maximum horizontal principal stress. The value of minimum principal horizontal stress is input from previous calculations. To perform this task, acousto-elastic parameter A_E in terms of the far-field shear moduli C_{55} and C_{66} is defined as:

$$A_E = \frac{C_{55} - C_{66}}{\sigma_v - \sigma_h} \tag{5.63}$$

In this equation it is assumed that the effects of permeability on the shear moduli (C_{55} and C_{66}) are similar and negligible [69].

Once the acousto-elastic parameter (A_E) has been determined for a given lithology interval, we can calculate the maximum horizontal principal stress magnitude (σ_{Hmax}) as a function of depth from the following equation:

$$\sigma_H = \sigma_h + \frac{C_{55} - C_{44}}{A_E} \tag{5.64}$$

Clay minerals have a huge impact on the difference between the Stoneley shear modulus C_{66} and the dipole shear moduli C_{44} or C_{55}. This generally makes shear modulus C_{66} in the isotropic plane of shale (along the clay minerals surfaces) to become larger than shear modulus C_{44} or C_{55} in the orthogonal planes; therefore

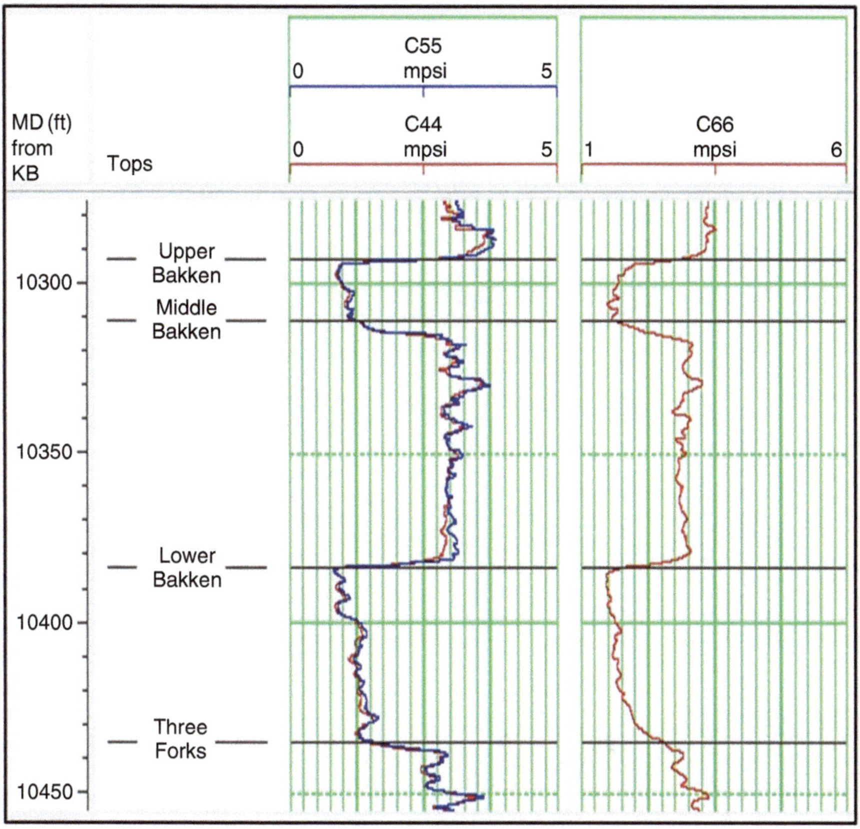

Fig. 5.18 Three shear moduli measured through the Bakken formation, [53]

C_{66} needed to be reduced by 40 % before combining it with the shear moduli C_{44} and C_{55} for stress magnitude estimation [71] in the upper and lower shale members.

5.4.1.7 Maximum Principal Horizontal Stress Orientation

The direction of maximum principal horizontal stress is found to be N65E by counting and recording the orientation of the fractures which have caused slowness or time based shear anisotropy greater than 20 % around the well. The results are plotted in a rose diagram showing the orientation of fracture planes with their frequencies through the Bakken interval (Fig. 5.20).

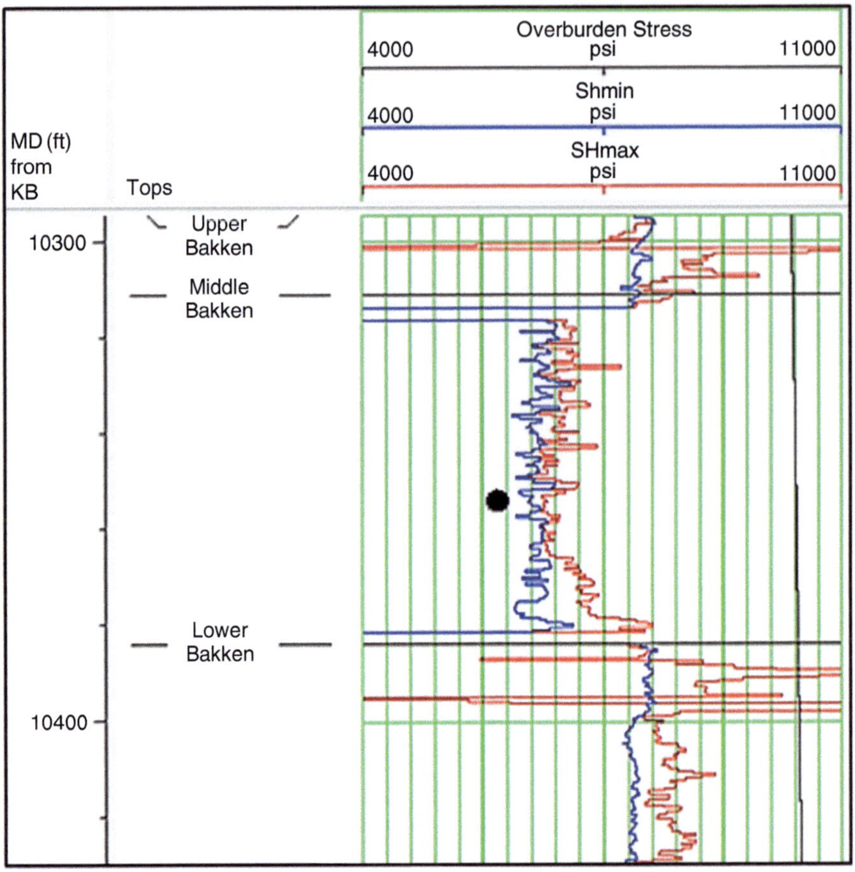

Fig. 5.19 Anisotropic horizontal principal stresses and overburden stress through the Bakken formation—the *black* star shows the DST measurement, [53]

5.4.2 3D Numerical Modeling

The numerical modeling of the inclined and vertical section of the wellbore through the VTI UB and isotropic MB is carried out by a series of 3D finite difference codes using FLAC3D (Fast Lagrangian Analysis of Continua in three dimensions).

5.4.2.1 Vertical Well (0° Deviation Angle)

The average elastic anisotropic mechanical properties that are assigned to the model were measured through integration of triaxial geomechanical testing and field data, summarized in Tables 5.2 and 5.3. The model was set under elastoplastic conditions.

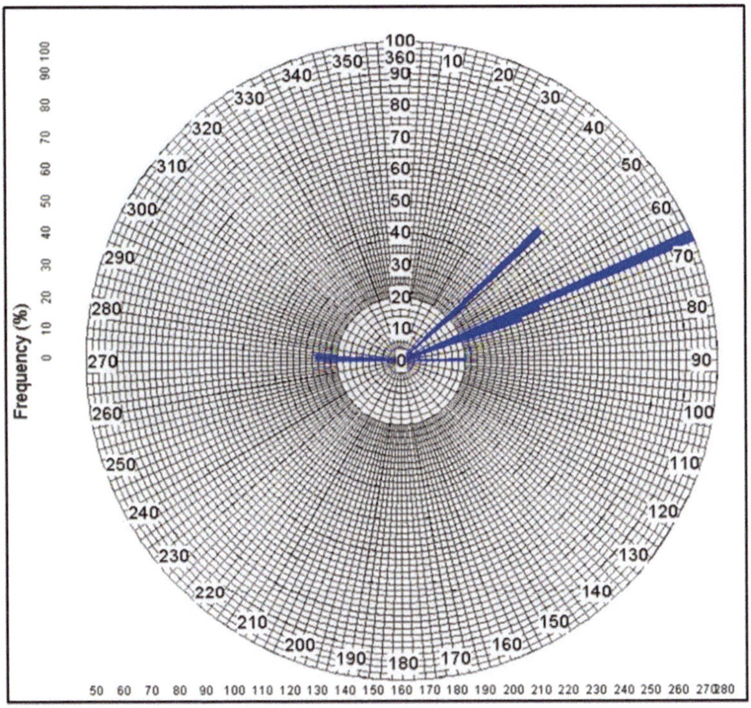

Fig. 5.20 Rose diagram of natural fractures that have caused more than 20 % shear anisotropy, frequency, and fracture plane orientation are presented, [53]

Table 5.2 Anisotropic elastic properties for the Bakken formation, [28]

	E_{vert} (GPa)	G_{vert} (GPa)	ν_{vert}
UB	5.01	2.53	0.18
MB	19.93	7.94	0.24
LB	5.83	2.77	0.16
	E_{Horz} (GPa)	G_{Horz} (GPa)	ν_{Horz}
UB	8.44	3.62	0.16
MB	19.92	7.97	0.24
LB	10.09	4.08	0.15

Table 5.3 Isotropic elastic parameters of the Bakken formation, [28]

	G (GPa)	E (GPa)	ν
UB	1.84	4.67	0.26
MB	8.02	19.90	0.24
LB	2.07	5.46	0.24

The following figure shows the total displacement magnitude in a 3D view for the Upper and Lower Bakken. From Fig. 5.21 it can be concluded that a higher degree of displacement takes place in the region adjacent to the wellbore in the

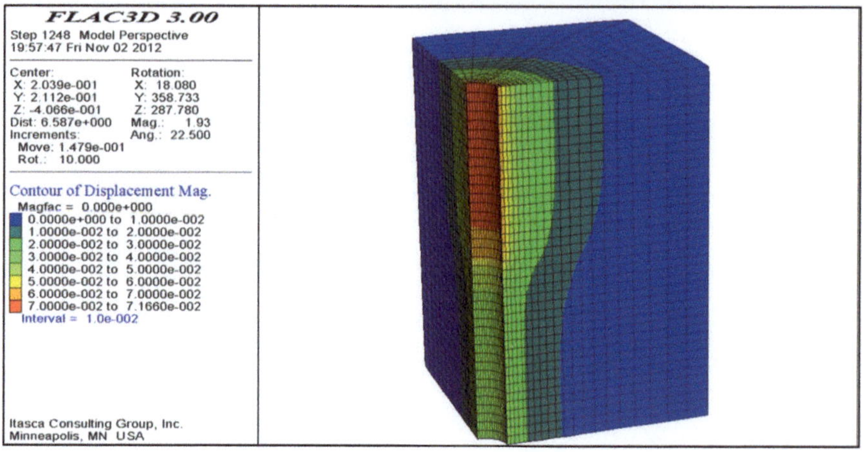

Fig. 5.21 Displacement around the borehole in the upper and lower Bakken under anisotropic assumptions, [53]

Upper Bakken, whereas in the middle member the displacement at the borehole wall is more moderate.

It can be observed that the displacements around the borehole in the Upper Bakken are much more severe than the displacement which takes place in the middle member. This difference is partially due to the lithology of the Upper Bakken compared to the Middle Bakken. Regarding the fact that shale is the main constituent of the upper member and deforms easier than carbonates and clastics with higher stiffness values (the main constituent of middle member) it was expected to see higher degrees of displacement in the upper member than the lower member.

The final step in geomechanical modeling was to run the elastoplastic Mohr–Coulomb failure criterion. To recall Mohr–Coulomb formulation: [3, 72]:

$$T = S_0 + \sigma \tan \varphi \tag{5.65}$$

$$\sigma_1 = C_0 + \sigma_3 \tan^2\left(45 + \frac{\varphi}{2}\right) \tag{5.66}$$

In Eq. (5.65), σ is the normal stress, T is the shear stress, S_0 is cohesion, and φ is the angle of internal friction. In Eq. (5.66), σ_1 is the maximum principal stress and σ_3 is the minimum principal stress. C_0, the unconfined compressive strength, given by [3, 54, 57]:

$$C_0 = 2S_0 \tan\left(45 + \frac{\varphi}{2}\right) \tag{5.67}$$

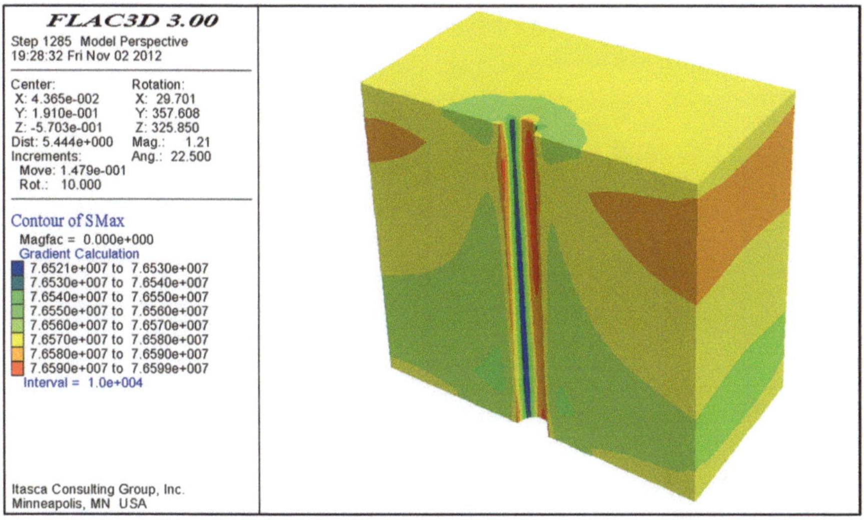

Fig. 5.22 (3D) view contours of maximum horizontal principal stress in UB under anisotropic assumptions, [53]

Figures 5.22 and 5.23 depict the variation of principal horizontal stresses (Mpa) (maximum and minimum principal horizontal stress) around the well in UB, MB, and UB-MB interface in the $X-Z$ plane under anisotropic assumptions.

Comparing two diagrams in Fig. 5.23 (right and left), which denotes the maximum horizontal principal stress magnitude in $X-Y$ plane, under anisotropic and isotropic assumptions, it was found that modeling the Upper Bakken under anisotropic assumptions can better define the stresses (hoop stress) around the borehole compared to the isotropic models. It can be seen in Fig. 5.23 that the hoop stress is more configured around the borehole. Hoop stress is compressive in the direction of minimum principal horizontal stress and tensile in the direction of maximum principal horizontal stress. This phenomenon is more visible when closer investigation is carried out in the vicinity of the wellbore wall (the blue contours). Hoop stress, if exceeds rock strength, will cause the formation to fail or tensile fractures to form. The concentration of hoop stress is a good indication of maximum and minimum principal horizontal stresses in nature. In the borehole, breakouts will occur where the hoop stress is concentrated, and reveals the direction of the minimum principal horizontal stress (Fig. 5.24).

Running the model under elastoplastic Mohr–Coulomb failure criterion will lead to the creation of plastic regions around the borehole. The following figures represent the position of various elastic and plastic regions which are developed around the borehole in a vertical well under anisotropic UB and isotropic MB, (Fig. 5.25).

Figure 5.25 (right) is generated when anisotropic parameters such as elastic mechanical properties and the stresses are incorporated in the geomechanical models for UB. This has caused the formation to behave elastically right after the

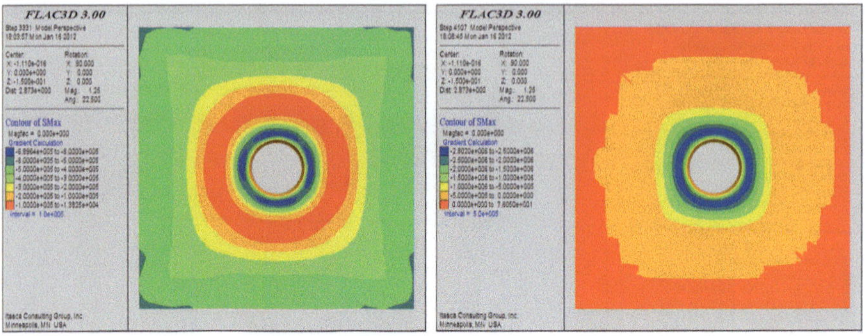

Fig. 5.23 Top view of maximum horizontal principal stress magnitude contours in UB under anisotropic assumptions (*left*), Top view of maximum horizontal principal stress magnitude contours in UB under isotropic assumptions (*right*), [53]

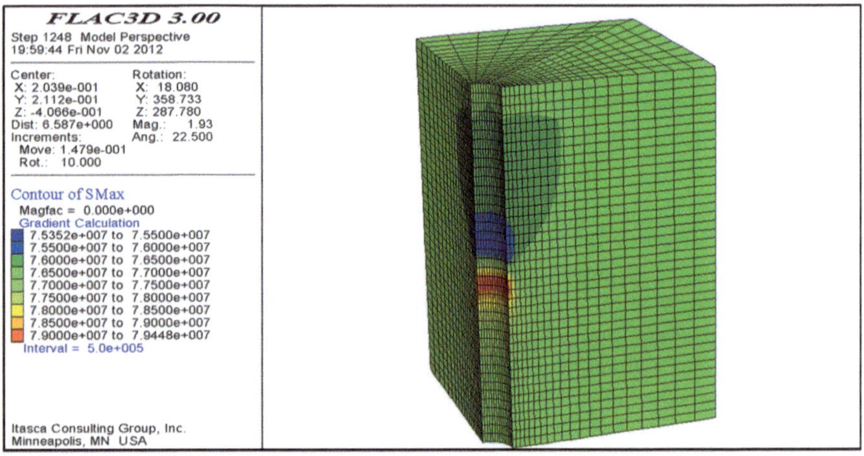

Fig. 5.24 Contours of maximum principal horizontal stress in a 3D view under anisotropic assumptions for the interface of UB-MB, [53]

formation is undergone a shear failure in the vicinity of the borehole (brown region) with better defined plastic areas matching the direction of principal horizontal stress (purple blocks).

5.4.2.2 Inclined Well (45° Attack Angle) (Figs. 5.26, 5.27, 5.28, and 5.29)

The following figure depicts how maximum total displacement versus well deviation angle decreases from a vertical well to a horizontal well in anisotropic conditions in the UB and isotropic MB (from 90° to 0° attack angle) (Fig. 5.30).

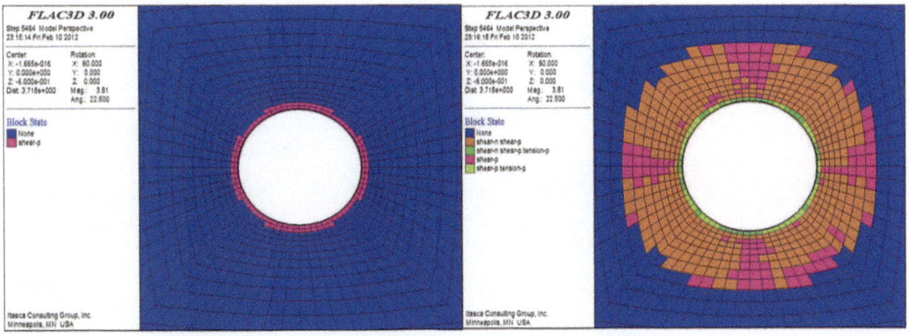

Fig. 5.25 Plastic regions around the borehole in isotropic MB (*left*). Plastic regions around the borehole in UB under anisotropic assumptions (*right*), [28]

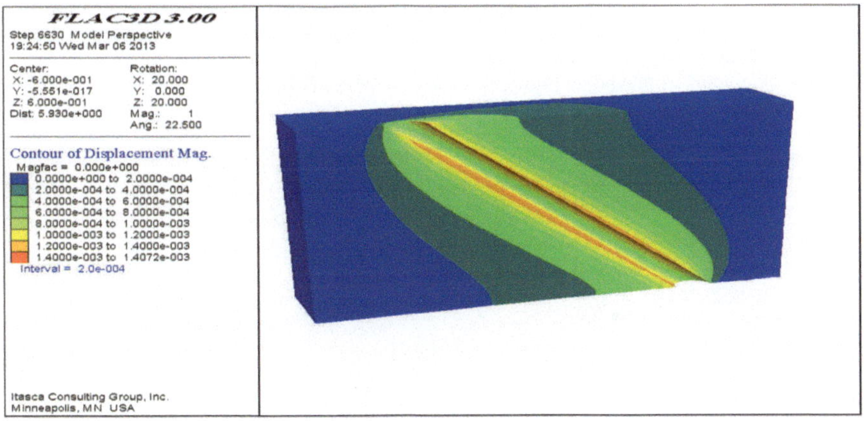

Fig. 5.26 Total displacement contours along the well in the UB, [53]

As expected and perceived, maximum displacement happens in a vertical well and decreases versus well deviation angle until a horizontal well is achieved. A horizontal well indicates the minimum value of total maximum displacement; at least half of which takes place in a vertical well.

Considering anisotropic behavior of upper member, the cohesion and internal friction angle plays an important role in the geomechanical modeling and they should be very well defined in vertical and horizontal directions (different planes of weakness or anisotropy) with respect to the well deviation angle. Thus, cohesion and internal friction angle vary along Z axis (the axis of symmetry). These two components can be defined through the following equations in two different directions, vertical and horizontal orientations or a combination of both in deviated wells. In the following relations, h denotes horizontal and v vertical directions. C is cohesion and φ is internal friction angle.

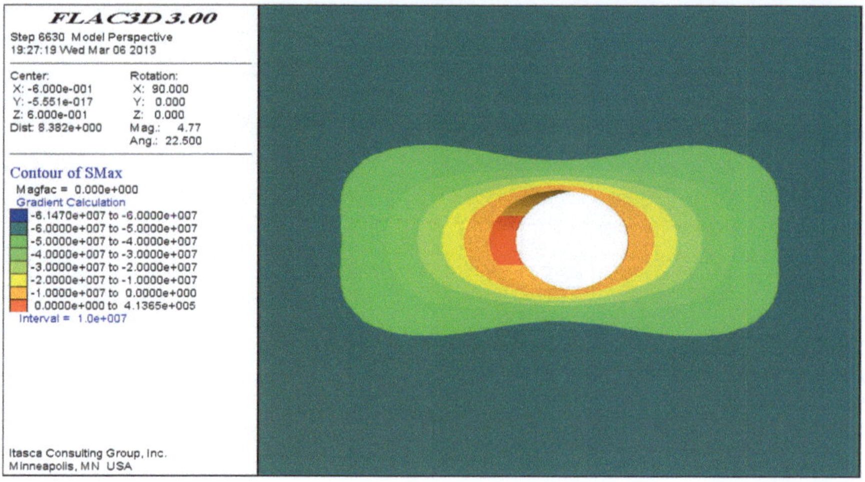

Fig. 5.27 Contours of maximum principal horizontal stress, [53]

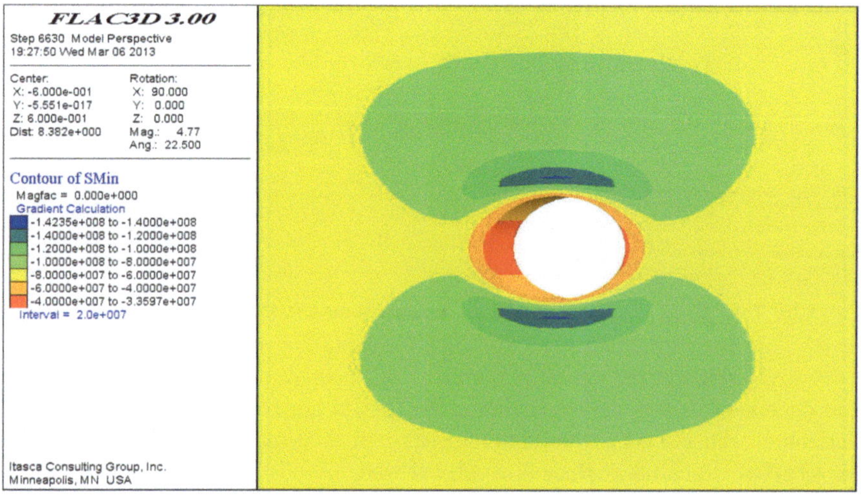

Fig. 5.28 Contours of minimum principal horizontal stress, [53]

$$C = C_{\mathrm{h}} \cos^2\theta + C_{\mathrm{v}} \sin^2\theta \tag{5.68}$$

$$\varphi = \varphi_{\mathrm{h}} \cos^2\theta + \varphi_{\mathrm{v}} \sin^2\theta \tag{5.69}$$

Table 5.4 summarizes the values for cohesion and internal friction angles used as input in the modeling of different well deviation angles in the Upper Bakken from laboratory measurements.

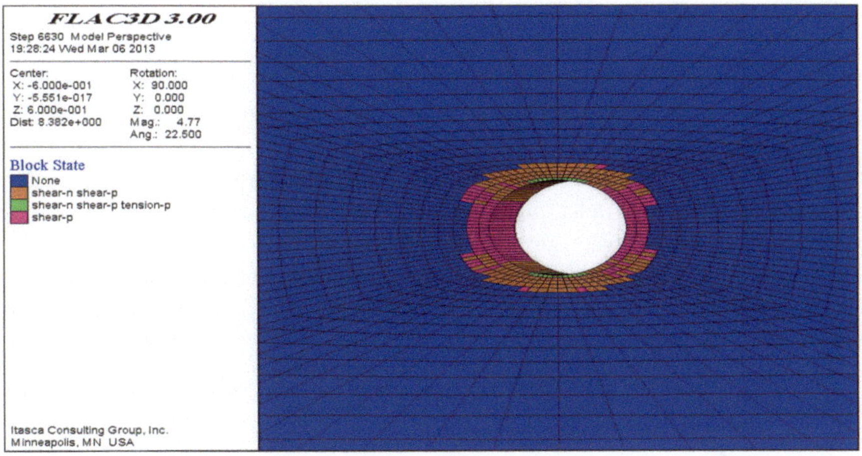

Fig. 5.29 Development of elastic and plastic regions of shear and tensile failures around the borehole, [53]

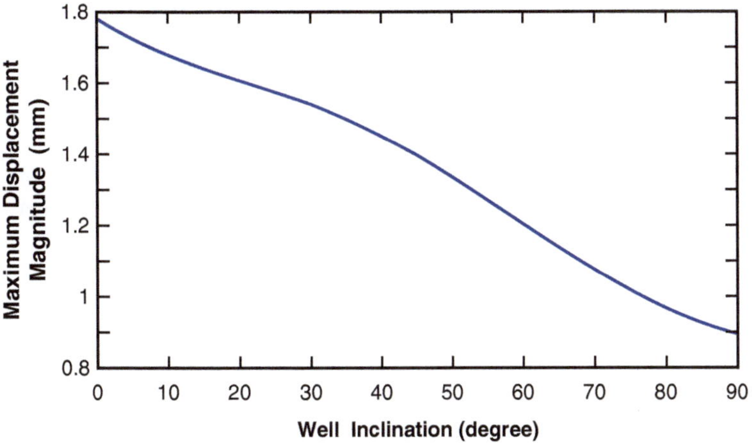

Fig. 5.30 Observed maximum total displacement versus well deviation angle, [53]

Table 5.4 Anisotropic internal friction angle (Φ) and cohesion (*C*) values used as input for geomechanical modeling in UB, [53]

Degree	C_h (Mpa)	C_v (Mpa)	Φ_h	Φ_v	θ	C	Φ
90	12	10	35	30	0	12.000	35.000
45	12	10	35	30	45	11.000	32.500
0	12	10	35	30	90	10.000	30.000

5.5 Summary and Recommendations

The anisotropic Mechanical Earth Model (MEM) for the Bakken Formation was generated. This task was performed by calculating the elastic parameters as well as the effective stresses in the UB (anisotropic) and MB (isotropic). Anisotropic behavior of the shale was determined through advanced sonic logging using dispersion analysis and radial profiling. The mechanical parameters to be input in the numerical models were generated through integration of field and lab data. These values were measured and calculated under isotropic and anisotropic formulae. It was found that UB is VTI with $E_{\text{horiz}} > E_{\text{vert}}$ and $\nu_{\text{vert}} > \nu_{\text{horiz}}$ and MB to be mechanically isotropic.

In situ stress and pore pressure profiles were generated under isotropic and anisotropic conditions. It is found that UB-LB is highly overpressured due to the kerogen to hydrocarbon transformation in low porosity-permeability shales. The Middle Bakken was found to be less overpressured due to the migration of hydrocarbon from UB and LB into the middle member. In addition, anisotropic stress model was found to better describe the stress conditions in the layers.

Geomechanical (numerical) modeling was performed under isotropic and anisotropic conditions in three main steps to evaluate the deformations and horizontal stress variations around the borehole. In the final step, the Mohr–Coulomb failure criterion was applied to the models and it was found that in anisotropic UB, the formation would undergo a shear failure followed by an elastic deformation.

In addition to conventional horizontal stress calculations, I introduced and applied a new approach to estimate the magnitude of anisotropic maximum horizontal principal stress. This method calculates the magnitude of maximum principal horizontal stress was through the measurement of three shear moduli of the formation and using the acousto-elastic parameter. The direction of maximum principal horizontal stress was found to be ~N65E from the direction of the fast shear azimuth (FSA). The direction of the fast shear azimuth was obtained by analyzing existing tensile fractures around the borehole that have caused shear anisotropy more than 20 %.

Biot's coefficient is an important parameter for poro-elastic geomechanical modeling and stress magnitude determination which should be measured with high precision. Elastic parameters such as Poisson's ratio, Young's modulus, and shear modulus in vertical and horizontal directions are better to be measured on preserved samples to better represent the formation mechanical properties. It is also important to understand how anisotropy parameters can be a function the TOC and kerogen content of the shale; Using X-ray diffraction (XRD) analysis can expand our knowledge regarding the chemical and elemental composition of the Bakken Formation and how they affect the anisotropy of the samples. Finally, it would be highly beneficial to run geomechanical models under different failure criterions rather than Mohr–Coulomb and compare the results in the lab.

References

1. EIA. (2010). Shale gas plays/lower 48 states.
2. EIA. (2011). Review of energy recourses: U.S. shale gas and shale oil plays.
3. Zimmerman, R. W., Jeager, J., & Cook, N. (2007). *Fundamentals of rock mechanics* (4th ed.). Malden, MA: Blackwell Publishing.
4. Tsvankin, I. (2005). *Seismic signatures and analysis of reflection data in anisotropic media* (2nd ed.). New York: Elsevier Science.
5. Pistre, V., Kinoshita, T., Endo, T., Schilling, K., Pabon, J., Sinha, B., Plona, T., Ikegamiand, T., & Johnson, D. (2005). A new modular wireline logging sonic tool for measurement of 3D (Azimuthal, radial and Axial) formation acoustic properties, In *Proceedings of SPWLA 46th Annual Logging Symposium*, New Orleans, June 26–29.
6. Winterstein, D. F. (1990). Velocity anisotropy terminology for geophysicists. *Geophysics, 55*, 1070–1088.
7. Tsvankin, I. (1997). Reflection move-out and parameter estimation for horizontal transverse isotropy. *Geophysics, 62*, 614–629.
8. Armstrong, P., Ireson, D., Chmela, B., Dodds, K., Esmersoy, C., Miller, D., et al. (1994). The promise of elastic anisotropy. *Oilfield Review, 6*(4), 36–47.
9. Hornby, B. E. (1994). The elastic properties of shales. PhD thesis, University of Cambridge.
10. Sayers, C. (2005). Seismic anisotropy of shales. *Geophysics, 64*, 93–98.
11. Sayers, C. (1994). The elastic anisotropy of shales. *Journal of Geophysical Research, 99*, 767–774. Solid Earth.
12. Vernik, L., & Liu, X. (1997). Velocity anisotropy in shales. A petrophysical study. *Geophysics, 62*, 521–532.
13. Walsh, J., Sinha, B. K., & Donald, A. (2006). Formation anisotropy parameters using borehole sonic data. *SPWLA 74th Annual Logging Symposium*. June 4–7.
14. Higgings, S., Goodwin, S., Donald, Q., Donald, A., Bratton, T., & Tracy, G. (2008). Anisotropic stress models improve completion design in the Baxter shale. In *Proceedings of SPE ATCE*, Denver, September 21–24, SPE 115736.
15. Nye, J. F. (1985). *Physical properties of crystals*. Oxford: Oxford University Press.
16. Hornby, B. E. (1998). Experimental laboratory determination of the dynamic elastic properties of wet, drained shales. *Journal of Geophysical Research, 103*(B12)
17. Johnston, J. E., & Christensen, N. I. (1995). Seismic anisotropy of shales. *Journal of Geophysical Research, 100*, 5991–6003.
18. Vernik, L., & Nur, A. (1992). Petrophysical classification of siliciclastics for lithology and porosity prediction from seismic velocities. *Bulletin of the American Association of Petroleum Geologists, 76*, 1295–1309.
19. Vernik, L., & Nur, A. (1990). Ultrasonic velocity and anisotropy of hydrocarbon source rocks. *Geophysics, 57*, 727–735.
20. Prasad, M., & Mukerji, T. (2003). Analysis of microstructural textures and wave propagation characteristics in shales. *73th Annual International Meeting*, SEG, Expanded Abstracts, (pp. 1648–1651).
21. Mukerji, T., & Prasad, M. (2004). Analysis of microstructural textures and wave propagation characteristics in shales. Retrieved February 9, 2010. http://www.osti.gov/bridge/servlets/purl/89053-yafcdC/890503.pdf.
22. Mukerji, T., & Prasad, M. (2007). Image processing of acoustic microscopy data to estimate textural scales and anisotropy in shales. *Acoustical Imaging, 28*, 21–29.
23. Arroyo Franco, J. L., Mercado Ortiz, M. A., De, G. S., Renlie, L., & Williams, S. (2006). Sonic investigation in and around the borehole. *Oilfield Review, 18*(1), 14–31.
24. Plona, T., Kane, M., Sinha, B., Walsh, J. & Viloria, O. (2000). Using acoustic anisotropy. *SPWLA 41th Annual Logging Symposium*. June 4–7.
25. Plona, T., Sinha, B., Kane, M., Shenoy, R., Bose, S., Walsh, J., Endo, T., Ikegami, T., & Skelton, O. (2002). Mechanical damage detection and anisotropy evaluation using dipole sonic dispersion analysis. *SPWLA 43th Annual Logging Symposium*. June 2–5.

26. Sinha, B., Vissapragada, B., Renlie, L., & Skomedal, E., (2006). Horizontal stress magnitude estimation using the three shear moduli—a Norwegian Sea case study. In *Proceedings of SPE ATCE*, San Antonio, 24–27 September, SPE 103079.
27. Sinha, B., Vissapragada, B., Renlie, L., & Tysse, S. (2006). Radial profiling of the three formation shear moduli and its application to well completions. *Journal of Geophysics, 71*(6), E65–E77.
28. Ostadhassan, M., Zeng Z., & Jabbari, H. Anisotropy analysis in shale by acquiring advanced sonic data-Bakken shale case study. *AAPG 2012 Annual Convention & Exhibition*, Long Beach, CA, USA.
29. Ostadhassan, M., Zeng Z., & Zamiran, S. Geomechanical modeling of an anisotropic formation-Bakken case study. *ARMA* 2012-221, Chicago, USA
30. Duseault, M. B. (1994). Analysis of borehole stability. *Computer Methods and Advances in Geomechanics Belkema* 125–137.
31. Fam, M. A., Dusseault, M. B., & Fooks, J. C. (2003). Drilling in mudrocks: Rock behavior issues. *Journal of Petroleum Science and Engineering, 38*, 155–166.
32. Fjaer, A., Holt, R., Raaen, A., Risnes, R., & Horsud, P. (1992). *Petroleum related rock mechanics*. Elsevier Publishing.
33. McLean, M. R. (1987). Wellbore stability analysis. PhD Thesis, University of London. LTK.
34. Xu, G. (2007). *Wellbore stability in geomechanics*. Ph.D. thesis, University of Nottingham.
35. Maury, V. (1994). Rock failure mechanisms identification: A key for wellbore stability and reservoir behavior problem, SPE 28049 presented at the SPE/ISRM Rock Mechanics in Petroleum Engineering Conference held in Delft, The Netherlands, August 29–31.
36. Li, X., Cui, L., & Rogegiers, J. C. (1998). Thermoporoelastic analyses of inclined borehole. SPE/ISRM, Eurock 98, Norway.
37. Charlez, P. A. (1997). *Rock mechanics. Vol 2: Petroleum applications*. Editions Technip.
38. Van Oort, E. (2003). On the physical and chemical stability of shales. *Journal of Petroleum Science & Engineering, 38*, 213–235.
39. Hale, A. H., Mody, F. K., & Salisbury, D. P. (1992). Experimental investigation of the influence of chemical potential on wellbore stability. IADC/SPE Paper 23885, presented at the 1992 IADC/SPE Drilling Conference in New Orleans, Louisiana, February 18–21.
40. Hubbert, M. K., & Willis, D. G. (1957). Mechanics of hydraulic fracturing. *Transactions of American Institute of Mining Engineering, 210*, 153–168.
41. Fairhurst, C. (1968). *Methods of determining in situ rock stress at great depth*. TR1-68 Missouri River Div. Corps of Engineering.
42. Bradley, W. B. (1979). Failure of inclined boreholes. *Journal of Energy Resources Technology, 101*, 232–239.
43. Santareli, F. J. & Brown, E. T. (1987). Performance of deep boreholes in rock with a confining pressure dependent elastic modulus. *Proceedings of 60th International Society of Rock Mechanics*. (Vol. 2. pp 1217–1222). Rotterdam: Balkema.
44. Aadnoy, B. S., & Chenevert, M. E. (1987). Stability of highly inclined boreholes. *SPE Drilling Engineering, 12*, 264–374.
45. Ong, S. H., & Roegiers, J. C. (1993). Influence of anisotropies in borehole stability. *Journal of Rock Mechanics and Mining Science & Geomechanics Abstracts, 30*(7), 1069–1075.
46. Roegiers, J. C. & Detournay, E. (1998). Considerations on failures initiation in inclined boreholes. *Proceedings of 29th U. S. Rock Symposium*. Rotterdam: Balkema.
47. Mody, F. K., & Hale, A. H. (1993). A borehole stability model to coupled the mechanics and chemistry of drilling fluid shale interaction. *10th SPE Drilling Conference*. Amsterdam.
48. Sherwood, J. D., & Bailey, L. (1994). Swelling of shale around a cylindrical wellbore. *Proceedings of the Royal Society of London*. 161–184.
49. Biot, M. A., & Willis, D. G. (1957). The elastic coefficient of the theory of consolidation. *Journal of Applied Mechanics, 24*, 594–601.
50. Gnirk, P. F. (1972). The mechanical behavior of the uncased wellbores situated in elastic/plastic media under hydrostatic stress. *SPEJ* 45–59.

51. Papamichos, E., & Vardoulakis, I. (1995). Shear band formation in sand according to non-coaxial plasticity model. *Geotechnique, 45*(5), 649–661.
52. Yu, H. S. (2000). *Cavity expansion methods in geomechanics*. Kluwer Academic Publishers.
53. Ostadhassan, M. (2013). *Geomechanics and elastic anisotropy of the Bakken Formation, Williston Basin*. Ph.D. thesis, University of North Dakota.
54. Jaeger, J. C., & Cook, N. W. G. (1979). *Fundamentals of rock mechanics* (Vol. 3). New York: Chapman & Hall.
55. Plumb, R. A., Edwards, S., Pidcock, G., Lee, D., & Stace, B. (2000). The mechanical earth model and its application to high risk well construction projects. In *Proceedings of IADC/SPE Drilling Conference*, New Orleans, 23–25 February, IADC/SPE 59128.
56. Plumb, R.A., Hooyman, P., Vineengen, D., Dutta, N., Ritchie, G., & Bennaceur, K. (2004). A new geomechanics process reduces operational risks from exploration to production. In *Proceedings of the NARMS*, Houston, June 5–9, ARMA/NARMS 04-616.
57. Sayers, C., Russel, C., Pelorosso, M., Adachi, J., Pastor, J., Singh, V., Tagbor, K. & Hooyman, P. (2009). Determination of rock strength using advanced sonic log interpretation techniques. In *Proceedings of the SPE ATCE*, New Orleans, 4–7 October, SPE 124161.
58. Sayers, C., Kisra, S., Tagbor, K., Taleghani, A. D. & Adachi, J. (2007). Calibrating the mechanical properties and in-situ stresses using acoustic radial profiles. In *Proceedings of the SPE ATCE*, Anaheim, November 11–14, SPE 110089.
59. Ostadhassan, M., Benson, S., & Zamiran, S. Stress analysis and wellbore stability in unconventional reservoirs. *ARMA* 2013-150, San Francisco, USA
60. Thiercelin, M. J., & Plumb, R. A. (1994). Core based predictions of lithologic stress contrasts in east Texas formations. *Journal of SPE Formation Evaluation, 9*(14), 251–258.
61. Gutierz, M., Braunsdor, N., & Couzens, B. (2006). Calibration and ranking of pore pressure prediction models. *TLE* 1516–1523.
62. Sayers, C. (2006). An introduction to velocity-based pore pressure estimation. *TLE*, 1496–1500.
63. Eaton, B. (1972). Graphical method predicts geopressures world wide. *World Oil, 182*, 51–56.
64. Mouchet, J., & Mitchell, A. (1989). Abnormal pressure while drilling. *Elf Acquitane Manuels Techniques 2*, Boussens, France.
65. Ruth, P., & Hillis, R. (2000). Estimating pore pressure in the Cooper Basin, South Australia: Sonic log method in an uplifted basin. *Journal of Exploration Geophysics, 31*, 441–447.
66. Waters, G., Lewis, R., & Bently, D. (2011). The effect of mechanical properties anisotropy in the generation of hydraulic fractures in organic shales. In *Proceedings of SPE ATCE*. Denver, 30 Oct–2 Nov, SPE 146776.
67. Amadei, B., Savage, W., & Swolfs, H. (1987). Gravitational stresses in anisotropic rock masses. *Journal of Rock Mechanics and Mining Science & Geomechanics Abstracts, 24*(1), 5–14.
68. Mavko, G., Mukerji, T., & Dvorkin, J. (1998). *The rock physics handbook, tools for seismic analysis in porous media*. New York: Cambridge University Press. 329 p.
69. Sinha, B., Vissapragada, B., Wendt, A., Kongslien, M., Eser, H., Skomedal, E., Renile, L., & Pedersen, E. (2007). Estimation of formation stresses using radial variation of three shear moduli- a case study from a high-pressure, high-temperature reservoir in Norwegian continental shelf. In *Proceedings of SPE ATCE*, Anaheim, November 11–14, SPE 109842.
70. Bowers, G. (1995). Pore pressure estimation from velocity data: Accounting for overpressure mechanisms besides undercompaction. *Journal of SPE Drilling & Completion, 10*(2), 89–95.
71. Sinha, B. K., Wang, J., Kisra, S., Li, J., Pistre, V., Bratton, T., Sanders, M. & Jun, C. (2008). Estimation of borehole stresses using sonic data. *49th Annual Logging Symposium*, May 25–28, Austin, TX.
72. Chang, C., Zoback, M., & Khaksar, A. (2006). Empirical relations between rock strength and physical properties in sedimentary rocks. *Journal of Petroleum Science and Engineering, 51*, 223–237.

Chapter 6
Nano-Scale Characterization of Organic-Rich Shale via Indentation Methods

Ange-Therese Akono and Pooyan Kabir

Abstract Gas shale or organic-rich shale is a porous multi-scale material that consists essentially of clay, silt inclusions, air voids, and kerogen, which is gaseous organic matter. Assessing the mechanical behavior of gas shale across several length scales is a challenging task due to the complex nature of the material. Therefore, the aim of this investigation is to introduce a novel framework based on nano-mechanics to characterize the elastic and plastic properties of gas shale using advanced techniques such as scanning electron microscopy (SEM), statistical nano-indentation, and micromechanical modeling. An indentation consists in pressing a diamond stylus against a soft material and measuring both the Young's modulus and hardness from the force and penetration depth measurements. Meanwhile, the grid indentation technique consists in carrying out a large array of indentation tests and applying statistical analysis so as to represent the overall behavior as the convolute response of several individual mechanical phases. The specimens analyzed in this study were extracted from major gas shale plays in the USA—Antrim shale from the Michigan Basin in Michigan State and Barnett shale from the Bend Arch-Fort Worth Basin in Texas—and in France—Toarcian shale from the Paris Basin. SEM reveals a heterogeneous granular microstructure with the grain size ranging from 30 to 100 μm; meanwhile, statistical indentation enables to identify the basic micro-constituents. Finally, micromechanics theory makes it possible to bridge the nanometer and macroscopic length scales. The field of applications is vast including major energy-related schemes such as hydrocarbon

The original version of this chapter was revised. An erratum to this chapter can be found at DOI 10.1007/978-3-319-40124-9_17

A.-T. Akono (✉)
Department of Civil and Environmental Engineering, University of Illinois at Urbana-Champaign, Urbana, IL, USA

Department of Mechanical Science and Engineering, University of Illinois at Urbana-Champaign, Urbana, IL, USA
e-mail: aakono@illinois.edu

P. Kabir
Department of Civil and Environmental Engineering, University of Illinois at Urbana-Champaign, Urbana, IL, USA

C. Jin, G. Cusatis (eds.), *New Frontiers in Oil and Gas Exploration*,
DOI 10.1007/978-3-319-40124-9_6

recovery for oil and gas wells, carbon dioxide geological sequestration, or nuclear waste store in depleted wells.

6.1 Introduction

Shale gas has radically transformed the energetic outlook of the USA. In 2011, shale gas shale represented 34 % of the domestic natural gas production compared to 1 % in 2000 [1]. Shale gas is expected to rise to 67 % of the overall production by 2035, generating lower natural gas prices [2]. Furthermore, the development of shale gas resources will also generate significant economic wealth as well as over 1.6 million jobs over the next 20 years [3]. Organic-rich shale, called black shale or unconventional shale, is a low permeability source rock in comparison to conventional reservoir source rock which exhibits a higher porosity and a higher permeability. The intrinsically porous and heterogeneous nature of the rock results in a challenge and makes it difficult to generate a well-connected network of fracture surfaces so as to be able to extract hydrocarbon.

On the other hand, natural gas harvesting from source rock is a complex operation that spans multiple length scales. The reservoir should be drilled both vertically and horizontally to be able to access an acceptable area for hydraulic fracturing. The directional drilling happens at depths of 5000–20,000 feet below the surface [4]. Next, lateral extensions ranging from 1000 to 10,000 feet are drilled horizontally so as to access a wider portion of the subsurface [5]. The basics of the hydraulic fracturing consists of injecting a mix of water, sand and chemical at a high pressure to propagate cracks in the source rock to extract the hydrocarbons. In order to optimize hydraulic fracturing schemes in unconventional reservoirs, it is imperative to get a fundamental understanding of the mechanical behavior of gas shale at both the nano/micrometer as well as the macroscopic length scales.

6.2 Multi-scale Thought Model for Shale

Shale is a highly complex and heterogeneous material with up to seven levels of hierarchy [6, 7] as illustrated on Fig. 6.1. At the macroscopic scale, levels 0–2, and mesoscopic levels, scales 3–4, it exists as a layered fabric resulting from the geological process of sedimentation. Conventional rock mechanical characterization using laboratory experiments has focused on scales 3–4 so far. Very recently, nano- and micro-indentation techniques have made it possible to probe the mechanical properties at the microscopic and nanometer length scales, respectively levels 5 and 6. At the microscopic level, scale 5, the microstructure is that of a porous clay composite with organic matter and quartz and feldspar grains. Advanced observational methods such as scanning electron microscopy or environmental

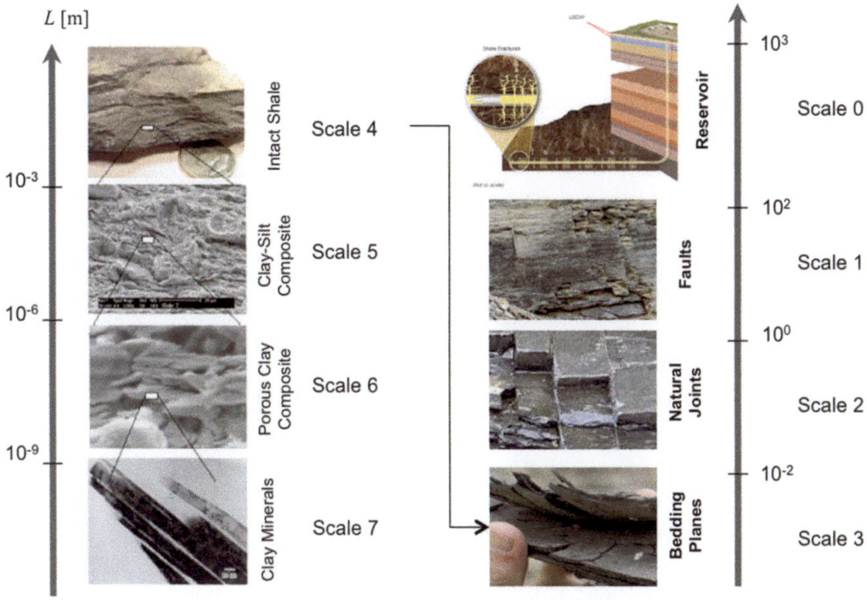

Fig. 6.1 Multi-scale structure thought model for unconventional shale spanning four levels, macroscopic (scale 0–2), mesoscopic (scale 3–4), microscopic level (scale 5–6), and nanometer level (scale 7). Source [7, 13]

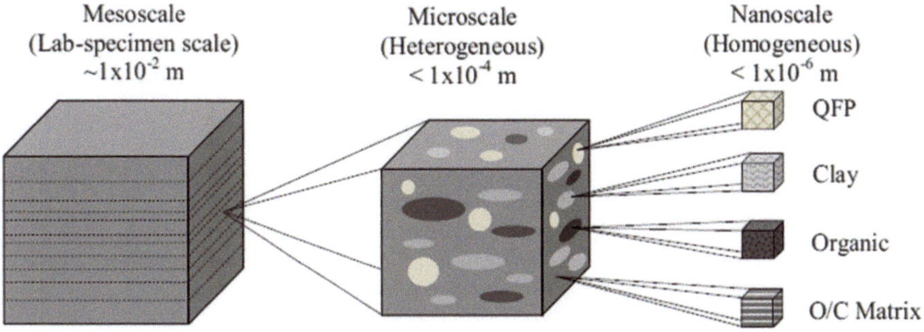

Fig. 6.2 Alternative thought model bridging scales 4–7. Courtesy of Bennet et al. [9]

scanning electron microscopy are used to study the structure at this scale. At level 6 it exists as a porous clay/organic matter composite. Finally, the nanometer length level, scale 7, is the fundamental building unit consisting of elementary clay particles bonded to kerogen molecules. This scale is the smallest scale present in the intrinsic nature of the material. Electron transmission microscopy should be used to reach this level of resolution [8].

Figure 6.2 illustrates an alternative thought model employed to connect the scales 7 to 4 and that was introduced by Bennett et al. [9]. This thought model postulates the existence of four phases at the fundamental scale: quartz, feldspar, organic matter, and clay. In particular, the microscopic examination of several specimens revealed that clay particles could be visualized using high-resolution SEM [10]. It was also found that the mineral particles and pockets of organic did not form a supporting skeleton with inter-granular contacts, which led to the introduction of the organic/clay micro-constituent [11, 12]. Although this model does not account for anisotropy, it will serve as a stepping stone to apply micromechanics in the later part of this chapter.

6.3 Experimental Procedure

6.3.1 Materials

Our research objective is to craft a novel characterization procedure anchored in nano-mechanics principles. We then implement our framework to gas shale specimens so as to obtain a fundamental understanding of the interplay between composition, microstructure and performance in gas shale. The specimens come from three gas shale plays: Antrim, Barnett, and Toarcian. Both Antrim shale and Barnett shale are major gas shale plays, respectively, in the Michigan state and in Texas in the USA, whereas the Toarcian shale is from the Paris Basin in France. Chemical characterization carried out using powder X-ray diffraction analysis and Leco TOC and rock evaluation revealed the presence of organic matter, clay minerals, and inorganic phases. Whereas Barnett and Antrim exhibit a high total organic content (TOC), respectively, 12.2 and 9.6 %, the TOC for Toarcian shale is only 1 %. Meanwhile for Toarcian shale, the volume fraction of quartz calculated using powder X-ray diffraction was found to be 78 % and the fraction of illite/smectite, 10.2 %. Other trace elements include chlorite, albite, microcline, pyrite, and kaolinite. The first step before applying small scale characterization method is to develop an adequate material preparation routine. This is the focus of the next paragraph.

6.3.2 Grinding and Polishing

A proper grinding and polishing procedure is the stepping stone to obtain flat and perfectly polished surfaces for microscopic observation as well as subsequent mechanical testing. Prior to grinding and polishing, a diamond saw, IsoMet® 5000 Linear Precision Saw (Buehler, Lake Bluff, IL), was used to machine flat cylindrical specimens. Various techniques were then designed and tested on both Toarcian B and Toarcian T specimens.

Protocol No. 1 consists in mounting the specimen onto aluminum disk followed by manual grinding and semi-automated wet polishing using an Ecomet 300/Automet 250 grinder/polisher. Manual grinding is carried out using a bubble

Fig. 6.3 (*Left*) Optical microscopy at 5× magnification of Toarcian B specimen with three steps of 8 min grinding using the grit sizes 240, 400, 600 and polishing in four steps with oil suspensions of 9, 3, 1, 0.25 μm for 40 min each step. (*Right*) SEM image under low vacuum with JOEL 6060LV

level on top of the aluminum disk so as to maintain a flat top surface. Three grit sizes of Carbimet abrasive paper are employed: consecutively 240, 400, and 600 for 8 min each and using ethanol as a lubricant. Between each grit size, the specimen is rinsed in an ultrasonic bath for two minutes to prevent cross-contamination. For polishing, Trident cloths are used along with diamond oil suspension, MetaDi® (Buehler, Lake Bluff, IL). The following solutions are used: 9, 3, 1, and 0.25 μm consecutively for 40 min each. This procedure works well on Toarcian B specimens and yields a mirror-like surface. Optical microscopy and scanning electron microscopy (SEM) images of the final result are shown in Fig. 6.3. The SEM image was obtained using a JOEL 6060 LV at low vacuum with uncoated specimens. The lubrication and ultrasonic baths are done at ethanol to prevent hydration and intermixing of water with the micro-constituents of organic-rich shale. Finally, after completing the protocol, the samples are air-dried and stored in a vacuum desiccator.

Protocol No. 2 was designed to correct the observed tilt of the final specimens. In fact, a major issue detected was the presence of a steep slope on the final polished surface, which results from the current design of the AutoMet 250 power head. To circumvent this problem, we embedded the specimens in epoxy resin, EpoThin 2. The specimens were wrapped in plastic before impregnation to prevent epoxy from reaching the pores. In other words, as shown in Fig. 6.4, the epoxy works as a support media for the samples to redistribute the applied load during grinding and polishing and thus prevent tilting.

We used EpoThin 2 which requires 9 h of curing at room temperature. The grinding and polishing steps described above were repeated on epoxy-embedded specimens. Grinding was performed using consecutively 240, 400 and 600 grit size for 8 min each. Meanwhile, ultrasonic baths were conducted in between each grit size for two minutes. Afterward, polishing ensued with consecutively the 9, 3, 1, and 0.25 μm diamond suspensions for 40 min each, the polishing cloth being the TriDent. The result can be seen in the optical microscopy image displayed in

Fig. 6.4 (*Left*) Epoxy embedding of Toarcian B specimens while grinding with three steps of 240, 400, 600 grit size each for 8 min, and polishing in four steps with oil suspension of 9, 3, 1, 0.25 μm for 40 min for each step. (*Right*) Optical microscopy of the specimen with 20× magnification

Fig. 6.4. The result for Toarcian B specimen was satisfactory but not as good as the previous method using an aluminum disk and no epoxy. The method did not yield a good result for Toarcian T specimens. Overall, casting the specimens in epoxy improved the flatness of the final surface. This approach also removes the need for manual grinding and as a result the whole process can be done in a semi-automated fashion. However, it leads to a poorer surface finish and significant cross-contamination of the specimens.

Protocol No. 3 was tested that consisted of manual grinding and manual polishing. It was found that the amount of relief could be significantly reduced by shortening the polishing time and using a different polishing cloth. Thus manual grinding was carried out using ethanol as a lubricant and with three grit sizes for 2 min each: 240, 400 and 600 consecutively. Afterward, polishing is carried out manually on a glass plate using a diamond oil suspension for 2 min per diamond particle size: 9, 3, 1, and 0.25 μm. The result of optical microscopy is shown in Fig. 6.5. Another variation of the same protocol consists in using FiberMet cloths with embedded abrasives, during the manual polishing phase. In order to ensure a flat polished surface, the polishing cloth was placed on a rigorously flat glass surface.

Protocol No. 3 was conducted on a virgin Toarcian T specimen that had not been mounted on any disc nor set in epoxy resin. The Toarcian T specimen cracked considerably during the semi-automated grinding intervals, crumbling apart into three pieces after the first polishing interval. The polishing was completed using one of the broken pieces but its top surface looked identical to the surface of the Toarcian T specimen that had undergone the long polishing times. In summary, Toarcian T specimens proved to be very difficult to polish as neither shorter polishing times nor different polishing media had led to a better surface finish.

A breakthrough was made by introducing two additional grinding steps: using the grit size 800 and 1200, leading to protocol No. 4. A considerable improvement

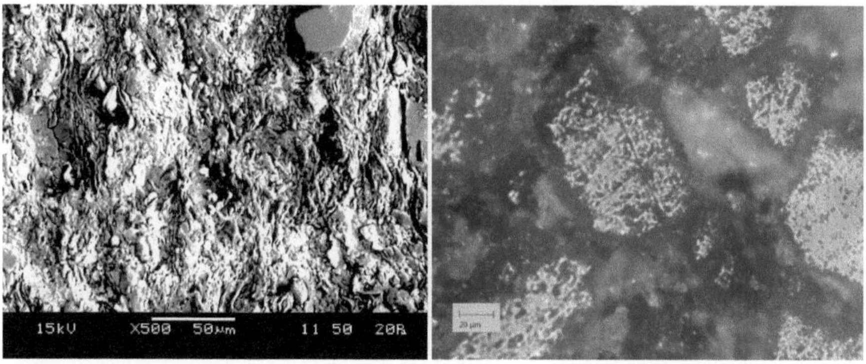

Fig. 6.5 (*Left*) SEM image of shale Toarcian T specimen. (*Right*) Optical microscopy of shale Toarcian T1 specimen with grinding with three steps of 240, 400, 600 grit size each for 2 min, and polishing four steps with oil suspension of 9, 3, 1, 0.25 μm for 2 min each step

in smoothness was finally achieved when a dry grinding and polishing procedure was tested, in which no ethanol or diamond suspension was used. In the first variation, the dry grinding was completed automatically with the base and head set at low revolution speeds of 200 rpm and 40 rpm, respectively. A minimal force of 1 lb. was applied to the specimen to prevent cracks from propagating while still providing enough force to hold it securely against the sanding papers. In the most recent variations, a dry, semi-automated grinding and polishing procedure and a dry, manual grinding and polishing procedures, both with time intervals varying from 2 to 5 min and minimal force, have been tried. Instead of moving to polishing with diamond abrasives after the 600 grit size paper, two additional grinding steps using 800 grit size and 1200 grit size paper were added. The resulting surface of this procedure was reflective under the light and no scratches were present when observed under the microscope.

As demonstrated through the use of up to four grinding and polishing protocols, there is not a unique protocol that will be applicable to all gas shale specimens in order to yield a flat and mirror-like surface. Instead, the protocol must be tailored to the specimen and typically results from several trials. Yet some principles remain. A longer polishing time will result in a better and smother finish. Moreover, cleaning the surface more often either with an ultrasonic bath and proper cleaning solution or using compressed air will reduce cross contamination.

Scanning Electron Microscopy was utilized to investigate the microstructure of organic-rich shale. Protocol No. 1 was applied to Toarcian B shale yielding a low level of magnification, 100×, as shown in Fig. 6.3. Meanwhile, protocol No. 4 was applied to Toarcian B shale, Antrim shale and Barnett shale as displayed in Figs. 6.6 and 6.7. A granular microstructure is revealed in Fig. 6.3 with the grain size ranging from 30 to 100 μm. In all SEM images, the composition is very heterogeneous where the quartz/feldspar inclusions are in white, the porous clay composite is in dark gray and the organic matter as well as the pores are in black. Finally, in Fig. 6.6 at a high level of magnification, 1700×, we observe a flaky

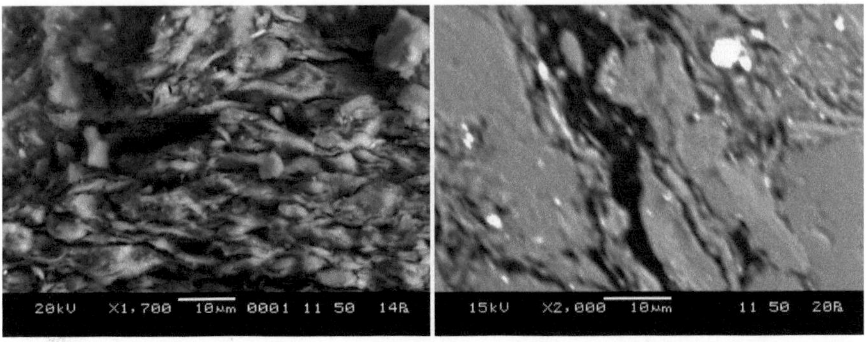

Fig. 6.6 Scanning electron microscopy of left image, Toarcian B specimen 1700× magnification, right image Antrim specimen 2000× magnification

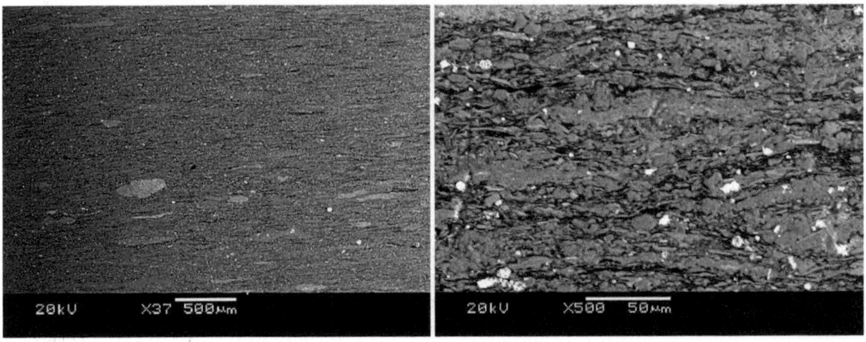

Fig. 6.7 The scanning electron microscopy of a Barnett shale specimen at different magnifications, 37×, 500×, respectively

microstructure which is characteristic of clay-rich materials. At the highest magnification reached, 2000×, the image is slightly blurred. In order to increase the resolution, we recommend coating the specimens using gold or carbon. An alternative method is to use focused ion beam milling.

6.3.3 Roughness Characterization

After grinding and polishing, it is important to assess the roughness of surface so as to quantitatively evaluate the quality of the surface polish. Moreover, the surface roughness can be used to design the nano-indentation testing by selecting the appropriate penetration depth and load levels. In a study by Miller et al. [13] the effect of different polishing times for surface roughness and criteria for cement paste nano-indentation was investigated. The criteria established in their study for cement paste was that the indentation depth should be higher than five times the

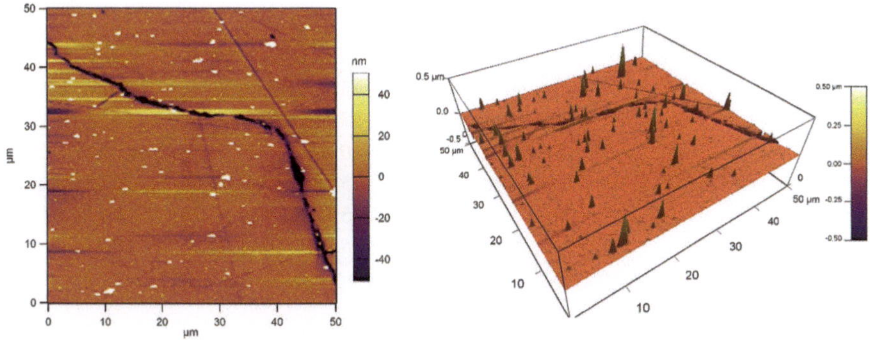

Fig. 6.8 (*Left*) Atomic force microscopy of polished Toarcian B specimen. (*Right*) 3-D representation topography of the Toarcian B specimen surface

root-mean-squared average roughness of the surface. The root-mean-squared average roughness of the specimens can be characterized by performing atomic force microscopy (AFM) on the samples. The topography images were obtained using the tapping mode imaging, non-contact mode, to minimize the effect of the imaging on the polished specimen for further investigation.

An AFM image of the surface of Toarcian B specimen is shown in Fig. 6.8, in this figure, the 3-D reconstruction of the surface also can be seen. The topography was recorded on a 50 × 50 μm area on the specimen. Atomic force microscopy topography images were obtained by using an Asylum Research MFP-3D instrument in the Frederick Seitz material research laboratory at University of Illinois at Urbana-Champaign. The average roughness is 30 nm. Therefore, in indentation testing, the normal load must be chosen so that the penetration depth exceeds 150 nm.

6.4 Mechanical Properties

6.4.1 Elastic Properties

A wide array of methods are commonly used to characterize the elastic properties of shale materials. These methods include uniaxial compressive tests, Brazilian tensile tests, and ultra-pulse velocity measurements. Yet, due to the intrinsically multi scale nature of shale, the mechanical behavior needs to be explored even at smaller scales: microscopic and nanometer length-scales. Some conventional methods to measure the elasticity of rocks are the Brazilian tensile tests, uniaxial compressive tests, as well as the ultrasonic pulse velocity (UPV) [14]. UPV testing has been reported as a useful reliable nondestructive tool for assessing the mechanical characteristics of concrete material [15, 16] as well as rocks. Another way to measure the mechanical properties is using a P-wave amplitude to obtain the parameters that fully describe the elastic behavior of shale. This method was used

Table 6.1 Elastic properties of different shale reservoirs

Sample group	*E* (GPa)	*N*
Barnett	33.0 [18]	0.2–0.3 [19]
Antrim	15.17 [20]	0.20 [20]
Marcellus	27.74 [16]	0.33 [21]
Allenwood	30.40 [22]	0.161 [22]
Elimsport	15.99 [22]	0.283 [22]
Mancos	15.17 [23]	0.35 [23]
New Albany	15.17–31.03 [24]	0.2 [24]

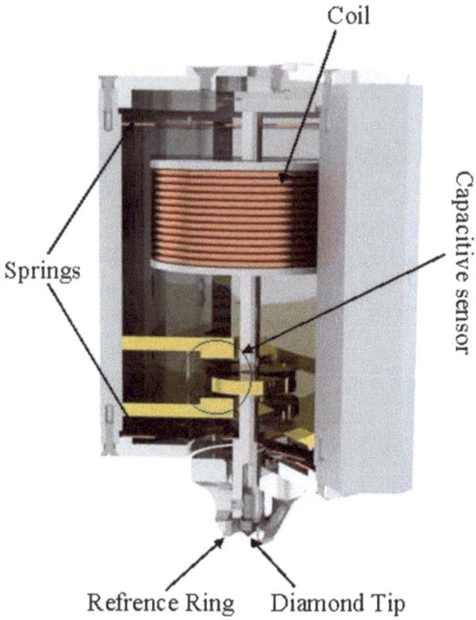

Fig. 6.9 A schematic of the nano-indentation head. Source [26]

to calculate the elastic properties of Marcellus shale [17]. Nano-indentation techniques were used to explore the mechanical properties of multi-scale organic-rich shale at the micrometer and nanometer scales [6, 18]. The elastic modulus and Poisson's ratio estimates for the rock source from different reservoirs throughout the USA are shown in Table 6.1.

6.4.2 Indentation Equipment

Nano-indentation experiments were performed on an Anton Paar TriTech multi-scale testing platform (Anton Paar, Ashland, VA). A schematic of the key components of the nano-indentation tester is provided in Fig. 6.9. The indentation load is applied electromagnetically by passing a current through a coil mechanically connected to the tip. The indentation depth with respect to the thermal calibrating ring is measured via the change in voltage of a parallel plate capacitor. The testing platform also houses an optical microscope that is used to select the location of the test and visualize the residual indent. Both the sample and the indenter are housed in an acoustic-emission enclosure to minimize the noise as well as thermal drift. The apparatus is extremely precise with a resolution on the prescribed load of 20 nN and a resolution on the penetration depth of 0.04 nm.

We applied the Oliver & Pharr method in indentation testing. Herein, the projected area of contact used in Eqs. (6.2) and (6.3) is a crucial parameter to ensure that subsequent data analysis provides meaningful results. As seen in these equations, the value of the projected area is crucial to calculate the indentation modulus and hardness. For a perfectly sharp Berkovich tip, the area function is well defined. In practice, however, the indenter probe exhibits some bluntness or imperfections that can significantly alter the area function. Therefore, it is imperative to accurately calibrate the area function. The calibration of the area function is typically performed indirectly by indentation on a reference material, usually fused silica. Using an inverse approach based on Eq. (6.3) and assuming a functional shape of A_c given by Eq. (6.1), one can compute the coefficients C_i. C_1 is representative of the ideal probe. Meanwhile, the remaining coefficients capture the bluntness of the tip:

$$A_c(h_c) = C_1 {h_c}^2 + C_2 h_c + C_3 h_c^{\frac{1}{2}} + C_4 h_c^{\frac{1}{4}} + \dots \tag{6.1}$$

6.4.3 Indentation Experiment

The indentation technique consists of establishing contact between an indenter of known geometry, mechanical property and the indented material for which the mechanical properties must be determined. In turn, the response of the material upon unloading provides access to the elastic properties of the indented material. A continuous load displacement measurement are carried out to extract the elastic material properties. Typically, the extraction of mechanical properties is achieved by applying a continuum scale mechanical model to derive the indentation modulus and indentation hardness. Equation (6.2) shows the definition of the hardness of a material with respect to the load applied and area affected by the load. The definition of material indentation modulus is shown in Eq. (6.3):

$$H \stackrel{\text{def}}{=} \frac{P}{A_c} \tag{6.2}$$

$$M \stackrel{\text{def}}{=} \frac{\sqrt{\pi}}{2} \frac{S}{\sqrt{A_c}} \tag{6.3}$$

The indentation modulus is related to the elastic properties of the material and indenter with Eq. (6.4), where E_i, υ_i are elastic modulus and Poisson's ratio of indenter:

$$\frac{1}{M} = \frac{(1 - \upsilon^2)}{E} + \frac{(1 - {\upsilon_i}^2)}{E_i} \tag{6.4}$$

To determine the indentation hardness H and the indentation modulus M, we analyze the load-penetration depth curves, $P - h$ curves. Figure 6.10 shows the method developed by Oliver & Pharr for extracting the information needed to calculate the elastic properties of the tested material based on the experimental data [27].

The area function A_c is determined by the calibration of the tip with a material of known elastic constants (mostly fused silica), which is a function of h_c. Where h_c is the vertical distance along which contact is made, which is shown in Fig. 6.10. Figure 6.11 illustrates a typical indentation response measured for the shale

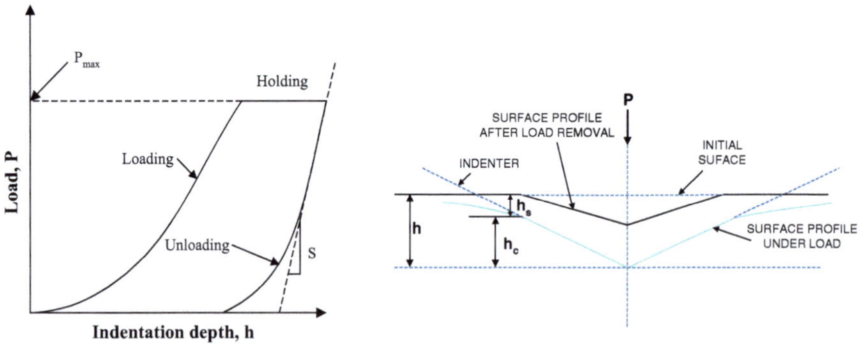

Fig. 6.10 A schematic representation of load versus indenter displacement showing quantities used in the analysis

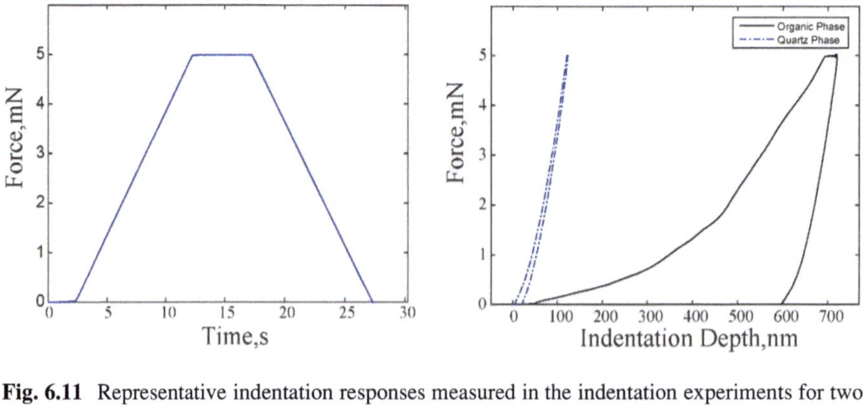

Fig. 6.11 Representative indentation responses measured in the indentation experiments for two different phase with 5 mN load threshold for both experiments

material for the loading pattern shown herein. The experiment consists of a loading part of 10 s with a 30 mN/min approach speed and 5 s of holding and 10 s of unloading with a 40 mN/min unloading rate. The magnitude of the load was 5mN with maximum indentation depth of 700 and 100 nm for two phases. The indentation modulus ad hardness inferred from these experimental curves are $M = 16.49$ GPa, $H = 0.36\,\text{GPa}$ for organic phase, $M = 105.22\,\text{GPa}, H = 11.99\,\text{GPa}$ for the quartz phase.

6.4.4 *Statistical Nano-Indentation*

The concept of the statistical indentation (grid indentation) was used by Ulm and coworkers [28–30] to address the heterogeneous nature of the microstructure by expanding the use of the classical indentation technique. The grid indentation technique consists in conducting a large grid of indentations over the surface of the heterogeneous medium of interest. Each indentation experiment could be regarded as statistically independent, which makes it feasible to apply statistical techniques. In turn, statistical analysis makes it possible to interpret the indentation results in terms of the mechanical properties of the individual material phases.

The concept of statistical indentation was developed based on the Gedanken experiment. Consider a material composed of two phases with different mechanical properties. The grid indentation principle can be introduced by considering an indentation test on an infinite half-space composed of two different material phases with contrasting properties, as shown in Fig. 6.12. Assume that the characteristic length of the phase are D and the indentation depth is much smaller than the characteristic size of the phases, $h << D$. In these conditions, a single indentation gives access to the material properties of either phase 1 or phase 2. In contrast, consider an indentation performed at a maximum indentation depth that is much larger than the characteristic size of the individual phases, $h >> D$. Such a test measures the response of the composite material, and the properties obtained from such an indentation experiment are representative of the average properties of the composite material. The grid indentation methods requires performing a large array of indentations as displayed in Fig. 6.12. As a result of the scale separability condition, most indentations will capture the properties of either one of the primary phases. For example, in the two-phase material two peaks are present in the frequency diagram, and those mean values represent the mechanical properties of the phases.

Figure 6.13 shows an indentation experiment of shale material with the thermal ring resting on the specimen during an indentation experiment. The right picture in Fig. 6.13 shows an optical microscopy of a grid of indentation performed on the surface of the shale specimen. The optical microscopy images of indentation provide information on the location of the indents on different phases.

The implementation of the deconvolution technique for the indentation analysis begins with the generation of the experimental probability density function (PDF) or cumulative distribution function (CDF) [18, 30, 31]. An assumption should be

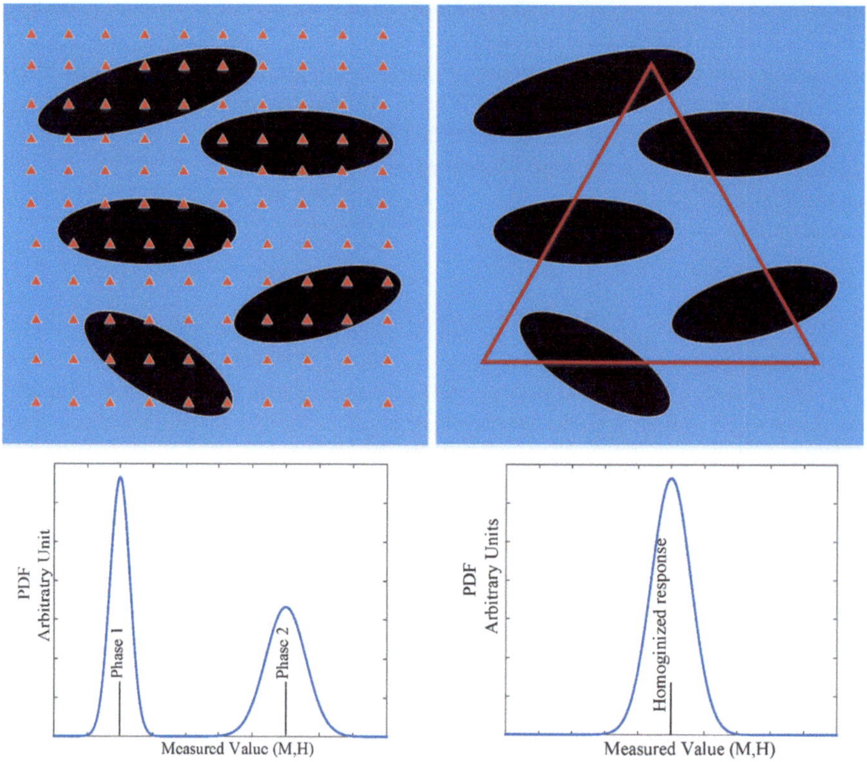

Fig. 6.12 (*Top*) Schematic of statistical indentation analysis performed on a two-phase heterogeneous material. (*Bottom*) Characterization of intrinsic phase properties from indentation at shallow depths, characterization of homogenized properties of two-phase material

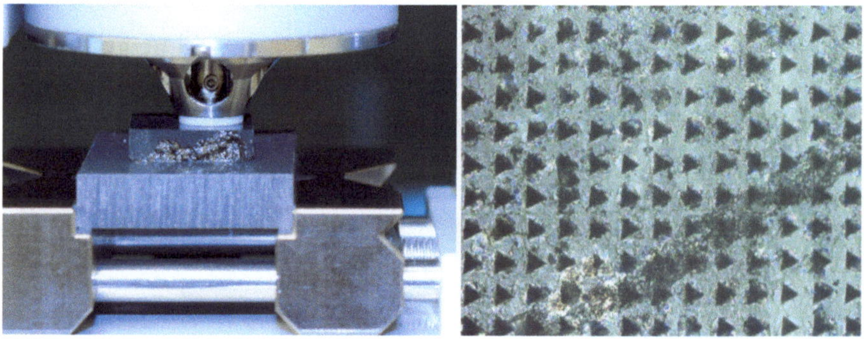

Fig. 6.13 Indentation experiments on gas shale. (*Left*) Digital photography of experimental setup. (*Right*) Optical microscopy image of the residual indentation grid on shale specimen

made on the form and shape of the distributions associated with the properties of each material phase. It is assumed that in statistical modeling of the heterogeneous material, each phase has its own mechanical (indentation) value for that phase. A spread of experimental measurement is expected because of the noise and inhomogeneity nature of phases. The second assumption is that the distribution obeys a Gaussian distribution, which assumes the variability is distributed evenly around the mean value. The application of the grid indentation on the heterogeneous material will always result in some measurements probing the composite response of two or more material phase at some location.

Each indentation test is considered as a single statistical event, with both extracted indentation modulus, hardness. The purpose of deconvolution is to determine the number of mechanical phases as well as their mean mechanical properties. Let N be the number of indentation test performed, $M_i, H_i (i = 1, N)$ are the sorted values of the measured indentation modulus and harness. The N points of the experimental CDF for indentation modulus and hardness shown in Eq. (6.5):

$$\left.\begin{aligned} F_M(M_i) &= \frac{i}{N} - \frac{1}{2N} \\ F_H(H_i) &= \frac{i}{N} - \frac{1}{2N} \end{aligned}\right\} \text{ for } i \in [1, N] \tag{6.5}$$

After constructing the experimental CDFs, we need to construct the model CDFs. Consider the material to be composed of $j = 1, n$ material phases with sufficient contrast in mechanical phase properties. Each phase occupies a surface fraction of f_j, of the indented surface. The distribution of the mechanical properties of each phase is assumed to be approximated by Gaussian distributions, described by the means of μ_j^M, μ_j^H and standard deviations of s_j^M, s_j^H. The CDF for each phase is given by Eq. (6.6):

$$\begin{aligned} F\left(X_i; \mu_j^X, s_j^X\right) &= \frac{1}{s_j^X} \int_{-\infty}^{X_i} \exp\left(\frac{-\left(u - \mu_j^X\right)^2}{2\left(s_j^X\right)^2}\right) du \\ X &= (M, H) \end{aligned} \tag{6.6}$$

The $n \times 5$ unknowns $\left\{f_i, \mu_j^M, s_j^M, \mu_j^H, s_j^H\right\}, j = 1, n$ are determined by minimizing the difference between the experimental and weighted sum of CDFs of different phases as shown in Eq. (6.7):

$$\begin{aligned} &\min \left[\sum_{i=1}^{N} \left(\sum_{j=1}^{n} f_j F(M_i; \mu_j^M, s_j^M) - F_M(M_i) \right)^2 + \sum_{i=1}^{N} \left(f_j F(H_i; \mu_j^H, s_j^H) - F_H(H_i) \right)^2 \right] \\ &\text{s.t.} \sum_{j=1}^{n} f_j = 1 \end{aligned} \tag{6.7}$$

where the constraint of the minimization problem requires that the surface fractions of the different phase sim to one. To ensure that individual phases exhibit sufficient contrast in properties, and avoid neighboring phases to overlap, the problem is additionally constrained with Eq. (6.8):

$$\mu_j^X + s_j^X \le {\mu_{j+1}}^X - {s_{j+1}}^X, X = (M, H) \tag{6.8}$$

The approach of statistical deconvolution works based on minimizing the error between the experimental data (M, H) and the model response. The principle is to compute the total error and find a set of parameters to minimize this error while satisfying other constraints. For a given set of data the error is due to the difference between the values of predicted modulus M_{pred} and hardness H_{pred}, and the experimental values of $M_{\text{exp}}, H_{\text{exp}}$. The error minimization can be carried out based on different definitions of the nominal error: absolute error, relative error, and normalized absolute error.

Absolute error: The absolute error is defined by Eq. (6.9) for each point:

$$S = \left(M_{\text{exp}} - M_{\text{pred}}(\mu, s)\right)^2 + \left(H_{\text{exp}} - H_{\text{pred}}(\mu, s)\right)^2 \tag{6.9}$$

The problem with this error is that it will give more dominance to points with high values compared to points with lower values. Since, in general the indentation modulus is much larger than the hardness (expressed in same units), the absolute error will lead to a good representation only for M and not of H.

Relative error: This error is defined by

$$S = \left(\frac{M_{\text{exp}} - M_{\text{pred}}(\mu, s)}{M_{\text{exp}}}\right)^2 + \left(\frac{H_{\text{exp}} - H_{\text{pred}}(\mu, s)}{H_{\text{exp}}}\right)^2 \tag{6.10}$$

This error has the advantage of not favoring high values of M and H. However, it will give somewhat more importance to points with low values of M, H.

Normalized absolute error: This error uses the same normalization factor except that the denominator is M_0, H_0 for the entire data set:

$$S = \left(\frac{M_{\text{exp}} - M_{\text{pred}}(\mu, s)}{M_0}\right)^2 + \left(\frac{H_{\text{exp}} - H_{\text{pred}}(\mu, s)}{H_0}\right)^2 \tag{6.11}$$

This definition of error helps to get the best minimization value, while fitting both M, H experimental data best.

One option to model the grid indentation is the expectation-maximization algorithm [32–34]. This methodology is well suited for the analysis of grid indentation data. In particular, normal mixture models are used to model multi-variate data based on efficient iterative solutions by maximum likelihood via the expectation–maximization algorithm. The mixture model for grid indentation data begins with considering each indentation event by a realization of a random two dimensional vector. The two dimensions correspond to the indentation modulus and hardness attributes. Mixture models are usually faced with the problem if multiple roots. The application of the maximum likelihood algorithm for the analysis of grid indentation data on shale can be done with the use of EEMIX algorithm developed by Mclachlan et al. [35]. The EMMIX algorithm automatically fits a range of normal mixture distributions with unrestricted variance-covariance matrices. The optimal number of mixture components is determined using the Bayesian information criterion. Ortega [6] has used the EMMIX algorithm to analyze the indentation modulus and hardness obtained from grid indentation experiments on shale specimens.

A second option for statistical analysis of grid indentation is cluster analysis. Cluster analysis is the automated identification of groups of related observations in a data set. The strength of this technique stems from its ability to determine the number of clusters in a data set and the uncertainty of experiments belonging to a cluster based on statistical criteria. The technique has been implemented in the R package Mclust, a contributed package for normal mixture modeling and model-based clustering [36]. The package provides functions for model based approaches assuming different data models and implement maximum likelihood estimation and Bayesian criteria to identify the most likely model and the number of clusters. The initialization of the model is done by hierarchical clustering for various parameterization of the Gaussian model [36–38]. The best model can be identified by fitting model with different parametrization and number of components to the data by maximum likelihood determined by the EM algorithm, while implementing a statistical criterion for model selection. Deirieh [8] used the Mclust package in R software to process the data obtained from wave dispersive spectroscopy for chemical assessment of shale by means of statistical grid spectroscopy on shale specimen.

The deconvolution is performed by a nonlinear least square solver using the computer programming software Matlab. The results of the deconvolution technique are estimates of the mean and standard deviation of indentation modulus and hardness for each mechanical phase. The result of deconvolution for a 20×20 grid indentation of a Toarcian B specimen is shown in Fig. 6.14. The load level in the experiment was 5 mN with loading time of 10 s, 5 s of holding and 10 s of unloading. The 400 data points yield three distinctive phases with different mean values for indentation modulus and hardness. The theoretical and experimental probability density function of both modulus and hardness are shown in Fig. 6.14, with an excellent agreement. To visualize the grid of indentation performed on the specimen a map of indentation hardness or indentation modulus can be made. DeJong et al. [39] used this visualization technique to understand the location of different phases with respect to other phases in cement paste.

The result of the deconvolution algorithm is presented in Table 6.2. The three phases presented in this study can be assigned to different phases by comparing the values of the Young's modulus estimates for the micro-constituents of organic-rich shale (clay, quartz/feldspar and organic matter) that are shown in Table 6.3. For instance, the Young's modulus of Phase 3, 90.60 GPa, is in the range of values expected for quartz. Therefore, it is reasonable to assume that Phase 3 represents quartz. Estimates of the Young's modulus for kerogen/organic matter are in the

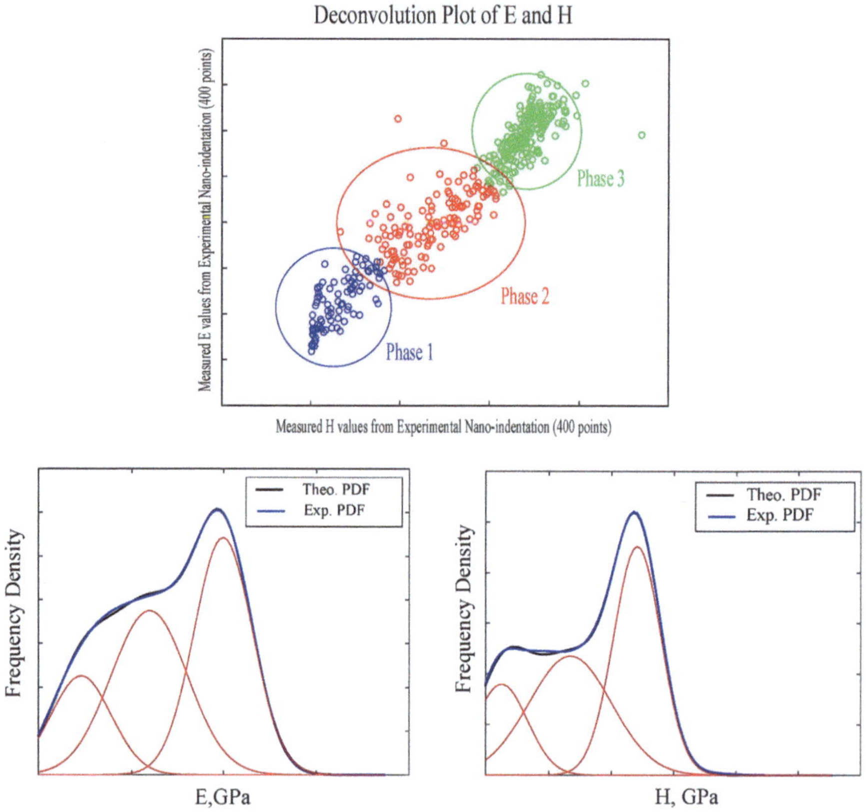

Fig. 6.14 (*Top*) Deconvolution graph of E, H for a Toarcian B specimen. (*Bottom*) PDF graphs of E, H for the three-phase material presented in the phase deconvolution graph

Table 6.2 Mean values and standard deviation for elastic mechanical properties of different phases from grid indentation

Phase	*E* (GPa)	Sd_E (GPa)	*H* (GPa)	Sd_H (GPa)	Volume fraction (percent)
Phase 1	20.43	5.02	2.04	0.21	7
Phase 2	60.68	12.92	4.3	0.44	45
Phase 3	90.60	10.61	9.4	0.29	48

range 7–16 GPa, slightly lower than the value for Phase 1, 20.43 GPa. One explanation is that Phase 1 is a composite made of clay, air voids and organic matter. Finally, the remaining phase can be interpreted as the clay phase in the material, based on the value of the modulus. Thus, statistical nano-indentation enables us to identify the micro-constituents of organic-rich shale. Furthermore, our findings agree with SEM observations and XRD analysis.

The maps created in Fig. 6.15 are the results of deconvolution technique plotted based on their *X* and *Y* positions in the grid. As discussed earlier the three-phase

Table 6.3 Reported elastic stiffness values of shale constituents

Material	E_1, Stiffness (GPa)	Source
Kaolinite	30.3	[45, 46]
Illite	73.9–84.3	[47]
Smectite/illite	51.5	[47]
Montmorillonite	44.7–85.5	[47]
Muscovite	118.1	[48]
Chlorite	82.2–214.0	[47]
Quartz	80–100	[7]
Pyrite	264–330	[49]
Feldspar	0.22	[50]
Organic phase/kerogen	7–16	[51]

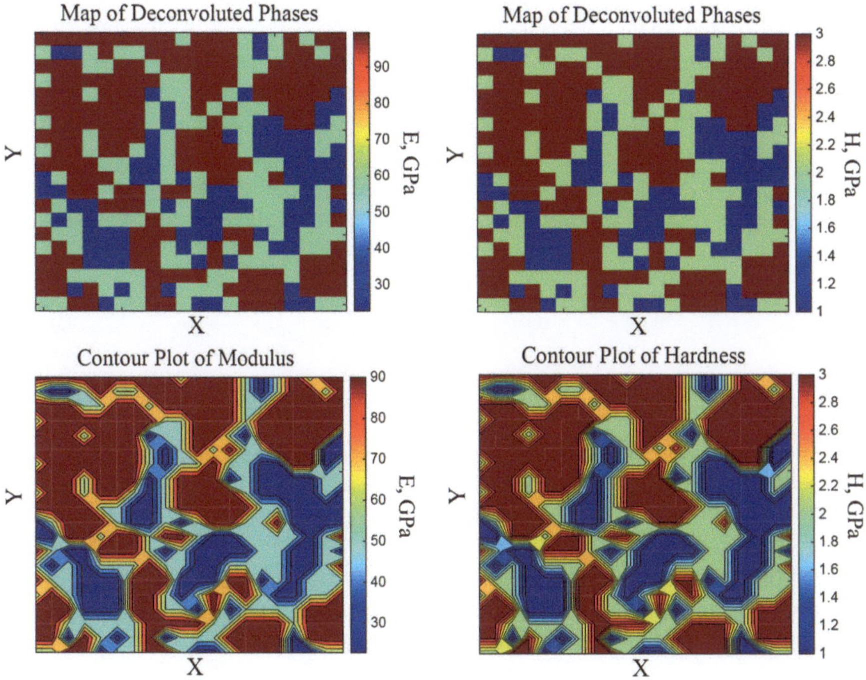

Fig. 6.15 Map of different phases for indentation modulus and hardness, also contour plot of indentation modulus and hardness, data of a 20-by-20 grid indentation experiment for Toarcian B specimen with spacing of 20 μm, the continuous map is drawn by linearly interpolating the discrete data over the whole surface

material can be observed in Fig. 6.15, while each phase corresponding to a distinct mechanical phase. The mechanical map can be overlapped with the microstructural observation obtained from optical microscopy to obtain useful information regarding microstructure. Moreover a continuous map of mechanical properties can be drawn to have a continuous mechanical map as opposed to a dotted map. The continuous map is constructed by averaging neighboring points, either indentation modulus or indentation hardness while the spacing from the points was chosen in the experiment, based on the load level, indentation depth, and characteristic size of different features in material.

6.4.5 Elastic Mechanical Homogenization

Having connected the mechanical response to constituents at the nanometer length-scale, the last step is to bridge scales 4–7 and predict the macroscopic behavior. To this end, we recall the thought model of Bennet et al. displayed in Fig. 6.2. Herein, we apply micromechanics theory to estimate the mechanical properties of the shale material at the macroscopic scale with the use of the information gained from nano-indentation results. Budiansky [47] developed a model to determine the elastic moduli of a composite material made of several constituents which are isotropic and elastic. The model is intended for heterogeneous materials composed of contiguous and spherical grains.

The model is developed by defining the effective shear modulus of the composite material in terms of a uniform macroscopic shear applied to the boundary and the average shear strain of the material as described in Eq. (6.12):

$$G^* = \frac{\tau_0}{\overline{\gamma}} \tag{6.12}$$

Then with the use of the Hill's Lemma [48] the elastic strain energy can be described as Eq. (6.13):

$$U = \frac{1}{2}\int_V \tau_0 \gamma_{xy} dV = \frac{V(\tau_0)^2}{2G^*} \tag{6.13}$$

where V is the volume of the material being considered. However, the equation for the elastic strain energy can be written in terms of individual constituent phases as Eq. (6.14):

$$U = \frac{1}{2}\int_V \frac{\tau_0 \tau_{xy}}{G_N} dV + \frac{1}{2}\int_V \tau_0 \left(\gamma_{xy} - \frac{\tau_{xy}}{G_N}\right) dV = \frac{V(\tau_0)^2}{2}\left[\frac{1}{G_N} + \sum_{i=1}^{n} c_i \left(1 - \frac{G_i}{G_N}\right)\left(\frac{\overline{\gamma}_i}{\tau_0}\right)\right] \tag{6.14}$$

where c_i is defined as the volume fraction of each phase. Solving for the composite shear modulus yields Eq. (6.15):

$$\frac{1}{G^*} = \frac{1}{G_N} + \sum_{i=1}^{n} c_i \left(1 - \frac{G_i}{G_N}\right)\left(\frac{\overline{\gamma}_i}{\tau_0}\right) \tag{6.15}$$

The next step is to assume a strain concentrations around spherical inclusion, which is obtained from Eshelby's inclusion derivations [49]. The final implicit expression of effective shear modulus can be obtained by use of Eq. (6.16):

$$\left[\begin{array}{l} \dfrac{1}{G^*} = \dfrac{1}{G_N} + \displaystyle\sum_{i=1}^{N-1} \left(1 - \dfrac{G_i}{G_N}\right) \dfrac{c_i}{G^* + \beta^*(G_i - G^*)} \\ \beta^* = \dfrac{2(4 - 5\upsilon^*)}{15(1 - \upsilon^*)} \end{array}\right. \tag{6.16}$$

where υ^* is the composite Poisson's ratio. Following the same derivation the bulk modulus is derived and presented in Eq. (6.17):

$$\left[\begin{array}{l} \dfrac{1}{K^*} = \dfrac{1}{K_N} + \displaystyle\sum_{i=1}^{N-1} \left(1 - \dfrac{K_i}{K_N}\right) \dfrac{c_i}{K^* + \alpha^*(K_i - K^*)} \\ \alpha^* = \dfrac{1 + \upsilon^*}{3(1 - \upsilon^*)} \end{array}\right. \tag{6.17}$$

The composite Poisson's ratio, υ^*, can be found from Eq. (6.18) using elastic theory:

$$\upsilon^* = \frac{3K^* - 2G^*}{6K^* + 2G^*} \tag{6.18}$$

An insight to the shale material and its composition will help the modeling scheme presented herein. We adopt the Bennet et al. thought model displayed in Fig. 6.2. Organic-rich shale material consists of clay minerals, quartz and feldspar inclusions and an organic phase. The clay minerals are mostly kaolinite, illite, smectite/illite, montmorillonite, muscovite, and chlorite. The inclusions are mostly quartz, pyrite, feldspar, and calcite. Table 6.3 shows some literature values for the elastic modulus of these primary constituents. The deformation and fracture properties of shale depend on the mechanical properties of its basic constituents. This suggests that an understanding of the overall macroscopic mechanical properties of shale can be obtained by studying the deformation and properties of these constituents and how they upscale to the overall behavior of the composite material [50, 51].

The thought model is borrowed from the study by Bennett et al. [9]. They carried out indentation tests with a maximum indentation depth of 200 nm to probe the mechanical behavior of individual phases. The four-phase thought model was used herein. The mechanical values tabulated in Table 6.4 are the results of a statistical deconvolution of Toarcian B specimens with a loading force of 5 mN, loading time

Table 6.4 Measured mechanical properties of different phases from indentation

Different phase	Phase 1	Phase 2	Phase 3	Phase 4
E (GPa)	108.28	33.48	10.12	20.38
υ	0.3	0.3	0.3	0.3
Volume fraction (percent)	16	10	18	56

of 10 s, holding time of 5 s, and unloading time of 10 s a well. The volume fractions are also calculated from the statistical indentation tests performed on specimens of Toarcian B, while the Poisson's ratio for all phases assumed to be 0.3. The QFP phase consists of quartz, feldspar, and pyrite because they have similar mechanical properties and they can be considered as one phase. The clay phase consists of different type of clays in the medium of shale. Furthermore, the organic matter phase appears in all of the organic-rich shale materials and typically exhibits very low stiffness values. Finally, there is a fourth phase consisting of both organic and clay phase, which is called the O/C phase.

The concept of the thought model used for elastic homogenization is demonstrated in Fig. 6.2. The interpretation of different phases shown in Table 6.4 are based on assumption of quartz having an indentation modulus value around 100 GPa, organic phase having the lowest elastic properties, and the dominant phase of the material being O/C matrix. This reasoning leads to phase 1, phase 2, phase 3, and phase 4 to be the quartz, clay, organic, and O/C matrix respectively. The homogenization approach based on the Budiansky's model was implemented in an implicit algorithm written using Matlab. The theoretical model predicts a macroscopic elastic modulus of shale material of 24 GPa. This theoretical value is in the range of the values of elastic modulus reported for gas shale materials and measured using macroscopic scale testing techniques, as shown in Table 6.1. Thus we have demonstrated the power of both nano-indentation methods and micromechanics in yielding a fundamental understanding of the mechanical response from the nanometer up to the macroscopic length-scales.

6.5 Conclusion and Future Perspectives

In this chapter, we provided a theoretico-experimental framework for the micro-structural and mechanical characterization of gas shale at the nanometer and micrometer length-scales. The starting point is to recognize the hierarchical nature of shale which calls for a thought model. Although organic-rich shale is composed of four basic elements, quartz/feldspar, gaseous kerogen, clay, and air voids, the exact composition and the local arrangement depend on the scale of observation. In turn this has strong influence on the mechanical performance as expressed in terms of elasticity, strength or fracture properties.

Accurate and precise small scale characterization relies on a rigorous material preparation technique. The basic preparation routine involves cutting with a diamond saw, grinding using a set of abrasive papers, and polishing using polishing solutions and/or polishing cloths. However, the exact sequence of steps must be tailored to each specific material. Nevertheless, we provided some samples of grinding and polishing routines using advanced methods such as optical microscopy, scanning electron microscopy and atomic force microscopy to quantitatively assess their efficiency on select gas shale specimens.

Nano-indentation consists in pushing a hard pyramidal probe into a soft material and it is an accurate means to measure the elastic-plastic properties and at the nanometer length-scale. Moreover, the grid indentation technique makes it possible to draw a compositional map based on the values of the indentation modulus and hardness. Applying statistical deconvolution methods to a large array of indentation tests enables to decompose the response and identify the primary constituents of the materials. The findings of the nano-indentation technique agreed with SEM observation and powder X-ray diffraction analysis.

Micromechanics hold the key to a bottom-up approach where the behavior at the larger scale is predicted from the composition and microstructure at a smaller scale. Herein we applied a 3D model using the phase decomposition results from statistical nano-indentation. The predicted macroscopic stiffness value agrees with macroscopic measurements reported in the scientific literature.

Thus we have demonstrated the power of nano-mechanics and micromechanics to shed light on the origin of the mechanical performance of gas shale by bridging the nanometer and macroscopic length scales. This an important development that will pave the way towards optimum hydraulic plant schemes for energy-harvesting from unconventional reservoir.

Acknowledgments The authors would like to thank Total S. A., Paris, France, for providing the gas shale specimens tested and analyzed in this investigation. The research was funded by Prof. Akono Start-up funds account which was provided by the Department of Civil and Environmental Engineering as well as the College of Engineering at University of Illinois at Urbana-Champaign. In addition, we acknowledge the Distinguished Structural Engineering Fellowship that supported Pooyan Kabir during his Ph.D. studies. The work was carried out in part in the Frederick Seitz Materials Research Laboratory Central Research Facilities, University of Illinois at Urbana-Champaign.

References

1. Annual Energy Outlook with Projections to 2035. (2010). U. S. Energy Information Administration. DOE/EIA-0383.
2. Boyer, C., et al. (2011). Shale gas: A global resource. *Oilfield Review, 23*(3), 28–39.
3. Bonakdarpour, M., Flanagan, B., Holling, C. and Larson J. W. (2011). The Economic and Employment Contributions of Shale Gas in the United States. IHS Global Insight (USA) Inc.
4. Abbas, S. Lecampion, B., and Prioul B. (2013). Competition between Transverse and Axial Hydraulic fractures in Horizontal Wells. Society of Petroleum Engineers. SPE-163848-MS.
5. Lhomme, T., Detournay, E., & Jeffrey, R. (2005). *Effect of fluid compressibility and borehole radius on the propagation of a fluid-driven fracture*. In *Proceedings of 11th International Conference on Fracture*. Turin, Italy.

6. Ortega, J. A. (2009). *Microporomechanical modeling of shale*. Massachusetts Institute of Technology.
7. Bobko, C., & Ulm, F.-J. (2008). The nano-mechanical morphology of shale. *Mechanics of Materials*, 40(4), 318–337.
8. Deirieh, A. A. M. (2011). *Statistical Nano-Chemo-Mechanical Assessment of Shale by Wave-Dispersive Spectroscopy and Nano-Indentation. Master Thesis*. Massachusetts Institute of Technology.
9. Bennett, K. C., et al. (2015). Instrumented nanoindentation and 3D mechanistic modeling of a shale at multiple scales. *Acta Geotechnica, 10*(1), 1–14.
10. Curtis, M. E., et al. (2011). Transmission and scanning electron microscopy investigation of pore connectivity of gas shales on the nanoscale. In *North American Unconventional Gas Conference and Exhibition*. Society of Petroleum Engineers.
11. Sone, H., & Zoback, M. D. (2013). Mechanical properties of shale-gas reservoir rocks—part 2: Ductile creep, brittle strength, and their relation to the elastic modulus. *Geophysics, 78*(5), D393–D402.
12. Vandamme, M., & Ulm, F.-J. (2006). Viscoelastic solutions for conical indentation. *International Journal of Solids and Structures, 43*(10), 3142–3165.
13. Miller, M., et al. (2008). Surface roughness criteria for cement paste nanoindentation. *Cement and Concrete Research, 38*(4), 467–476.
14. Vasconcelos, G., et al. (2008). Ultrasonic evaluation of the physical and mechanical properties of granites. *Ultrasonics, 48*(5), 453–466.
15. Hassan, M., Burdet, O., & Favre, R. (1995). Ultrasonic measurements and static load tests in bridge evaluation. *NDT & E International, 28*(6), 331–337.
16. Sack, D. A., & Olson, L. D. (1995). Advanced NDT methods for evaluating concrete bridges and other structures. *NDT & E International, 28*(6), 349–357.
17. Far, M. E., Hardage, B., & Wagner, D. (2013). Inversion of elastic properties of fractured rocks from AVOAZ data Marcellus Shale example. In *2013 SEG Annual Meeting*. Society of Exploration Geophysicists.
18. Bobko, C. P. (2008). Assessing the Mechanical Microstructure of Shale by Nano-Indentation: The link between Mineral Composition and Mechanical Properties. Ph. D. Thesis. Massachusetts Institute of Technology.
19. Gale, J. F., Reed, R. M., & Holder, J. (2007). Natural fractures in the Barnett Shale and their importance for hydraulic fracture treatments. *AAPG Bulletin, 91*(4), 603–622.
20. Hill, R. R. C. C. (1992). *Analysis of natural fractures in the Barnett Shale, in Mitchell Energy Corporation T. P. Sims no. 2, Wise County, Texas* (p. 50). Chicago, IL: Gas Research Institute.
21. Kim, K., & Mubeen, A. (1980). *Fracture toughness of Antrim shale*. Houghton, MI, USA: Michigan Technological Univ., Dept. of Mining Engineering.
22. Koesoemadinata, A., et al. (2011). Seismic reservoir characterization in Marcellus Shale. In *2011 SEG Annual Meeting*. Society of Exploration Geophysicists.
23. McGinley, M., Zhu, D., & Hill, A. D. (2015). The effects of fracture orientation and elastic property anisotropy on hydraulic fracture conductivity in the Marcellus Shale. In *SPE Annual Technical Conference and Exhibition*. Society of Petroleum Engineers.
24. Hustrulid, W., & Johnson, G. A. (1990). *Rock Mechanics Contributions and Challenges: Proceedings of the 31st US Symposium on Rock Mechanics* (Vol. 31). CRC Press.
25. Salehi, I. (2010). *New Albany shale gas project*. Des Plaines, IL: Gas Technology Institute.
26. Instruments, C. (2014). *Technical features 2014- nanoindentation tester (NHT2)*. Switzerland.
27. Oliver, W. C., & Pharr, G. M. (1992). An improved technique for determining hardness and elastic modulus using load and displacement sensing indentation experiments. *Journal of Materials Research, 7*(06), 1564–1583.
28. Constantinides, G., et al. (2006). Grid indentation analysis of composite microstructure and mechanics: Principles and validation. *Materials Science and Engineering A, 430*(1), 189–202.
29. Constantinides, G., & Ulm, F.-J. (2007). The nanogranular nature of C–S–H. *Journal of the Mechanics and Physics of Solids, 55*(1), 64–90.

30. Constantinides, G., Ulm, F.-J., & Van Vliet, K. (2003). On the use of nanoindentation for cementitious materials. *Materials and Structures, 36*(3), 191–196.
31. Vandamme, M. (2008). *The nanogranular origin of concrete creep: A nanoindentation investigation of microstructure and fundamental properties of calcium-silicate-hydrates.* Cambridge, MA: Massachusetts Institute of Technology.
32. Dempster AP, Laird N, Rubin DB. Maximum Likelihood from Incomplete Data via the EM Algorithm. Journal of the Royal Statistical Society Series B (Methodological). 1977;39(1):1–38.
33. McLachlan G, Basford KE. Mixture models: Inference and applications to clustering. Applied statistics: Textbooks and monographs. New York: Dekker; 1988. doi:10.2307/2348072.
34. McLachlan, G., & Peel, D. (2004). *Finite mixture models.* New York: John Wiley & Sons.
35. McLachlan, G. J., et al. (1999). The EMMIX software for the fitting of mixtures of normal and t-components. *Journal of Statistical Software, 4*(2), 1–14.
36. Fraley, C., & Raftery, A. E. (1999). MCLUST: Software for model-based cluster analysis. *Journal of Classification, 16*(2), 297–306.
37. Fraley, C., & Raftery, A. E. (2007). Model-based methods of classification: Using the mclust software in chemometrics. *Journal of Statistical Software, 18*(6), 1–13.
38. Fraley, C., & Raftery, A. E. (2002). Model-based clustering, discriminant analysis, and density estimation. *Journal of the American Statistical Association, 97*(458), 611–631.
39. DeJong, M. J., & Ulm, F.-J. (2007). The nanogranular behavior of CSH at elevated temperatures (up to 700 °C). *Cement and Concrete Research, 37*(1), 1–12.
40. Hornby, B. E., Schwartz, L. M., & Hudson, J. A. (1994). Anisotropic effective-medium modeling of the elastic properties of shales. *Geophysics, 59*(10), 1570–1583.
41. Marion, D., et al. (1992). Compressional velocity and porosity in sand-clay mixtures. *Geophysics, 57*(4), 554–563.
42. Wang, Z., Wang, H., & Cates, M. E. (2001). Effective elastic properties of solid clays. *Geophysics, 66*(2), 428–440.
43. Mavko, G., Mukerji, T., & Dvorkin, J. (1998). *The rock physics handbook: Tools for seismic analysis in porous media.* Cambridge, UK: Cambridge Univ. Press. 329 pp.
44. Simmons G, Wang H. Single crystal elastic constants and calculated aggregate properties. Cambridge, MA, USA: The MIT Press; 1971.
45. Gandais, M., & Willaime, C. (1984). Mechanical Properties of Feldspars. In Feldspars and Feldspathoids: Structures, Properties and Occurrences. NATO ASI Series, 137, 207–246.
46. Kumar, V., et al. (2012). Estimation of elastic properties of organic matter in Woodford Shale through nanoindentation measurements. In *SPE Canadian Unconventional Resources Conference.* Society of Petroleum Engineers.
47. Budiansky, B. (1965). On the elastic moduli of some heterogeneous materials. *Journal of the Mechanics and Physics of Solids, 13*(4), 223–227.
48. Hill, R. (1967). The essential structure of constitutive laws for metal composites and polycrystals. *Journal of the Mechanics and Physics of Solids, 15*(2), 79–95.
49. Eshelby, J. D. (1957) The determination of the elastic field of an ellipsoidal inclusion, and related problems. In *Proceedings of the Royal Society of London A: Mathematical, Physical and Engineering Sciences.* The Royal Society
50. Arson, C., & Pereira, J. M. (2013). Influence of damage on pore size distribution and permeability of rocks. *International Journal for Numerical and Analytical Methods in Geomechanics, 37*(8), 810–831.
51. Barthélémy, J. F., Souque, C., & Daniel, J. M. (2013). Nonlinear homogenization approach to the friction coefficient of a quartz-clay fault gouge. *International Journal for Numerical and Analytical Methods in Geomechanics, 37*(13), 1948–1968.

Chapter 7
On the Production Analysis of a Multi-Fractured Horizontal Well

Erfan Sarvaramini and Dmitry I. Garagash

Abstract This paper investigates the post fracture transient analysis of multi-fractured horizontal wells under the assumption of infinitely large fracture conductivity. Most of the existing studies of multi-fractured wells have considered finite fracture conductivity, when the dynamic fluid pressure drop in the flow within fractures is a part of the solution. This led to computationally intensive solution methods, particularly when a reasonably large number of fractures representative of current field applications is considered. In this work, we limit our consideration to low-permeability, tight (e.g., shale) reservoirs, when pressure losses in propped fractures can be neglected. This assumption allows to develop a rigorous, accurate, and computationally efficient solution method based on the fundamental problem of a unit step pressure decline in an array of identically sized and equally spaced fractures. The study of this fundamental problem is analogous to the well testing analysis of a fractured well produced at constant bottom-hole pressure conditions. The solution for a unit step pressure decline is used within the Green's function framework to formulate and solve for the transient pressure response of a multi-fracture array produced at a constant volumetric flow rate. We also explore two simplified approaches to the production analysis of multi-fractured wells based on (1) the infinite fracture array approximation for finite arrays, and (2) an extension of the ad hoc method of Gringarten et al. (Soc Pet En J 14(04):347–360), respectively. We show that both methods lead to very good approximations of the rigorous solution for a finite fracture array problem, thus allowing to further simplify the transient analysis of multi-fracture wells.

E. Sarvaramini • D.I. Garagash (✉)
Department of Civil and Resource Engineering, Dalhousie University,
1360 Barrington Street, Halifax, NS, Canada, B3H 4R2
e-mail: erfan.sarvaramini@dal.ca; garagash@dal.ca

C. Jin, G. Cusatis (eds.), *New Frontiers in Oil and Gas Exploration*,
DOI 10.1007/978-3-319-40124-9_7

7.1 Introduction

Hydraulic fracturing is the primary technology used in petroleum industry to extract hydrocarbon from unconventional reservoirs (e.g., shale gas, tight gas sandstone, oil shale) [1]. In this technique, a fluid is pumped from a borehole under high pressure to create and propagate fractures in rock formation. The preferred method of completion during the hydraulic fracturing treatment is multistage fracturing in which an array of fractures is created from a horizontal borehole.

The performance of horizontal wells is usually monitored using production well tests, such as pressure drawdown or build-up tests. In a horizontal, multi-fractured producing well, the interaction between fractures through the reservoir pore fluid flow and fluid-solid mechanical coupling may result in multiple flow regimes [2]. In order to accurately characterize the reservoir, it is important to identify each of the flow regimes and time-dependent transition between them. The latter can be done using the pressure transient data analysis.

Boundary element problem of distributed sources/sinks of fluid along a fracture is a common way to analyze the pressure transient data. It appears that this approach was first used in the context of fluid production from a fractured well by Gringarten et al. [3], who identified the various flow regimes developing during production from a *single, infinite-conductivity fracture* in a homogenous reservoir. Gringarten et al. [3] carried out a rigorous numerical analysis of the boundary element problem and evaluated fluid pressure response $p(t)$ in the fracture (uniform along the fracture due its infinite conductivity) and the point-wise fluid exchange flux between the fracture and the rock $\bar{g}(x,t)$. It was shown that the flux distribution is approximately uniform along the fracture only at early times, when the pore pressure diffusion in the surrounding rock is essentially one-dimensional, while the flux distribution becomes strongly nonuniform along the fracture at some extended production times coincidental with pore pressure diffusion in the surrounding rock on a scale comparable to the fracture length or larger. Furthermore the flux distribution is essentially singular at the fracture tips [4, 5]. Although an accurate evaluation of the pressure response in an infinitely conductive fracture requires the rigorous analysis of the governing equations, Gringarten et al. [3] proposed an alternative, approximate approach which considerably simplifies the pressure transient analysis of a single fracture. In this approach, the fluid-exchange flux distribution is assumed to be uniform along the fracture at all times (i.e., $\bar{g}^{\mathrm{app}}(t)$, where "app" refers to the approximate nature of the assumption). This assumption results in a now nonuniform pressure distribution along the fracture $p^{\mathrm{app}}(x,t)$. The latter is then evaluated at an intermediate location ($x=0.732\ell$) selected by Gringarten et al. [3] to best approximate the uniform pressure $p(t)$ in the rigorous solution.

The pressure transient analysis of a fractured well was later extended to the case of a single finite-conductivity hydraulic fracture by Cinco et al. [6]. They combined the Green's function technique with Laplace transform to calculate the transient pressure response. This combination then became a powerful tool in analyzing the pressure transient data, allowing to extend the previous seminal studies [3, 6] to the case of a heterogeneous reservoir (e.g., naturally fractured formations) intersected by a hydraulic fracture.

With advances in horizontal drilling and multistage hydraulic fracturing technologies, the well testing of multi-fractured wells have also attracted a great attention. Raghavan et al. [7] carried out a rigorous study of the behavior of an array of multiple finite-conductivity fractures produced at a constant volumetric flow rate. They investigated the effects of fracture conductivity, number and location of fractures in the array on the pressure response in homogeneous reservoirs with relatively high permeability. Although the analysis of Raghavan et al. is comprehensive, the numerical solution method can be computationally intensive even for small number of fractures. This issue necessitated further investigations attempting to simplify the analysis of finite conductivity fracture arrays. For instance, Al-Ahmadi et al. [8] considered production from multi-fractured wells at their early life when the interaction between the fractures is negligible. Brown et al. [9] and Ozkan et al. [10] presented mathematical models which only considered the pore fluid flow in the region intermediate to hydraulic fractures and a linear flow was assumed at all times.

The pressure transient analysis of a finite-conductivity hydraulic fracture can be of great interests in a number of practical reservoir applications. For example, it can be used in well testing analysis of a fractured well characterized by a poor fracture conductivity (e.g., un-propped fracture). Another application is the analysis of the short time transient data of a fractured well when the fluid pressure has not yet equilibrated along the crack. Cinco et al. [6] established a parametric range when an approximate pressure uniformity condition holds along a crack, expressed as a condition that the dimensionless fracture conductivity defined by $C_{fD} = k_f \bar{w} / k\ell$ is large enough, $C_{fD} \geq 300$. Here k_f and k are the fracture and formation values of permeability, $\bar{w}$ and ℓ are average fracture aperture and half-length, respectively. A more universal criterion of pressure uniformity assumption was recently established in [11]. Based on the latter criteria, the condition of pressure uniformity is likely to be reached for hydraulic fractures in the low-permeability, tight reservoirs, particularly in shale formations. For example, for a typical hydraulic fracture of half-length $\ell = 50$ m and average fracture conductivity $k_f \bar{w} = 0.5\,$d.m (e.g., [12]) in the Bakken shale formation characterized by the representative value of the formation permeability $k = 0.01$ md, the dimensionless fracture conductivity is $C_{fD} = 1000$. The condition of pressure uniformity or, conversely, infinite fracture conductivity allows for a simpler and more accurate solution method for multiple-fracture arrays.

In this study, the pressure transient analysis of infinite-conductivity, multiple-fracture arrays is investigated. Section 7.2 summarizes the mathematical model and underlying assumptions. In Sect. 7.3, we formulate and solve an auxiliary problem in which all fractures in the array are subjected to a unit step pressure decline. This solution is an extension of an auxiliary problem for a single fracture studied by

Detournay and Cheng [4] and Sarvaramini and Garagash [5] in the injection context (i.e., for a unit pressure rise). The auxiliary problem solution can be applied to well testing analysis of multi-fractured horizontal wells producing at constant pressure conditions. The latter has not received enough attention despite its importance in reservoir applications. For example, production from a hydrocarbon reservoir during production decline period normally takes place under the constant pressure conditions. Other applications of constant pressure production are in geothermal reservoirs and water artisan wells [13]. In Sect. 7.4, the auxiliary solution is used within the Green's function framework to formulate the transient response of a multiple fracture array producing at a constant volumetric flow rate. Finally, we examine the validity of the uniform leak-in method of Gringarten et al. [3] when applied to a multiple-fracture array. This approximation can considerably improve the computational time involved in calculations of the fracture response of the multi-fractured horizontal wells. The summary of this study is presented in Sect. 7.5.

7.2 Mathematical Formulation

We consider a problem of a horizontal borehole fractured with a linear array of equally spaced cracks of the Perkins–Kern–Nordgren (PKN) geometry. Reservoir rock is characterized by permeability k, bulk compressibility c_t, porosity ϕ, and initial pore pressure p_0. Fractures in the array are characterized by the half-length ℓ, height h, and uniform spacing d along the well (Fig. 7.1). Pore fluid with viscosity μ is produced from the borehole under either constant volumetric flow rate Q_0 or constant pressure decline $p - p_0 < 0$ conditions.

We make use of the following assumptions. (1) Hydraulic fractures are propped, i.e., filled with proppant during placement (the multistage hydraulic fracturing treatment) in order to prevent/minimize the fracture closure during production pressure decline period. (2) Changes of the fracture volume during the pressure decline period (i.e., fracture storage effects) are neglected. This assumption is valid in the transient analysis of propped hydraulic fractures characterized by negligible

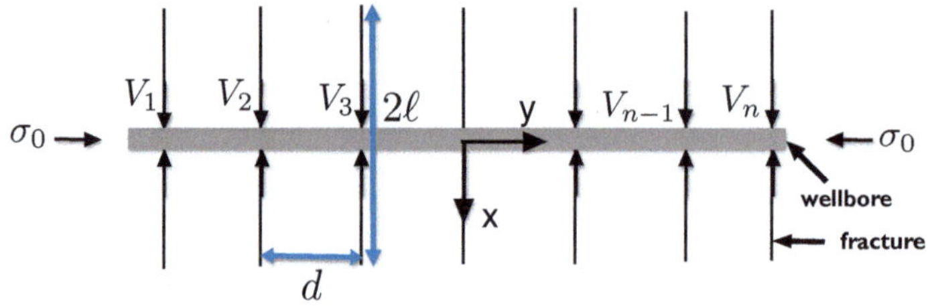

Fig. 7.1 Horizontal well with a multiple (n) fracture array

compressibility in the reservoir applications [14]. (3) The fluid pressure along each propped fracture is assumed to be spatially uniform (infinite fracture conductivity assumption). (4) Fractures are confined between two horizontal impermeable (cap-rock) layers, which, together with the assumption of pressure equilibrium within a vertical crack cross-section, suggests a 2-D fluid diffusion (in the horizontal x-y plane) within the permeable (reservoir) rock layer. (5) A single phase flow is assumed at all times. (6) Reservoir properties are assumed to be homogenous.

To aid the transient pressure analysis of multiple hydraulic fractures, we first consider the fundamental solution to an auxiliary problem of a unit step pressure decline in a fracture array. The latter was previously developed for a single fracture in the injection context (i.e., unit step pressure rise) in [4, 5]. We extend this single fracture solution to the case of a multi-fracture array, now in the production context (i.e., unit step pressure decline). The auxiliary solution is used to formulate and solve a convolution integral equation governing the transient pressure decline in a multi-fracture array for a given production rate history. Moreover, the auxiliary problem can be applied directly to the analysis of wells producing at constant pressure conditions. [In this case, the constant pressure conditions have to be maintained from $t=0$ (beginning of the production) for the auxiliary problem to be directly applicable.]

7.3 Auxiliary Problem (Unit Step Pressure Decline)

Consider that each fracture in a multi-fracture array is subjected to a unit step pressure decline

$$p - p_0 = -1, \quad (|x| < \ell,\ y = y_i\ (i = 1, \ldots, n),\ t > 0), \tag{7.1}$$

where y_i is the ith crack position along the well.

Pore pressure diffusivity equation together with the leak-in boundary condition

$$\frac{\partial p}{\partial t} = \alpha \nabla^2 p, \quad -\frac{k}{\mu}\left[\left.\frac{\partial p}{\partial y}\right|_{y=y_i^+} - \left.\frac{\partial p}{\partial y}\right|_{y=y_i^-}\right] = \bar{g}_i(|x| < \ell, t), (i = 1, \ldots, n), \tag{7.2}$$

can be used to obtain the pore pressure distribution surrounding producing fractures. Here $\alpha = k/(\mu\phi c_t)$ is the diffusivity coefficient, and $\bar{g}_i$ is the ith fracture fluid leak-in rate (positive for fluid production from a fracture). The solution of (7.2) can be obtained by superposition of distributed point-sinks [7, 15], which yields along the ith fracture

$$p(|x| < \ell, y_i, t) - p_0 = \sum_{j=1}^{n} \int_0^t \int_{-\ell}^{\ell} \frac{-\bar{g}_j(x', t')}{4\pi S\alpha(t - t')} \times \exp\left(-\frac{|x - x'|^2 + |y_i - y_j|^2}{4\alpha(t - t')}\right) dx' dt', \ (i = 1, \ldots, n), \tag{7.3}$$

where $S = \phi\ c_t$ is the fluid storage coefficient.

In the following, we solve the set of equations (7.3) for the rates of fluid leak-in for each fracture in the array ($\bar{g}_i(x, t)$, $i = 1, \ldots, n$) subjected to a unit step pressure decline (7.1). The corresponding cumulative produced volumes from each fracture in the array can be obtained by integration

$$V_i(t) = h \int_0^t \int_{-\ell}^{\ell} \bar{g}_i(x', t') dx' dt' \ (i = 1, \ldots, n). \tag{7.4}$$

To facilitate the numerical solution of (7.3) with (7.1), we introduce the following normalized parameters: the non-dimensional time (τ), x coordinate (ξ), y coordinate (η), leak-in rate (ψ), and cumulative leak-in volume (Φ):

$$\tau = \frac{t}{t_*}, \quad \xi = \frac{x}{\ell}, \quad \eta = \frac{y}{\ell}, \quad \psi = \frac{\bar{g}}{\bar{g}_*}, \quad \Phi = \frac{V}{V_*}, \tag{7.5}$$

defined in terms of the respective scales,

$$t_* = \frac{\ell^2}{4\alpha}, \quad \ell, \quad \ell, \quad \bar{g}_* = \frac{4\alpha S}{\ell}, \quad V_* = Sh\ell^2. \tag{7.6}$$

The normalized form of (7.3) with (7.1) is expressed as

$$1 = \frac{1}{\pi} \sum_{j=1}^{n} \int_0^{\tau} \int_{-1}^{1} \psi_j(\xi', \tau') \exp\left(-\frac{|\xi - \xi'|^2 + |\eta_i - \eta_j|^2}{\tau - \tau'}\right) \frac{d\xi' d\tau'}{\tau - \tau'}, \ (i = 1, \ldots, n). \tag{7.7}$$

The corresponding normalized form of the cumulative fracture leak-in volume (7.4) is given by

$$\Phi_i(\tau) = \int_0^{\tau} \int_{-1}^{1} \psi_i(\xi', \tau') d\xi' d\tau', \ (i = 1, \ldots, n). \tag{7.8}$$

Applying Laplace transform to (7.7) results in

$$\frac{1}{s}=\frac{2}{\pi}\sum_{j=1}^{n}\int_{-1}^{1}\psi_j(\xi',s)K_0\left(2\sqrt{s}\sqrt{|\xi-\xi'|^2+|\eta_i-\eta_j|^2}\right)d\xi',\ (i=1,\ldots,n),\quad (7.9)$$

where s is the transform variable, $\psi(\xi,s)$ is the Laplace image of $\psi(\xi,\tau)$, K_0 is the modified Bessel function of the second kind. We use the Gauss–Chebyshev method to discretize the integral for $\psi_j(\xi,s)$, $(j=1,\ldots,n)$ along each fracture (see Appendix D of [5] for details of integral discretization). The resulting system of linear equations is solved for $\psi_j(\xi,s)$, $(j=1,\ldots,n)$ using the Newton–Raphson method and then numerically inverted to the time domain using the Stehfest algorithm [16].

Before we carry out the numerical analysis of the auxiliary problem for particular examples of production from a finite fracture array, we first consider the limiting cases corresponding to the production from a single fracture ($n=1$) and an infinite fracture array ($n=\infty$) due to a unit step pressure decline.

7.3.1 Single Fracture

The solution of the auxiliary problem of a single crack subjected to a unit step pressure rise (i.e., in the injection context) was developed in [4, 5]. The transient solution for the production problem can be obtained from that for the injection problem by reversing the direction (sign) of the flow rate.

Evolution of the crack-average of the fluid leak-in rate $\langle\psi\rangle(\tau)=(1/2)\int_{-1}^{1}\psi(\xi',\tau)d\xi'$ with normalized time $\tau=t/t_*$ for a single fracture is shown by a black dotted line in Fig. 7.3a. Note that the solution of the fluid leak-in rate $\psi(\xi,\tau)$ was numerically tabulated in [5]. The cumulative produced volume is reproduced in Fig. 7.3b after [5].

7.3.2 Infinite Fracture Array ($n=\infty$)

We define a problem in which an infinite linear array of fractures is hydraulically connected to a horizontal borehole (Fig. 7.1 with $n=\infty$). The solution of fluid leak-in rate ψ_j, $(j=1,..,\infty)$ in the infinite array problem due to a unit step pressure decline can be obtained from (7.3) with $n=\infty$. Note that since number of fractures are assumed to be infinite, the rate of fluid leak-in for each fracture is identical (i.e., $\psi_i=\psi$ for all integer i's). A good approximation of the solution of the infinite array problem is afforded by truncating the series to a finite odd number of terms m (i.e., effectively approximating infinite array by a finite one) and solving truncated (7.3) for the central crack, $i=(m+1)/2$, (i.e., at $y_i=0$). The truncation number m in this approximate solution would have to scale with production time,

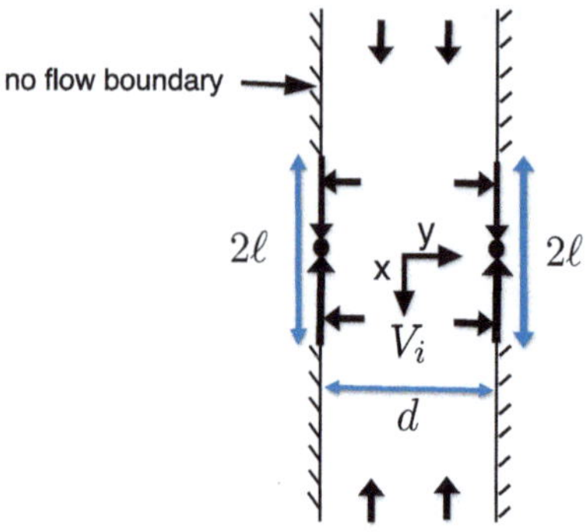

Fig. 7.2 Problem of a partially permeable strip equivalent to the infinite fracture array problem (Fig. 7.1 with $n = \infty$)

such that the influence of the first ($i = 1$) and last ($i = m$) cracks in the truncated array on the central crack ($i = (m + 1)/2$) can be reasonably neglected; i.e., $\sqrt{4\alpha t} \ll (m - 1)d/2$.

Alternatively, we can define a problem of fluid production from a strip (infinite in the fracture length direction x) of thickness d from two line sinks of length 2ℓ located on each side of the strip (Fig. 7.2). The rest of the strip boundary is a no-flow boundary. The pressure depletion solutions of the permeable strip and the infinite linear array of fractures are identical. For the strip problem, we can explicitly discretize the diffusivity equation and leak-in boundary conditions (7.2) to obtain the fluid leak-in rate.

In this study, we use the truncated series approximation of (7.3) to obtain the fluid production rate and cumulative production volume for the infinite fracture array. Since the spacing between fractures is assumed to be equal, the normalized solution of (7.9) with $n = m$, $\psi_j = \psi$, and $i = (m + 1)/2$ depends on a single parameter, the non-dimensional crack spacing d/ℓ (note that $(\eta_i - \eta_j) = (i - j)d/\ell$ for equally spaced fractures). Figure 7.3 illustrates the evolution of the crack-averaged normalized leak-in rate for various values of normalized crack spacing d/ℓ. As shown in Fig. 7.3a, the solution of the fluid leak-in rate for the infinite fracture array is well approximated by that of a single fracture at early production time, before the interference effects between fractures become significant. The latter causes the departure of the interacting infinite array solution from the single fracture solution with the onset of the interaction dependent upon the fracture spacing.

The large-time asymptote of the crack-averaged-leak-in rate can be approximated by the solution to 1-D diffusion in the along-crack direction (x) due to a unit pressure decline [15]:

$$\langle\psi\rangle(\tau) = \frac{\langle\bar{g}\rangle(t)}{\bar{g}_*} = \frac{d}{\ell}\frac{1}{\sqrt{16t/t_*}}. \tag{7.10}$$

This asymptote is shown by dashed line on Fig. 7.3a.

The cumulative produced volume from a fracture within an infinite fracture array is shown in Fig. 7.3b for various values of normalized crack spacing d/ℓ.

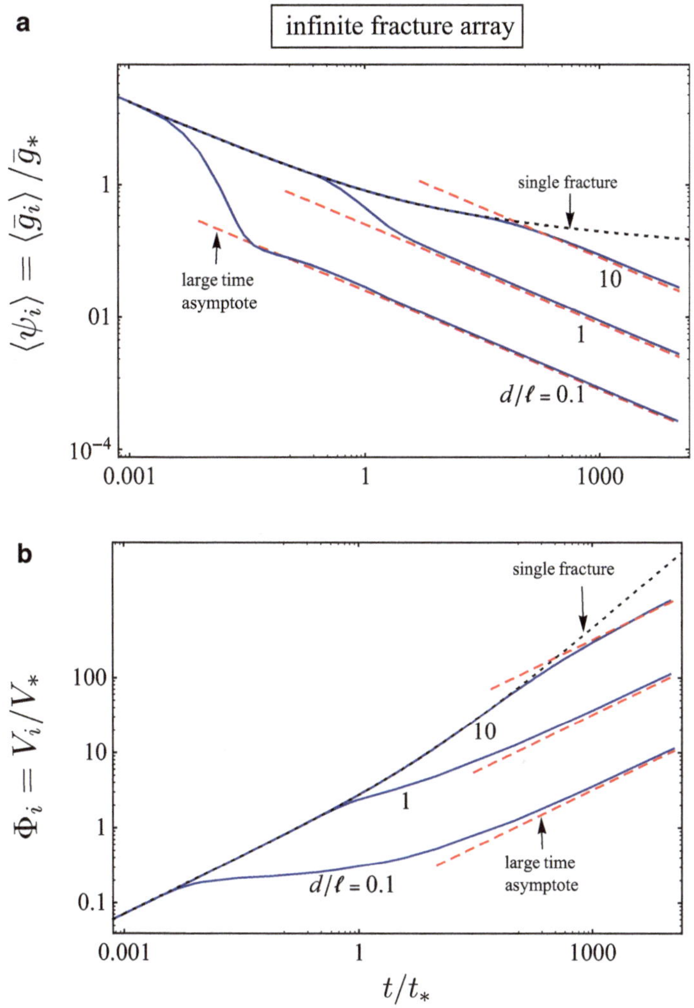

Fig. 7.3 Auxiliary problem of a unit step pressure decline in an infinite fracture array: Evolution of (**a**) the crack-averaged normalized leak-in rate, and (**b**) the cumulative produced volume from a fracture within the array with the normalized time and for various values of fracture spacing d/ℓ. The solution of a single fracture is shown by a *black dotted line* after [5]. The large-time asymptotic solution is shown by a *red dashed line* [Eq. (7.10)]. Time, leak-in rate, and cumulative volume scales are $t_* = \ell^2/4\alpha$, $\bar{g}_* = 4S\alpha/\ell$, and $V_* = Sh\,\ell^2$, respectively

7.3.3 Finite Fracture Array Problem

The analysis of the auxiliary problem of a unit step pressure decline in a finite fracture array is carried out for a particular example of production from a typical multi-fractured horizontal well in Bakken shale formation.

Hydrocarbon with viscosity $\mu = 0.64$ cP is produced from 1 km long interval of a horizontal well aligned in the direction of the minimum confining stress. The horizontal well has been stimulated with 21 hydraulic fractures (Fig. 7.1 with $n = 21$) oriented perpendicular to the direction of the minimum confining stress (and, hence, the well axis). Hydraulic fractures of length $2\ell = 100$ m and height $h = 10$ m are uniformly distributed with spacing $d = 50$ m in a reservoir formation characterized by the formation permeability $k = 0.01$ md, bulk compressibility $c_t = 2 \times 10^{-9}\ Pa^{-1}$, and porosity $\phi = 0.12$ [17].

The calculated values of relevant problem parameters are $S = 2.4 \times 10^{-10}\ Pa^{-1}$ (storage parameter), $\alpha = 6.5 \times 10^{-5}\ m^2/s$ (diffusivity coefficient), $t_* = 111.1$ days (diffusion time scale), $\bar{g}_* = 1.33 \times 10^{-15}$ m^3/m^2 s/unit of pressure decline (leak-in scale), $V_* = 6 \times 10^{-6}$ m^3/unit of pressure decline (volume scale), and dimensionless fracture spacing $d/\ell = 1$.

7.3.3.1 Production Rate

Figure 7.4 illustrates the evolution of the averaged-normalized leak-in rate $\langle \psi_i \rangle(\tau) = (1/2)\int_{-1}^{1} \psi_i(\xi', \tau) d\xi'$ with normalized time for outer ($i = 1$), intermediate ($i = 2$, $i = 5$), and central ($i = 11$) fractures in the array. The rate of production in the

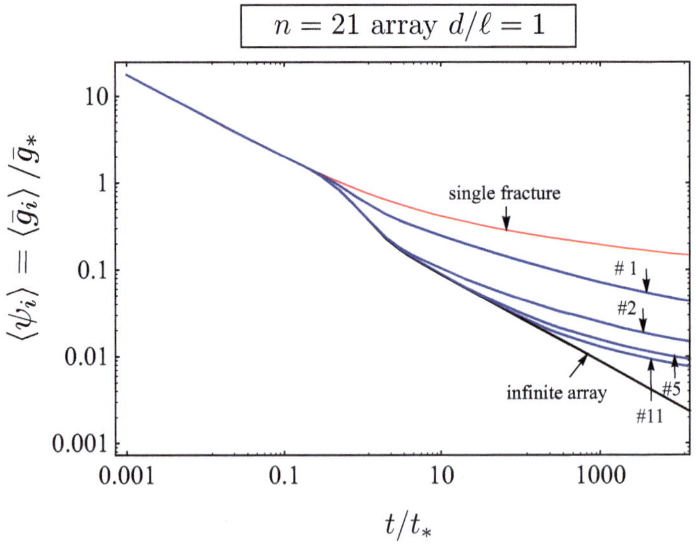

Fig. 7.4 Auxiliary problem of a unit step pressure decline: Evolution of the crack-average normalized leak-in rate with normalized time for crack 1 (outer), 2, 5 (intermediate), and 11 (central). The solutions of limiting cases of a single fracture and infinite fracture array are shown for comparison. The time and leak-in rate scales are $t_* = \ell^2/4\alpha = 111.1$ days and $\bar{g}_* = 4\ S\alpha/\ell = 1.33 \times 10^{-15}$ m^3/m^{2}s/unit of pressure decline, respectively

single fracture and the infinite fracture array solutions are shown on the same graph for comparison.

As shown in Fig. 7.4, the production rates from different fractures in the finite array are approximately the same at early production times (when the hydraulic interference effects between fractures are negligible), and are approximated equally well by that of the single fracture and the infinite fracture array solutions. However, the contribution of each fracture to the total flow rate in the finite fracture array is no longer equal at later production times. As expected, the outer crack produces at the highest rate, while the central crack produces at the lowest rate.

We also notice that the production rate from the *outer* crack cannot be described by the single fracture solution nor the infinite array solution after the interference effects between fractures become important. However, the production rate from *inner* fractures are well approximated by the infinite fracture array solution up to $t/t_* \sim 30$, which corresponds to entire range of realistic production times up to $t \sim 10$ years for the considered example ($t_* = 111$ days). Note that the rate of production from individual fractures in the finite fracture array always remains in between the single fracture solution (upper bound) and infinite array solution (lower bound) at any given time.

An example of the numerical solution for the leak-in distribution along the outer ($i = 1$) and central ($i = 11$) cracks in the finite fracture array at various normalized time is shown in Fig. 7.5. The comparison with the single fracture [5] and infinite fracture array solutions (shown by dashed line) reinforces the previous observation that the infinite array solution provides an excellent approximation for the inner cracks in a finite array for normalize time up to ~ 100.

7.3.3.2 Cumulative Production

The cumulative produced volumes from the crack #1 (outer), #2, #5 (intermediate), and #11 (central) are shown in Fig. 7.6a. We also plot the total cumulative production from the finite fracture array in Fig. 7.6b.

As illustrated in Fig. 7.6b, the interference effects between fractures result in significant departure of the cumulative produced volume solution from the single fracture solution. However, the cumulative produced volume from the array is closely approximated by the infinite fracture array solution until $t = 3.6$ years (or, $t/t_* = 12$) when the latter under-predicts the cumulative production by 10 %.

7.3.4 Uniform Leak-in Approximation

Calculation of the fracture responses such as fluid leak-in rate and cumulative leak-in volume for a finite fracture array in the auxiliary problem requires the rigorous solution of (7.7). As we showed in Fig. 7.5, the distribution of the fluid leak-in rate is only approximately uniform along the fracture at early times. The objective of

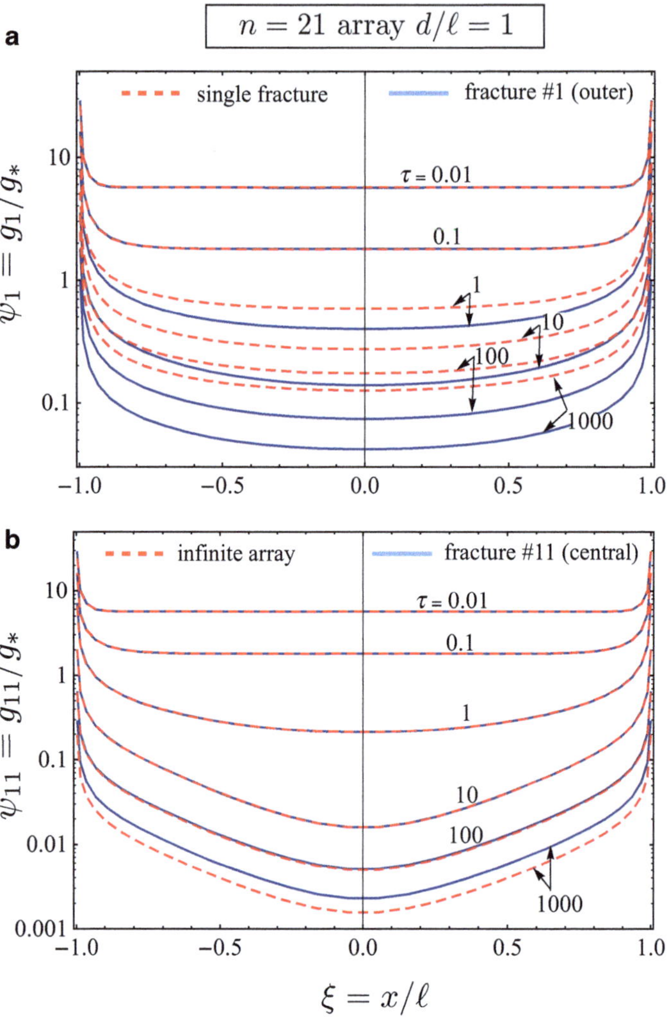

Fig. 7.5 Auxiliary problem of a unit step pressure decline: Comparison of the numerical solutions of the leak-in rate distribution for (**a**) the #1 (outer) crack with that of a single (after [5]) fracture solution shown by the *dashed line*, and (**b**) the #11 (central) crack with that of an infinite fracture array solution shown by the *dashed line*. Time and leak-in rate scales are $t_* = \ell^2/4\alpha = 111.1$ days and $\bar{g}_* = 4S\alpha/\ell = 1.33 \times 10^{-15}$ m^3/m^{2}s/unit of pressure decline, respectively

this section is to investigate if the uniform leak-in assumption applied indiscriminately at all times can still provide an acceptable approximation for the fracture responses as obtained in the complete (rigorous) solution. This approach was originally introduced by Gringarten et al. [3] for the case of a single fracture produced at a constant volumetric rate. They showed that a simple, yet accurate approximate solution for the fluid pressure $p(t)$ can be obtained by assuming uniform leak-in distribution along the crack (i.e., $\bar{g}^{\mathrm{app}}(t)$) and evaluating the now

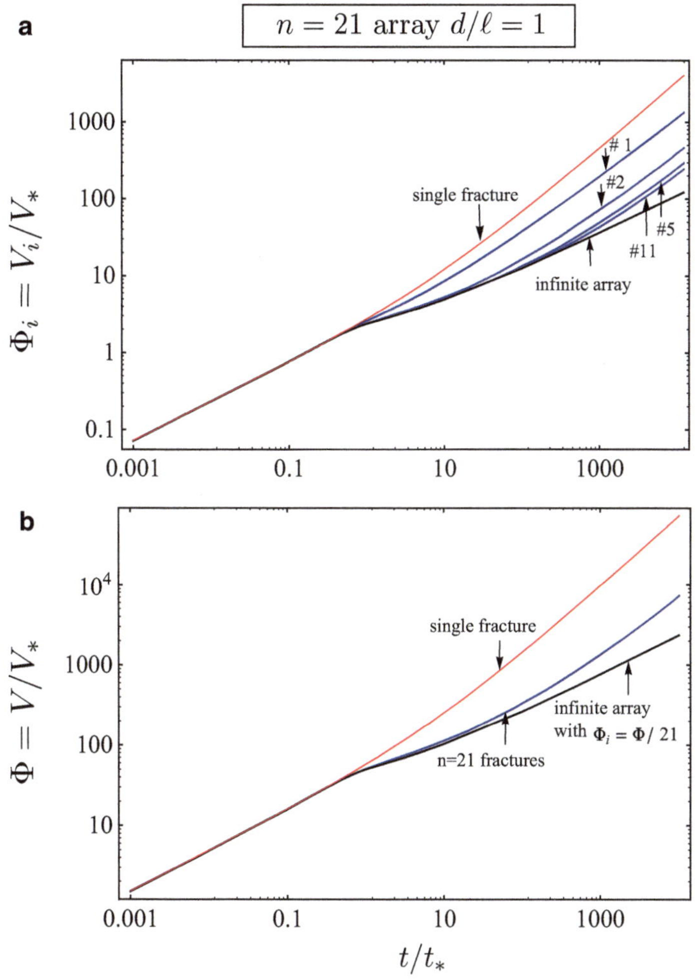

Fig. 7.6 Auxiliary problem of a unit step pressure decline: Evolution of (**a**) the cumulative leak-in volume for crack #1 (outer), #2, #5 (intermediate), and #11 (central), and (**b**) the total cumulative leak-in volume from the finite fracture array with normalized time. Time and cumulative leak-in volume scales are $t_* = \ell^2/4\alpha = 111.1$ days, and $V_* = 6 \times 10^{-6}$ m^3/unit of pressure decline, respectively

nonuniform pressure distribution $p^{\mathrm{app}}(x,t)$ at an ad hoc intermediate location $x = 0.732\ell$ along the crack. In this approach $p^{\mathrm{app}}(x = 0.732\ell, t)$ is used to approximate the evolution of the uniform pressure $p(t)$ in the original problem. Here we extend this ad hoc approximation approach to the case of multiple fractures subjected to a unit step pressure decline. In this case, set of equations (7.3) with (7.1) is solved for the individual fracture leak-in rates assuming $\bar{g}_i(x,t) \approx \bar{g}_i^{\mathrm{app}}(t)$ $(i = 1, \ldots, n)$ while the integral in (7.3) is evaluated at $x = 0.732\ell$.

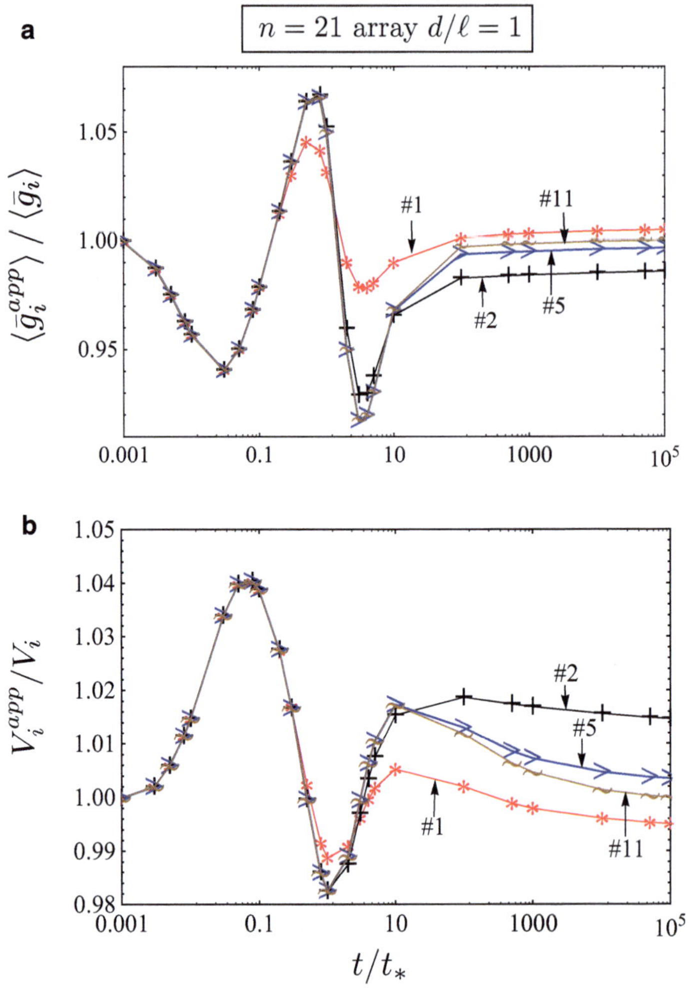

Fig. 7.7 Auxiliary problem of a unit step pressure decline: Comparison between (**a**) fracture fluid-leak-in rates and (**b**) fracture cumulative leak-in volumes obtained by using the Gringarten's-like uniform leak-in approximation (superscript "app") and the rigorous solution for #1 (outer), #2, #5 (intermediate), and #11 (central) cracks. Diffusion time scale is $t_* = \ell^2/4\alpha = 111.1$ days

Figure 7.7 shows a comparison between the values of the fluid-leak-in rates and cumulative leak-in volumes for #1 (outer), #2, #5 (intermediate), and #11 (central) cracks obtained using the uniform leak-in approximate solution and the rigorous solution. As evidenced by Fig. 7.7, the uniform leak-in solution provides an excellent (within 5 %) approximation for the crack-averaged-fluid-leak-in rates and cumulative produced volumes.

7.4 Transient Pressure Decline: Constant Rate of Production from Fractured Well

The cumulative production from an individual crack (V_i) for a given pressure history $p(t)$ can be obtained by application of the Duhamel's theorem [15]

$$V_i(t) = \int_0^t V_i^{\text{aux}}(t-t')\left(-\frac{dp(t')}{dt'}\right)dt', \quad (i = 1, \ldots, n), \tag{7.11}$$

where $V_i^{\text{aux}}(t) = \ell^2 hS\Phi_i^{\text{aux}}(4\alpha t/\ell^2)$ is the cumulative produced volume for the ith fracture in the auxiliary problem of a unit step pressure decline, with Φ_i^{aux} given by (7.8). Global cumulative volume balance equation for a finite fracture array produced at a total constant production rate Q_0,

$$\sum_{i=1}^{n} V_i = Q_0 t, \tag{7.12}$$

and system of equations (7.11) are simultaneously solved for pressure evolution $p(t)$ and fracture cumulative production volumes $V_i(t)$ $(i = 1,\ldots,n)$.

We introduce the non-dimensional time $\tau = t/t_2$, pressure change $\pi = (p - p_0)/p_2$, and volume $\Phi = V/V_2$ defined in terms of the respective "2-D leak-in" time, pressure, and volume scales of [5]

$$t_2 = t_*, \quad p_2 = -\frac{Q_{I0}t_2}{\ell^2 hS}, \quad V_2 = \ell^2 hSp_2, \tag{7.13}$$

where $Q_{I0} = Q_0/n$ is the nominal fracture production rate. The normalized forms of (7.11) and (7.12) in terms of non-dimensional scales (7.13) are given by

$$\Phi_i(\tau) = \int_0^{\tau} \Phi_i^{aux}(\tau-\tau')\frac{d\pi}{d\tau'}d\tau'(i = 1, \ldots, n), \quad \sum_{i=1}^{n}\Phi_i(\tau) = n\tau, \tag{7.14}$$

respectively.

After applying Laplace transform, (7.14) becomes

$$\Phi_i(s) = s\pi(s)\Phi_i^{aux}(s), \quad \sum_{i=1}^{n}\Phi_i(s) = \frac{n}{s^2}. \tag{7.15}$$

In the following, we first consider the end member cases of production from a single fracture ($n = 1$) and an infinite fracture array ($n = \infty$), respectively, followed by an example of a finite fracture array solution.

7.4.1 Single Fracture

The pressure evolution in a single fracture produced at the nominal rate Q_{I0} is shown in Fig. 7.8 after [5]. This solution corresponds to the early-time, non-interacting regime for the n-fracture array when $Q_{I0} = Q_0/n$.

7.4.2 Infinite Fracture Array

Figure 7.8 illustrates the evolution of the normalized pressure with normalized time in an infinite fracture array characterized by the individual fracture production rate of Q_{I0}. We use (7.11) to obtain the pressure evolution in the infinite array problem where $V_i^{\text{aux}}(t) = \ell^2 hS\Phi_i^{\text{aux}}(4\alpha t/\ell^2)$ is given in Fig. 7.3b.

Similar to the step pressure decline (auxiliary) problem, the large-time asymptotic solution of the pressure response can be obtained from the 1-D diffusion (in the along-fracture direction) solution for a continuous sink,

$$\frac{p - p_0}{p_2} \approx \frac{\ell}{d}\sqrt{\frac{t}{t_2}}. \tag{7.16}$$

This asymptote is shown by a dashed line in Fig. 7.8.

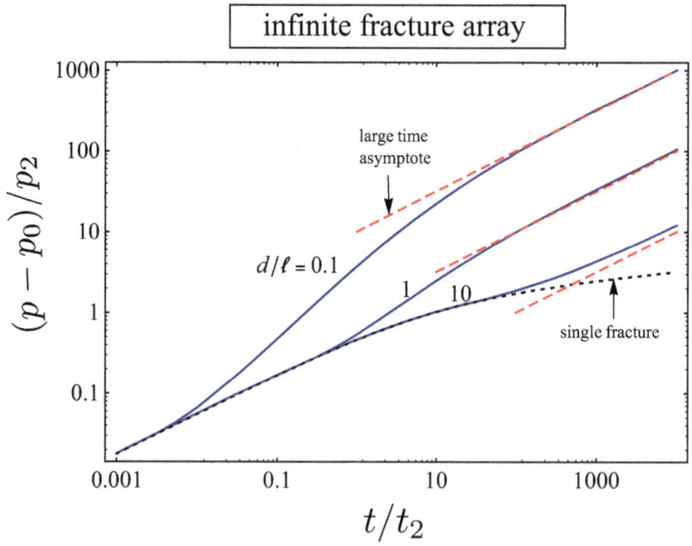

Fig. 7.8 Evolution of the normalized pressure with time in an infinite fracture array characterized by the individual fracture production rate of Q_{I0} and various fracture spacing d/ℓ. The single fracture solution is reproduced after [3, 5] (*black dotted line*). The large time asymptotic solution for the infinite fracture array [Eq. (7.16)] is shown by a *red dashed line*. Time and pressure scales are $t_2 = \ell^2/4\alpha$ and $p_2 = -Q_{I0}t_2/\ell^2 hS$, respectively

7.4.3 *Finite Fracture Array*

We investigate the pressure evolution and production from individual fractures in the $n = 21$ fracture array, previously considered in the context of step pressure decline (Sect. 7.3), and now considered to be produced at a constant volumetric flow rate of $Q_0 = 10^{-4}\ m^3/s$. The additional data (beyond that provided in Sect. 7.3) are $p_0 = 15.6$ MPa (initial reservoir pressure), and $p_b = 10.81$ MPa (pore fluid bubble point pressure) [17].

The calculated values of the relevant problem parameters are $t_2 = t_* = 111.1$ days (diffusion time scale), $p_2 = -7.14$ MPa (pressure scale), $V_2 = 6 \times 10^{-8}\ m^3$ (cumulative produced volume scale), and $(p_b - p_0)/p_2 = 0.67$ (pressure change at the fluid bubble point).

7.4.3.1 Pressure Evolution

The evolution of the fracture pressure in the finite fracture array problem is illustrated in Fig. 7.9. The solutions for a single fracture and an infinite fracture array produced at the nominal rate $Q_{I0} = Q_0/n$ per fracture are also shown for comparison. We locate the point with $(p_b - p_0)/p_2 = 0.032$ on Fig. 7.9 (point A) when the fluid pressure declined to the bubble point value. At this point the two-phase flow begins to develop, therefore invalidating the single phase flow assumption used in this study.

The onset of the two phase flow is forecasted at $t_b/t_2 = 1.45$, or in $t_b = 161.1$ days from the beginning of the production. As shown in Fig. 7.9a, the pressure evolution in the infinite and finite fracture array solutions is identical throughout the one-phase flow window.

7.4.3.2 Cumulative Produced Volume

Cumulative produced volumes from #1 (outer), #2, #5 (intermediates), and #11 (central) cracks are shown in Fig. 7.9b. Note that the outer and central cracks produce the largest and the smallest fractions of the total produced volume, respectively.

7.4.4 *Uniform Leak-in Approximation*

In this section, we investigate the applicability of the Gringarten et al. [3] approximate method, described earlier in Sect. 7.3.4 in the context of step pressure decline production, to the case of a constant global rate of production. The approximate solution method is essentially the same as described in the beginning of this section

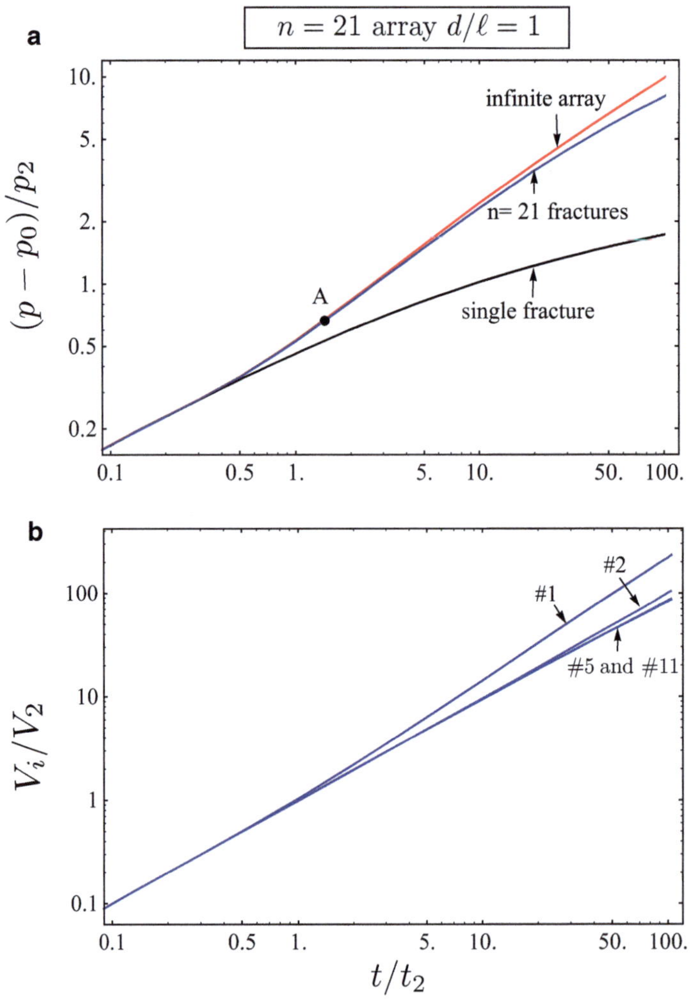

Fig. 7.9 Evolution of (**a**) the fracture pressure and (**b**) the cumulative leak-in volume for #1 (outer), #2, #5 (intermediate), and #11 (central) cracks in the $n = 21$ finite fracture array problem produced at a constant global production rate Q_0. The fracture pressure solutions for a single fracture and infinite fracture array produced at $Q_{I0} = Q_0/n$ per fracture are also shown. Time, pressure, and volume scales are $t_2 = \ell^2/4\alpha = 111.1$ days, $p_2 = -Q_{0I}t_2/\ell^2 hS = -7.14$ MPa, and $V_2 = \ell^2 hSp_2 = 6 \times 10^{-8}\ m^3$, respectively

when solving global volume balance (7.12) together with convolution integral expressions for volume produced from individual fractures, (7.11), with the only change corresponding to the replacement of the set of convolution kernels $V_i^{\text{aux}}(t)$ (corresponding to the fracture produced volumes in the auxiliary, step pressure decline problem) with the ones obtained through Gringarten's-like method in Sect. 7.3.4, denoted here by $V_i^{\text{app. aux}}(t)$. (Recall that the ratios of the approximate and the rigorous solutions, $V_i^{\text{app. aux}}(t)/V_i^{\text{aux}}(t)$, have been evaluated in Fig. 7.7b).

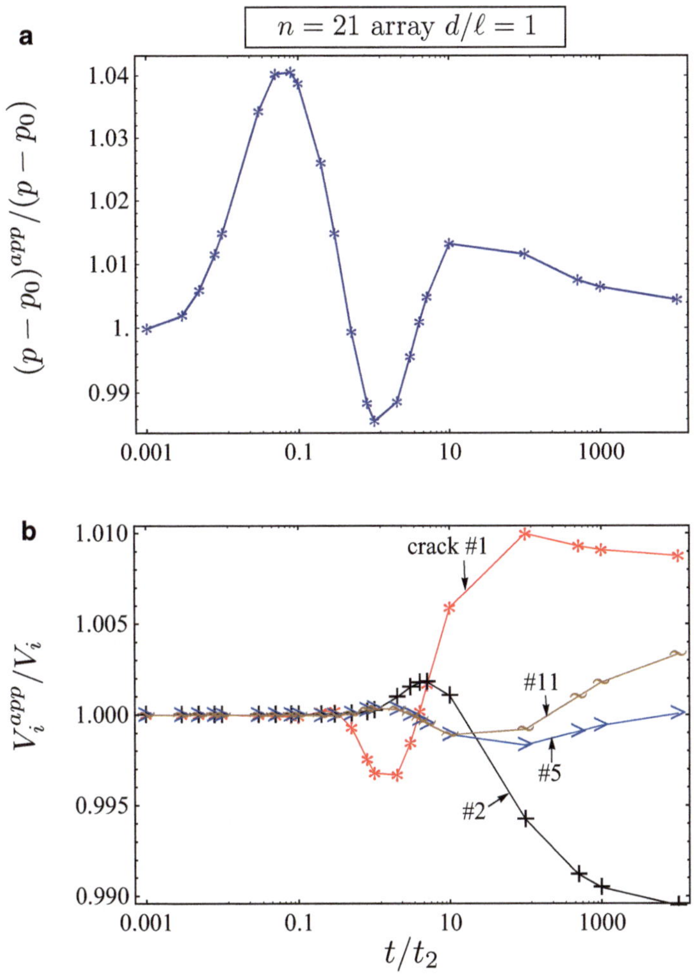

Fig. 7.10 Comparison between the approximate solution using the Gringarten's-like method and the rigorous solution for a finite fracture array produced at a constant volumetric rate: (**a**) the pressure evolution comparison and (**b**) fracture cumulative produced volumes comparison [cracks #1 (outer), #2, #5 (intermediate), and 11 (central)]. Time scale is $t_2 = \ell^2/4\alpha = 111.1$ days

Figure 7.10a compares the pressure response of the finite fracture array evaluated by using the full rigorous solution and the Gringarten's-like approximation. A similar comparison is also carried out for the fracture cumulative leak-in volumes for fractures #1, #2, #5, and #11 in Fig. 7.10b. As evident from Fig. 7.10, the uniform leak-in (Gringarten's) assumption provides a very good approximation of the pressure response (within 4 %) and of the fracture cumulative produced volumes (within 10 %) for the particular example of a finite fracture array considered here.

7.5 Summary

In this study, we addressed the pressure transient analysis of a multi-fractured horizontal well in the limit of large fracture hydraulic conductivity, corresponding to reasonably negligible fluid pressure drop in the flow within a fracture. This condition, leading to the uniform fluid pressure along a fracture, is representative of low-permeability, tight reservoirs (e.g., shale, tight sandstone). This, in turn, allows for a simpler and more accurate solution methods for reservoir fluid flow to a multi-fracture array.

We begun this study by presenting the solution for an auxiliary problem in which each individual fracture in a finite fracture array undergoes a unit step pressure decline (e.g., production at a constant bottom-hole pressure conditions). The general method of solution for the production rates and the cumulative produced volumes from fractures within the array due to a step pressure decline was presented and used to evaluate a particular example representative of the oil field. It was shown that in the finite number fracture array, the fluid production rates from individual fractures are approximately identical only at early production times when the effects of hydraulic interference between fractures are insignificant. An unequal partition of the produced fluid volume between the fractures in the array becomes noticeable at some extended production time, with the two outer, array-bounding fractures producing at ever increasing rate relative to the inner fractures in the array. The solutions to the limiting cases of production from a single fracture and an infinite fracture array were also presented. We showed that the total and individual fracture cumulative production volumes in the finite fracture array cannot be described by that of a single fracture past the onset of hydraulic interference effects between fractures. However, the infinite fracture array solution provides an excellent approximation for the total and individual fracture cumulative production volumes (with the exception of the two outer cracks) over realistic production times, i.e., up to 10 years.

The auxiliary solution for a step pressure decline was then used within the Green's function framework to formulate the pressure transient analysis of a multi-fractured well produced at a constant volumetric flow rate. We applied the analysis to a particular field example, and showed how it can be used to quantify the onset of the two-phase flow in the reservoir. We also showed that the corresponding infinite fracture array solution can be confidently used to approximate the production from a finite fracture array at least until the onset of the two-phase reservoir flow.

Finally, we considered an ad hoc simplification of the rigorous analysis of production from a multi-fractured well described in the above, following the framework previously studied by Gringarten et al. [3] in the case of a single crack. In this approach, the potentially intensive numerical treatment of the rigorous solution of a coupled system of convolution integral equations governing the time evolution of the spatially nonuniform rates of fluid exchange between the fractures and the surrounding rock (local fracture leak-in rates) are circumvented by ad hoc assumptions of a uniform leak-in along a fracture and of a particular choice for the evaluation of the fracture fluid pressure, as described in the main text. We show that the Gringarten et al. approach applied to finite fracture arrays provides an excellent alternative to the rigorous solution method, allowing to considerably simplify the transient analysis of multi-fractured horizontal wells at least within the considered geometrical constraints of identically sized, equally spaced fractures.

References

1. Economides, M. J., & Nolte, K. G. (Eds.). (2000). *Reservoir stimulation* (3rd ed.). Chichester, UK: Wiley.
2. Clarkson, C. R. (2013). Production data analysis of unconventional gas wells: Review of theory and best practices. *International Journal of Coal Geology, 109*, 101–146.
3. Gringarten, A. C., Ramey, H. J., & Raghavan, R. (1974). Unsteady-state pressure distributions created by a well with a single infinite-conductivity vertical fracture. *Society of Petroleum Engineers Journal, 14*(04), 347–360.
4. Detournay, E., & Cheng, A. H.-D. (1991). Plane strain analysis of a stationary hydraulic fracture in a poroelastic medium. *International Journal of Solids and Structures, 27*(13), 1645–1662.
5. Sarvaramini, E., & Garagash, D. I. (2015). Breakdown of a pressurized fingerlike crack in a permeable solid. *Journal of Applied Mechanics, 82*(6), 061006.
6. Cinco, L. H., Samaniego, V. F., & Dominguez, A. N. (1978). Transient pressure behavior for a well with a finite-conductivity vertical fracture. *Society of Petroleum Engineers Journal, 18* (4), 253–264.
7. Raghavan, R., Chen, C. C., & Agarwal, B. (1997). An analysis of horizontal wells intercepted by multiple fractures. *SPE Journal, 2*(03), 235–245.
8. Al-Ahmadi, H. A., Almarzooq, A. M., & Wattenbarger, R. A. (2010). Application of linear flow analysis to shale gas wells-field cases. In *SPE Unconventional Gas Conference*. Society of Petroleum Engineers. Pittsburgh, Pennsylvania, USA.
9. Brown, M., Ozkan, E., Raghavan, R., & Kazemi, H. (2011). Practical solutions for pressure-transient responses of fractured horizontal wells in unconventional shale reservoirs. *SPE Reservoir Evaluation & Engineering, 14*(06), 663–676.
10. Ozkan, E., Brown, M., Raghavan, R., & Kazemi, H. (2011). Comparison of fractured-horizontal-well performance in tight sand and shale reservoirs. *SPE Reservoir Evaluation & Engineering, 14*(02), 248–259.
11. Sarvaramini, E. (2015, August). *Reactivation of a Hydraulic Fracture in Permeable Rock*. PhD thesis, Dalhousie University, Halifax, Canada.
12. Montgomery, C. T., & Steanson, R. E. (1985). Proppant selection: The key to successful fracture stimulation. *Journal of Petroleum Technology, 37*(12), 2163–2172.
13. Ehlig-Economides, C. A. (1979). *Well Test Analysis for Wells Produced at a Constant Pressure*. PhD thesis, Stanford University, Stanford, CA.
14. Cinco-Ley, H., & Samaniego, V. F. (1981). Transient pressure analysis for fractured wells. *Journal of Petroleum Technology, 33*, 1749–1766.
15. Carslaw, H., & Jaeger, J. C. (1959). *Conduction of heat in solids* (2nd ed.). London, England: Oxford University Press.
16. Stehfest, H. (1970). Algorithm 368: Numerical inversion of laplace transforms [D5]. *Communications of the ACM, 13*(1), 47–49.
17. Clarkson, C. R., & Pedersen, P. K. (2011). Production analysis of Western Canadian unconventional light oil plays. In *Canadian Unconventional Resources Conference*. Society of Petroleum Engineers. Calgary, Alberta, Canada

Chapter 8
Interfacial Engineering for Oil and Gas Applications: Role of Modeling and Simulation

Kshitij C. Jha, Vikram Singh, and Mesfin Tsige

Abstract Interfaces control the functional performance of advanced materials used in the oil and natural gas industry for applications ranging from oil recovery, and flow assurance to gas separation, and carbon capture and utilization. The interactions that govern such functional performance are extremely challenging to obtain empirically. This is partly because of the instability at fluid interfaces, but also due to the intrinsic complexity in quantification of the behavior of a large number of components and interactions. Molecular modeling offers a pathway to examine confined wettability, specific adsorption, and cooperative network formation with changes in chemical structure that act as a design platform for custom functional performance. This is especially important in oil and natural gas processing because of the large number of variations introduced through changes in environment from one location to another. This chapter highlights the iterative design of injection fluids, kinetic inhibitors, separation membranes, and conversion technologies through mechanistic insight gained from simulations primarily based on molecular dynamics and density functional theory approaches.

8.1 Introduction

Innovative material design to improve process efficiencies and integration common in the oil and natural gas industry, such as separation, purification, extraction, transport, capture, and conversion, requires a mechanistic overview of the physico-chemical factors contributing to functional performance. Numerical

K.C. Jha • M. Tsige (✉)
Department of Polymer Science, The University of Akron, Akron, OH, USA
e-mail: kcj4@uakron.edu; mtsige@uakron.edu

V. Singh
Department of Chemical Engineering, Indian Institute of Technology Delhi, New Delhi, India
e-mail: vs225@chemical.iitd.ac.in

C. Jin, G. Cusatis (eds.), *New Frontiers in Oil and Gas Exploration*,
DOI 10.1007/978-3-319-40124-9_8

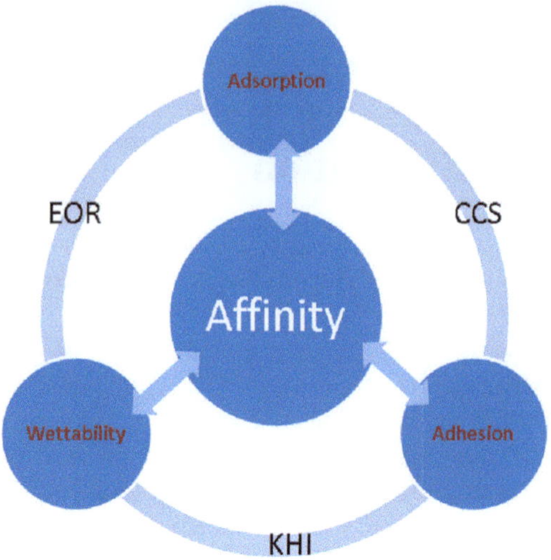

Fig. 8.1 Affinity controls adhesion, wettability, and adsorption which in turn guide design of materials for Carbon Capture and Separation (CCS), Enhanced Oil Recovery (EOR), and Hydrate Inhibition (KHI)

models and process simulation approaches have existed for decades, while material design through molecular modeling is gaining increasing importance in recent years [1–36]. For example, gas separation processes are being refined for improvements in permeability and selectivity, through rational design of robust hybrid membranes that incorporates insight from molecular modeling [37–49].

In key areas of oil recovery, flow assurance, carbon capture, conversion, and CO_2 utilization, controlled interaction at interfaces (fluid–fluid, membrane–gas mixture, hydrate-inhibitor, etc.) can only be achieved with insight into the molecular mechanisms driving key behaviors. Mechanisms that influence these behaviors include confined wettability, selective adsorption, adhesion, and cooperative network formation, all of which can be related to differentials in affinity (Fig. 8.1).

Molecular dynamics (MD) based approaches, because of their ability to access nanoscale regimes and buried interfaces, that are experimentally challenging to characterize, afford quantification that both aids design of experiments and leads innovation through discovery of novel materials and hybrids. Density functional theory (DFT) is commonly used in conjunction with MD for chemically accurate information on binding and activation energies that feed into facilitated transport and catalytic design. Differentials in affinity can especially be approached through relative breakdown of interactions, which has led to the recent increase in interests in molecular modeling approaches for oil and gas applications.

We highlight such approaches for the specific areas of enhanced oil recovery, hydrate formation and disruption, CO_2 capture, separation, and utilization in the following sections.

8.2 Enhanced Oil Recovery

In enhanced oil recovery, typically a fluid is injected to improve the recovery of oil. The overall effort through chemical injection is to reduce interfacial tension (between oil and water) and increase mobility of oil to reduce viscous fingering. In a conventional reservoir, large amount of oil is trapped in micropores. It is estimated that such trapping leads to a 50 % under-utilization in reservoir extraction.

Surfactants and their behavior in emulsions (oil-in-water, water-in-oil) have been studied extensively and are the most common chemical injection methods. Additionally, supercritical CO_2 approaches rival the brine/surfactant traditional processes, due to advantages of process integration and efficiencies due to amphiphilic behavior of CO_2, that has been hypothesized as acting like a surfactant. Both approaches are discussed in the following sections. Post recovery, the utilization of produced water is a concern that is being addressed through design of innovative membranes, discussed in the last subsection (Sect. 8.2.3).

8.2.1 Surfactants and Additives

Surfactant flooding has been a traditional means of enhanced oil recovery [50–53], due to the tunability offered by moiety design and solution concentration. There has been considerable research interest in both design of surfactants [54–63] and modeling of the flooding process [64–71] to optimize oil recovery.

While macroscopic models exist, a micro- or atomistic view of surfactant aided oil recovery is hindered by the plethora of interactions at play at the buried fluid interfaces. Understanding the molecular mechanism behind surfactant aided oil recovery entails quantifying the effect of two factors: hydrophobicity and confined wettability. Tail matching between the surfactant and oil molecules allows for greater miscibility and adsorption at the oil–water interface [73, 74]. The surfactant tail-oil molecule hydrophobic interactions initiate the process of recovery, which involves formation of a water gate [72] (Fig. 8.2). The mechanism illustrated in Fig. 8.2 shows a system of alkanes (dodecane) deposited on a silica surface that was penetrated by water molecules through long-range interactions aided by surfactants. It was shown that the hydrogen bond of the water molecules with silica surface and between themselves allowed formation and penetration of "water channel" through relative affinity [72].

Further investigation of the water gate mechanism through a Lattice Monte Carlo (LMC) approach [75] that takes into account surfactant shape and charge proposed three regimes of surfactant aided carrying, stripping, and diffusion of oil molecules from the substrate. It is to be noted that the recovery process would be greatly aided by an atomistic view since the changes in environment and

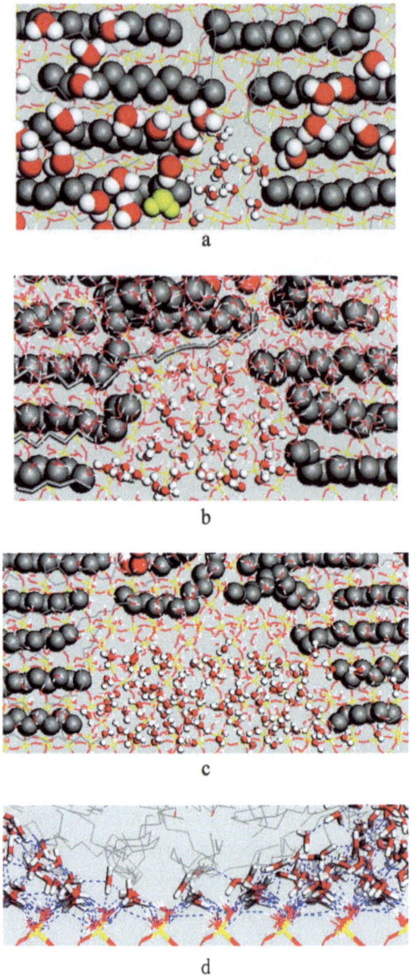

Fig. 8.2 Snapshots showing water gate formation at (**a**) 150 ps, (**b**) 700 ps, and (**c**) 1.5 ns. Water molecules in the channel have been highlighted and shown in ball-and-stick display mode, while the dodecane (C12) molecules are shown in CPK display mode. (**d**) Highlights the H-bonding between the water molecules and the -OH group of the silica surface. The snapshot in (**d**) has been taken at 4 ns, post detachment of the oil molecules from the silica surface. O- is shown in *red*, Si- in *yellow*, H- in *white*, CH_2- in *gray*, and N- in *blue*. Adapted with permission from [72]

geographies need mechanistic insight that feeds into the designed composition of injected fluids.

Reservoir wettability is also being explored as an additional parameter that can be affected by choice of surfactants and additives [76–83]. The effect of additives such as nano-silica [84–89] and metal oxide nanoparticles [90–98] have been observed to aid oil recovery, and been the subject of renewed research interest [99]. Nanoscale wettability needs statistical models that take into account the effect of confinement, as well as interactions resulting from cooperative affinity, charge screening, and dispersion forces. Efforts to compute line tension on chemically modified surfaces for multiscale models need careful calibration of long range interactions, especially when computing contact angles. Validated models would provide a path to controlled wettability alterations sought through introduction of functional nanoparticles and mesoporous materials [100–103].

8.2.2 Supercritical CO_2

The excellent solubility of CO_2 in oil and its ability to maintain reservoir pressure has made it a popular choice for oil recovery. At higher pressures, above 50 atm, the solubility increases and CO_2 is said to be in a supercritical phase. Utilizing all-atom molecular dynamics (MD), de Lara et al. [104] probed the interface of CO_2-crude oil and found that in the supercritical phase, CO_2 ($scCO_2$) becomes so miscible that it is hard to demarcate the interface. There is preferred interfacial accumulation of aromatics and the orientational order of various components at the interface can differ sharply (Fig. 8.3). With increase in alkane chain length (component of crude oil), there was an exhibited increase in the order parameter (Fig. 8.3b). Twin orientation peaks are observed for CO_2 in the supercritical state. The high mobility coupled with miscibility of both CO_2 and $scCO_2$ was found to exceed the performance of gases such as N_2 and CH_4, as well as the traditional usage of brine solutions.

Analysis of orientational order, diffusion dynamics, and interaction energies provides information that is hard to glean at the nanoscale level for fluid interfaces. Specifically, the diffusion of "like" species (molecules with similar molecular structure or solubility) in the oil-reagent mixture can be obtained through monitoring validated MD trajectories. It is important to understand the cooperative motion in such mixtures to gain insight into recovery solutions that would work for a given oil composition, and a given environmental (temperature, pressure, humidity, salt concentration, etc.) regime. Further, quantification of minimum miscible pressure (MMP), through trends in diffusion and solubility, would provide valuable operational guidelines.

The mechanism behind reduction of interfacial tension due to $scCO_2$ was examined by Liu et al. [105] for a water-decane(oil) system. They concluded that comparable affinities of water–CO_2 and oil–CO_2, due to amphiphilic nature of CO_2, causes accumulation and reorientation of oil moieties at the water-oil interface. The reduction in interfacial tension through this mechanism favors EOR. Amphiphilic behavior of CO_2 was also observed by Zhao et al. [106] in systems of CO_2–hexane/water–NaCl. In these ternary (hexane–CO_2–brine) systems, the roughness of the interface, and CO_2 diffusion increased with increase in CO_2 concentration, both of which contribute to reduction in interfacial tension. Interestingly, a maximum concentration of 46 % by weight fraction leads to surface excess of CO_2, with its behavior approaching that of a surfactant.

8.2.3 Produced Water Demulsification and Treatment

Industrial waste water generated from injecting mixtures of alkane–water–surfactant in various proportions in oil wells for enhanced recovery needs to be treated to reduce the environmental footprint. It can also be re-injected if the oil has been

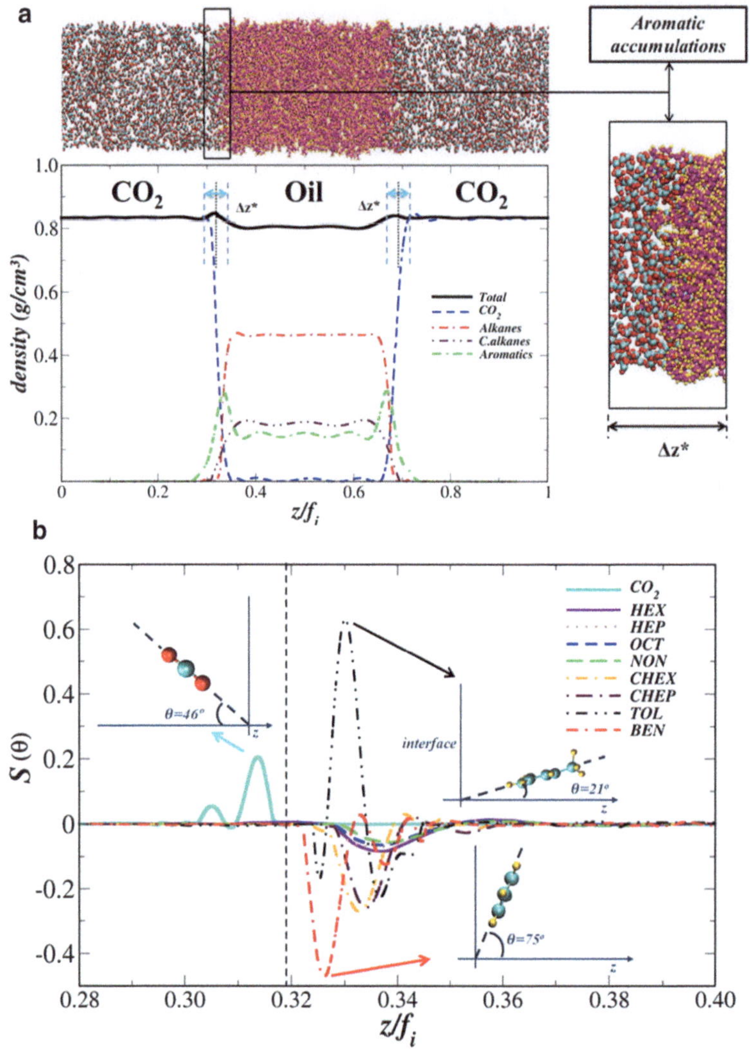

Fig. 8.3 The behavior of the crude oil interface in the presence of CO_2 (330 K, 150 atm). (**a**) Shows accumulation of the aromatics at the interface. (**b**) Shows the average orientational order of components, measured with respect to the *Z*-axis (surface normal). The *broken line* near 0.32 represents the Gibbs dividing surface. *Insets* in (**b**) also show the orientation of CO_2, toluene, and benzene molecules with respect to the *Z*-axis. The *Z*-axis has been scaled by varying amounts for individual components (factor f_i). Abbreviations are: *HEX* hexane, *HEP* heptane, *OCT* octane, *NON* nonane, *CHEX* cyclohexane, *CHEP* cycloheptane, *TOL* toluene, *BEN* benzene. Adapted with permission from [104]

remedied from the effluent. Micro-, ultra, and nano-filtration membranes have been developed over the decades [108, 109] with ceramic materials such as SiC [110, 111], which are both more expensive and durable when compared to polymer membranes such as PVDF.

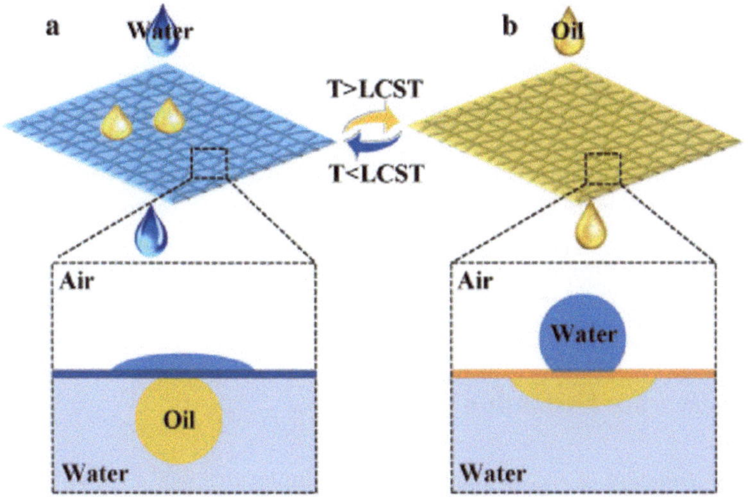

Fig. 8.4 Temperature dependent response of a PMMA-b-PNIPAAm membrane. (**a**) Shows an oleophobic/hydrophilic which switches to (**b**), an oleophilic/hydrophobic state upon heating. Adapted with permission from [107]

Switchability and super-wettability allow for a high degree of discrimination in separation of oil-in-water and water-in-oil emulsions [112–118]. To control the wettability of oil and water on the membrane surface, Tuteja and co-workers [119] deployed a novel approach wherein the membranes were hygro-responsive through incorporation of POSS functionalized nanoparticles. Control of "surface porosity" and "breakthrough pressure" (defined as maximum pressure below which the second phase does not permeate), led to continuous flow membrane design with larger than 99.9 % separation efficiency. Varanasi and co-workers proposed a more scalable version [120], based on the same principle of superhydrophilicity and superoleophobicity through a cold spray coat method of TiO_2. Various mesh [121, 122] and nanofiber [123, 124] based approaches that take advantage of hierarchy and chemistry towards designed wettability have also been proposed. For example, pH [125] and temperature responsive [107] (Fig. 8.4) membranes were synthesized, with near surface morphology varying with change in external stimuli [126–129]. Sensitive changes, with external stimuli, of modified surfaces such as self-assembled monolayers, have been quantified by work in the Tsige lab [130] through validated models that track tilt angle differentials, which are difficult to probe empirically. Such an approach provides an additional tool towards design of responsive wettable surfaces.

Incorporation of membranes and meshes along-with design of novel magnetic sorbents [131, 132] that take into account a large number of pollutants presents the next step in the evolution of EOR with reduced footprint and increased processing efficiency (through injection of treated water). Problems of flux, selectivity, robustness, and fouling are being addressed for oil-in-water and water-in-oil emulsion separation membranes through hybrid materials [123, 133, 134], hierarchical design [135], and biomimicry [136, 137].

8.3 Flow Assurance

The large methane concentration in pipelines is ideal for hydrate formation. Clogging of pipelines through hydrates leads to downtime and/or replacement, which are both expensive propositions. Additionally, it is a safety hazard. Methane hydrates on seabeds and in sediments constitute untapped energy, which by most estimates amount to more than twice all current traditional reserves. Controlled disruption of hydrate formation is industrially important and needs mechanistic insights that would feed into inhibitor design and possible alternate dissociation routes.

8.3.1 Hydrate Formation Mechanisms

A typical structure of gas hydrate constitutes molecules that are non-polar (gas) trapped (enclathrated) within polar water molecules with frozen, cage like, structure. Of particular interest is methane hydrate, which contains large amounts of untapped energy, and is present as sediments in oceans and deep lakes and on seabeds. Decomposition of methane hydrate is also a concern because of the potential release of CH_4, a more potent greenhouse gas than CO_2. Accelerated decomposition of methane hydrate may lead to geohazards. Thus the interest in the mechanism of methane hydrate formation is not only a scientific issue, with implications in understanding phobic–philic interactions and order–disorder transitions, but a compelling technical challenge.

Empirically, the spatial and temporal regimes of methane hydrate nucleation present a difficulty in quantification as they are on the nano-scale. Since MD trajectories can access such scales, considerable research has been done in modeling of methane hydrate formation and dissociation. Particularly, for gas pipelines, prevention of hydrate agglomeration would need a clear molecular scale understanding of the drivers behind hydrate formation. The hydrate nucleation itself qualifies as a rare event, which means to capture the mechanism, either independent dynamic runs for different starting configurations have to be done for large time periods (orders of μs) or sampling methods such as "transition path" and "forward flux" have to be employed [139–146].

Utilizing classical MD trajectories of μs, Walsh et al. [147] were able to show that the cooperative nature of bonding is a key driver for hydrate formation. They also found that while there was initial formation of the kinetically favored sII structure ($5^{12}6^4$ cage with face sharing), the mechanism eventually directed towards an increase in the thermodynamically favored sI structure ($5^{12}6^2$ cage without face sharing), through linkage from the uncommon $5^{12}6^3$ cages. The unusual occurrence of $5^{12}6^3$ cages was first reported by Vatamanu et al. [148] using atomistic simulations and shows the probative value of modeling in determining unpredictable structural formations and molecular ordering. For supersaturated mixtures of

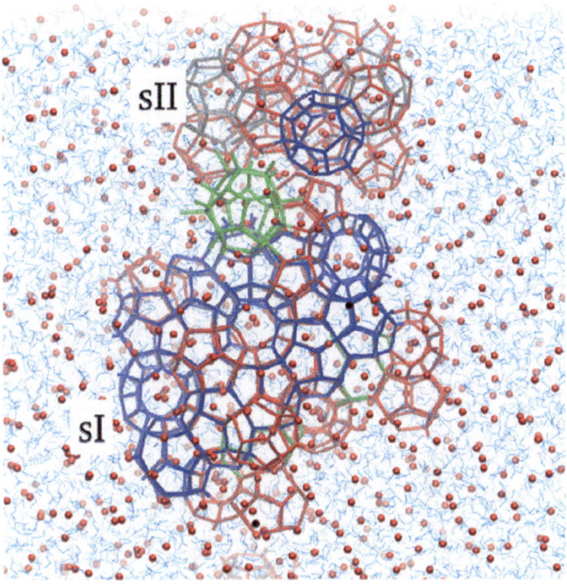

Fig. 8.5 Two crystalline structures of methane hydrate coexist for a supersaturated water–methane mixture. Adapted with permission from [138]

water–methane at moderate temperatures, it was found that the two crystalline structures coexist (Fig. 8.5).

The Dhinojwala research group at The University of Akron has found that the behavior of cations at the interface regulates formation of hydrates [149, 150]. To shed light on this behavior, ongoing collaborative research in the Tsige lab examines the modulation in interfacial water structure through change in surface charge and nature of cation [151]. Knowledge of hydrate structure and formation mechanism is key to their disruption, and the design of hydrate inhibitors.

8.3.2 *Kinetic Inhibitor Design*

Usage of inhibitors has traditionally been of the thermodynamic type, with a recent shift in interest towards kinetic hydrate inhibitors (KHIs) that allow for biodegradability while also utilizing lower quantities. The mechanism of hydrate formation disruption revolves around strengthening and weakening of local water structure around the inhibitor. Sa et al. [152] examined the disruption behavior of amino acids and correlated their inhibition efficacy to hydrophobicity. It was found that decrease in hydrophobicity of the amino acid disrupts the hydrogen bond network while an increase in hydrophobicity serves to strengthen it (Fig. 8.6). Hence, amino acids with lower hydrophobicity are better KHIs. The design principles from local disruption could lead to scalable synthesis of biodegradable, high efficiency KHIs.

One such approach through peptidomimetic structures, with modular incorporation of amino acid mimics as pendant groups that have been synthesized by our

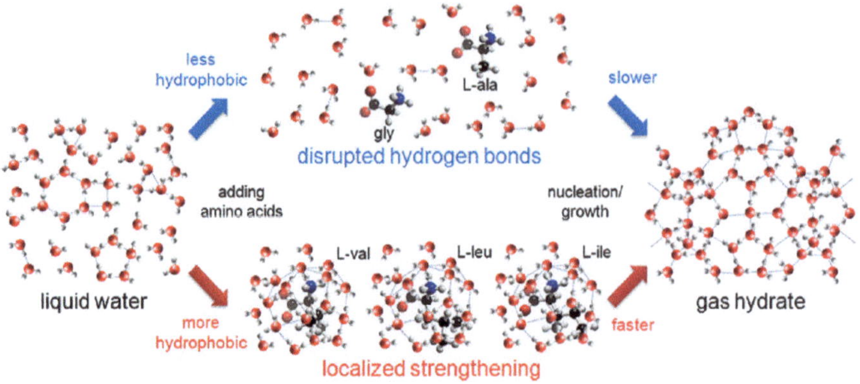

Fig. 8.6 Local water structure for less hydrophobic amino acids (glycine and L-alanine) showing disruption, and more hydrophobic amino acids (L-valine, L-leucine, and L-isoleucine) showing network formation. Adapted with permission from [152]

collaborators [153, 154], is currently under investigation in the Tsige lab, to examine their hydrophobic behavior and transient properties as a function of framework design. Computational screening for a range of inhibitors, with classical MD, has shown statistically significant variations in hydrate formation even for small explorative runs of less than 10 ns [155].

8.4 Carbon Capture and Separation

Large amounts of CO_2 produced from gas powered plants, and also purification needs of natural gas, make capture and separation of CO_2 a key area for materials development. Hybrid and novel material design has taken off exponentially, and ranges from hierarchical to enzymatic design. In the following sections, some approaches to design of adsorbents and membranes, and the challenges to field applications are discussed.

8.4.1 Adsorbents

Novel adsorbents and hierarchies for carbon capture and sequestration are constantly being developed with increased adsorption efficacy and high discrimination for enhanced selectivity [157–167]. The focus has shifted from amine based chemistries that have large energetic footprint for adsorption–desorption cycle to affinity directed, and porosity controlled covalent organic frameworks (COFs),

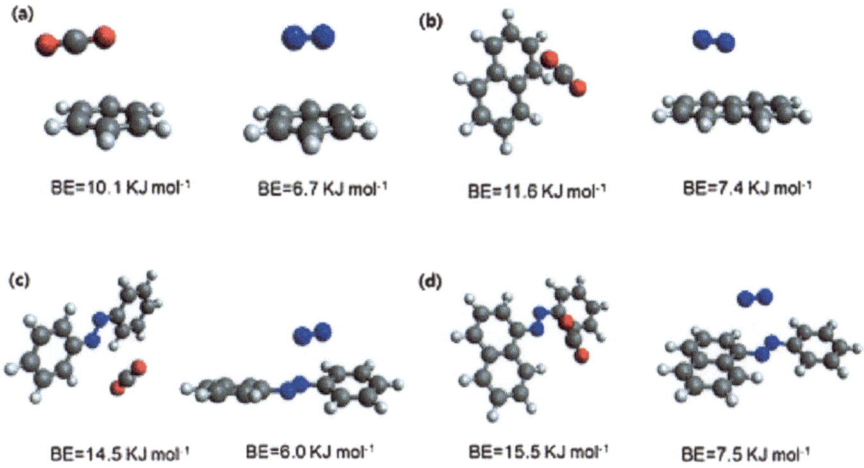

Fig. 8.7 Binding energy and optimized structures using DFT-D2 and a PBE functional for: (**a**) benzene, (**b**) naphthalene, (**c**) azo-benzene, and (**d**) azo-naphthalene. Adapted with permission from [156]

covalent organic polymers (COPs) [156, 168–171], and metal organic frameworks (MOFs).

Control in affinity, through change in interaction with the aromatic moiety bridged by azo groups, has allowed synthesis of COPs with the highest CO_2/N_2 selectivity reported to date [156, 168]. These COPs have the additional advantage of increase in selectivity with temperature increase, a key to post-combustion applications. Scalability and low-cost of azo-bridged COPs are afforded through catalyst free coupling reactions. The binding energy through DFT calculations show a mechanism of N_2 phobicity (Fig. 8.7). Monte Carlo simulations showed that the azo-bridge enthalpically favors CO_2 adsorption, but binding brings an entropic loss. Changes in chemistry (bridge groups and moieties) lead to different porosities, CO_2 uptake and selectivity [169–171] and points to molecular understanding being critical to adsorbent design and functionality.

8.4.2 *Membranes*

Polymeric membranes have a long history of utilization as gas separation media, with hollow fiber membranes being the most commonly deployed solution [172–175]. The desired properties of permeability and selectivity for a mixture of gases have been advanced through molecular engineering of backbones and pendant functional groups. For example, it was shown that for microporous organic polymers (MOPs), introducing rigidity in the backbone through a class of compounds classified as polymers with intrinsic microporosity (PIMs) [176–182], allowed greater control

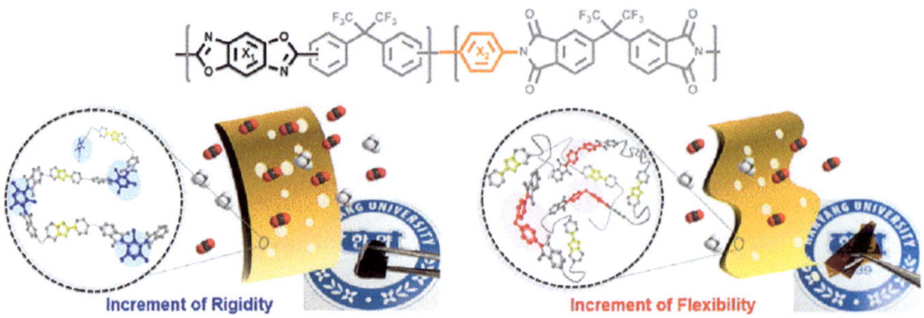

Fig. 8.8 Modification of thermally rearranged membranes with controlled ratios of polybenzoxazoles and polyimide leads to increase in mechanical strength and fractional free volume. The modified membranes show both high permeability and selectivity for CO_2/N_2 separations. Adapted with permission from [183]

over pore size and hence permeability. A similar design principle was applied to thermally rearranged polybenzoxazoles, where differentials in flexibility and rigidity led to increased robustness of the gas separation membrane (Fig. 8.8).

The formation of scalable MOF thin films and membranes [184–189], due to their high discrimination and CO_2 uptake would lead to advanced functional separation hierarchies that provide degrees of improvement in energetics. Robustness of MOFs is a concern that limits post-combustion applications. To overcome this limitation, mixed matrix membranes that incorporate MOFs have been fabricated and show promise in industrial application. Affinity based computational screening has been applied to both MOFs [190–200] and supported ionic liquid membranes (SILMs) [201–206].

An additional effort has been to electrospin membranes for carbon capture that incorporate functional moieties, such as adsorbents discussed in Sect. 8.4.1. Crucial to such design is the need for control over hierarchies. The Jana research group in the College of Polymer Science and Polymer Engineering at The University of Akron has developed a scalable method (gas jet nanofiber—GJF) [207] to morphologically controlled (side by side, encapsulated) nanofibers that are being explored in a collaborative effort with Tsige's research group as a possible means for functional nanofiber mat fabrication. Crucial amongst the design principles for such mats would be understanding the surface migration in common polymers, which has been extensively studied by the Tsige lab for stereoregular PMMA in a series of papers [208–210]. Stereocomplexation, coupled with functional incorporation, provides a polymeric design space that has also sparked increasing interest from a number of research groups. We believe that applications in selective separation membranes is a specific area that would benefit from stereocomplexation aided functional fabrication.

8.5 CO_2 Conversion and Utilization

The large amount of CO_2 available through capture, separation, and sequestration means pathways to utilization and conversion have been topics of vigorous scientific interest. While conversion to urea has been the industrially most relevant process, along-with use in EOR in the supercritical state outlined in Sect. 8.2.2, innovative routes to conversion towards polymers have emerged in recent years. Among the compounds, green-methanol synthesis and conversion to pharmaceutical compounds such as salicylic acid through the Kolbe–Schmitt reaction are significant in terms of current organochemical syntheses. The utilization for material manufacturing with research timescales is shown in Fig. 8.9.

Conversion of CO_2 to polycarbonates, linear and cyclic, is a viable economic pathway. Figure 8.10 shows the general conversion scheme, the free energies for which have been computed through chemically accurate DFT based computations [212]. CO_2 based polymers are currently an under-utilized industrial process under rapid development. Use of innovative chemistries and frameworks such as outlined in Fig. 8.11, which combine capture and conversion, would advance material science towards streamlined utilization. Catalytic efficiency would depend both on selective adsorption and consequent catalytic conversion-quantification on which can be obtained through dual employment of classical MD and DFT with validated force fields and functionals.

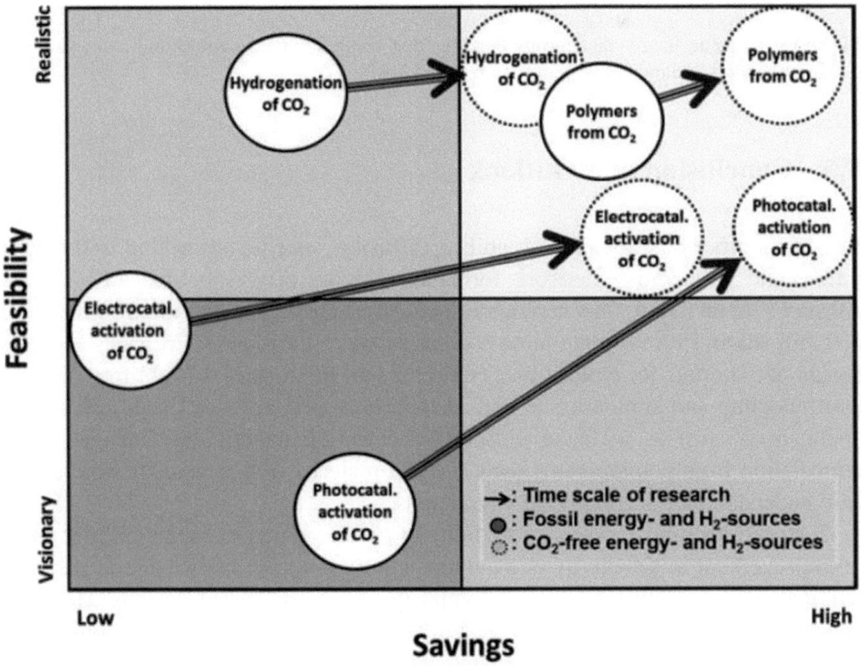

Fig. 8.9 Technologies for CO_2 utilization and their potential gauged through a qualitative assessment. Adapted with permission from [211]

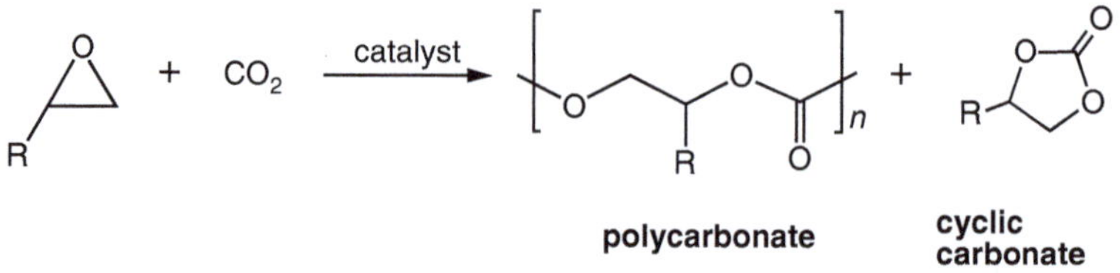

Fig. 8.10 CO_2 reaction with cyclic ethers to produce polycarbonate and cyclic carbonate. Adapted with permission from [212]

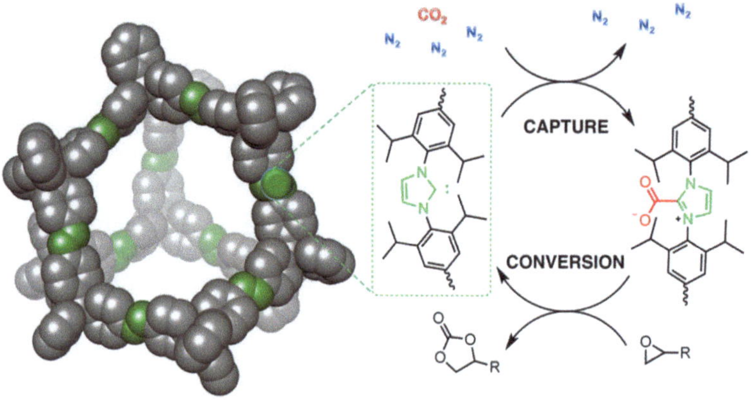

Fig. 8.11 Carbene based nanoporous polymer that combines CO_2 capture and conversion to cyclic carbonates. Adapted with permission from [213]

8.6 Conclusion and Outlook

Interfacial processes, such as wettability, diffusion, specific adsorption, and recognition, and cooperative network formation are the drivers behind advances in enhanced oil recovery, flow assurance, and CO_2 capture, sequestration, conversion, and utilization. Process integration such as between EOR and CCS needs bespoke design of materials for multifarious environments and locales. Insight from molecular modeling and simulation is critical to custom performance. Additionally, for applications such as methane sensors for detecting natural gas leakage, water remediation from contaminants such as TCE, material design benefits from atomistic understanding of interfacial interactions.

The need for robustness and scalability of advanced materials that offer degrees of improvement in selectivity or catalytic activity involves knowledge of parameters that would allow for layered, composite, or matrix based designs. Here the development of force fields that accurately predict the behavior of conjugated organics as well as organic–inorganic interactions would be keys to in silico development of functionality. Rapid prototyping, such as through additive manufacturing or roll to roll processes, typically involves extrusion or coating of gels under stress, where understanding of the role of functional fillers and their shear alignment, surface migration, and molecular ordering would be increasingly important. Statistical models that take into account both host–matrix interactions and external fields would need careful iterative development in conjunction with empirical results.

Through examples discussed above we have shown that advanced functional materials for a number of oil and gas applications have the need for understanding of molecular mechanisms not only for improvements in efficiency but also robustness in performance and scalability in manufacturing. A broader impact of interfacial engineering principles gained from the efforts outlined above has found use in the fields of thin film, membrane, coatings, and device design for separation, sensing, and catalysis.

Acknowledgements This work was made possible by funding from ACS Petroleum Research Fund (ACS PRF 54801- ND5) and National Science Foundation (NSF DMR-1410290).

References

1. Aman, Z. M., & Koh, C. A. (2016). Interfacial phenomena in gas hydrate systems. *Chemical Society Reviews, 45*(6), 1678–1690.
2. Pathak, M., Pawar, G., Huang, H., & Deo, M. D. (2015). Carbon dioxide sequestration and hydrocarbons recovery in the gas rich shales: An insight from the molecular dynamics simulations. In *Carbon Management Technology Conference*.
3. Fouad, W. A., Yarrison, M., Song, K. Y., Cox, K. R., & Chapman, W. G. (2015). High pressure measurements and molecular modeling of the water content of acid gas containing mixtures. *AIChE Journal, 61*(9), 3038–3052.
4. Jin, Z., & Firoozabadi, A. (2015). Thermodynamic modeling of phase behavior in shale media. *SPE Journal*.
5. Alvarez-Majmutov, A., Chen, J., & Michal Gieleciak, R. (2015). Molecular-level modeling and simulation of vacuum gas oil hydrocracking. *Energy & Fuels, 30*(1), 138–148.
6. Garcia-Ratés, M., de Hemptinne, J.-C., Avalos, J. B., & Nieto-Draghi, C. (2012). Molecular modeling of diffusion coefficient and ionic conductivity of CO2 in aqueous ionic solutions. *The Journal of Physical Chemistry B, 116*(9), 2787–2800.
7. Fouad, W. A., Song, K. Y., & Chapman, W. G. (2015). Experimental measurements and molecular modeling of the hydrate equilibrium as a function of water content for pressures up to 40 MPa. *Industrial & Engineering Chemistry Research, 54*(39), 9637–9644.
8. Pathak, M., Deo, M. D., Panja, P., & Levey, R. A. (2015). The effect of kerogen-hydrocarbons interaction on the pvt properties in liquid rich shale plays. In *SPE/CSUR Unconventional Resources Conference*. Society of Petroleum Engineers
9. Yu, W., Sepehrnoori, K., & Patzek, T. W. (2016). Modeling gas adsorption in Marcellus shale with Langmuir and bet isotherms. *SPE Journal, 21*(02), doi:10.2118/170801-PA.
10. Dehghanpour, H., Lan, Q., Saeed, Y., Fei, H., & Qi, Z. (2013). Spontaneous imbibition of brine and oil in gas shales: Effect of water adsorption and resulting microfractures. *Energy & Fuels, 27*(6), 3039–3049.

11. Kuznetsova, T., Kvamme, B., Parmar, A., Simos, T. E., & Maroulis, G. (2012). Molecular dynamics simulations of methane hydrate pre-nucleation phenomena and the effect of PVCap kinetic inhibitor. In *AIP Conference Proceedings-American Institute of Physics* (Vol. 1504, p. 776).
12. Firouzi, M., & Wilcox, J. (2012). Molecular modeling of carbon dioxide transport and storage in porous carbon-based materials. *Microporous and Mesoporous Materials, 158*, 195–203.
13. Welch, W. R. W., & Piri, M. (2016). Pore diameter effects on phase behavior of a gas condensate in graphitic one- and two-dimensional nanopores. *Journal of Molecular Modeling, 22*(1), 1–9.
14. Taylor, C. D., Chandra, A., Vera, J., & Sridhar, N. (2015). Design and prediction of corrosion inhibitors from quantum chemistry ii. A general framework for prediction of effective oil/water partition coefficients and speciation from quantum chemistry. *Journal of the Electrochemical Society, 162*(7), C347–C353.
15. Alvarez-Majmutov, A., Gieleciak, R., & Chen, J. (2015). Deriving the molecular composition of vacuum distillates by integrating statistical modeling and detailed hydrocarbon characterization. *Energy & Fuels, 29*(12), 7931–7940.
16. Alvarez-Majmutov, A., Chen, J., Gieleciak, R., Hager, D., Heshka, N., & Salmon, S. (2014). Deriving the molecular composition of middle distillates by integrating statistical modeling with advanced hydrocarbon characterization. *Energy & Fuels, 28*(12), 7385–7393.
17. Al-Arfaj, M., Sultan, A., & Abdulraheem, A. (2015). Understanding shale-fluid interactions using molecular modeling techniques for drilling applications: A literature review. In *SPE Kuwait Oil and Gas Show and Conference*. Society of Petroleum Engineers.
18. Khoshghadam, M., Lee, W. J., & Khanal, A. (2015). Numerical study of impact of nano-pores on gas-oil ratio and production mechanisms in liquid-rich shale oil reservoirs. In *Unconventional Resources Technology Conference (URTEC)*.
19. Bui, K., Akkutlu, I. Y., Zelenev, A., Saboowala, H., Gillis, J. R., Silas, J. A., et al. (2016). Insights into mobilization of shale oil by use of microemulsion. *SPE Journal*.
20. Liyana-Arachchi, T. P., Zhang, Z., Ehrenhauser, F. S., Avij, P., Valsaraj, K. T., & Hung, F. R. (2014). Bubble bursting as an aerosol generation mechanism during an oil spill in the deep-sea environment: Molecular dynamics simulations of oil alkanes and dispersants in atmospheric air/salt water interfaces. *Environmental Science: Processes & Impacts, 16*(1), 53–64.
21. Collell, J., Ungerer, P., Galliero, G., Yiannourakou, M., Montel, F., & Pujol, M. (2014). Molecular simulation of bulk organic matter in type ii shales in the middle of the oil formation window. *Energy & Fuels, 28*(12), 7457–7466.
22. Ungerer, P., Rigby, D., Leblanc, B., & Yiannourakou, M. (2014). Sensitivity of the aggregation behaviour of asphaltenes to molecular weight and structure using molecular dynamics. *Molecular Simulation, 40*(1–3), 115–122.
23. Firouzi, M., Alnoaimi, K., Kovscek, A., & Wilcox, J. (2014). Klinkenberg effect on predicting and measuring helium permeability in gas shales. *International Journal of Coal Geology, 123*, 62–68.
24. Jiménez-Ángeles, F., & Firoozabadi, A. (2014). Induced charge density and thin liquid film at hydrate/methane gas interfaces. *The Journal of Physical Chemistry C, 118*(45), 26041–26048.
25. Aimoli, C. G., Maginn, E. J., & Abreu, C. R. A. (2014). Force field comparison and thermodynamic property calculation of supercritical CO 2 and CH 4 using molecular dynamics simulations. *Fluid Phase Equilibria, 368*, 80–90.
26. Sun, M., & Firoozabadi, A. (2013). New surfactant for hydrate anti-agglomeration in hydrocarbon flowlines and seabed oil capture. *Journal of Colloid and Interface Science, 402*, 312–319.
27. Stukan, M., & Abdallah, W. (2015). Nano-confined adsorbed and free gas in shale reservoirs: A molecular dynamic study. In *SPE Middle East Oil & Gas Show and Conference*. Society of Petroleum Engineers.

28. Morimoto, M., Boek, E. S., Hibi, R., Matsuoka, T., Uetani, T., Murata, S., et al. (2014). Investigation of asphaltene-asphaltene association and aggregation for compositional reservoir simulators by quantitative molecular representations. In *International Petroleum Technology Conference*.
29. Kaya, S., Tüzün, B., Kaya, C., & Obot, I. B. (2016). Determination of corrosion inhibition effects of amino acids: Quantum chemical and molecular dynamic simulation study. *Journal of the Taiwan Institute of Chemical Engineers, 58*, 528–535.
30. Forte, E., Galindo, A., & Trusler, J. P. M. (2013). Experimental and molecular modelling study of the three-phase behaviour of (propane+ carbon dioxide+ water) at reservoir conditions. *The Journal of Supercritical Fluids, 75*, 30–42.
31. Sæthre, B. S., van der Spoel, D., & Hoffmann, A. C. (2012). Free energy of separation of structure ii clathrate hydrate in water and a light oil. *The Journal of Physical Chemistry B, 116*(20), 5933–5940.
32. dos Santos, E. S., de Souza, L. C. V., de Assis, P. N., Almeida, P. F., & Ramos-de Souza, E. (2014). Novel potential inhibitors for adenylylsulfate reductase to control souring of water in oil industries. *Journal of Biomolecular Structure and Dynamics, 32*(11), 1780–1792.
33. Metin, C. O., Baran, J. R., Jr., & Nguyen, Q. P. (2012). Adsorption of surface functionalized silica nanoparticles onto mineral surfaces and decane/water interface. *Journal of Nanoparticle Research, 14*(11), 1–16.
34. Kvamme, B., Kuznetsova, T., & Kivelæ, P.-H. (2012). Adsorption of water and carbon dioxide on hematite and consequences for possible hydrate formation. *Physical Chemistry Chemical Physics, 14*(13), 4410–4424.
35. Hu, Y., Devegowda, D., Striolo, A., Phan, A., Ho, T.A., Civan, F., et al. (2015). The dynamics of hydraulic fracture water confined in nano-pores in shale reservoirs. *Journal of Unconventional Oil and Gas Resources, 9*, 31–39.
36. Tenney, C. M., & Cygan, R. T. (2014). Molecular simulation of carbon dioxide, brine, and clay mineral interactions and determination of contact angles. *Environmental Science & Technology, 48*(3), 2035–2042.
37. Low, B. T., Xiao, Y., & Chung, T. S. (2009). Amplifying the molecular sieving capability of polyimide membranes via coupling of diamine networking and molecular architecture. *Polymer, 50*(14), 3250–3258.
38. Mahurin, S. M., Hillesheim, P. C., Yeary, J. S., Jiang, D.-E., & Dai, S. (2012). High CO 2 solubility, permeability and selectivity in ionic liquids with the tetracyanoborate anion. *RSC Advances, 2*(31), 11813–11819.
39. Kim, J., Abouelnasr, M., Lin, L.-C., & Smit, B. (2013). Large-scale screening of zeolite structures for CO2 membrane separations. *Journal of the American Chemical Society, 135*(20), 7545–7552.
40. Adibi, M., Barghi, S.H., & Rashtchian, D. (2011). Predictive models for permeability and diffusivity of CH 4 through imidazolium-based supported ionic liquid membranes. *Journal of Membrane Science, 371*(1), 127–133.
41. Yuanyan, G., Cussler, E. L., & Lodge, T. P. (2012). Aba-triblock copolymer ion gels for CO 2 separation applications. *Journal of Membrane Science, 423*, 20–26.
42. Bahukudumbi, P, & Ford, D. M. (2006). Molecular modeling study of the permeability-selectivity trade-off in polymeric and microporous membranes. *Industrial & Engineering Chemistry Research, 45*(16), 5640–5648.
43. Comesaña-Gándara, B., José, G., Hernández, A., Jo, H. J., Lee, Y. M., de Abajo, J., et al. (2015). Gas separation membranes made through thermal rearrangement of ortho-methoxypolyimides. *RSC Advances, 5*(124), 102261–102276.
44. Fried, J. R. (2006). Gas diffusion and solubility in poly (organophosphazenes): Results of molecular simulation studies. *Journal of Inorganic and Organometallic Polymers and Materials, 16*(4), 407–418.
45. Song, Q., Cao, S., Pritchard, R. H., Ghalei, B., Al-Muhtaseb, S. A., Terentjev, E. M., et al. (2014). Controlled thermal oxidative crosslinking of polymers of intrinsic

microporosity towards tunable molecular sieve membranes. *Nature Communications, 5*:4813, doi:10.1038/ncomms5813.
46. Diestel, L., Liu, X. L., Li, Y. S., Yang, W. S., & Caro, J. (2014). Comparative permeation studies on three supported membranes: Pure ZIF-8, pure polymethylphenylsiloxane, and mixed matrix membranes. *Microporous and Mesoporous Materials, 189*, 210–215
47. Au, H. (2012). *Molecular Dynamics Simulation of Nanoporous Graphene for Selective Gas Separation*. PhD thesis, Massachusetts Institute of Technology.
48. Bera, D., Bandyopadhyay, P., Ghosh, S., Banerjee, S., & Padmanabhan, V. (2015). Highly gas permeable aromatic polyamides containing adamantane substituted triphenylamine. *Journal of Membrane Science, 474*, 20–31.
49. Ohs, B., Lohaus, J., & Wessling, M. (2016). Optimization of membrane based nitrogen removal from natural gas. *Journal of Membrane Science, 498*, 291–301.
50. Gale, W. W., & Sandvik, E. I. (1973). Tertiary surfactant flooding: Petroleum sulfonate composition-efficacy studies. *Society of Petroleum Engineers Journal, 13*(04), 191–199.
51. Fathi, Z., & Ramirez, F. W. (1984). Optimal injection policies for enhanced oil recovery: Part 2-surfactant flooding. *Society of Petroleum Engineers Journal, 24*(03), 333–341.
52. Shuler, P., Blanco, M., Jang, S. S., Lin, S.-T., Maiti, P., Wu, Y., et al. (2004). Lower cost methods for improved oil recovery (IOR) via surfactant flooding. DoE Project Report.
53. Rosen, M. J., Wang, H., Shen, P., & Zhu, Y. (2005). Ultralow interfacial tension for enhanced oil recovery at very low surfactant concentrations. *Langmuir, 21*(9), 3749–3756.
54. Al-Amodi, A. O., Al-Mubaiyedh, U. A., Sultan, A. S., Kamal, M. S., & Hussein, I. A. (2015). Novel fluorinated surfactants for enhanced oil recovery in carbonate reservoirs. *The Canadian Journal of Chemical Engineering, 94*(3), 454–460.
55. Babu, K., Maurya, N. K., Mandal, A., & Saxena, V. K. (2015). Synthesis and characterization of sodium methyl ester sulfonate for chemically-enhanced oil recovery. *Brazilian Journal of Chemical Engineering, 32*(3), 795–803.
56. Yuan, C.-D., Pu, W.-F., Wang, X.-C., Sun, L., Zhang, Y.-C., & Cheng, S. (2015). Effects of interfacial tension, emulsification, and surfactant concentration on oil recovery in surfactant flooding process for high temperature and high salinity reservoirs. *Energy & Fuels, 29*(10), 6165–6176.
57. Co, L., Zhang, Z., Ma, Q., Watts, G., Zhao, L., Shuler, P.J., et al. (2015). Evaluation of functionalized polymeric surfactants for EOR applications in the Illinois basin. *Journal of Petroleum Science and Engineering, 134*, 167–175.
58. Babu, K., Pal, N., Bera, A., Saxena, V. K., & Mandal, A. (2015). Studies on interfacial tension and contact angle of synthesized surfactant and polymeric from castor oil for enhanced oil recovery. *Applied Surface Science, 353*, 1126–1136.
59. Husin, H., Ibrahim, M. N., Hassan, Z., Taib, N. M., Hamid, K. H. K., Ab Lah, N. K. I. N., & Shayuti, M. S. (2015). Overview on chemical-based, bio-based and natural-based surfactants in EOR applications. In *ICGSCE 2014* (pp. 3–9). Singapore: Springer.
60. Taiwo, O. A., & Olafuyi, O. A. (2015). Surfactant and surfactant-polymer flooding for light oil: A gum arabic approach. *Petroleum & Coal, 57*(3), 205–215.
61. Kittisrisawai, S., & Beatriz Romero-Zerón, L. (2015). Complexation of surfactant/β-cyclodextrin to inhibit surfactant adsorption onto sand, kaolin, and shale for applications in enhanced oil recovery processes. part iii: Oil displacement evaluation. *Journal of Surfactants and Detergents, 18*(5), 797–809.
62. Zhan, W., Pengyuan, Z., Guangwen, C., Haikui, Z., Yun, J., & Jianfeng, C. (2015). Synthesis of petroleum sulfonate surfactant with ultra-low interfacial tension in rotating packed bed reactor. *China Petroleum Processing & Petrochemical Technology, 17*(1), 59–68.
63. Luan, H., Wu, Y., Wu, W., Zhang, W., Chen, Q., Zhang, H., et al. (2015). Study on cardanolbetaine surfactants for ultralow interfacial tension in a low range of surfactant concentration and wide range of temperature applied in compound flooding. *Tenside Surfactants Detergents, 52*(1), 29–34.

64. Fletcher, P. D., Savory, L. D., Woods, F., Clarke, A., & Howe, A. M. (2015). Model study of enhanced oil recovery by flooding with aqueous surfactant solution and comparison with theory. *Langmuir, 31*(10), 3076–3085.
65. Chen, X., Feng, Q., Sepehrnoori, K., Goudarzi, A., & Bai, B. (2015). Mechanistic modeling of gel microsphere surfactant displacement for enhanced oil recovery after polymer flooding. In *SPE/IATMI Asia Pacific Oil & Gas Conference and Exhibition*. Society of Petroleum Engineers.
66. Jin, L., Jamili, A., Harwell, J. H., Shiau, B. J., & Roller, C. (2015). Modeling and interpretation of single well chemical tracer tests (SWCTT) for pre and post chemical EOR in two high salinity reservoirs. In *SPE Production and Operations Symposium*. Society of Petroleum Engineers.
67. Keshtkar, S., Sabeti, M., & Mohammadi, A. H. (2015). Numerical approach for enhanced oil recovery with surfactant flooding. *Petroleum, 2*(1), 98–107.
68. Danaev, N., Akhmed-Zaki, D., Mukhambetzhanov, S., & Imankulov, T. (2015). Mathematical modelling of oil recovery by polymer/surfactant flooding. In *Mathematical Modeling of Technological Processes* (pp. 1–12). Springer International Publishing.
69. Alshehri, A. J., & Khatib, A. M. (2015). Implementation time of chemical flood and its impact on ultimate recovery. In *IOR 2015—18th European Symposium on Improved Oil Recovery*.
70. Hosseini-Nasab, S. M., Padalkar, C., Battistutta, E., & Zitha, P. L. (2015). Mechanistic modelling of alkaline/surfactant/polymer flooding process at under-optimum salinity condition for enhanced oil recovery. In *SPE Asia Pacific Enhanced Oil Recovery Conference*. Society of Petroleum Engineers.
71. Tavassoli, S., Korrani, A. K. N., Pope, G. A., & Sepehrnoori, K. (2015). Low salinity surfactant flooding—A multi-mechanistic enhanced oil recovery method. In *SPE International Symposium on Oilfield Chemistry*. Society of Petroleum Engineers.
72. Liu, Q., Yuan, S., Yan, H., & Zhao, X. (2012). Mechanism of oil detachment from a silica surface in aqueous surfactant solutions: Molecular dynamics simulations. *The Journal of Physical Chemistry B, 116*(9), 2867–2875.
73. Tokiwa, Y., Sakamoto, H., Takiue, T., Aratono, M., & Matsubara, H. (2015). Effect of alkane chain length and counterion on the freezing transition of cationic surfactant adsorbed film at alkane mixture-water interfaces. *The Journal of Physical Chemistry B, 119*(20), 6235–6241.
74. Tamam, L., Pontoni, D., Sapir, Z., Yefet, S., Sloutskin, E., Ocko, B. M., et al. (2011). Modification of deeply buried hydrophobic interfaces by ionic surfactants. *Proceedings of the National Academy of Sciences, 108*(14), 5522–5525.
75. Wang, S., Li, Z., Liu, B., Zhang, X., & Yang, Q. (2015). Molecular mechanisms for surfactant-aided oil removal from a solid surface. *Applied Surface Science, 359*, 98–105.
76. Wang, Y., Xu, H., Yu, W., Bai, B., Song, X., & Zhang, J. (2011). Surfactant induced reservoir wettability alteration: Recent theoretical and experimental advances in enhanced oil recovery. *Petroleum Science, 8*(4), 463–476.
77. Giraldo, J., Benjumea, P., Lopera, S., Cortés, F. B., & Ruiz, M. A. (2013). Wettability alteration of sandstone cores by alumina-based nanofluids. *Energy & Fuels, 27*(7), 3659–3665.
78. Roustaei, A., Saffarzadeh, S., & Mohammadi, M. (2013). An evaluation of modified silica nanoparticles' efficiency in enhancing oil recovery of light and intermediate oil reservoirs. *Egyptian Journal of Petroleum, 22*(3), 427–433.
79. Castro Dantas, T. N., Soares A, P. J., Wanderley Neto, A. O., Dantas Neto, A. A., & Barros Neto, E. L. (2014). Implementing new microemulsion systems in wettability inversion and oil recovery from carbonate reservoirs. *Energy & Fuels, 28*(11), 6749–6759.
80. Hou, B.-F., Wang, Y.-F., & Huang, Y. (2015). Study of spontaneous imbibition of water by oil-wet sandstone cores using different surfactants. *Journal of Dispersion Science and Technology, 36*(9), 1264–1273.

81. Hou, B.-F., Wang, Y.-F., & Huang, Y. (2015). Mechanistic study of wettability alteration of oil-wet sandstone surface using different surfactants. *Applied Surface Science, 330*, 56–64.
82. Mohammed, M., & Babadagli, T. (2015). Wettability alteration: A comprehensive review of materials/methods and testing the selected ones on heavy-oil containing oil-wet systems. *Advances in Colloid and Interface Science, 220*, 54–77.
83. Hou, B., Wang, Y., Cao, X., Zhang, J., Song, X., Ding, M., et al. (2015). Mechanisms of enhanced oil recovery by surfactant-induced wettability alteration. *Journal of Dispersion Science and Technology, 37*(9), 1259–1267.
84. Ragab, A. M. S., & Hannora, A. E. (2015). An experimental investigation of silica nano particles for enhanced oil recovery applications. In *SPE North Africa Technical Conference and Exhibition*. Society of Petroleum Engineers.
85. Zargartalebi, M., Kharrat, R., & Barati, N. (2015). Enhancement of surfactant flooding performance by the use of silica nanoparticles. *Fuel, 143*, 21–27.
86. Umar, M., Novriansyah, A., Rita, N., & Husbani, A. (2015). Effect of nanosilica injection to oil recovery factor in low porosity and permeability reservoir. *Jurnal Intelek, 9*(2), 11–13.
87. Ahmadi, M. A., & Shadizadeh, S. R. (2013). Induced effect of adding nano silica on adsorption of a natural surfactant onto sandstone rock: Experimental and theoretical study. *Journal of Petroleum Science and Engineering, 112*, 239–247.
88. Zeyghami, M., Kharrat, R., & Ghazanfari, M. H. (2014). Investigation of the applicability of nano silica particles as a thickening additive for polymer solutions applied in EOR processes. *Energy Sources, Part A: Recovery, Utilization, and Environmental Effects, 36*(12), 1315–1324.
89. Dai, C., Wang, S., Li, Y., Gao, M., Liu, Y., Sun, Y., et al. (2015). The first study of surface modified silica nanoparticles in pressure-decreasing application. *RSC Advances, 5*(76), 61838–61845.
90. Bayat, A. E., Junin, R., Shamshirband, S., & Chong, W. T. (2015). Transport and retention of engineered Al2O3, TiO2, and SiO2 nanoparticles through various sedimentary rocks. *Scientific Reports, 5*:14264, doi:10.1038/srep14264.
91. Bayat, A. E., Junin, R., Mohsin, R., Hokmabadi, M., & Shamshirband, S. (2015). Influence of clay particles on Al2O3 and TiO2 nanoparticles transport and retention through limestone porous media: Measurements and mechanisms. *Journal of Nanoparticle Research, 17*(5), 1–14.
92. Kazemzadeh, Y., Eshraghi, S. E., Sourani, S., & Reyhani, M. (2015). An interface-analyzing technique to evaluate the heavy oil swelling in presence of nickel oxide nanoparticles. *Journal of Molecular Liquids, 211*, 553–559.
93. Nazari Moghaddam, R., Bahramian, A., Fakhroueian, Z., Karimi, A., & Arya, S. (2015). Comparative study of using nanoparticles for enhanced oil recovery: Wettability alteration of carbonate rocks. *Energy & Fuels, 29*(4), 2111–2119.
94. Cheraghian, G., & Hendraningrat, L. (2015). A review on applications of nanotechnology in the enhanced oil recovery part b: Effects of nanoparticles on flooding. *International Nano Letters, 6*(1), 1–10.
95. Bhuiyan, M. H. U., Saidur, R., Mostafizur, R. M., Mahbubul, I. M., & Amalina, M. A. (2015). Experimental investigation on surface tension of metal oxide–water nanofluids. *International Communications in Heat and Mass Transfer, 65*, 82–88.
96. Bayat, A. E., Junin, R., Derahman, M. N., & Samad, A. A. (2015). TiO 2 nanoparticle transport and retention through saturated limestone porous media under various ionic strength conditions. *Chemosphere, 134*, 7–15.
97. Sygouni, V., & Chrysikopoulos, C. V. (2015). Characterization of TiO 2 nanoparticle suspensions in aqueous solutions and TiO 2 nanoparticle retention in water-saturated columns packed with glass beads. *Chemical Engineering Journal, 262*, 823–830.
98. Al-Marshed, A., Hart, A., Leeke, G., Greaves, M., & Wood, J. (2015). Optimization of heavy oil upgrading using dispersed nanoparticulate iron oxide as a catalyst. *Energy & Fuels, 29* (10), 6306–6316.

99. Tang, J., Quinlan, P. J., & Tam, K. C. (2015). Stimuli-responsive pickering emulsions: Recent advances and potential applications. *Soft Matter, 11*(18), 3512–3529.
100. de Lara, L. S., Michelon, M. F., Metin, C. O., Nguyen, Q. P., & Miranda, C. R. (2012). Interface tension of silica hydroxylated nanoparticle with brine: A combined experimental and molecular dynamics study. *The Journal of Chemical Physics, 136*(16), 164702.
101. Metin, C. O., Lake, L. W., Miranda, C. R., & Nguyen, Q. P. (2011). Stability of aqueous silica nanoparticle dispersions. *Journal of Nanoparticle Research, 13*(2), 839–850.
102. Rigo, V. A., de Lara, L. S., & Miranda, C. R. (2014). Energetics of formation and hydration of functionalized silica nanoparticles: An atomistic computational study. *Applied Surface Science, 292*, 742–749.
103. Wu, D., Guo, X., Sun, H., & Navrotsky, A. (2015). Energy landscape of water and ethanol on silica surfaces. *The Journal of Physical Chemistry C, 119*(27), 15428–15433.
104. de Lara, L. S., Michelon, M. F., & Miranda, C. R. (2012). Molecular dynamics studies of fluid/oil interfaces for improved oil recovery processes. *The Journal of Physical Chemistry B, 116*(50), 14667–14676.
105. Liu, B., Shi, J., Wang, M., Zhang, J., Sun, B., Shen, Y., et al. (2015). Reduction in interfacial tension of water-oil interface by supercritical CO 2 in enhanced oil recovery processes studied with molecular dynamics simulation. *The Journal of Supercritical Fluids*, 111, 171–178.
106. Zhao, L., Tao, L., & Lin, S. (2015). Molecular dynamics characterizations of the supercritical CO2-mediated hexane–brine interface. *Industrial & Engineering Chemistry Research, 54*(9), 2489–2496.
107. Li, J.-J., Zhu, L.-T., & Luo, Z.-H. (2016). Electrospun fibrous membrane with enhanced swithchable oil/water wettability for oily water separation. *Chemical Engineering Journal, 287*, 474–481.
108. Alzahrani, S., & Mohammad, A. W. (2014). Challenges and trends in membrane technology implementation for produced water treatment: A review. *Journal of Water Process Engineering, 4*, 107–133.
109. Drioli, E., Ali, A., Lee, Y. M., Al-Sharif, S. F., Al-Beirutty, M., & Macedonio, F. (2015). Membrane operations for produced water treatment. *Desalination and Water Treatment, 57* (31), 1–19.
110. Bakshi, A. K., Ghimire, R., Sheridan, E., & Kuhn, M. (2015). Treatment of produced water using silicon carbide membrane filters. *Advances in Bioceramics and Porous Ceramics VIII: Ceramic Engineering and Science Proceedings, 36*(5), 91.
111. Frisk, S., Lim, H. S., Bates, L. C., Andrin, P., & El-Borno, B. (2014). Produced water treatment in oil recovery, October 8 2014. US Patent App. 14/509,201.
112. Tao, M., Xue, L., Liu, F., & Jiang, L. (2014). An intelligent superwetting PVDF membrane showing switchable transport performance for oil/water separation. *Advanced Materials, 26* (18), 2943–2948.
113. Yuan, T., Meng, J., Hao, T., Wang, Z., & Zhang, Y. (2015). A scalable method toward superhydrophilic and underwater superoleophobic PVDF membranes for effective oil/water emulsion separation. *ACS Applied Materials & Interfaces, 7*(27), 14896–14904.
114. Ou, R., Wei, J., Jiang, L., Simon, G. P., & Wang, H. (2015). Robust thermo-responsive polymer composite membrane with switchable superhydrophilicity and superhydrophobicity for efficient oil-water separation. *Environmental Science & Technology, 50*(2), 906–914.
115. Sasaki, K., Tenjimbayashi, M., Manabe, K., & Shiratori, S. (2015). Asymmetric superhydrophobic/superhydrophilic cotton fabrics designed by spraying polymer and nanoparticles. *ACS Applied Materials & Interfaces, 8*(1), 651–659.
116. Zhu, H., & Guo, Z. (2016). Understanding the separations of oil/water mixtures from immiscible to emulsions on super-wettable surfaces. *Journal of Bionic Engineering, 13*(1), 1–29.
117. Si, Y., & Guo, Z. (2015). Superwetting materials of oil-water emulsion separation. *Chemistry Letters, 44*(7), 874–883.

118. Zhu, H., & Guo, Z. (2015). Order separation of oil/water mixtures by superhydrophobic/ superoleophilic Cu (OH) 2-thioled films. *Chemistry Letters, 44*(10), 1431–1433.
119. Kota, A. K., Kwon, G., Choi, W., Mabry, J. M., & Tuteja, A. (2012). Hygro-responsive membranes for effective oil–water separation. *Nature Communications, 3*, 1025.
120. Gondal, M. A., Sadullah, M. S., Dastageer, M. A., McKinley, G. H., Panchanathan, D., & Varanasi, K. K. (2014). Study of factors governing oil–water separation process using TiO2 films prepared by spray deposition of nanoparticle dispersions. *ACS Applied Materials & Interfaces, 6*(16), 13422–13429.
121. Deng, D., Prendergast, D. P., MacFarlane, J., Bagatin, R., Stellacci, F., & Gschwend, P. M. (2013). Hydrophobic meshes for oil spill recovery devices. *ACS Applied Materials & Interfaces, 5*(3), 774–781.
122. Lin, X., Lu, F., Chen, Y., Liu, N., Cao, Y., Xu, L., et al. (2015). One-step breaking and separating emulsion by tungsten oxide coated mesh. *ACS Applied Materials & Interfaces, 7* (15), 8108–8113.
123. Si, Y., Fu, Q., Wang, X., Zhu, J., Yu, J., Sun, G., et al. (2015). Superelastic and superhydrophobic nanofiber-assembled cellular aerogels for effective separation of oil/water emulsions. *ACS Nano, 9*(4), 3791–3799.
124. Huang, M., Si, Y., Tang, X., Zhu, Z., Ding, B., Liu, L., et al. (2013). Gravity driven separation of emulsified oil–water mixtures utilizing in situ polymerized superhydrophobic and superoleophilic nanofibrous membranes. *Journal of Materials Chemistry A, 1*(45), 14071–14074.
125. Xiang, Y., Shen, J., Wang, Y., Liu, F., & Xue, L. (2015). A pH-responsive PVDF membrane with superwetting properties for the separation of oil and water. *RSC Advances, 5*(30), 23530–23539.
126. Cao, P.-F., Mangadlao, J. D., & Advincula, R. C. (2015). Stimuli-responsive polymers and their potential applications in oil-gas industry. *Polymer Reviews, 55*(4), 706–733.
127. Che, H., Huo, M., Peng, L., Fang, T., Liu, N., Feng, L., et al. (2015). CO2-responsive nanofibrous membranes with switchable oil/water wettability. *Angewandte Chemie International Edition, 54*(31), 8934–8938.
128. Wang, X., Yu, J., Sun, G., & Ding, B. (2015). Electrospun nanofibrous materials: A versatile medium for effective oil/water separation. *Materials Today*, doi:10.1016/j.mattod.2015.11. 010.
129. Kwon, G., Post, E., & Tuteja, A. (2015). Membranes with selective wettability for the separation of oil–water mixtures. *MRS Communications, 5*(03), 475–494.
130. Yimer, Y. Y., Jha, K. C., & Tsige, M. (2014). Epitaxial transfer through end-group coordination modulates the odd–even effect in an alkanethiol monolayer assembly. *Nanoscale, 6*(7), 3496–3502.
131. Li, S., Li, N., Yang, S., Liu, F., & Zhou, J. (2014). The synthesis of a novel magnetic demulsifier and its application for the demulsification of oil-charged industrial wastewaters. *Journal of Materials Chemistry A, 2*(1), 94–99.
132. Flores, J. A., Pavía-Sanders, A., Chen, Y., Pochan, D. J., & Wooley, K. L. (2015). Recyclable hybrid inorganic/organic magnetically active networks for the sequestration of crude oil from aqueous environments. *Chemistry of Materials, 27*(10), 3775–3782.
133. Hu, L., Gao, S., Zhu, Y., Zhang, F., Jiang, L., & Jin, J. (2015). An ultrathin bilayer membrane with asymmetric wettability for pressure responsive oil/water emulsion separation. *Journal of Materials Chemistry A, 3*(46), 23477–23482.
134. Yang, Y., Tong, Z., Ngai, T., & Wang, C. (2014). Nitrogen-rich and fire-resistant carbon aerogels for the removal of oil contaminants from water. *ACS Applied Materials & Interfaces, 6*(9), 6351–6360.
135. Solomon, B. R., Hyder, Md. N., & Varanasi, K. K. (2014). Separating oil-water nanoemulsions using flux-enhanced hierarchical membranes. *Scientific Reports, 4*:5504, doi:10.1038/srep05504.

136. Wang, Z.-X., Lau, C.-H., Zhang, N.-Q., Bai, Y.-P., & Shao, L. (2015). Mussel-inspired tailoring of membrane wettability for harsh water treatment. *Journal of Materials Chemistry A, 3*(6), 2650–2657.
137. Wang, B., Liang, W., Guo, Z., & Liu, W. (2015). Biomimetic super-lyophobic and super-lyophilic materials applied for oil/water separation: A new strategy beyond nature. *Chemical Society Reviews, 44*(1), 336–361.
138. Jiménez-Ángeles, F., & Firoozabadi, A. (2014). Nucleation of methane hydrates at moderate subcooling by molecular dynamics simulations. *The Journal of Physical Chemistry C, 118* (21), 11310–11318.
139. Peters, B., Zimmermann, N. E. R., Beckham, G. T., Tester, J. W., & Trout, B. L. (2008). Path sampling calculation of methane diffusivity in natural gas hydrates from a water-vacancy assisted mechanism. *Journal of the American Chemical Society, 130*(51), 17342–17350.
140. Sarupria, S., & Debenedetti, P. G. (2012). Homogeneous nucleation of methane hydrate in microsecond molecular dynamics simulations. *The Journal of Physical Chemistry Letters, 3* (20), 2942–2947.
141. Barnes, B. C., & Sum, A. K. (2013). Advances in molecular simulations of clathrate hydrates. *Current Opinion in Chemical Engineering, 2*(2), 184–190.
142. Sum, A. K., Wu, D. T., & Yasuoka, K. (2011). Energy science of clathrate hydrates: Simulation-based advances. *MRS Bulletin, 36*(03), 205–210.
143. Bi, Y., & Li, T. (2014). Probing methane hydrate nucleation through the forward flux sampling method. *The Journal of Physical Chemistry B, 118*(47), 13324–13332.
144. Małolepsza, E., & Keyes, T. (2015). Pathways through equilibrated states with coexisting phases for gas hydrate formation. *The Journal of Physical Chemistry B, 119*(52), 15857–15865.
145. Yagasaki, T., Matsumoto, M., & Tanaka, H. (2015). Adsorption mechanism of inhibitor and guest molecules on the surface of gas hydrates. *Journal of the American Chemical Society, 137*(37), 12079–12085.
146. Yagasaki, T., Matsumoto, M., & Tanaka, H. (2015). Effects of thermodynamic inhibitors on the dissociation of methane hydrate: A molecular dynamics study. *Physical Chemistry Chemical Physics, 17*(48), 32347–32357.
147. Walsh, M. R., Koh, C. A., Sloan, E. D., Sum, A. K., & Wu, D. T. (2009). Microsecond simulations of spontaneous methane hydrate nucleation and growth. *Science, 326*(5956), 1095–1098.
148. Vatamanu, J., & Kusalik, P. G. (2006). Unusual crystalline and polycrystalline structures in methane hydrates. *Journal of the American Chemical Society, 128*(49), 15588–15589.
149. Anim-Danso, E., Zhang, Y., & Dhinojwala, A. (2013). Freezing and melting of salt hydrates next to solid surfaces probed by infrared–visible sum frequency generation spectroscopy. *Journal of the American Chemical Society, 135*(23), 8496–8499.
150. Zhang, Y., Anim-Danso, E., & Dhinojwala, A. (2014). The effect of a solid surface on the segregation and melting of salt hydrates. *Journal of the American Chemical Society, 136*(42), 14811–14820.
151. Jha, K. C., Anim-Danso, E., Bekele, S., Eason, G., & Tsige, M. (2016). On modulating interfacial structure towards improved anti-icing performance. *Coatings, 6*(1), 3.
152. Sa, J.-H., Kwak, G.-H., Lee, B. R., Park, D.-H., Han, K., & Lee, K.-H. (2013). Hydrophobic amino acids as a new class of kinetic inhibitors for gas hydrate formation. *Scientific Reports, 3*:2428, doi:10.1038/srep02428.
153. Gokhale, S., Xu, Y., & Joy, A. (2013). A library of multifunctional polyesters with "peptide-like" pendant functional groups. *Biomacromolecules, 14*(8), 2489–2493.
154. Swanson, J. P., Monteleone, L. R., Haso, F., Costanzo, P. J., Liu, T., & Joy, A. (2015). A library of thermoresponsive, coacervate-forming biodegradable polyesters. *Macromolecules, 48*(12), 3834–3842.

155. Oluwunmi, P. A., Finney, A. R., & Rodger, P. M. (2015). Molecular dynamics screening for new kinetic inhibitors of methane hydrate. *Canadian Journal of Chemistry, 93*(9), 1043–1049.
156. Patel, H. A., Je, S. H., Park, J., Jung, Y., Coskun, A., & Yavuz, C. T. (2014). Directing the structural features of N2-phobic nanoporous covalent organic polymers for CO2 capture and separation. *Chemistry-A European Journal, 20*(3), 772–780.
157. Qian, D., Lei, C., Hao, G.-P., Li, W.-C., & Lu, A.-H. (2012). Synthesis of hierarchical porous carbon monoliths with incorporated metal–organic frameworks for enhancing volumetric based CO2 capture capability. *ACS Applied Materials & Interfaces, 4*(11), 6125–6132.
158. Bali, S., Chen, T. T., Chaikittisilp, W., & Jones, C. W. (2013). Oxidative stability of amino polymer–alumina hybrid adsorbents for carbon dioxide capture. *Energy & Fuels, 27*(3), 1547–1554.
159. Chaikittisilp, W., Khunsupat, R., Chen, T. T., & Jones, C. W. (2011). Poly (allylamine)–mesoporous silica composite materials for CO2 capture from simulated flue gas or ambient air. *Industrial & Engineering Chemistry Research, 50*(24), 14203–14210.
160. Yang, H., Xu, Z., Fan, M., Gupta, R., Slimane, R. B., Bland, A. E., et al. (2008). Progress in carbon dioxide separation and capture: A review. *Journal of Environmental Sciences, 20*(1), 14–27.
161. Goeppert, A., Czaun, M., May, R. B., Surya Prakash, G. K., Olah, G. A., & Narayanan, S. R. (2011). Carbon dioxide capture from the air using a polyamine based regenerable solid adsorbent. *Journal of the American Chemical Society, 133*(50), 20164–20167.
162. Drage, T. C., Snape, C. E., Stevens, L. A., Wood, J., Wang, J., Cooper, A. I., et al. (2012). Materials challenges for the development of solid sorbents for post-combustion carbon capture. *Journal of Materials Chemistry, 22*(7), 2815–2823.
163. Choi, S., Drese, J. H., & Jones, C. W. (2009). Adsorbent materials for carbon dioxide capture from large anthropogenic point sources. *ChemSusChem, 2*(9), 796–854.
164. Chatti, R., Bansiwal, A. K., Thote, J. A., Kumar, V., Jadhav, P., Lokhande, S. K., et al. (2009). Amine loaded zeolites for carbon dioxide capture: Amine loading and adsorption studies. *Microporous and Mesoporous Materials, 121*(1), 84–89.
165. Herm, Z. R., Swisher, J. A., Smit, B., Krishna, R., & Long, J. R. (2011). Metal-organic frameworks as adsorbents for hydrogen purification and precombustion carbon dioxide capture. *Journal of the American Chemical Society, 133*(15), 5664–5667.
166. Li, P., Zhang, S., Chen, S., Zhang, Q., Pan, J., & Ge, B. (2008). Preparation and adsorption properties of polyethylenimine containing fibrous adsorbent for carbon dioxide capture. *Journal of Applied Polymer Science, 108*(6), 3851–3858.
167. Wang, Q., Luo, J., Zhong, Z., & Borgna, A. (2011). CO2 capture by solid adsorbents and their applications: Current status and new trends. *Energy & Environmental Science, 4*(1), 42–55.
168. Patel, H. A., Je, S. H., Park, J., Chen, D. P., Jung, Y., Yavuz, C. T., et al. (2013). Unprecedented high-temperature CO2 selectivity in N2-phobic nanoporous covalent organic polymers. *Nature Communications, 4*, 1357.
169. Byun, J., Je, S.-H., Patel, H. A., Coskun, A., & Yavuz, C. T. (2014). Nanoporous covalent organic polymers incorporating tröger's base functionalities for enhanced CO 2 capture. *Journal of Materials Chemistry A, 2*(31), 12507–12512.
170. Patel, H. A., Karadas, F., Byun, J., Park, J., Deniz, E., Canlier, A., et al. (2013). Highly stable nanoporous sulfur-bridged covalent organic polymers for carbon dioxide removal. *Advanced Functional Materials, 23*(18), 2270–2276.
171. Patel, H. A., Karadas, F., Canlier, A., Park, J., Deniz, E., Jung, Y., et al. (2012). High capacity carbon dioxide adsorption by inexpensive covalent organic polymers. *Journal of Materials Chemistry, 22*(17), 8431–8437.
172. Khalilpour, R., Abbas, A., Lai, Z., & Pinnau, I. (2012). Modeling and parametric analysis of hollow fiber membrane system for carbon capture from multicomponent flue gas. *AIChE Journal, 58*(5), 1550–1561.

173. Zhang, Y.-T., Zhang, L., Chen, H.-L., & Zhang, H.-M. (2010). Selective separation of low concentration CO 2 using hydrogel immobilized ca enzyme based hollow fiber membrane reactors. *Chemical Engineering Science, 65*(10), 3199–3207.
174. Ren, J., Wang, R., Zhang, H.-Y., Li, Z., Liang, D. T., & Tay, J. H. (2006). Effect of PVDF dope rheology on the structure of hollow fiber membranes used for CO 2 capture. *Journal of Membrane Science, 281*(1), 334–344.
175. Kim, S., Han, S. H., & Lee, Y. M. (2012). Thermally rearranged (TR) polybenzoxazole hollow fiber membranes for CO 2 capture. *Journal of Membrane Science, 403*, 169–178.
176. Guiver, M. D., & Lee, Y. M. (2013). Polymer rigidity improves microporous membranes. *Science, 339*(6117), 284–285.
177. McKeown, N. B., Budd, P. M., Msayib, K. J., Ghanem, B. S., Kingston, H. J., Tattershall, C. E., et al. (2005). Polymers of intrinsic microporosity (PIMs): Bridging the void between microporous and polymeric materials. *Chemistry-A European Journal, 11*(9), 2610–2620.
178. Budd, P. M., Msayib, K. J., Tattershall, C. E., Ghanem, B. S., Reynolds, K. J., McKeown, N. B., et al. (2005). Gas separation membranes from polymers of intrinsic microporosity. *Journal of Membrane Science, 251*(1), 263–269.
179. Budd, P. M., McKeown, N. B., & Fritsch, D. (2005). Free volume and intrinsic microporosity in polymers. *Journal of Materials Chemistry, 15*(20), 1977–1986.
180. Budd, P. M., Makhseed, S. M., Ghanem, B. S., Msayib, K. J., Tattershall, C. E., & McKeown, N. B. (2004). Microporous polymeric materials. *Materials Today, 7*(4), 40–46.
181. Peng, F., Lu, L., Sun, H., Wang, Y., Liu, J., & Jiang, Z. (2005). Hybrid organic-inorganic membrane: Solving the tradeoff between permeability and selectivity. *Chemistry of Materials, 17*(26), 6790–6796.
182. Swaidan, R., Ghanem, B., Litwiller, E., & Pinnau, I. (2015). Physical aging, plasticization and their effects on gas permeation in "rigid" polymers of intrinsic microporosity. *Macromolecules, 48*(18), 6553–6561.
183. Jo, H. J., Soo, C. Y., Dong, G., Do, Y. S., Wang, H. H., Lee, M. J., et al. (2015). Thermally rearranged poly (benzoxazole-co-imide) membranes with superior mechanical strength for gas separation obtained by tuning chain rigidity. *Macromolecules, 48*(7), 2194–2202.
184. Li, J.-R., Sculley, J., & Zhou, H.-C. (2011). Metal–organic frameworks for separations. *Chemical Reviews, 112*(2), 869–932.
185. Chen, Z., Xiang, S., Arman, H. D., Mondal, J. U., Li, P., Zhao, D., et al. (2011). Three-dimensional pillar-layered copper (ii) metal-organic framework with immobilized functional oh groups on pore surfaces for highly selective CO2/CH4 and C2H2/CH4 gas sorption at room temperature. *Inorganic Chemistry, 50*(8), 3442–3446.
186. Andirova, D., Cogswell, C. F., Lei, Y., & Choi, S. (2016). Effect of the structural constituents of metal organic frameworks on carbon dioxide capture. *Microporous and Mesoporous Materials, 219*, 276–305.
187. Venna, S. R., & Carreon, M. A. (2015). Metal organic framework membranes for carbon dioxide separation. *Chemical Engineering Science, 124*, 3–19.
188. Buchan, I., Ryder, M. R., & Tan, J.-C. (2015). Micromechanical behavior of polycrystalline metal–organic framework thin films synthesized by electrochemical reaction. *Crystal Growth & Design, 15*(4), 1991–1999.
189. Seth, S., Savitha, G., & Moorthy, J. N. (2015). Carbon dioxide capture by a metal–organic framework with nitrogen-rich channels based on rationally designed triazole-functionalized tetraacid organic linker. *Inorganic Chemistry, 54*(14), 6829–6835.
190. Keskin, S., & Sholl, D. S. (2008). Assessment of a metal-organic framework membrane for gas separations using atomically detailed calculations: CO2, CH4, N2, H2 mixtures in MOF-5. *Industrial & Engineering Chemistry Research, 48*(2), 914–922.
191. Erucar, I., & Keskin, S. (2012). Computational screening of metal organic frameworks for mixed matrix membrane applications. *Journal of Membrane Science, 407*, 221–230.

192. Watanabe, T., & Sholl, D. S. (2012). Accelerating applications of metal–organic frameworks for gas adsorption and separation by computational screening of materials. *Langmuir, 28*(40), 14114–14128.
193. Krishna, R., & van Baten, J. M. (2011). Investigating the potential of MgMOF-74 membranes for CO 2 capture. *Journal of Membrane Science, 377*(1), 249–260.
194. Pillai, R. S., Benoit, V., Orsi, A., Llewellyn, P. L., Wright, P. A., & Maurin, G. (2015). Highly selective CO2 capture by small pore scandium-based metal–organic frameworks. *The Journal of Physical Chemistry C, 119*(41), 23592–23598.
195. Yang, Q., Liu, D., & Zhong, C. (2015). Molecular modeling of gas separation in metal–organic frameworks. In *Metal-Organic Frameworks: Materials Modeling Towards Engineering Applications* (p. 295). Pan Stanford Publishing.
196. Erucar, I., & Keskin, S. (2013). High CO2 selectivity of an amine-functionalized metal organic framework in adsorption-based and membrane-based gas separations. *Industrial & Engineering Chemistry Research, 52*(9), 3462–3472.
197. Adatoz, E., Avci, A. K., & Keskin, S. (2015). Opportunities and challenges of MOF-based membranes in gas separations. *Separation and Purification Technology, 152*, 207–237.
198. Shah, M., McCarthy, M. C., Sachdeva, S., Lee, A. K., & Jeong, H.-K. (2012). Current status of metal–organic framework membranes for gas separations: Promises and challenges. *Industrial & Engineering Chemistry Research, 51*(5), 2179–2199.
199. Jeazet, H. B. T., Staudt, C., & Janiak, C. (2012). Metal–organic frameworks in mixed-matrix membranes for gas separation. *Dalton Transactions, 41*(46), 14003–14027.
200. Ismail, A. F., Khulbe, K. C., & Matsuura, T. (2015). Gas separation membrane materials and structures. In *Gas Separation Membranes* (pp. 37–192). Springer International Publishing.
201. Hojniak, S. D., Khan, A. L., Hollóczki, O., Kirchner, B., Vankelecom, I. F. J., Dehaen, W., et al. (2013). Separation of carbon dioxide from nitrogen or methane by supported ionic liquid membranes (SILMs): Influence of the cation charge of the ionic liquid. *The Journal of Physical Chemistry B, 117*(48), 15131–15140.
202. Chen, Y., Hu, Z., Gupta, K. M., & Jiang, J. (2011). Ionic liquid/metal–organic framework composite for CO2 capture: A computational investigation. *The Journal of Physical Chemistry C, 115*(44), 21736–21742.
203. Dai, Z., Noble, R. D., Gin, D. L., Zhang, X., & Deng, L. (2016). Combination of ionic liquids with membrane technology: A new approach for CO 2 separation. *Journal of Membrane Science, 497*, 1–20.
204. Bara, J. E., Gabriel, C. J., Carlisle, T. K., Camper, D. E., Finotello, A., Gin, D. L., et al. (2009). Gas separations in fluoroalkyl-functionalized room-temperature ionic liquids using supported liquid membranes. *Chemical Engineering Journal, 147*(1), 43–50.
205. Ramdin, M. (2015). *CO2 Capture with Ionic Liquids: Experiments and Molecular Simulations*. PhD thesis, TU Delft, Delft University of Technology.
206. Tomé, L. C., Isik, M., Freire, C. S. R., Mecerreyes, D., & Marrucho, I. M. (2015). Novel pyrrolidinium-based polymeric ionic liquids with cyano counter-anions: High performance membrane materials for post-combustion CO 2 separation. *Journal of Membrane Science, 483*, 155–165.
207. Benavides, R. E., Jana, S. C., & Reneker, D. H. (2012). Nanofibers from scalable gas jet process. *ACS Macro Letters, 1*(8), 1032–1036.
208. Zhu, H., Jha, K. C., Bhatta, R. S., Tsige, M., & Dhinojwala, A. (2014). Molecular structure of poly (methyl methacrylate) surface. i. Combination of interface-sensitive infrared–visible sum frequency generation, molecular dynamics simulations, and ab initio calculations. *Langmuir, 30*(39), 11609–11618.
209. Jha, K. C., Zhu, H., Dhinojwala, A., & Tsige, M. (2014). Molecular structure of poly (methyl methacrylate) surface ii: Effect of stereoregularity examined through all-atom molecular dynamics. *Langmuir, 30*(43), 12775–12785.
210. Jha, K. C., Dhinojwala, A., & Tsige, M. (2015). Local structure contributions to surface tension of a stereoregular polymer. *ACS Macro Letters, 4*(11), 1234–1238.

211. Müller, T. E., Leitner, W., Markewitz, P., & Kuckshinrichs, W. (2015). Opportunities for utilizing and recycling CO2. In *Carbon Capture, Storage and Use* (pp. 67–100). Springer International Publishing.
212. Darensbourg, D. J., & Yeung, A. D. (2012). Thermodynamics of the carbon dioxide–epoxide copolymerization and kinetics of the metal-free degradation: A computational study. *Macromolecules, 46*(1), 83–95.
213. Talapaneni, S. N., Buyukcakir, O., Je, S. H., Srinivasan, S., Seo, Y., Polychronopoulou, K., et al. (2015). Nanoporous polymers incorporating sterically confined n-heterocyclic carbenes for simultaneous CO2 capture and conversion at ambient pressure. *Chemistry of Materials, 27*(19), 6818–6826.

Chapter 9
Petroleum Geomechanics: A Computational Perspective

Maurice B. Dusseault, Robert Gracie, Dipanjan Basu, Leo Rothenburg, and Shunde Yin

Abstracts Petroleum geomechanics is concerned with rock and fracture behavior in reservoir, drilling, completion, and production engineering. Typical problems in petroleum geomechanics include subsidence, borehole stability, and hydraulic fracturing. All are coupled problems that involve heat transfer, fluid flow, rock/fracture deformation, and/or solute transport. Numerical solutions through modeling are desired for such complicated systems. In this chapter, we present the mathematical descriptions of these typical problems in petroleum geomechanics, point out the challenges in solving these problems, and address those challenges by a variety of classical and emerging numerical techniques.

9.1 Introduction

Petroleum geomechanics is concerned with rock and fracture behavior in reservoir, drilling, completion, and production engineering. Typical problems in petroleum geomechanics include subsidence, borehole stability, and hydraulic fracturing. All are coupled problems that involve heat transfer, fluid flow, rock/fracture deformation, and/or solute transport. Numerical solutions through modeling are desired for such complicated systems. In this chapter, we present the mathematical descriptions of these typical problems in petroleum geomechanics, point out the challenges in solving these problems, and address those challenges by a variety of classical and emerging numerical techniques.

M.B. Dusseault
Department of Earth and Environmental Sciences, University of Waterloo, Waterloo, ON, Canada, N2L 3G1

R. Gracie • D. Basu • L. Rothenburg
Department of Civil and Environmental Engineering, University of Waterloo, Waterloo, ON, Canada, N2L 3G1

S. Yin (✉)
Department of Petroleum Engineering, University of Wyoming, Laramie, WY 82071, USA
e-mail: syin@uwyo.edu

C. Jin, G. Cusatis (eds.), *New Frontiers in Oil and Gas Exploration*,
DOI 10.1007/978-3-319-40124-9_9

9.2 Subsidence

In thick, high-porosity reservoirs, compaction during pressure depletion can lead to surface subsidence great enough to cause increased risk to facilities. In the case of the Ekofisk Field, over 9 m of reservoir compaction has led to 8 m of seafloor subsidence. The economic consequences were vast: inserting additional 6 m leg sections in the platforms at a cost of $\$485 \times 10^6$ in 1988–1990, abandonment of all the seafloor storage that made Ekofisk the hub of the southern North Sea, and a total field redevelopment in 1999–2001 at a cost of $\$2 \times 10^9$ (OGJ) [1]. The real technical shock of Ekofisk on the petroleum geomechanics community is that despite many signs of high compaction potential before decisions were made about platform installation in the 1970s, engineers failed to predict and mitigate the consequences of subsidence. One hopes that geomechanics has since progressed enough that subsidence can be accurately predicted and misinterpretations do not recur. Geertsma [2] developed an analytical solution of subsidence in half-space in response to certain pressure depletion in disc-shaped reservoirs. To deal with scenarios of more complicated boundary conditions, numerical models have proved more practical. However, most models failed to relate pressure depletion to subsidence in a half-space domain, which has been suggested necessary.

Field observations suggest that while pressure depletion is a local process that occurs mainly inside aquifers or reservoirs, through induced volume changes it triggers a redistribution of effective stress in a more extensive domain that may involve not just the surrounding strata, but the entire overburden and surrounding rocks out to considerable distances. In an analytical solution presented by Rothenburg et al. [3] for transient two-dimensional radial flow of a compressible fluid into a fully penetrating line well, it is shown that the stiffness of the overburden is an essential coupling element; that is, redistribution of stresses depends on the relative stiffness of the reservoir surroundings. Hettema et al. [4] also show that depletion-induced subsidence modeling requires incorporating the surrounding strata mechanical response.

To simulate the subsidence in half-space, we developed a 3D fully-coupled single-multiphase model of a reservoir and surrounding rocks (referred later to as the DDFEM model). In this model, the displacement discontinuity method (DDM) from seam mining simulation [5] is introduced for the surrounding rocks and combined with the finite element method (FEM) approach for the reservoir. A 3-D, large-scale reservoir is thereby simulated to exist in a vastly larger semi-infinite poroelastic domain through a 3-D FEM model accounting for the reservoir (based on poroelasticity) and a 3-D DDM model to incorporate the semi-infinite outer domain (based on elasticity).

The DDM—displacement discontinuity method—is an indirect boundary element method for solving problems in solid mechanics. This method proved especially useful for simulating large-scale mining activity in tabular ore bodies which extend at most a few meters in one direction and hundreds or thousands of meters in the other two [5, 6]. It is also used for analyzing other geomechanical cases involving displacements along faults or joints, and in fracture mechanics [7]. An advantage of the DDM for problems in geomechanics, like any boundary element method, is that the boundary conditions at infinity are automatically satisfied. Hence, full domain discretization and stipulation of boundary conditions on non-infinite boundaries can be avoided, reducing computation time in dealing with elastic rocks surrounding an opening or a reservoir.

In mining problems, the displacement discontinuity is defined as the relative displacement between the roof and floor of a small area of a seam-like deposit. Similarly, for a petroleum reservoir, the displacement discontinuity components can be defined as the relative displacement components between the top and bottom of a small area of a tabular reservoir, inclined or horizontal.

Consider a displacement discontinuity as a plane crack with a normal in the x_3 direction; its two faces can be distinguished by specifying one in the positive side ($x_3=0^+$) and the other is in the negative side ($x_3=0^-$). In crossing from one side to the other, the displacements undergo a specified change in value $D_i=(D_1, D_2, D_3)$ given by

$$D_1(x_1,x_2,0) = u_1(x_1,x_2,0^-) - u_1(x_1,x_2,0^+) \tag{9.1}$$

$$D_2(x_1,x_2,0) = u_2(x_1,x_2,0^-) - u_2(x_1,x_2,0^+) \tag{9.2}$$

$$D_3(x_1,x_2,0) = u_3(x_1,x_2,0^-) - u_3(x_1,x_2,0^+) \tag{9.3}$$

The general form solution for a displacement discontinuity element in an isotropically elastic space can be expressed as [5, 7]

$$u_1 = \left\{\left[2(1-v)\phi_{1,2} - x_3\phi_{1,13}\right] - x_3\phi_{2,12} - \left[(1-2v)\phi_{3,1} + x_3\phi_{1,13}\right]\right\} \tag{9.4}$$

$$u_2 = \left\{\left[2(1-v)\phi_{2,3} - x_3\phi_{2,22}\right] - x_3\phi_{1,12} - \left[(1-2v)\phi_{3,2} + x_3\phi_{3,23}\right]\right\} \tag{9.5}$$

$$u_3 = \left\{\left[2(1-v)\phi_{3,3} - x_3\phi_{3,33}\right] + \left[(1-2v)\phi_{1,1} - x_3\phi_{1,13}\right] - \left[(1-2v)\phi_{2,2} - x_3\phi_{2,23}\right]\right\} \tag{9.6}$$

$$\sigma_{11} = 2G\left\{\left[2\phi_{1,13} - x_3\phi_{1,111}\right] + \left[2v\phi_{2,23} - x_3\phi_{2,112}\right] + \left[\phi_{3,33} + (1-2v)\phi_{3,22} - x_3\phi_{3,113}\right]\right\} \tag{9.7}$$

$$\sigma_{22} = 2G\left\{\left[2v\phi_{1,13} - x_3\phi_{1,122}\right] + \left[2\phi_{2,23} - x_3\phi_{2,222}\right] + \left[\phi_{3,33} + (1-2v)\phi_{3,11} - x_3\phi_{3,223}\right]\right\} \tag{9.8}$$

$$\sigma_{33} = 2G\left\{-x_3\phi_{1,133} - x_3\phi_{2,233} + \left[\phi_{2,33} - x_3\phi_{2,333}\right]\right\} \tag{9.9}$$

$$\begin{aligned}\sigma_{12} = 2G\{&\left[(1-v)\phi_{1,23} - x_3\phi_{1,112}\right] + \left[(1-v)\phi_{2,13} - x_3\phi_{2,122}\right] \\ &-\left[(1-2v)\phi_{3,12} + x_3\phi_{3,123}\right]\}\end{aligned} \tag{9.10}$$

$$\sigma_{23} = 2G\{[-\nu\phi_{1,12} - x_3\phi_{1,123}] + [\phi_{2,23} + \nu\phi_{2,11} - x_3\phi_{2,223}] - x_3\phi_{3,123}\} \tag{9.11}$$

$$\sigma_{13} = 2G\{[\phi_{1,33} + \nu\phi_{1,22} - x_3\phi_{1,113}] + [-\nu\phi_{2,12} - x_3\phi_{2,123}] - x_3\phi_{3,133}\} \tag{9.12}$$

where $\varphi_{i,j}, \varphi_{i,jk}, \varphi_{i,jkl}(j,k,l = 1,2,3)$ are the derivatives of the kernel function

$$\varphi_i(x_1,x_2,x_3) = \frac{1}{8\pi(1-\nu)} \iint_{\Re} D_i\left[(x_1-\xi)^2 + (x_2-\eta)^2 + x_3^2\right]^{-1/2} d\xi d\eta \tag{9.13}$$

in which $\Re$ is the area of the element, D_i ($i = 1, 2, 3$) are the displacement discontinuities, (x_1, x_2, x_3) is the coordinate system originated at the element, and (ξ, η, 0) are the coordinates of the loading point. For the most commonly used constant displacement discontinuity element, the displacement discontinuities can be taken out of the integration formula. The last equation is in terms of the basic kernel function:

$$I(x_1,x_2,x_3) = \iint_{\Re} \left[(x_1-\xi)^2 + (x_2-\eta)^2 + x_3^2\right]^{-1/2} d\xi d\eta \tag{9.14}$$

which depends on the geometry of the element. The kernel functions were derived for the rectangular element by Salamon [5] for an isotropic elastic case.

Poroelastic theory [8, 9] is the basis for the simulation and prediction of reservoir compaction and the induced surface subsidence. Using Biot's poroelastic theory and Darcy's law [10] for a compressible fluid flowing through a saturated porous medium, the governing equations for the problem of oil flow in deforming reservoir rock can be described as (the body force is ignored)

$$G\nabla^2\mathbf{u} + (G+\lambda)\nabla div\mathbf{u} - \left(1 - \frac{K}{K_m}\right)\nabla p = 0 \tag{9.15}$$

$$\left(1 - \frac{K}{K_{\mathrm{m}}}\right) div\mathbf{u}_{\mathrm{t}} + \left(\frac{1-\phi}{K_{\mathrm{m}}} + \frac{\phi}{K_{\mathrm{f}}} - \frac{1}{(3K_{\mathrm{m}})^2}\mathbf{i}^T\mathbf{D}\mathbf{i}\right)p_{\mathrm{t}} + \frac{k}{\mu}\nabla^2 p = 0 \tag{9.16}$$

where G and λ are Lamé constants. k is the permeability of the porous medium; μ is the viscosity of the fluid; $\mathbf{u}$ and p denote the displacement of the porous medium and the pore pressure, respectively; the subscript t denotes time derivative; φ is the porosity of the porous medium (assumed constant hereafter as a first-order approximation); and K, K_{f}, and K_{m} are the bulk modulus of the skeleton, fluid, and matrix, respectively. Furthermore, $\mathbf{i}^{\mathrm{T}} = [1, 1, 1, 0, 0, 0]$, and $\mathbf{D}$ is the elastic stiffness matrix expressed using Young's modulus E and Poisson's ratio ν:

$$D = \frac{E(1-v)}{(1+v)(1-2v)} \begin{bmatrix} 1 & \frac{v}{1-v} & \frac{v}{1-v} & 0 & 0 & 0 \\ \frac{v}{1-v} & 1 & \frac{v}{1-v} & 0 & 0 & 0 \\ \frac{v}{1-v} & \frac{v}{1-v} & 1 & 0 & 0 & 0 \\ 0 & 0 & 0 & \frac{1-2v}{2(1-v)} & 0 & 0 \\ 0 & 0 & 0 & 0 & \frac{1-2v}{2(1-v)} & 0 \\ 0 & 0 & 0 & 0 & 0 & \frac{1-2v}{2(1-v)} \end{bmatrix} \tag{9.17}$$

To mathematically describe multiphase fluid flow through a deformable porous medium, it is necessary to determine functional expressions that best define the relationship among the hydraulic properties of the porous medium, i.e., saturation, relative permeability, and capillary pressure. The capillary pressure relationship is required to couple phase pressures, and relative permeability values are required to calculate phase velocity. The porous medium voids are assumed to be filled with water, gas, and oil; thus the sum of their saturations will be unity, i.e.,

$$S_{\mathrm{o}} + S_{\mathrm{w}} + S_{\mathrm{g}} = 1 \tag{9.18}$$

where S_π is the saturation of the fluid phase π, with o, w, and g representing oil, water, and gas phases, respectively. When more than one fluid exists in a porous medium, the pressure exerted by the fluids may be evaluated using the effective average pore pressure, $\overline{p}$, which is calculated from

$$\overline{p} = S_{\mathrm{o}}P_{\mathrm{o}} + S_{\mathrm{w}}P_{w} +^{w} S_{\mathrm{g}}P_{\mathrm{g}} \tag{9.19}$$

The water pressure P_{w}, gas pressure P_{g}, and oil pressure P_{o} are related through the capillary pressure, and the three capillary terms are defined as

$$P_{\mathrm{cow}}(S_{\mathrm{o}}, S_{\mathrm{w}}) = P_{\mathrm{o}} - P_{\mathrm{w}} \tag{9.20}$$

$$P_{\mathrm{cgo}}(S_{\mathrm{g}}, S_{\mathrm{o}}) = P_{\mathrm{g}} - P_{\mathrm{o}} \tag{9.21}$$

$$P_{\mathrm{cgw}}(S_{\mathrm{g}}, S_{\mathrm{w}}) = P_{\mathrm{g}} - P_{\mathrm{w}} \tag{9.22}$$

where P_{cgw} is the capillary pressure between the gas and water phases, P_{cow} is the capillary pressure between the oil and the water phases, and P_{cgo} is the capillary pressure between the gas and oil phases. In general, for a multiphase system, the saturation of any of the three phases is a function of three capillary pressure relationships, i.e., oil-water, gas-oil, and gas-water, respectively:

$$S_{\mathrm{p}} = f(P_{\mathrm{cgw}}, P_{\mathrm{cow}}, P_{\mathrm{cgo}}) \tag{9.23}$$

The gas-water capillary pressure, expressed in terms of the other two capillary pressures, yields the following:

$$P_{\mathrm{cgo}} = P_{\mathrm{cgw}} - P_{\mathrm{cow}} \tag{9.24}$$

and we can rewrite the equation as

$$S_{\pi} = f(P_{\mathrm{cgw}}, P_{\mathrm{cow}}) \tag{9.25}$$

In a multiphase flow model of a porous medium, the simultaneous flow of the fluid phases, water, oil, and gas, depends primarily on the pressure gradient, the gravitational force and the capillary pressures between the multiphase fluids. The fluid pressures and the displacement values are used as the primary dependent variables.

A general equilibrium equation incorporating the concept of effective stress can be written as follows:

$$G\nabla^2 \mathbf{u} + (G + \lambda)\nabla \operatorname{div} \mathbf{u} - \left(1 - \frac{K}{K_{\mathrm{m}}}\right)\nabla \overline{p} = 0 \tag{9.26}$$

A general form of the continuity equation for each flowing phase π, incorporating Darcy's law, can be expressed as follows:

$$\begin{aligned} &-\nabla^T\left[\frac{kk_{r\pi}\rho_\pi}{\mu_\pi B_\pi}\nabla(P_\pi + \rho_\pi gh)\right] + \phi\frac{\partial}{\partial t}\left(\frac{\rho_\pi S_\pi}{B_\pi}\right) \\ &+ \rho_\pi\frac{S_\pi}{B_\pi}\left[\left(\mathbf{i}^{\mathrm{T}} - \frac{\mathbf{i}^{\mathrm{T}}\mathbf{D}}{3K_{\mathrm{m}}}\right)\frac{\partial \varepsilon}{\partial t} + \frac{\mathbf{i}^{\mathrm{T}}\mathbf{D}c}{3K_{\mathrm{m}}} + \left(\frac{1-\phi}{K_{\mathrm{m}}} - \frac{\mathbf{i}^T\mathbf{D}\mathbf{i}}{(3K_{\mathrm{m}})^2}\right)\frac{\partial \overline{p}}{\partial t}\right] + \rho_\pi Q_\pi \\ &= 0 \end{aligned} \tag{9.27}$$

where Q_π represents external sinks and sources, k is the absolute permeability, $k_{r\pi}$ is the relative permeability, and B_π is the formation volume factor.

The FEM has been effectively used to solve the above equations for the Biot poroelastic formulation since Sandhu and Wilson [11] first applied the FEM to single-phase poroelasticity; later, refinements and more extensive applications were achieved [12–14].

We emphasize that because the reservoir is discretized with finite elements, various types of nonlinearity can be incorporated, although each additional degree of nonlinearity requires additional iterations of the FEM zone to achieve convergence.

The relationship between the DD element and the FE element is shown in Fig. 9.1. From experience in mining problems, the interface shear stress is of a

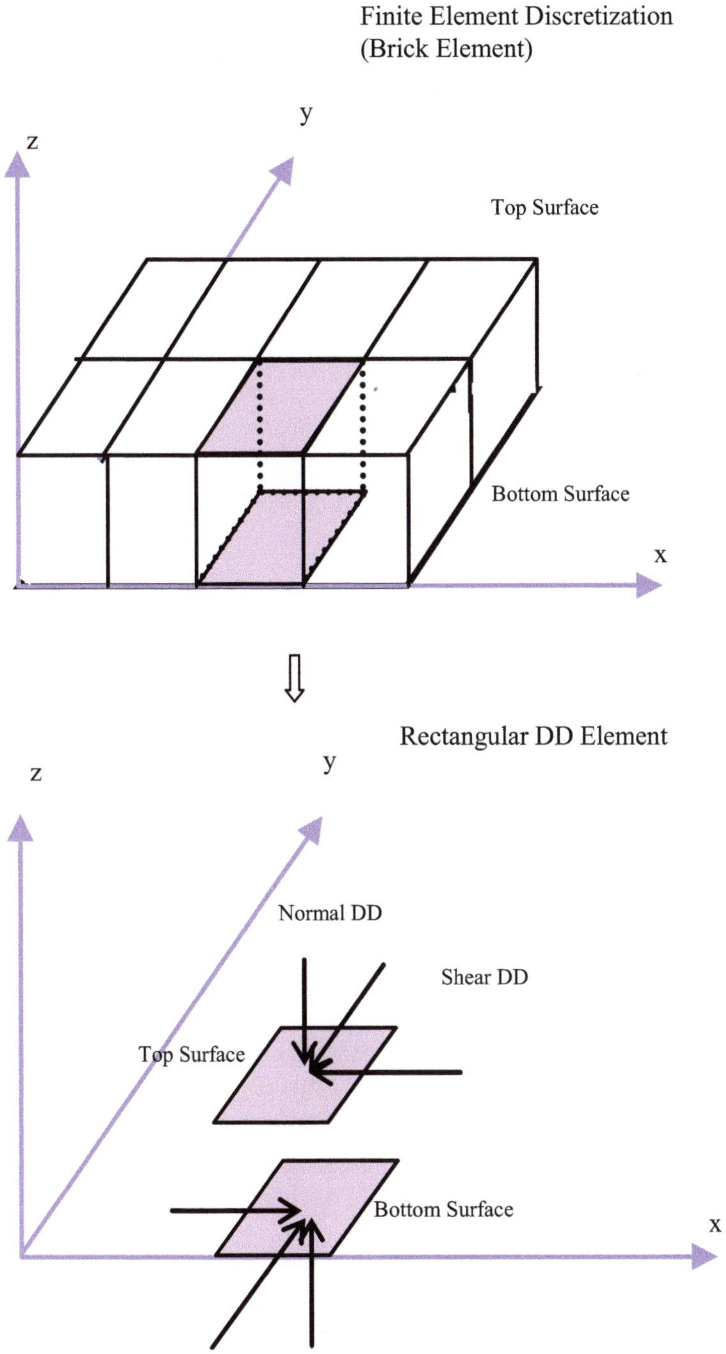

Fig. 9.1 Sketch of relationship between finite element and displacement discontinuity element

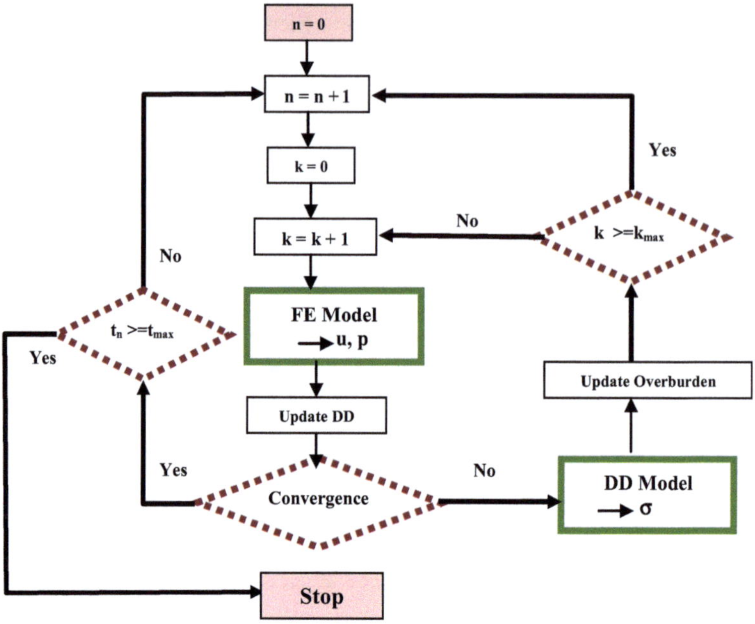

Fig. 9.2 Flowchart of the iterative calculations to couple the DD and FE models

lower order of importance compared to the normal stresses, and is generally ignored. In the present reservoir problem, for simplicity we provide only for the continuity of vertical displacements and do not match other components. In other words, we assume that the shear displacements along the coupling interface between DD and FE elements are unimportant (this may not always be the case, and can be accounted for in more complex formulations).

An exchange of information between the reservoir FE model and the DD model is necessary. The information that the FEM model provides is the deformation of the reservoir, which is then converted into a displacement discontinuity provided to the DD model; the information that the DD model provides is the stress distribution acting upon the reservoir, which is then converted into overburden loads provided to the FEM model. We use an iterative method to implement data exchange between the DD and FEM models; the procedure is as follows (shown in Fig. 9.2):

1. Start with the FEM (reservoir) model to calculate the displacement and pressure under prescribed external loads and fluid discharge conditions within a specified time period.
2. Convert displacements obtained from the FEM model into displacement discontinuities which are applied to the DD elements defining the surrounding strata.
3. Execute the DD model, from which the local stresses can be computed.
4. Apply the induced stresses calculated from the DD model, along with the difference between the stresses in FEM and DDM, into the external loadings

to be applied to the FEM model in the next iteration. This ensures that the stresses retain continuity across the coupling interface:

$$q^{(k+1)} = q^{(k)} + \left(\sigma_{33}^{(k)} - \sigma_{v}^{(k)}\right) \tag{9.28}$$

where k denotes the iteration number, $\sigma_{33}^{(k)}$ is the vertical stress calculated from the DDM model in last step, and $\sigma_{v}^{(k)}$ is the vertical total stress calculated from the FEM model by

$$\sigma_{v}^{(k)} = \sigma_{v}^{'\,(k)} - p^{(k)} \tag{9.29}$$

where $\sigma_{v}^{'\,(k)}$ is the effective stress (sign convention: compression as negative in FE model), and p is the pore water pressure.

Based on our experience, to accelerate the convergence, we introduce a constant, χ, to multiply the stress difference between the two models:

$$q^{(k+1)} = q^{(k)} + \chi\left(\sigma_{33}^{(k)} - \sigma_{v}^{(k)}\right) \tag{9.30}$$

The formulation for χ is recommended as

$$\chi = \frac{E_{\mathrm{r}}}{E_{\mathrm{r}} + E_{\mathrm{o}}} \tag{9.31}$$

where E_{r} represents the Young's modulus of the reservoir, and E_{o} represents the Young's modulus of the surroundings.

To examine the computational efficiency of the DDFEM method mentioned above, we consider a $25\,\mathrm{m} \times 25\,\mathrm{m} \times 4\,\mathrm{m}$ reservoir at depth of 300 m with the following parameters (Fig. 9.3): $E = 2.0 \times 10^{4}\mathrm{kPa}$, $v = 0.25$, $\phi = 0.30$, $K_{\mathrm{f}} = 1 \times 10^{6}\mathrm{kPa}$, $K_{\mathrm{m}} = 1 \times 10^{8}\mathrm{kPa}$, and $k = 1.0$ D, $\mu = 22.0$ cP.

The elastic deformation parameters for the surrounding rock are $E = 2.0 \times 10^{5}$ kPa and $v = 0.25$. The reservoir has an impermeable boundary, and is subjected to a production rate of $Q = 720.0\ \mathrm{m}^3$/day with uniform pumping. The time step is set as $\Delta t = 0.1$ days. We want to predict the subsidence with time.

When we use the DDFEM model to solve this problem, in the FEM mesh, the domain is discretized into 50 elements, 360 nodes; in DDM mesh, the domain is discretized into 25 DD elements.

Execution time of 2 min was noted for the DDFEM program to finish the problem. The ground surface subsidence evolution curves produced by the DDFEM model along the diagonal direction are shown in Fig. 9.4. The lateral ground surface movement curves produced by DDFEM model along the diagonal direction are shown in Fig. 9.5. It should be noted that we did not put the complete solution provided by the DDFEM model into these two figures. Clearly, values are approaching zero with increasing distance from the center.

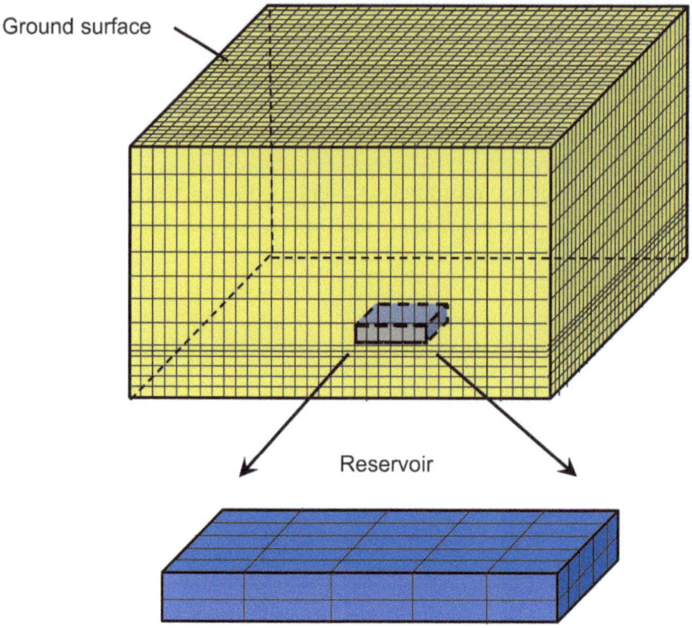

Fig. 9.3 Fine finite element discretization of the reservoir and surroundings

When we use a FEM approach alone to solve the problem, the outer boundary of the domain must be extended reasonably far from the reservoir, and it needs to be meshed inside the entire domain. If we still use $5 \times 5 \times 2$ finite elements to account for the reservoir domain, there are in total $35 \times 35 \times 14$ finite elements for the reservoir and its surroundings (Fig. 9.3) in a typical discretization. The reservoir is in layers 5 and 6 in the vertical direction, and from grid blocks 16–20 in the horizontal directions. The sideburden, underburden, and overburden are obtained by extending 75, 40, and 300 m from the side, bottom, and top of the reservoir, respectively, and discretized uniformly in each zone. To solve the same problem, the full finite element model took 241 min, ~120 times as long as the DDFEM model.

The ground surface subsidence evolution curves produced by FEM model along the diagonal direction are shown in Fig. 9.4. The lateral ground surface movement curves produced by FEM model along the diagonal direction are shown in Fig. 9.5. The DDFEM model is excellent in computing efficiency compared to the full FEM model, and furthermore we can see significant differences between the outcomes of the two models from Figs. 9.4 and 9.5. It was calculated that the subsidence 'volume' taking place at the ground surface produced by the DDFEM model and FEM model are close. Obviously, the limited reservoir outer domain included in the FEM model led to inaccuracies which may be called "wrong predictions." However, if the outer domain of the FEM model is made extremely large, leading to

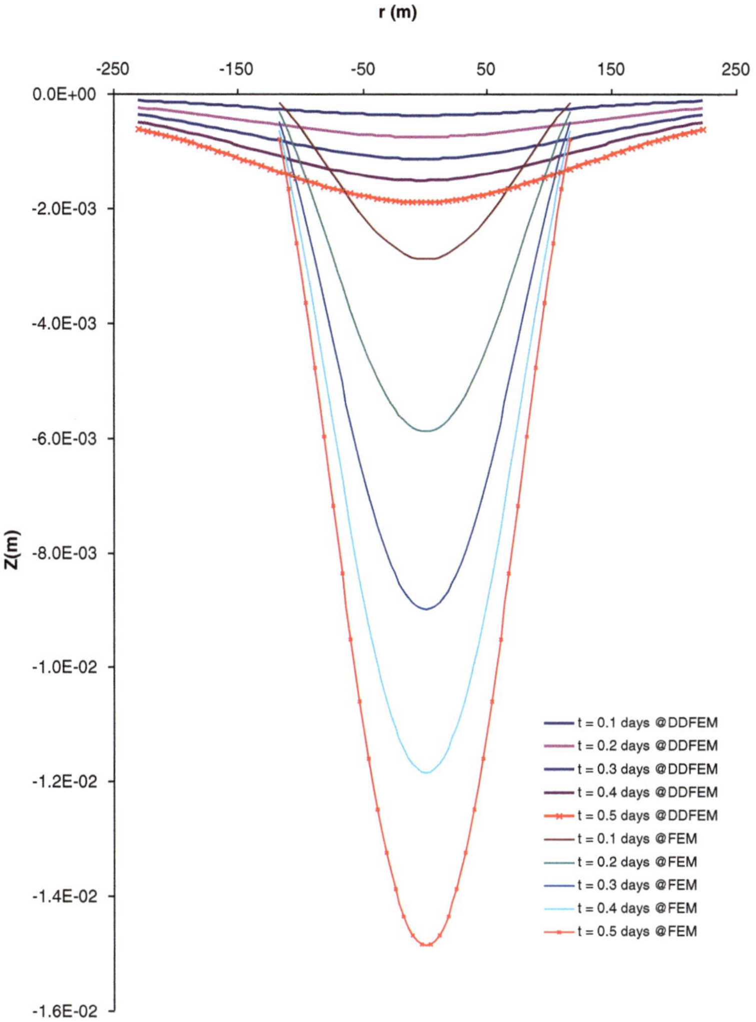

Fig. 9.4 Comparison of ground surface subsidence profiles predicted by DDFEM model and FEM model

"correct predictions," the computing efforts will be accordingly larger, even with elements of different size.

In the second numerical experiment, two-phase flow is considered in a production-induced compaction/subsidence problem. Scenarios including reservoirs with different sets of reservoir surrounding stiffness and reservoir depth are studied.

Assume a 1000 m × 1000 m × 20 m reservoir (Fig. 9.6) with $E = 1.0 \times 10^5$ kPa, $v = 0.3$, $\phi = 0.35$, $K_m = 1.4 \times 10^8$ kPa, $k = 10.0$ D, $\mu_w = 1$ cP, $\mu_{oil} = 1$ cP,

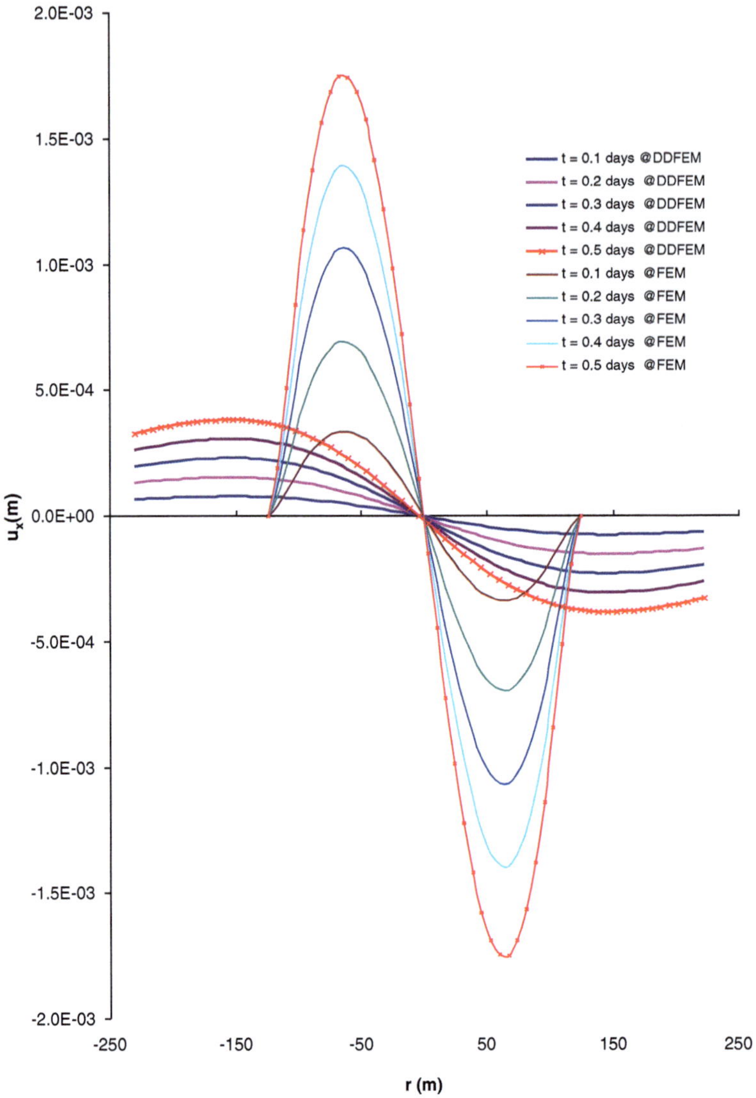

Fig. 9.5 Comparison of lateral ground surface movements predicted by DDFEM model and FEM model

$\rho_w = 1.0 \times 10^3$ kg/m^3, $\rho_o = 1.0 \times 10^3$ kg/m^3, $B_o = 1.0$, $B_w = 1.0$. The relative permeability-saturation functions are

$$k_{rw} = 0.4\left(\frac{S_w - S_{wc}}{1 - S_{or} - S_{wc}}\right)^2 \tag{9.32}$$

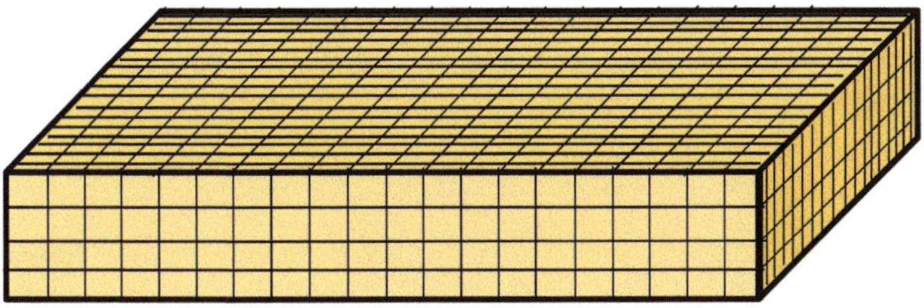

Fig. 9.6 FEM mesh for a 1000 m × 1000 m × 20 m reservoir in numerical examples

$$k_{\mathrm{ro}} = \left(\frac{1 - S_{\mathrm{w}} - S_{\mathrm{or}}}{1 - S_{\mathrm{or}} - S_{\mathrm{wc}}}\right)^2 \tag{9.33}$$

The capillary curve is

$$P_{\mathrm{c}} = 0.016\ln\left(\frac{1 - S'}{S'}\right) \tag{9.34}$$

$$S' = \frac{S_{\mathrm{w}} - S_{\mathrm{wc}}}{1 - S_{\mathrm{wc}} - S_{\mathrm{or}}} \tag{9.35}$$

where the units of P_{c} are MPa. $S_{\mathrm{wc}} = 0.10$ and $S_{\mathrm{or}} = 0.20$ are assumed, as well as an initial water saturation of $S_{\mathrm{w}} = 0.20$. The reservoir is subjected to a no-flow boundary and a production well is established in the center with a pumping rate of 1000 m^3/day.

In the first case, a softer surrounding rock mass with $E = 1.0 \times 10^4$ kPa and $\nu = 0.30$ is assumed, and the reservoir is located at a depth of $Z = 300$ m from the ground surface. The subsidence and compaction profiles at $t = 300$ days are shown in Fig. 9.7.

In the second case, a softer surrounding rock mass with $E = 1.0 \times 10^4$ kPa and $\nu = 0.30$ is assumed, and the reservoir is located at a depth of $Z = 3000$ m. The subsidence and compaction profiles at $t = 300$ days are shown in Fig. 9.8.

In the third case, a stiffer surrounding rock mass with $E = 1.0 \times 10^6$ kPa and $\nu = 0.30$ is assumed, and reservoir is located at a depth of $Z = 300$ m. The subsidence and compaction profiles at $t = 300$ days are shown in Fig. 9.9.

In the fourth case, a stiffer surrounding rock mass with $E = 1.0 \times 10^6$ kPa and $\nu = 0.30$ is assumed, and the reservoir is at a depth of $Z = 3000$ m. The subsidence and compaction profiles at $t = 300$ days are shown in Fig. 9.10.

Figure 9.7 shows more extensive subsidence in the area adjacent to the reservoir center, compared with Fig. 9.8; the deeper reservoir is located, the larger the area of surface subsidence. The same effect is seen in comparing Figs. 9.9 and 9.10, which also show a more concave compaction profile compared with Figs. 9.7 and 9.8. Thus, the constraining effect from the sideburden is more significant with stiffer surroundings. Figures 9.8 and 9.10 show that Δz at $x = \pm 2500$ m (reservoir area

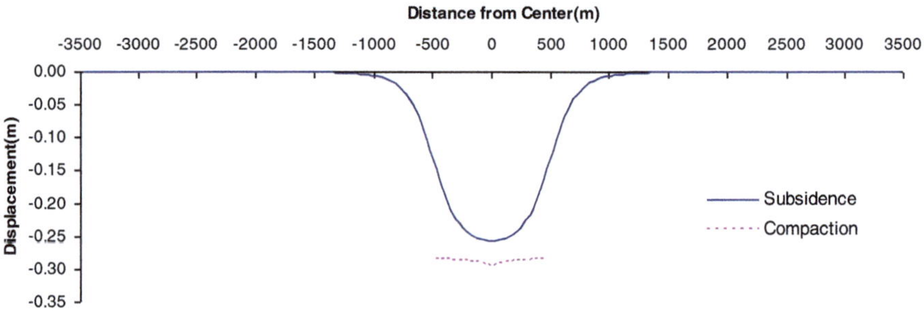

Fig. 9.7 Subsidence and compaction profiles at $t = 300$ days for a shallow reservoir with softer surrounding

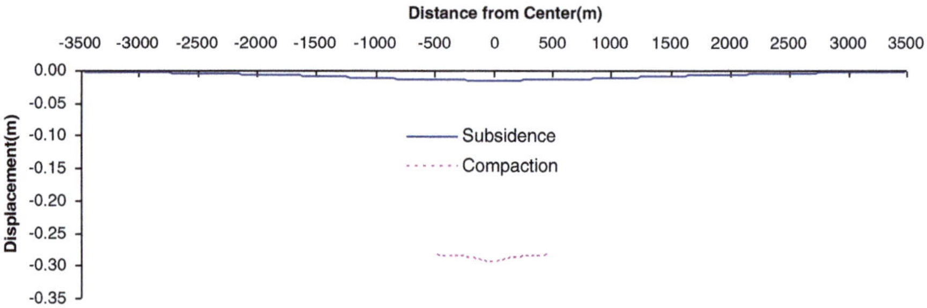

Fig. 9.8 Subsidence and compaction profiles at $t = 300$ days for a deep reservoir with softer surrounding

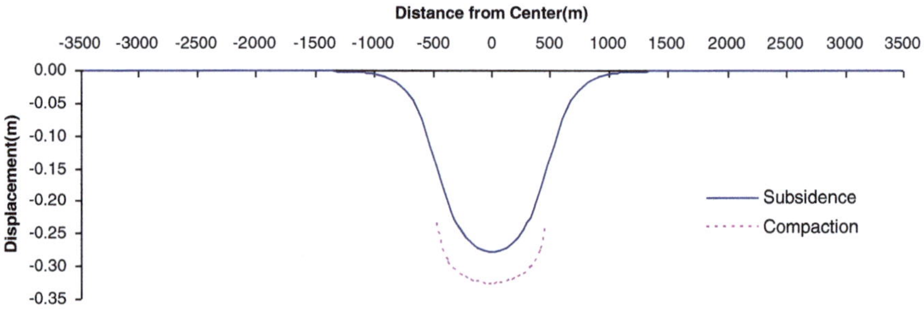

Fig. 9.9 Subsidence and compaction profiles at $t = 300$ days for a shallow reservoir with stiffer surrounding

from −500 to +500) is still significant compared to the center, showing that the discretized domain has to be large to obtain a locally accurate solution at the surface. With the DDFEM model, this is handled naturally.

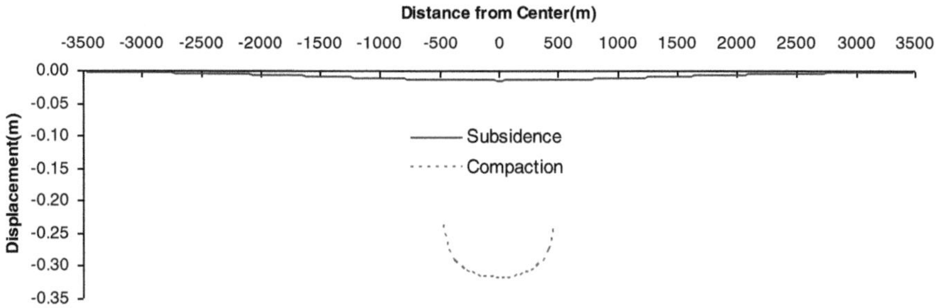

Fig. 9.10 Subsidence and compaction profiles at $t = 300$ days for a deep reservoir with stiffer surrounding

9.3 Borehole Stability

Development of oil, gas, and geothermal energy, as well as deep geological storage of CO_2 and disposal of liquid and slurried solid wastes, requires access boreholes [15, 16]. Massive sloughing and borehole wall fracturing are frequently observed instability phenomena arising from removal of the original supporting rock and the interaction between the drilling fluids and shale formations. Strong predictive models for these processes are of great practical value to drilling planning, as well as to *postmortem* analysis of problem cases.

Analysis of wellbore instability issues is a basic subject in petroleum geomechanics and engineering [17–19]. Mechanical behavior, pore pressure evolution, heat flux, and chemical changes all influence wellbore stability. Analysis involves studying the interactions among changes of pore pressure (Δp), temperature (ΔT), effective stress ($\Delta\sigma'_{ij}$), and geochemistry (ΔC) during drilling shale formations, a typical coupled thermal-hydraulic-mechanics-chemical (THMC) process in geomechanics [20]. Analytical and numerical analyses of wellbore stability issues have focused more and more on coupling mechanisms [17, 20–28]. Most of these studies focused on extremely low permeability shale in which thermal and solute convection are not included. However, for porous media with intermediate to high permeability, the accuracy of analysis could be impaired if advective processes are neglected [29]. Analytical solutions considering thermal and chemical convection factors have been occasionally proposed [27], but in general, complex coupling implies mathematical simulation.

The basis for fully coupled THMC modeling of wellbore stability is the combination of theories of poroelasticity, thermoelasticity, elastoplasticity, and chemo-thermoporoelasticity [8–10, 30, 31]. This combination allows us to address situations of strong coupling among thermal flux, solute flux, fluid flow, and the deformations of the porous media.

Based on chemo-thermoporoelastic theory for non-isothermal fluid flow and solute transport through a deformable wellbore region, a general equilibrium equation incorporating the concept of effective stress can be written (for simplicity, body forces are ignored):

$$G\nabla^2\mathbf{u} + (G+\lambda)\nabla\operatorname{div}\mathbf{u} - \left(1-\frac{K}{K_{\mathrm{m}}}\right)\nabla p + \frac{\xi RT}{M}\nabla C - K\beta_{\mathrm{s}}\nabla T = 0 \quad (9.36)$$

The symbols G and λ denote the Lamé elastic constants, and $\mathbf{u}, p, C$, and T represent displacement, pore pressure, solute concentration, and temperature, respectively. ξ is the ratio of the coefficients of volume change due to chemical consolidation and mechanical consolidation, R is the universal gas constant, and M is the solute molar mass. β_{s} is the volumetric thermal expansion coefficient of the skeleton, and K and K_{m} are bulk moduli for the rock skeleton and for the matrix mineral, respectively. In this version of the general equilibrium equation, issues such as non-isotropic elastic properties or nonlinearities are not addressed explicitly; these can be included through a more general tensorial formulation to account for fabric, and through use of iterative solutions to solve nonlinear cases.

Equations representing mass conservation and energy conservation are expressed below. The general form of the continuity equation for fluid flow, incorporating Darcy's Law, can be expressed as follows:

$$\nabla^T\left(\frac{k}{\mu}\nabla p\right) + \left(\frac{\alpha-\phi}{K_{\mathrm{m}}} + \frac{\phi}{K_{\mathrm{w}}}\right)\frac{\partial p}{\partial t} - \nabla^T\left(\frac{k}{\mu}\frac{\eta RT}{M}\nabla C\right) + \left(\phi\beta_{\mathrm{c}} + \frac{\omega}{K_{\mathrm{m}}}\right)\frac{\partial C}{\partial t} + \alpha\frac{\partial \varepsilon}{\partial t} - [(\alpha-\phi)\beta_{\mathrm{s}} + \phi\beta_{\mathrm{w}}]\frac{\partial T}{\partial t} = 0 \quad (9.37)$$

where α is Biot's coefficient, equal to $1.0-K/K_{\mathrm{m}}$, k is the permeability of the porous media, ϕ is the porosity, and μ is the viscosity of the fluid. The membrane efficiency of the shale formation is η, and ω is the swelling coefficient. β_{c} is the coefficient that relates the density to the solute mass concentration in the fluid, and β_{w} is the volumetric thermal expansion coefficient of the fluid.

Next, the general form of the continuity equation for solute transport, including both the solute diffusion and convection terms, is written as follows:

$$\nabla^T(D_{\mathrm{e}}\nabla C) + \phi R_{\mathrm{d}}\frac{\partial C}{\partial t} + \left(\frac{k}{\mu}p_i - \frac{k}{\mu}\frac{\eta RT}{M}C_i\right)\nabla C = 0 \quad (9.38)$$

where D_{e} is the coefficient of molecular diffusion, and R_{d} is the retardation coefficient. The term with the concentration gradient is the Darcy velocity, v, which is related to both the pressure gradient and the solute concentration gradient.

Finally, the general form of the energy balance equation, including the thermal convection and thermal conduction terms, can be expressed as follows:

$$\nabla^T[\lambda_T \nabla \mathbf{T}] + \rho_w c_w v \nabla \mathbf{T} + T\left[(1-\phi)c_s \frac{\rho_s}{K_m} + \phi c_w \frac{\rho_w}{K_w}\right]\frac{\partial \mathbf{p}}{\partial t} + [-\phi c_w \rho_w \beta_w T - (1-\phi)\rho_s c_s \beta_s T + (1-\phi)\rho_s c_s + \phi \rho_w c_w]\frac{\partial \mathbf{T}}{\partial t} + Q_h = 0 \tag{9.39}$$

Here, λ_T is the porous medium thermal conductivity, c_l is the specific heat capacity (the subscript l represents the solid, s, and water, w), ρ_l is the density, Q_h is an external sink or source, and, as before, v is the Darcy velocity.

Because of its power and flexibility, the FEM has been widely applied in numerical modeling in geomechanics and reservoir simulations [32–34]. The final matrix form of the Galerkin FEM solution for the above equations after finite element discretization is expressed as follows:

$$\begin{bmatrix} \mathbf{M} & -\mathbf{C}_{sw} & -\mathbf{C}_{sc} & -\mathbf{C}_{sT} \\ 0 & \mathbf{H}_{ww} & \mathbf{H}_{wc} & 0 \\ 0 & 0 & \mathbf{H}_{cc} & 0 \\ 0 & 0 & 0 & \mathbf{H}_{TT} \end{bmatrix} \begin{Bmatrix} \mathbf{u} \\ \mathbf{p} \\ \mathbf{C} \\ \mathbf{T} \end{Bmatrix} + \begin{bmatrix} 0 & 0 & 0 & 0 \\ \mathbf{C}_{ws} & \mathbf{R}_{ww} & \mathbf{C}_{wc} & \mathbf{C}_{wT} \\ 0 & 0 & \mathbf{R}_{cc} & 0 \\ 0 & \mathbf{C}_{Tw} & 0 & \mathbf{R}_{TT} \end{bmatrix} \begin{Bmatrix} \mathbf{u}_t \\ \mathbf{p}_t \\ \mathbf{C}_t \\ \mathbf{T}_t \end{Bmatrix} = \begin{Bmatrix} \mathbf{f}^u \\ \mathbf{f}^w \\ \mathbf{f}^c \\ \mathbf{f}^T \end{Bmatrix} \tag{9.40}$$

where $[\mathbf{u}, \mathbf{p}, \mathbf{C}, \mathbf{T}]^T$ and $[\mathbf{u}_t, \mathbf{P}_t, \mathbf{C}_t, \mathbf{T}_t]^T$ are the vectors of unknown variables and corresponding time derivatives. $[\mathbf{f}^u, \mathbf{f}^w, \mathbf{f}^c, \mathbf{f}^T]^T$ is the vector for the nodal loads ($\mathbf{f}^u$), the flow source ($\mathbf{f}^w$), the solute source ($\mathbf{f}^c$), and the heat source. The explicit expressions of the above matrices can be found in Yin et al. [34].

To integrate the above equations with respect to time, linear interpolation in time using a finite difference method is introduced, so the equations can be written as

$$\begin{bmatrix} \theta\mathbf{M} & -\theta\mathbf{C}_{sw} & -\theta\mathbf{C}_{sc} & -\theta\mathbf{C}_{sT} \\ \mathbf{C}_{ws} & \mathbf{R}_{ww} + \theta\Delta t\mathbf{H}_{ww} & \mathbf{C}_{wc} + \theta\Delta t\mathbf{H}_{wc} & \mathbf{C}_{wT} \\ 0 & 0 & \mathbf{R}_{cc} + \theta\Delta t\mathbf{H}_{cc} & 0 \\ 0 & \mathbf{C}_{Tw} & 0 & \mathbf{R}_{TT} + \theta\Delta t\mathbf{H}_{TT} \end{bmatrix} \begin{Bmatrix} \Delta\mathbf{u} \\ \Delta\mathbf{p} \\ \Delta\mathbf{C} \\ \Delta\mathbf{T} \end{Bmatrix} = \begin{Bmatrix} \Delta\mathbf{f}^u \\ \Delta t\mathbf{f}^w - \Delta t\mathbf{H}_{ww}\mathbf{p} - \Delta t\mathbf{H}_{wc}\mathbf{C} \\ \Delta t\mathbf{f}^c - \Delta t\mathbf{H}_{cc}\mathbf{C} \\ \Delta t\mathbf{f}^T - \Delta t\mathbf{H}_{TT}\mathbf{T} \end{Bmatrix} \tag{9.41}$$

Standard FE approximations are based upon the Galerkin formulation of the method of weighted residuals. This formulation has proven highly successful for problems in solid/structural mechanics and in other situations, such as heat conduction, governed by diffusion-type equations. The reason for this success is that, when applied to problems governed by self-adjoint elliptic or parabolic partial differential equations, the Galerkin FE method leads to symmetric stiffness matrices. In this case the difference between the FE solution and the exact solution is minimized with respect to the energy norm (e.g., [35]).

The success of the Galerkin FEM in solid/structural mechanics and heat conduction problems is not replicated for the case of fluid flow simulations, especially for modeling convection-dominated transport phenomena. The main difficulty is due to the presence of convection operators in the formulation of flow problems based on non-Lagrangian kinematical descriptions. Convection operators are in fact non-symmetric; thus the best approximation property in the energy norm of the Galerkin method, which is the basis for success in symmetric cases, is lost when convection dominates transport [36].

The energy conservation equation and the chemical solute transport equation mentioned above are essentially transient advection-diffusion equations. When the equation is diffusion dominated, the traditional Galerkin FEM is adequate to handle it; however, in advection-dominated cases, this usually leads to spurious oscillations.

These features are common with central-difference-type finite difference methods. In the finite difference field, upstream weighting has been employed to mitigate the oscillations, but this may also severely degrade accuracy because of excessive numerical diffusion [37].

Solving the steady-state advection-diffusion problem by FE methods has been extensively studied, and many stabilized methods, such as the streamline upwind Petrov-Galerkin (SUPG) and the Galerkin/least-squares (GLS) methods, have helped make FE modeling suitable for such flow problems. These approaches stabilize the numerical scheme by adding an additional stabilization term to the original Galerkin formulation. The magnitude of the stabilization parameters in the additional term can be determined by the dimensionless Peclet number (Pe), which describes the rate of advective flux to the rate of diffusive flux. When it comes to the unsteady advection-diffusion problem, additional numerical oscillations take place at small time steps, and this problem is much less tractable.

In time-dependent advection-diffusion equations, after the FEM approximation, a mass matrix containing the time derivative terms is formed. If the mass matrix were diagonal-dominant (or diagonalized), one could consider the use of explicit techniques for integrating the system in time, whereas the consistent mass (non-diagonal dominant) matrix essentially demands the use of implicit methods. The relative speed and simplicity of explicit methods has led to the sometimes compromising and always ad hoc concept of 'mass lumping' wherein the mass matrix is converted to a diagonal form. Mass lumping is known to add significant numerical diffusion, so if it is employed, the FE method accuracy can be severely compromised, although the FEM solution to the advection-diffusion equation can nevertheless be more accurate than solutions generated via conventional finite difference methods [38]. Therefore, mass lumping is not an ideal technique in FEM modeling of advection-diffusion flow problems.

Next, the performance of different FE methods in dealing with the transient advection-diffusion problem will be discussed through some examples.

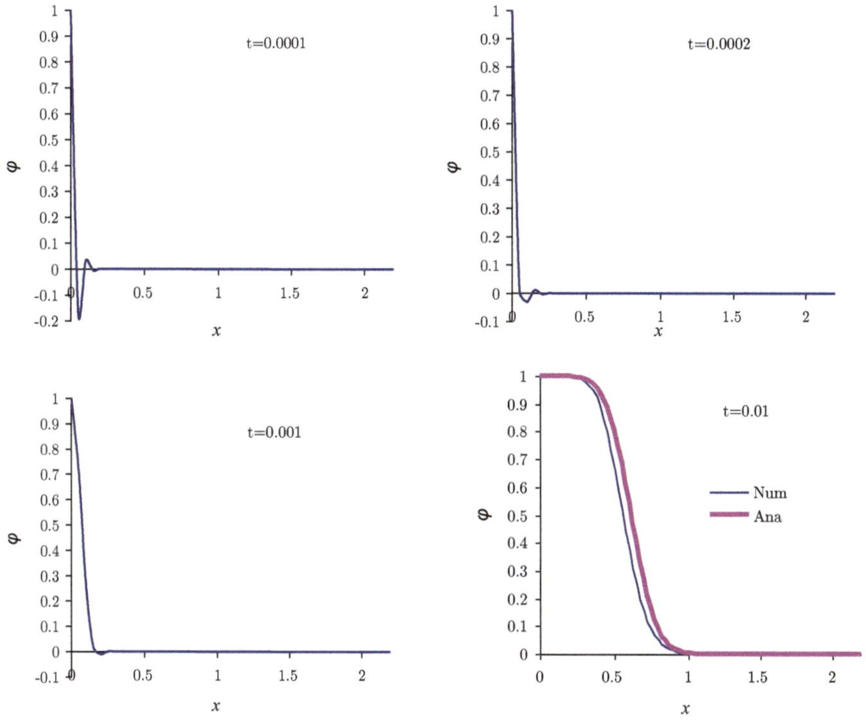

Fig. 9.11 1-D advection-diffusion problem solved by a classic Galerkin method

First let us observe the performance of the traditional Galerkin FEM. Consider a 1-D problem seeking a numerical solution for the equation:

$$\begin{aligned} &\varphi_t + u\varphi_x - \kappa\varphi_{xx} = 0 \\ &\varphi(0,t) = 1, \quad \varphi(x,0) = 0 \end{aligned} \tag{9.42}$$

where u is the convective velocity and k is the diffusion coefficient. φ can refer to temperature, concentration, saturation, etc. in different physical processes and applications. One hundred equal-size linear elements are used for spatial discretization between $x=0$ and $x=5$, where an adiabatic boundary condition exists, and the generalized trapezoidal method (or θ-method) in direct time-integration is used.

Assume $k=1$, and $u=60$, noting that the spurious oscillations appear at small time steps and disappear at later, larger time steps (see Fig. 9.11).

For a multidimensional case, suppose a 2-D problem where a numerical solution is sought for the equation:

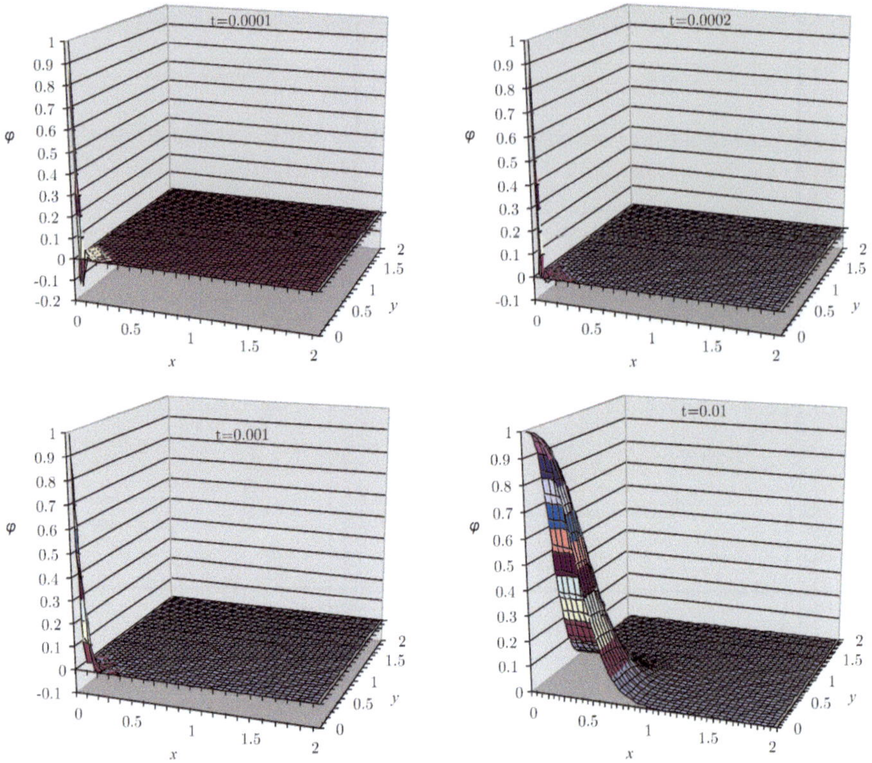

Fig. 9.12 2-D advection-diffusion problem solved by a classic Galerkin method

$$\begin{aligned} &\varphi_t + \mathbf{u} \cdot \nabla \varphi - \nabla \cdot (\kappa \nabla \varphi) = 0 \\ &\varphi(0,0,t) = 1, \quad \varphi(x,y,0) = 0 \end{aligned} \tag{9.43}$$

where **u** is the convective velocity vector and k is the diffusion coefficient. We assume $k = 1$ and $\|\mathbf{u}\| = 60$ at a direction of (cos 45°, sin 45°). We find phenomena similar (see Fig. 9.12) to those observed in the 1-D case.

Harari [39] gives a solution that can incorporate the time-dependent factor naturally into the determination of the stabilizing parameters. He suggests that one may transform the transient term into the reaction term by first discretizing the time domain, instead of the conventional method of first discretizing the spatial domain. For example, a transient diffusion problem can thus be converted to a steady diffusion-reaction problem. This suggested the conversion of the transient advection-diffusion problem to a steady advection-diffusion-reaction problem.

Once a transient diffusion-convection problem is converted to a steady diffusion-convection-reaction problem, the latter can be addressed by the subgrid scale/gradient subgrid scale (SGS/GSGS) method [40]. The SGS/GSGS method combines two types of stabilization integrals and presents appealing traits for accurately solving the steady diffusion-convection-reaction equation.

Consider a steady diffusion-convection-reaction equation such as

$$\begin{aligned} u \cdot \nabla \varphi - \nabla \cdot (\kappa \nabla \varphi) - s\varphi = 0 \quad \text{in } \Omega, \\ \varphi = g \quad \text{on } \Gamma, \end{aligned} \tag{9.44}$$

where u is the velocity field, k is the diffusion coefficient ($k > 0$), and s is the source parameter where $s > 0$ for production and $s < 0$ for injection (or dissipation or absorption. . .). The function g is the Dirichlet datum on the boundary Γ. Of particular interest for this problem are the advective-diffusive-reactive operator and its adjoint, which are, respectively,

$$\begin{aligned} L\varphi &= u \cdot \nabla \varphi - \nabla \cdot (\kappa \nabla \varphi) - s\varphi, \\ L^* \varphi &= -u \cdot \nabla \varphi - \nabla \cdot (\kappa \nabla \varphi) - s\varphi. \end{aligned} \tag{9.45}$$

Now one may apply the combined SGS/GSGS method. Consider a mesh formed by n_{el} non-overlapping elements Ω^{e}, so that $\Omega = \cup_{e=1}^{n_{el}} \Omega^e$. Let S^{h} be the associated FE solution space and V^{h} be the weighting space. The integral form of the stabilized FEM is defined as follows: find $\varphi^{\mathrm{h}} \in S^{\mathrm{h}}$ such that for all $w^h \in V^{\mathrm{h}}$:

$$\begin{aligned} &\int_{\Omega} \left(w^{\mathrm{h}} u \cdot \nabla \varphi^{\mathrm{h}} + \nabla w^{\mathrm{h}} \cdot \left(\kappa \nabla \varphi^{\mathrm{h}} \right) - w^{\mathrm{h}} s \varphi^{\mathrm{h}} \right) d\Omega \\ &\quad + \sum_{e=1}^{n_{el}} \int_{\Omega^e} - L^{*} w^{\mathrm{h}} \tau_{00}^{e} L \varphi^{\mathrm{h}} d\Omega \\ &\quad + \sum_{e=1}^{n_{el}} \int_{\Omega^e} \nabla \left(-L^{*} w^{\mathrm{h}} \right) \cdot \tau_{11}^{e} \nabla L \varphi^{\mathrm{h}} d\Omega = 0, \end{aligned} \tag{9.46}$$

where the stabilization parameters are defined as

$$\begin{aligned} \tau_{00}^{e} &= \frac{h}{u} \left(-2D_K + \frac{{D_K}^2 \sin h(Pe)}{- \cos h(Pe) + \cos h(\gamma) + D_K \sin h(Pe)} \right)^{-1}, \\ \tau_{11}^{e} &= \frac{h^3}{6u{D_K}^3} \left\{ -3 - {D_K}^2 + \frac{3D_K}{Pe} \right. \\ &\quad \left. + \frac{D_K \left[3D_K \cos h(\gamma) + \left(-3 + {D_K}^2 \right) \sin h(Pe) \right]}{-2 \cos h(Pe) + 2 \cos h(\gamma) + D_K \sin h(Pe)} \right\}, \end{aligned} \tag{9.47}$$

in which

$$\begin{aligned} &Pe = uh/2\kappa \quad \text{Peclet number}, \\ &D_K = sh/u \quad \text{Damköhler number}, \\ &\gamma = \sqrt{Pe(-2D_K + Pe)}. \end{aligned} \tag{9.48}$$

For the transient advection-diffusion Eq. (9.42), based on the Rothe approach [39], time discretization is performed before the space discretization, and this leads to the following by generalized trapezoidal method (θ-method):

$$\begin{aligned}&(\theta\Delta tu)\varphi_x^{n+1} - (\theta\Delta t\kappa)\varphi_{xx}^{n+1} - (-1)\varphi^{n+1}\\&= -(1-\theta)\Delta tu\varphi_x^n + (1-\theta)\Delta t\kappa\varphi_{xx}^n + \varphi^n\end{aligned} \tag{9.49}$$

where the superscripts n and $n+1$ represent the evaluation at the nth and $n+1$th time step, respectively.

Now the link between the semi-discretized Eqs. (9.49) and (9.44) can be seen to be as follows:

$$\begin{array}{lcll} \kappa \text{ in Eq.(9.44)} & \sim & \theta\Delta t\kappa & \text{in Eq.(9.49)} \\ u \text{ in Eq.(9.44)} & \sim & \theta\Delta tu & \text{in Eq.(9.49)} \\ s \text{ in Eq.(9.44)} & \sim & -1 & \text{in Eq.(9.49)} \end{array} \tag{9.50}$$

and the stabilizing parameters can be calculated according to Eq. (9.47). Thence, solution procedures can just follow Eq. (9.46) to incorporate the stabilizing terms. The transient advection-diffusion problem has now been converted into a steady advection-diffusion-reaction problem framework, the time factor is taken into account naturally for determining the stabilizing parameters, and the transition between the advection-dominant, diffusion-dominant, and small time step-dominant situations is determined naturally through the stabilizing parameters.

The numerical solution of the SGS/GSGS method is shown in Fig. 9.13; the spurious oscillations at small time are successfully circumvented. To compare the accuracy of the SGS/GSGS method, a comparison of the results with the analytical solution is also provided (Fig. 9.13). The analytical solution [41] for the specific problem is as follows:

$$\varphi = \frac{1}{2}\left[erfc\frac{x-ut}{2(\kappa t)^{1/2}} + e^{ux/\kappa}erfc\frac{x+ut}{2(\kappa t)^{1/2}}\right] \tag{9.51}$$

In the 2-D case, spurious oscillations at small time steps are also successfully circumvented (Fig. 9.14).

Comparing the solute transport Eq. (9.38) after first applying discretization in time, we can find the following links to Eq. (9.48) to calculate the stabilization parameters for our problem:

$$\begin{aligned}Pe &\leftarrow \frac{\left|\frac{k}{\mu}p_i - \frac{k}{\mu}\frac{\sigma RT}{M}C_i\right|h}{2D_\mathrm{e}},\\ D_K &\leftarrow \frac{\phi R_\mathrm{d}}{\left|\frac{k}{\mu}p_i - \frac{k}{\mu}\frac{\sigma RT}{M}C_i\right|\Delta t}.\end{aligned} \tag{9.52}$$

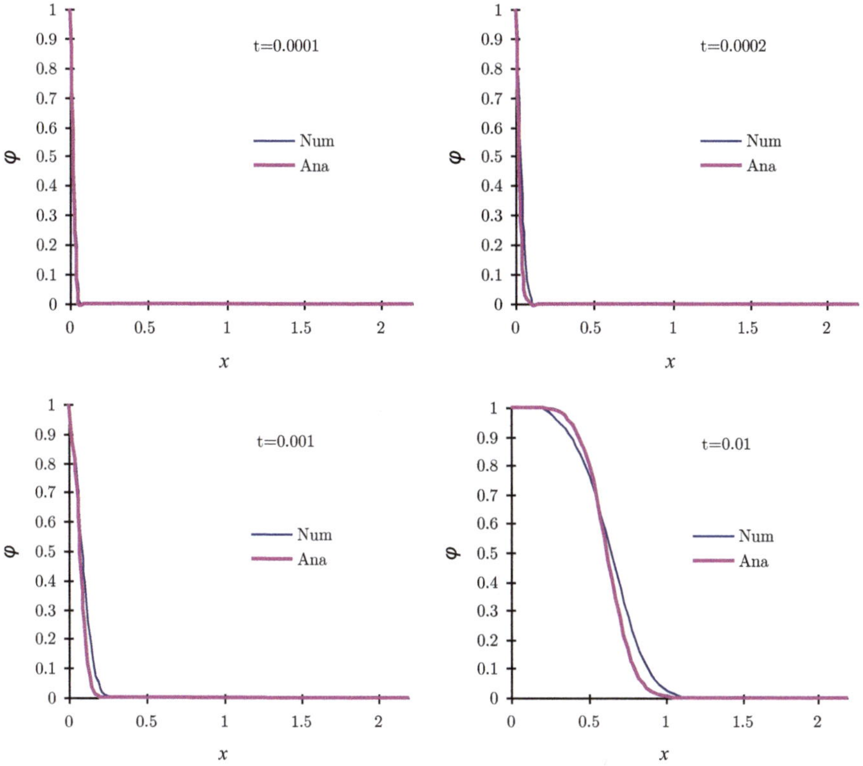

Fig. 9.13 1-D advection-diffusion problem solved by the SGS/GSGS method

Similarly, comparing the energy conservation Eq. (9.39) after first applying discretization in time, we can find the following links to Eq. (9.48) to calculate the stabilization parameters for our problem:

$$\begin{aligned} Pe &\leftarrow \frac{\left|\rho_{\mathrm{w}} c_{\mathrm{w}} \left(\frac{k}{\mu} p_i - \frac{k}{\mu}\frac{\sigma RT}{M} C_i\right)\right| h}{2\lambda_{\mathrm{T}}}, \\ D_K &\leftarrow \frac{-\phi c_{\mathrm{w}} \rho_{\mathrm{w}} \beta_{\mathrm{w}} T - (1-\phi)\rho_{\mathrm{s}} c_{\mathrm{s}} \beta_{\mathrm{s}} T + \phi \rho_{\mathrm{w}} c_{\mathrm{w}} + (1-\phi)\rho_{\mathrm{s}} c_{\mathrm{s}}}{|\rho_{\mathrm{w}} c_{\mathrm{w}} v| \Delta t}. \end{aligned} \tag{9.53}$$

After the stabilizing parameters are determined by Eq. (9.47), the solution procedure is just to follow Eq. (9.46) to append the stabilizing terms.

A numerical experiment is performed here based on the abovementioned model. We suppose a vertical wellbore (Fig. 9.15) of radius 0.127 m drilled in shale in an anisotropic stress field: $\sigma_x = 2.8 \times 10^4$ kPa, $\sigma_y = 3.2 \times 10^4$ kPa, $\sigma_z = 4.0 \times 10^4$ kPa. The initial pore pressure is $p_0 = 2.0 \times 10^4$ kPa, maintained constant at the far-field boundary. The initial solute concentration is zero and is also maintained constant at

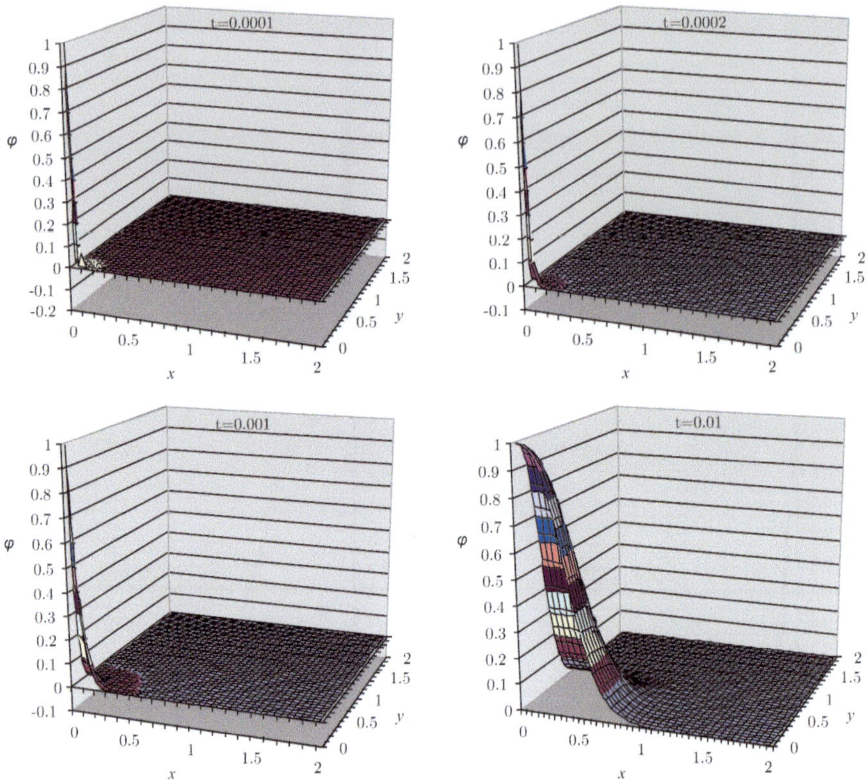

Fig. 9.14 2-D advection-diffusion problem solved by the SGS/GSGS method

the far-field boundary, as is the initial temperature of the shale formation, 77 °C. In the wellbore, the water-based drilling fluid is maintained at a pressure of $p_i = 2.2 \times 10^4$ kPa, a temperature $T_i = 32$ °C, and a solute concentration (NaCl) of $C_i = 200$ kg/m^3. Values of the other parameters are $E = 1.5 \times 10^6$ kPa, $\nu = 0.30$, $\phi = 0.30$, $K_m = 1.4 \times 10^6$ kPa, $K_w = 1.0 \times 10^6$ kPa, $k = 9.87 \times 10^{-18}$ m^2, and $\mu = 1$ cP. Also, the membrane efficiency parameter $\eta = 0.5$, the gas constant $R = 8.314$ J/mol K, the solute molar mass $M = 0.0585$ kg/mol, and the coefficient of molecular diffusion $D_e = 1.5 \times 10^{-10}$ m^2/s.

9.3.1 Case 1: Impact of the FEM Schemes (SGS/GSGS and Galerkin FEM)

In this case, we want to explore the impact of the stabilized FEM scheme on the results. The basic parameters of the system are the same as mentioned above.

Fig. 9.15 Geometry of the wellbore

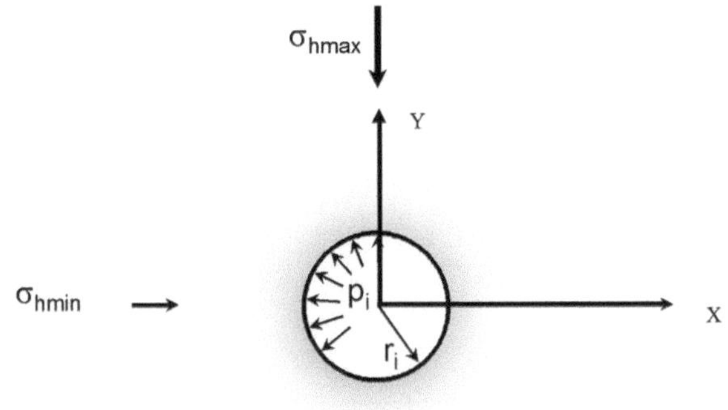

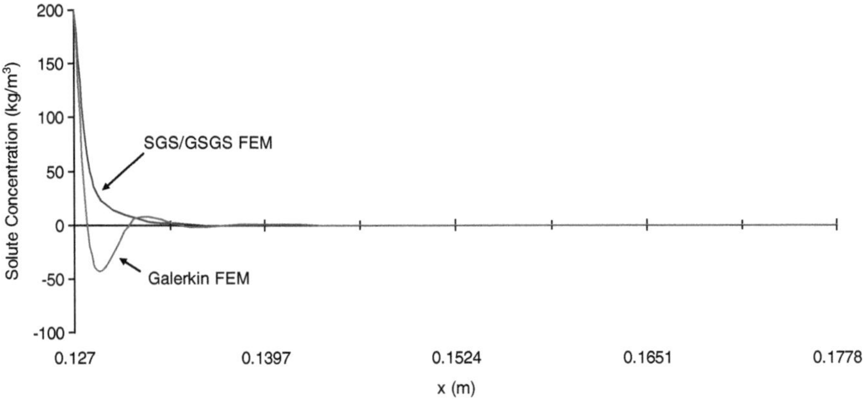

Fig. 9.16 Concentration profiles by SGS/GSGS FEM and Galerkin FEM

Figure 9.16 shows the solute concentration profiles (all the profiles in this section are along the *X*-axis as shown in Fig. 9.15) at 0.024 h computed by the stabilized SGS/GSGS FEM and the traditional Galerkin FEM. It can be seen that the Galerkin FEM leads to numerical oscillations whereas our stabilized FEM approach avoids such oscillations. Figure 9.17 shows the pore pressure profiles at 0.024 h by SGS/GSGS FEM and Galerkin FEM; the Galerkin FEM solution again evidences oscillations, whereas the SGS/GSGS FEM solution does not.

9.3.2 *Case 2: Impact of Thermal and Solute Convection in Lower Permeability Formations*

In this case, we want to explore the effect of thermal and solute convection on the pressure change in formations with lower permeability.

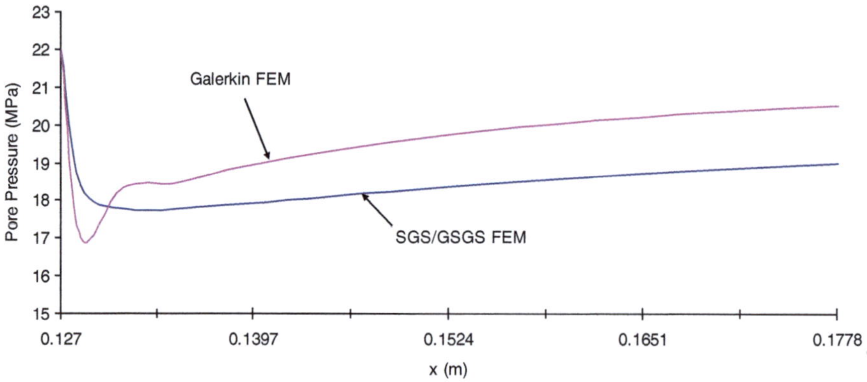

Fig. 9.17 Pore pressure profiles by SGS/GSGS FEM and Galerkin FEM

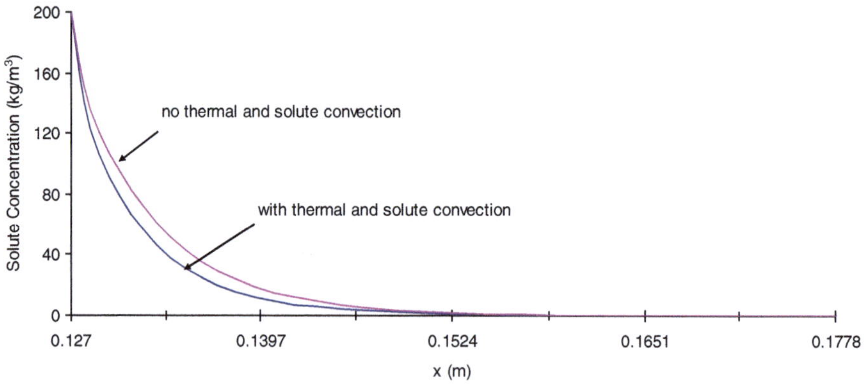

Fig. 9.18 Solute concentration profiles with and without thermal and solute convection

Figure 9.18 shows the solute concentration profiles at 2.4 h using the models with and without considering thermal and solute convection (i.e., the second term in Eq. (9.39) and the third term in Eq. (9.38)). Figure 9.19 shows temperature profiles and Fig. 9.17 shows pressure profiles, also at 2.4 h, and also with and without considering thermal and solute convection. As expected, the cooler drilling fluid with a higher solute concentration will reduce the pore pressure in the formation and contribute to wellbore stability enhancement. However, the solute concentration in the case with advective terms propagates more slowly than in the case without advective terms, leading to the difference in the pore pressure evolution profiles shown in Fig. 9.20. From Figs. 9.18 and 9.19, we conclude that temperature propagation is less influenced than solute concentration propagation in cases including advective terms.

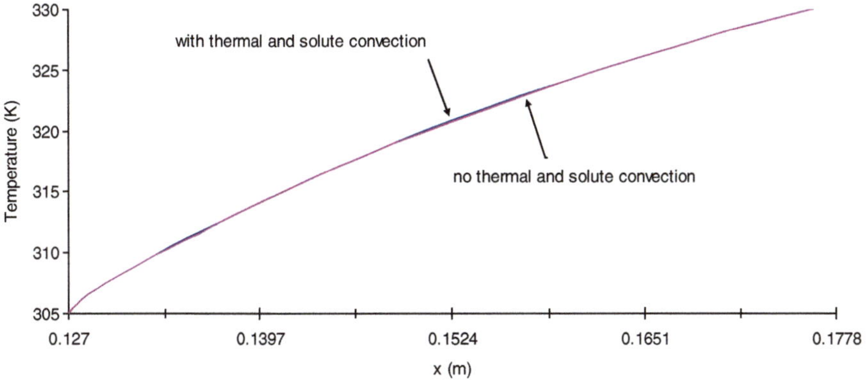

Fig. 9.19 Temperature profiles with and without thermal and solute convection

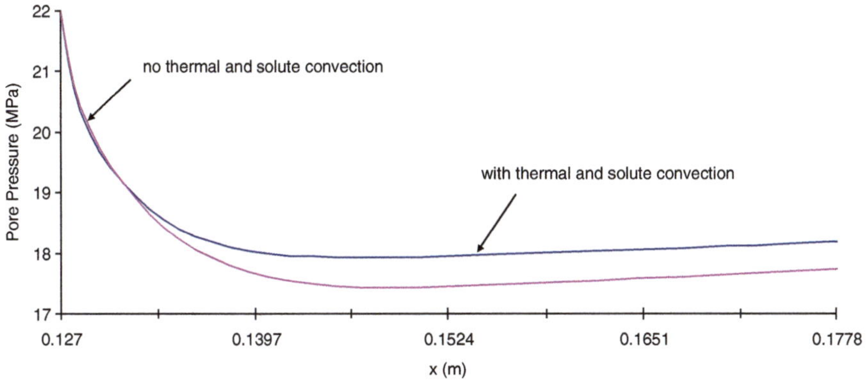

Fig. 9.20 Pore pressure profiles with and without thermal and solute convection

9.3.3 *Case 3: Impact of Thermal and Solute Convection in Higher Permeability Formations*

For this case, we want to explore the effects of thermal and solute convection on the pressure change in formations with higher permeability (keep the value of other parameters the same as those in case 2).

Figures 9.21, 9.22, and 9.23 show the solute concentration profiles, temperature, and pore pressure profiles of a case with a higher permeability of 9.87×10^{-16} m^2, corresponding to the profiles of case 2 with the lower permeability of 9.87×10^{-18} m^2. The effect of thermal and solute convection is more significant in the case with higher permeability than that in the case with lower permeability, as expected. The fact that the temperature and solute concentration propagate more slowly in the case including thermal and solute convection shows that the Darcy velocity that appears in Eqs. (9.38) and (9.39) is mostly directed toward the wellbore.

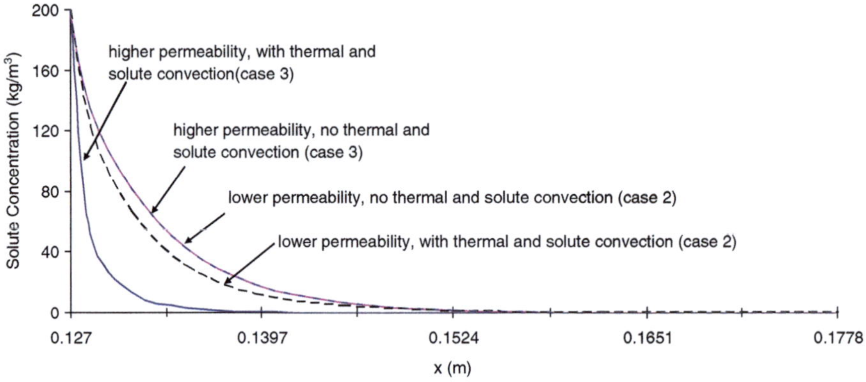

Fig. 9.21 Solute concentration profiles with and without thermal and solute convection

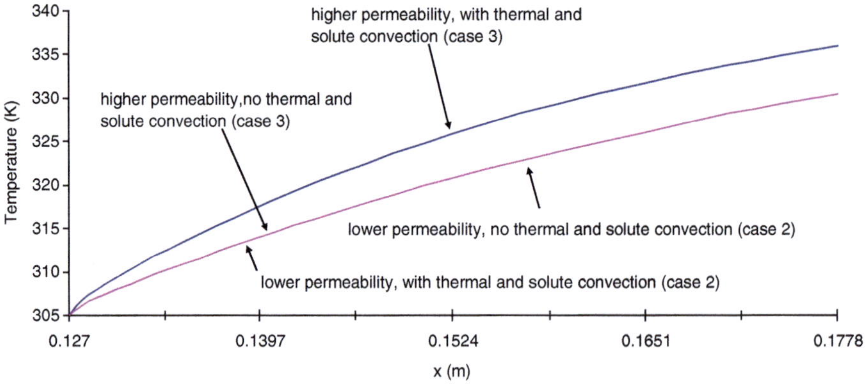

Fig. 9.22 Temperature profiles with and without thermal and solute convection

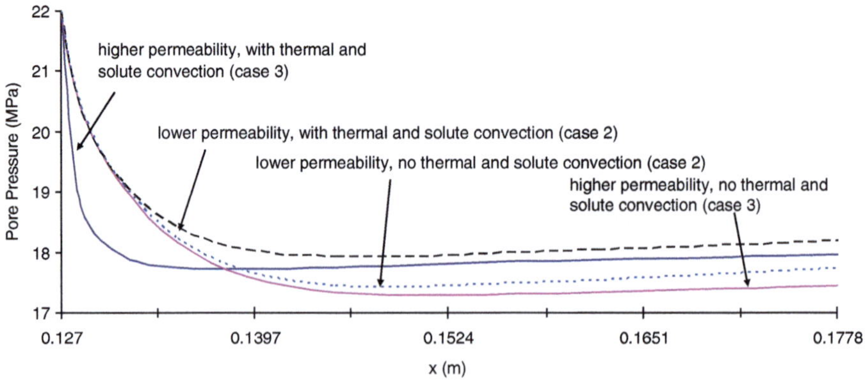

Fig. 9.23 Pore pressure profiles with and without thermal and solute convection

9.3.4 Case 4: Impact of the Membrane Efficiency

The effect of membrane efficiency on the pressure change at formations is now explored with a lower membrane efficiency (keep the value of other parameters the same as case 3).

Figures 9.24, 9.25, and 9.26 show the solute concentration profiles, temperature, and pore pressure profiles with lower membrane efficiency of 0.05, as well as for the higher membrane efficiency of 0.5 (Case 3). When the membrane efficiency is very low, the pore pressure gradient dominates the Darcy flow velocity, and therefore the temperature and solute concentration propagate more rapidly in the case considering advective terms than in the case that does not consider advective terms; when the membrane efficiency is very high, the Darcy flow velocity will be significantly reduced by the counter-acting osmotic pressure gradient, and this leads to less advection, and therefore the temperature and solute concentration propagate

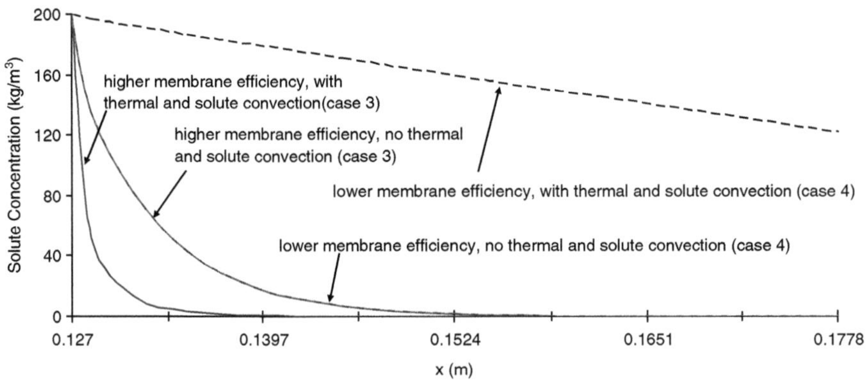

Fig. 9.24 Solute concentration profiles with and without thermal and solute convection

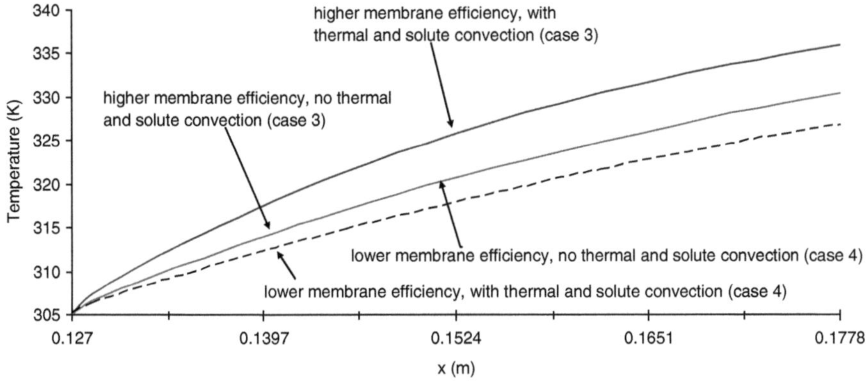

Fig. 9.25 Temperature profiles with and without thermal and solute convection

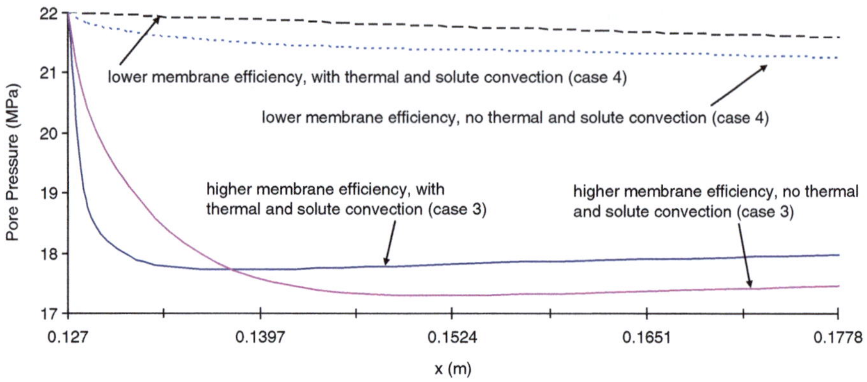

Fig. 9.26 Pore pressure profiles with and without thermal and solute convection

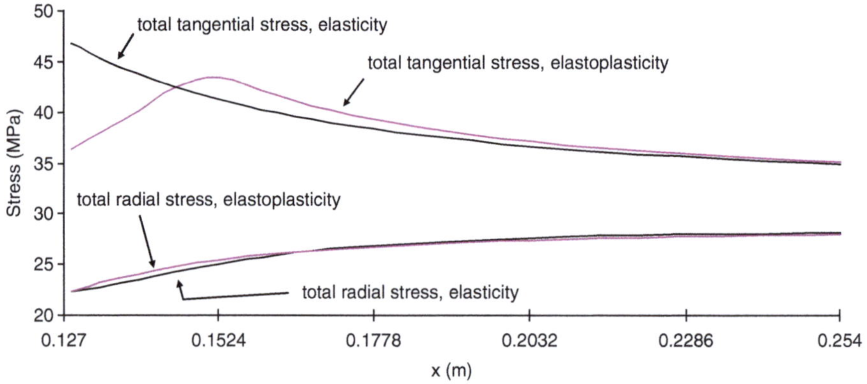

Fig. 9.27 Profiles of radial and tangential stresses with and without plasticity considered

more slowly in the situation considering advective terms than in the situation that does not consider advective terms.

Figure 9.27 shows the total tangential and radial stresses around the borehole with and without plasticity being considered. It can be seen that the geomechanics constitutive relationship of the rock (i.e., the elastoplastic behavioral law) also plays an important role on the stresses redistribution during the drilling process.

The cases presented show that the effect of thermal and solute convection is an essential element to be considered in fully coupled THMC wellbore modeling, at least for drilling formations of a particular intermediate permeability range. Furthermore, a stabilized FEM approach is necessary to obtain a stable solution. The numerical experiments demonstrate that that the permeability and membrane efficiency both play an important role in the impact of thermal and solute convection on pressure change, and thus are key elements for wellbore stability modeling in a range of circumstances.

9.4 Hydraulic Fracturing

Along with hydraulic fracture applications in areas such as conventional well stimulation, acid injection into carbonates, and restressing unconsolidated sandstones to reduce the risk of sand influx, other emerging activities associated with hydraulic fracturing include massive slurried solid waste disposal (including biosolids), produced-water reinjection, CO_2 geo-sequestration, injectivity enhancement in compressed air energy storage systems, geothermal energy exploitation, and shale gas development. All these activities involve injection of fluid at lower temperature into deep subsurface porous media at a significantly higher temperature, where hydraulic fracture might not be induced without contribution of significant temperature differences. Accumulation of cold fluid around an injection well leads to formation contraction and induced thermal stresses reduce effective stresses substantially. When the reduced effective stresses become tensile with magnitude exceeding the tensile strength of the formation rock, hydraulic fractures are initiated at the wellbore in the plane mostly perpendicular to the minimum horizontal in situ stress. The induced fractures provide higher injectivity for fluid flow, make cooled regions larger through convection-dominated heat transfer, and thus result in continuing propagation of the fractures. Thermal contribution to hydraulic fracturing mostly is beneficial in these activities in enhancing the injectivity, whereas it may be detrimental when fractures develop excessively into undesired zones. Therefore hydraulic fracturing modeling with thermal contribution considered is important for accurate estimation of fracture length, width, and height to avoid potential environmental risks. The model involves the coupling of heat transfer, mass transport, stress change as well as the fracture propagation. Researchers have been making efforts to simulate the coupled system by different tools, including analytical methods, finite element methods, finite difference methods, and boundary element methods [42–45].

Among these methods, the extended finite element method (XFEM) enables the fracturing problem to be simulated in a continuous fashion by introducing additional degrees of freedom to the existing nodes of the traditional finite elements when containing discontinuity [46]. This leads to the advantage that the finite element mesh does not need to be aligned with the fracture path and remeshing is not needed. Extended finite element method has been widely applied in modeling fracturing processes in solid, and it has also been used to model the fracturing processes in porous media in recent years by introducing different techniques in treating the fluid exchange between the fracture and the matrix [47]. These techniques include a regularized Delta Dirac function added for a discontinuous fluid field [48], a moisturizing model without enhancing the fluid phase [49], a weak discontinuity model for the fluid phase with Couette flow in the crack [50], and a double-porosity model [51]. Among these techniques, the double-porosity model seems to be convenient to be combined with XFEM and recently has been applied to the displacement and pressure analysis in fractured porous media [52]. The other advantage of this technique is that thermal fracturing in porous media can be modeled naturally based on the existing double porosity model of deformable porous media with temperature coupled [53]. XFEM combined with double-porosity model to simulate non-isothermal fluid flow in fracturing and fractured porous media can be described as follows.

The general equilibrium equation for non-isothermal single phase water flow through deformable porous medium incorporating the concept of effective stress can be written as

$$\nabla \cdot (\mathbf{D}\boldsymbol{\varepsilon} - \alpha\mathbf{i}p_{\mathrm{w}} - K\beta_{\mathrm{s}}\mathbf{i}T) = 0 \tag{9.54}$$

where $\mathbf{D}$ is the tangential stiffness matrix, $\boldsymbol{\varepsilon}$ is the strain of the matrix, $\mathbf{i}$ is the identity vector, and p_{w} and T denote pore pressure and temperature, respectively. α is Biot's coefficient, β_{s} is the volumetric thermal expansion coefficient of solid phase, and K is bulk modulus for the skeleton. It's straightforward to extended to multiphase flow system with p_{w} replaced by average pore pressure, $p = S_{\mathrm{n}}p_{\mathrm{n}} + S_{\mathrm{w}}p_{\mathrm{w}}$ where S_{n}, S_{w}, p_{n}, and p_{w} are the saturation and pore pressure with respect to non-wetting and wetting phases, respectively.

The general form of the continuity equation for single phase water flow in the matrix and fractured zone according to the double-porosity model can be expressed as equations below:

$$\nabla^T\left(-\frac{\mathbf{k}_1}{\mu}\nabla p_{\mathrm{w}}\right) + \frac{\bar{\alpha}\mathbf{k}_1}{\mu}(p_{\mathrm{w}} - p_{\mathrm{f}}) + \left(\frac{\alpha - \phi_1}{K_{\mathrm{s}}} + \frac{\phi_1}{K_{\mathrm{w}}}\right)\frac{\partial p_{\mathrm{w}}}{\partial t} + \alpha\frac{\partial \boldsymbol{\varepsilon}}{\partial t} - [(\alpha - \phi_{\mathrm{T}})\beta_{\mathrm{s}} + \phi_1\beta_{\mathrm{w}}]\frac{\partial T}{\partial t} + Q_1 = 0 \tag{9.55}$$

$$\nabla^T\left(-\frac{\mathbf{k}_2}{\mu}\nabla p_{\mathrm{f}}\right) - \frac{\bar{\alpha}\mathbf{k}_1}{\mu}(p_{\mathrm{w}} - p_{\mathrm{f}}) + \left(\frac{\phi_2}{K_{\mathrm{w}}}\right)\frac{\partial p_{\mathrm{f}}}{\partial t} - \phi_2\beta_{\mathrm{w}}\frac{\partial T}{\partial t} + Q_2 = 0 \tag{9.56}$$

where $\mathbf{k}_1$ and $\mathbf{k}_2$ are the permeability of the matrix and fractured zone, μ is the viscosity of water, ϕ_1 and ϕ_2 are the porosity of matrix and fractured zone, ϕ_{T} is the sum of ϕ_1 and ϕ_2, p_{f} is the pressure of the fractured zone, K_{s} and K_{w} are the bulk moduli of solid grains and water, $\bar{\alpha}$ is a coefficient depending on the fracture width and geometry, and Q_1 and Q_2 are the flow source/sink terms.

Here only one thermodynamic continuum is assumed to be representative of both the matrix and fractured zone rather than a double temperature approach, considering that it is difficult to arrive at an appropriate external entropy supply for the matrix and fractured zone and it is also an extremely difficult experimental procedure to determine the heat transfer coefficient in a fractured porous material. Then the general form of the energy balance equation, including the thermal convection and thermal conduction terms, can be expressed as follows:

$$
\begin{aligned}
&\nabla^{T}[-\lambda\nabla T]+\rho_{\mathrm{w}}c_{\mathrm{w}}(v_1+v_2)\nabla T+T\left[(1-\phi_{\mathrm{T}})c_{\mathrm{s}}\frac{\rho_{\mathrm{s}}}{K_{\mathrm{s}}}+\phi_1 c_{\mathrm{w}}\frac{\rho_{\mathrm{w}}}{K_{\mathrm{w}}}\right]\frac{\partial p_{\mathrm{w}}}{\partial t}\\
&+\left(T\phi_2 c_{\mathrm{w}}\frac{\rho_{\mathrm{w}}}{K_{\mathrm{w}}}\right)\frac{\partial p_{\mathrm{f}}}{\partial t}+[(1-\phi_{\mathrm{T}})\rho_{\mathrm{s}}c_{\mathrm{s}}+\phi_{\mathrm{T}}\rho_{\mathrm{w}}c_{\mathrm{w}}\\
&-(1-\phi_{\mathrm{T}})\rho_{\mathrm{s}}c_{\mathrm{s}}\beta_{\mathrm{s}}T-\phi_{\mathrm{T}}\rho_{\mathrm{w}}c_{\mathrm{w}}\beta_{\mathrm{w}}T]\frac{\partial T}{\partial t}+Q_{\mathrm{h}}=0
\end{aligned}
\tag{9.57}
$$

where λ is the porous medium thermal conductivity, c_{w} and c_{s} are the specific heat capacity of water and solid phase, ρ_{s} is the density of solid phase, Q_{h} is heat source/sink, and v_1 and v_2 are the velocity of flow in matrix and fractured zone, respectively.

To simulate the fracturing process in aforementioned double porosity system, the strong discontinuity will be introduced in displacement field by employing the extended finite element method with remeshing avoided. In addition, the formation rock is considered as quasi-brittle material and discrete traction-displacement constitutive model based on the cohesive crack concept can be employed [54]. Further, modeling of the fracture propagation in extended finite elements is carried out by applying elements that are entirely cut by fracture. Therefore, in the extended finite element method described below, no fracture tip enhancement is needed, and the discontinuous displacement field can be decomposed into a continuous part represented by degrees of freedom in traditional finite element framework, **a**, and an enhanced part represented by extra degrees of freedom, **b**, and Heaviside function H (**x**) (H = 1 if $\mathbf{x} \in \Omega^{+}$, H = −1 if $\mathbf{x} \in \Omega^{-}$, where Ω^{+} and Ω^{-} are the sub-bodies of whole body Ω separated by the discontinuity Γ_{d}):

$$
\mathbf{u}(\mathbf{x})=\mathbf{N}(\mathbf{x})\mathbf{a}+\mathcal{H}(\mathbf{x})\mathbf{N}_{\mathrm{b}}(\mathbf{x})\mathbf{b}
\tag{9.58}
$$

where $\mathbf{N}$ and $\mathbf{N}_{\mathrm{b}}$ are the conventional shape functions for the regular displacement and displacement jump. The strain field in terms of nodal displacements can be written as

$$
\varepsilon=\mathbf{Ba}+\mathcal{H}\mathbf{B}_{\mathrm{b}}\mathbf{b}+2(\delta\Gamma_{\mathrm{d}}\mathbf{n})\mathbf{Nb}
\tag{9.59}
$$

where $\mathbf{B}$ and $\mathbf{B}_{\mathrm{b}}$ are the conventional strain-displacement matrices for the regular displacement and displacement jump, $\delta\Gamma_{\mathrm{d}}$ is the Dirac-delta distribution centered on the discontinuity, and $\mathbf{n}$ is the unit normal vector to the discontinuity.

Pressures and temperature fields in the double-porosity model framework are approximated in the traditional way by $\mathbf{N}_{\mathrm{p}}\mathbf{p}_{\mathrm{w}}$, $\mathbf{N}_{\mathrm{p}}\mathbf{p}_{\mathrm{f}}$, and $\mathbf{N}_{\mathrm{p}}\mathbf{T}$, respectively. After applying Galerkin's FEM of weighted residual to the abovementioned system equations, the finite element formulations for the above equations after discretization can be expressed as follows:

$$
\begin{aligned}
&\int_{\Omega}\mathbf{B}^{\mathrm{T}}\mathbf{DB}\mathrm{d}\Omega\mathbf{a}+\int_{\Omega^{+-}}\mathcal{H}\mathbf{B}^{\mathrm{T}}\mathbf{DB}_{\mathrm{b}}\mathrm{d}\Omega\mathbf{b}-\int_{\Omega}\left(\mathbf{B}^{\mathrm{T}}\mathbf{i}-\mathbf{B}^{\mathrm{T}}\mathbf{D}\frac{\mathbf{i}}{3K_{\mathrm{s}}}\right)\mathbf{N}_{\mathrm{p}}\mathrm{d}\Omega\mathbf{p}_{w}\\
&-\int_{\Omega}\mathbf{B}^{\mathrm{T}}\mathbf{Di}\frac{\beta_{\mathrm{s}}}{3}\mathbf{N}_{\mathrm{p}}\mathrm{d}\Omega\mathbf{T}=\mathbf{f}^{\mathrm{a}}
\end{aligned}
\tag{9.60}
$$

$$\begin{aligned}&\int_{\Omega^{+-}} \mathcal{H}\mathbf{B}_b^T\mathbf{D}\mathbf{B}\mathrm{d}\Omega\mathbf{a} + \int_{\Omega^{+-}} \mathcal{H}\mathbf{B}_b^T\mathbf{D}\mathbf{B}_b\mathcal{H}\mathrm{d}\Omega\mathbf{b} + 2\int_{\Gamma d} \mathbf{N}_b{}^T\mathbf{T}_d\mathbf{N}_b\mathrm{d}\Gamma\mathbf{b}\\ &- \int_{\Omega^{+-}} \mathcal{H}\left(\mathbf{B}_b^T\mathbf{i} - \mathbf{B}_b^T\mathbf{D}\frac{\mathbf{i}}{3K_s}\right)\mathbf{N}_p\mathrm{d}\Omega\mathbf{p}_w - 2\int_{\Gamma d} \mathbf{N}_b{}^T\mathbf{n}_s\mathbf{N}_p\mathrm{d}\Gamma\mathbf{p}_f\\ &- \int_{\Omega^{+-}} \mathcal{H}\mathbf{B}_b^T\mathbf{D}\mathbf{i}\frac{\beta_s}{3}\mathbf{N}_p\mathrm{d}\Omega\mathbf{T} = \mathbf{f}^b\end{aligned} \tag{9.61}$$

$$\begin{aligned}&\int_{\Omega} \mathbf{N}_p^T\left(\mathbf{i}^T - \frac{\mathbf{i}^T\mathbf{D}}{3K_s}\right)\mathbf{B}\mathrm{d}\Omega\dot{\mathbf{a}} + \int_{\Omega^{+-}} \mathcal{H}\mathbf{N}_p^T\left(\mathbf{i}^T - \frac{\mathbf{i}^T\mathbf{D}}{3K_s}\right)\mathbf{B}_b\mathrm{d}\Omega\dot{\mathbf{b}}\\ &+ \int_{\Omega_1} (\nabla\mathbf{N}_p)^T\frac{\mathbf{k}_1}{\mu}\nabla\mathbf{N}_p\mathrm{d}\Omega\mathbf{p}_w + \int_{\Omega_2} \mathbf{N}_p^T\overline{\alpha}\frac{\mathbf{k}_1}{\mu}\mathbf{N}_p\mathrm{d}\Omega\mathbf{p}_w + \int_{\Omega_1} \mathbf{N}_p^T\left(\frac{\alpha - \phi_1}{K_s} + \frac{\phi_1}{K_w}\right)\mathbf{N}_p\mathrm{d}\Omega\dot{\mathbf{p}}_w\\ &- \int_{\Omega_2} \mathbf{N}_p^T\overline{\alpha}\frac{\mathbf{k}_1}{\mu}\mathbf{N}_p\mathrm{d}\Omega\mathbf{p}_f + \int_{\Omega_1} \mathbf{N}_p^T[(\alpha - \phi_T)\beta_s + \phi_1\beta_w]\mathbf{N}_p\mathrm{d}\Omega\dot{\mathbf{T}} = \mathbf{f}^w\end{aligned} \tag{9.62}$$

$$\begin{aligned}&-\int_{\Omega_2} \mathbf{N}_p^T\overline{\alpha}\frac{\mathbf{k}_1}{\mu}\mathbf{N}_p\mathrm{d}\Omega\mathbf{p}_w + \int_{\Omega_2} (\nabla\mathbf{N}_p)^T\frac{\mathbf{k}_2}{\mu}\nabla\mathbf{N}_p\mathrm{d}\Omega\mathbf{p}_f + \int_{\Omega_2} \mathbf{N}_p^T\overline{\alpha}\frac{\mathbf{k}_1}{\mu}\mathbf{N}_p\mathrm{d}\Omega\mathbf{p}_f\\ &+ \int_{\Omega_2} \mathbf{N}_p^T\frac{\phi_2}{K_w}\mathbf{N}_p\mathrm{d}\Omega\dot{\mathbf{p}}_f + \int_{\Omega_2} \mathbf{N}_p^T\phi_2\beta_w\mathbf{N}_p\mathrm{d}\Omega\dot{\mathbf{T}} = \mathbf{f}^f\end{aligned} \tag{9.63}$$

$$\begin{aligned}&\int_{\Omega} \mathbf{N}_p^T T\left[(1-\phi_T)c_s\frac{\rho_s}{K_s} + \phi_1 c_w\frac{\rho_w}{K_w}\right]\mathbf{N}_p\mathrm{d}\Omega\dot{\mathbf{p}}_w + \int_{\Omega} \mathbf{N}_p^T T\phi_2 c_w\frac{\rho_w}{K_w}\mathbf{N}_p\mathrm{d}^p\Omega\,\dot{\mathbf{p}}_f\\ &+ \int_{\Omega}\left((\nabla\mathbf{N}_p)^T\lambda^T\nabla\mathbf{N}_p + \mathbf{N}_p^T\rho_w c_w\left(\frac{\mathbf{k}_1}{\mu}P_{w,i} + \frac{\mathbf{k}_2}{\mu}P_{f,i}\right)\nabla\mathbf{N}_p\right)\mathrm{d}\Omega\mathbf{T}\\ &+ \int_{\Omega}\mathbf{N}_p^T[(1-\phi_T)\rho_s c_s + \phi_T\rho_w c_w\\ &- (1-\phi_T)\rho_s c_s\beta_s T - \phi_T\rho_w c_w\beta_w T]\mathbf{N}_p\mathrm{d}\Omega\dot{\mathbf{T}}\\ &= \mathbf{f}^T\end{aligned} \tag{9.64}$$

The final matrix form for the above equations can be expressed as

$$\begin{bmatrix}\mathbf{M}_{aa} & \mathbf{M}_{ab} & -\mathbf{C}_{aw} & 0 & -\mathbf{C}_{aT}\\ \mathbf{M}_{ba} & \mathbf{M}_{bb} & -\mathbf{C}_{bw} & 0 & -\mathbf{C}_{bT}\\ 0 & 0 & \mathbf{H}_{ww} & -\mathbf{H}_{wf} & 0\\ 0 & 0 & -\mathbf{H}_{fw} & \mathbf{H}_{ff} & 0\\ 0 & 0 & 0 & 0 & \mathbf{H}_{TT}\end{bmatrix}\begin{Bmatrix}\mathbf{a}\\ \mathbf{b}\\ \mathbf{p}_w\\ \mathbf{p}_f\\ \mathbf{T}\end{Bmatrix} + \begin{bmatrix}0 & 0 & 0 & 0 & 0\\ 0 & 0 & 0 & 0 & 0\\ \mathbf{C}_{wa} & \mathbf{C}_{wb} & \mathrm{R}_{ww} & 0 & \mathbf{C}_{wT}\\ 0 & 0 & 0 & \mathrm{R}_{ff} & \mathbf{C}_{fT}\\ 0 & 0 & \mathbf{C}_{Tw} & \mathbf{C}_{Tf} & \mathrm{R}_{TT}\end{bmatrix}\begin{Bmatrix}\mathbf{a}_t\\ \mathbf{b}_t\\ \mathbf{p}_{wt}\\ \mathbf{p}_{ft}\\ \mathbf{T}_t\end{Bmatrix} = \begin{Bmatrix}\mathbf{f}^a\\ \mathbf{f}^b\\ \mathbf{f}^w\\ \mathbf{f}^f\\ \mathbf{f}^T\end{Bmatrix} \tag{9.65}$$

where $[\mathbf{a}, \mathbf{b}, \mathbf{p}_{\mathrm{w}}, \mathbf{p}_{\mathrm{f}}, \mathbf{T}]^{\mathrm{T}}$ and $[\mathbf{a}_{\mathrm{t}}, \mathbf{b}_{\mathrm{t}}, \mathbf{p}_{\mathrm{wt}}, \mathbf{p}_{\mathrm{ft}}, \mathbf{T}_{\mathrm{t}}]^{\mathrm{T}}$ are the vectors of unknown variables and corresponding time derivatives. $[\mathbf{f}^{\mathrm{a}}, \mathbf{f}^{\mathrm{b}}, \mathbf{f}^{\mathrm{w}}, \mathbf{f}^{\mathrm{f}}, \mathbf{f}^{\mathrm{T}}]^{\mathrm{T}}$ is the vector for the nodal loads, the flow source, the heat source, and the solute source. The explicit expressions of the above matrices can be found in Yin [45].

To integrate the above equations with respect to time, linear interpolation in time using a finite difference method is needed. For example, in the classic theta method, $\theta = 0$ reflects the explicit Euler method, $\theta = 1$ implies the implicit Euler method, whereas setting $\theta = 0.5$ results in Crank-Nicolson method. Using the general theta method, the equations can be written as

$$\begin{bmatrix} \theta\mathbf{M}_{\mathrm{aa}} & \theta\mathbf{M}_{\mathrm{ab}} & -\theta\mathbf{C}_{\mathrm{aw}} & 0 & -\theta\mathbf{C}_{\mathrm{aT}} \\ \theta\mathbf{M}_{\mathrm{ba}} & \theta\mathbf{M}_{\mathrm{bb}} & -\theta\mathbf{C}_{\mathrm{bw}} & 0 & -\theta\mathbf{C}_{\mathrm{bT}} \\ \mathbf{C}_{\mathrm{wa}} & \mathbf{C}_{\mathrm{wb}} & \mathrm{R}_{\mathrm{ww}} + \theta\Delta t\mathbf{H}_{\mathrm{ww}} & -\theta\Delta t\mathbf{H}_{\mathrm{wf}} & \mathbf{C}_{\mathrm{wT}} \\ 0 & 0 & -\theta\Delta t\mathbf{H}_{\mathrm{fw}} & \mathrm{R}_{\mathrm{ff}} + \theta\Delta t\mathbf{H}_{\mathrm{ff}} & \mathbf{C}_{\mathrm{fT}} \\ 0 & 0 & \mathbf{C}_{\mathrm{Tw}} & \mathbf{C}_{\mathrm{Tf}} & \mathrm{R}_{\mathrm{TT}} + \theta\Delta t\mathbf{H}_{\mathrm{TT}} \end{bmatrix} \begin{Bmatrix} \Delta\mathbf{a} \\ \Delta\mathbf{b} \\ \Delta\mathbf{p}_{\mathrm{w}} \\ \Delta\mathbf{p}_{\mathrm{f}} \\ \Delta\mathrm{T} \end{Bmatrix} = \begin{Bmatrix} \Delta\mathbf{f}^{\mathrm{a}} \\ \Delta\mathbf{f}^{\mathrm{b}} \\ \Delta t\mathbf{f}^{\mathrm{w}} - \Delta t\mathbf{H}_{\mathrm{ww}}\mathbf{p}_{\mathrm{w}} + \Delta t\mathbf{H}_{\mathrm{wf}}\mathbf{p}_{\mathrm{f}} \\ \Delta t\mathbf{f}^{\mathrm{f}} - \Delta t\mathbf{H}_{\mathrm{ff}}\mathbf{p}_{\mathrm{f}} + \Delta t\mathbf{H}_{\mathrm{fw}}\mathbf{p}_{\mathrm{w}} \\ \Delta t\mathbf{f}^{\mathrm{T}} - \Delta t\mathbf{H}_{\mathrm{TT}}\mathbf{T} \end{Bmatrix} \tag{9.66}$$

The implementation of the three-dimensional finite element model is based on 20-node brick elements. All degrees of freedom are solved on the corner nodes of the brick element, whereas the degree of freedom a are solved on the midside nodes as well in order to make its shape function one order higher than the rest of the degrees of freedom and thus to ensure no spurious oscillations occurring in the numerical solution. The fracture is assumed to fully propagate through an element if the minor principal stress (tensile) in the element exceeds the tensile strength of the material, and the fracture plane is normal to the direction of the minor principal stress. The extent of success of the assumption may be subject to mesh size but it has proven satisfactory for 3D fracturing problem [55]. For a fractured element, the hexahedron is decomposed into tetrahedra (see Fig. 9.28), and the fracture may separate a tetrahedron into two sub-elements in shape of either tetrahedron or prismatic by a triangular or quadrilateral plane. Both sub-elements need to be considered in numerical integration for each tetrahedron separated by a fracture [56]. The volume integral of the fractured element takes account into the contribution from all the sub-elements. The surface integral of the fracture plane in the fracture element takes account into the contribution from all the sub-elements that are separated by the fracture.

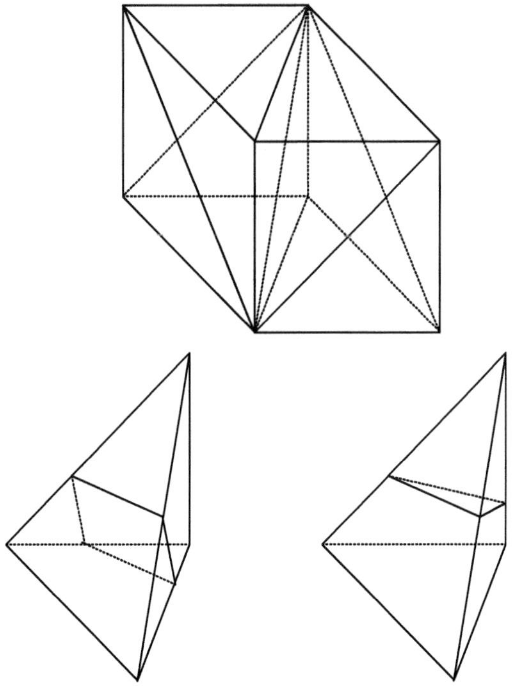

Fig. 9.28 The decomposition of a hexahedron into tetrahedral and the possible fracture planes crossing the tetrahedral

The discrete traction-displacement constitutive model adopted is an exponential softening curve focusing on mode-I fracture opening. The normal traction t_n relates the history parameter k (equal to the largest crack opening, $[\![u]\!]_n$, reached in loading history) as

$$t_n = f_t \exp\left(-\frac{f_t}{G_f}\kappa\right) \tag{9.67}$$

where f_t is the tensile strength and G_f is the fracture energy of the material. This equation can be differentiated with respect to time to form a consistently linearized tangent [54]:

$$\dot{t}_n = -\frac{f_t^2}{G_f}\exp\left(-\frac{f_t}{G_f}\kappa\right)\dot{u}_n \tag{9.68}$$

Following the approach in works [57] concerning the fracture permeability of double-porosity model, the permeability for an element in fractured zone can be determined by

$$\mathbf{k}_2 = \frac{V_f}{V_e}\mathbf{k}_2^{'} \tag{9.69}$$

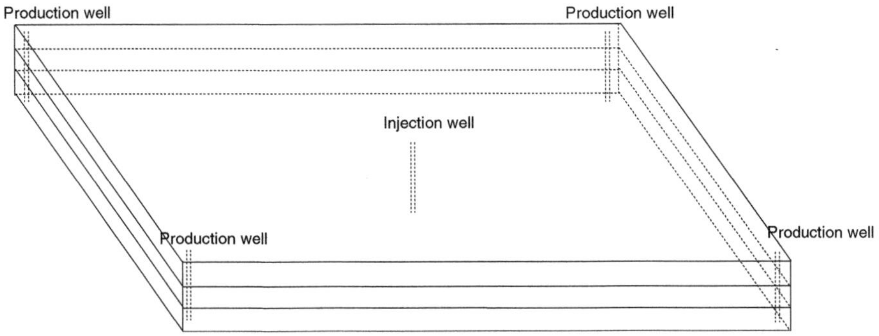

Fig. 9.29 Geometry of the problem domain

where V_f and V_e are the volume of the fracture and element, and $\mathbf{k}_2'$ is the effective fracture permeability and its absolute value can be determined based on the parallel plate model:

$$\|\mathbf{k}_2'\| = \frac{w^2}{12} \tag{9.70}$$

where w is the width of the fracture.

The energy balance equation of heat transfer is a transient diffusion-advection equation. Traditional finite element method has difficulty overcoming the numerical oscillation in solving in this type of equation. The stabilized finite element scheme, the subgrid scale/gradient subgrid scale (SGS/GSGS) method, needs to be employed to address the problem based on its previous success [33, 34].

Based on the abovementioned numerical model, numerical experiments are conducted for scenarios of cold-water injection into a hypothetical deep warm aquifer. Suppose a 180 m × 180 m × 4 m aquifer (see Fig. 9.29), 2000 m deep, with $E = 1.5 \times 10^7$ kPa, $\nu = 0.3$, $\phi = 0.3$, $K_s = 1.4 \times 10^8$ kPa, $k_1 = 100$ mD, $\mu = 1$ cP, $\rho_s = 2.5 \times 10^3$ kg/m^3, $\rho_w = 1.0 \times 10^3$ kg/m^3, $\beta_s = 2.5 \times 10^{-5}$ K^{-1}, $\beta_w = 2.0 \times 10^{-4}$ K^{-1}, $c_s = 2.5$ kJ/kg K, $c_w = 4.2$ kJ/kg K, and $\lambda = 2.65$ J/m s K. The tensile strength of the formation rock is 1.0×10^3 kPa, and the fracture energy is 0.1 N/mm. A Cartesian coordinate system is used where the top and bottom of the aquifer are normal to the Z-axis, and all sides are vertical and normal to the $X-Y$ plane. Extending 4 m from the outer aquifer boundary in vertical directions are the overlying and underlying formations. In these formations, the following properties apply: $E = 1.0 \times 10^7$ kPa, $\nu = 0.30$, $\varphi = 0.30$, $k = 1.0$ mD, $\mu = 1.0$ cP, $\beta_s = 1.0 \times 10^{-6}$ K^{-1}, $\rho_s = 2.5 \times 10^3$ kg/m^3, $c_s = 1.25$ kJ/kg K, and $\lambda = 2.3$ J/m s K. The tensile strength and the fracture energy are the same as those of the aquifer formation rock.

The aquifer in situ stresses consist of a vertical stress $\sigma_v = 6.0 \times 10^4$ kPa and horizontal far-field stresses of $\sigma_{hx} = 5.0 \times 10^4$ kPa, $\sigma_{hy} = 3.0 \times 10^4$ kPa, aligned with Z, X, and Y coordinates, respectively. The initial pore pressure is $p_0 = 2.0 \times 10^4$ kPa. Initial temperature of the aquifer is 100 °C. The in situ stresses

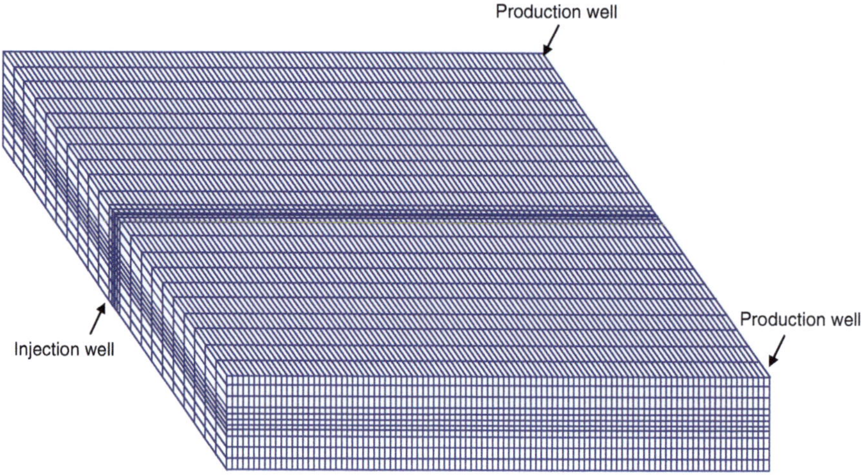

Fig. 9.30 Finite element mesh of the half domain

overlying and underlying formation consist of vertical stress $\sigma_v = 6.0 \times 10^4$ kPa and horizontal far-field stresses of $\sigma_{hx} = 5.0 \times 10^4$ kPa, $\sigma_{hy} = 4.5 \times 10^4$ kPa, aligned with *Z*, *X*, and *Y* coordinates, respectively. The initial pressure and temperature are the same as those of the aquifer formation.

Water at 20 °C is centrally injected at a rate of 240 m^3/day, and initial pressure is maintained as constant at production wells at the four corners of the aquifer zone. The outer boundary of the aquifer, overlying and underlying forms system is a no heat or mass flux boundary. In the numerical model, finite element mesh is shown in Fig. 9.30, and implicit Euler method has been chosen for time integration.

9.4.1 Case 1: Fully Coupled XFEM Solution

In this section, fully coupled XFEM solution is presented including pore pressure, temperature changes, as well as fracture development.

Results show that vertical fracture perpendicular to minimum horizontal in situ stress, σ_{hy}, is initiated from the injection well and propagates along the *X* direction, as expected. Due to the large stress contrast between the aquifer formation and its overlying and underlying formations, the fracture is contained within the aquifer formation, with height equal to 4 m.

Figure 9.31 shows the matrix pressure profiles along *X* direction from the injection well at different time. Figure 9.32 shows the fracture pressure profiles along *X* direction from the injection well at different time. It can be seen that relatively little variance exists in matrix pressure in the area penetrated by fracture,

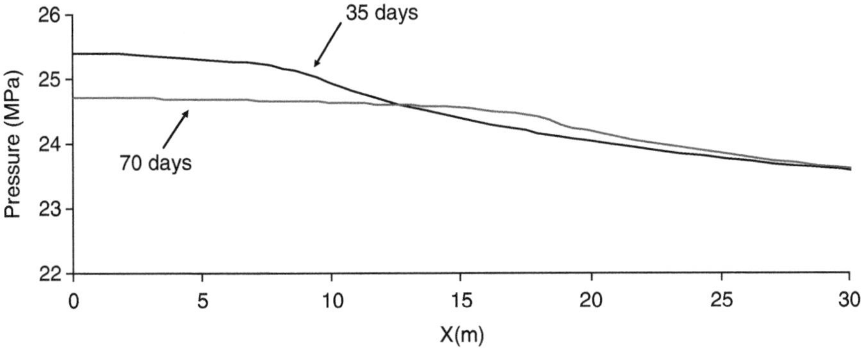

Fig. 9.31 Matrix pressure profiles at different time

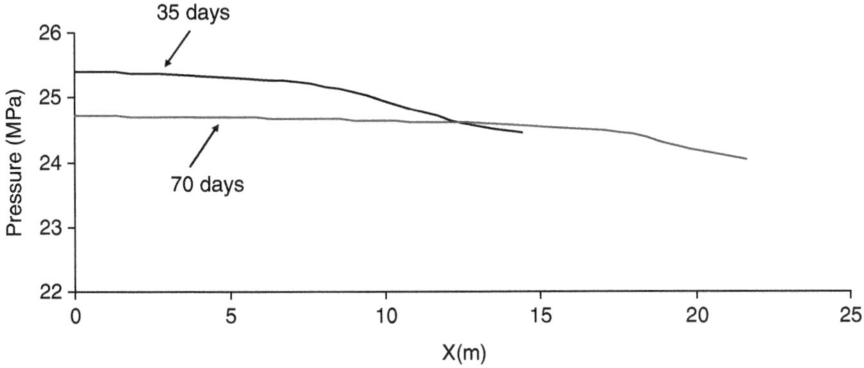

Fig. 9.32 Fracture pressure profiles at different time

while greater pressure gradient exists in intact area. This can be explained by the double porosity model in which there is fluid exchange between the matrix and fracture. Because the permeability of the fracture is much higher, the pressure difference within the fracture is much smaller at given injection rate. Hence the pressure difference within the matrix in the area close to the two opposing surfaces of the fracture is insignificant because of the fluid exchange between the fractured zone and matrix in the double porosity model. While in the intact zone, there is no such mechanism existing, and the pressure difference displays as usual depending on the flow rate, formation permeability and diffusion process. Figure 9.33 shows the temperature change profiles along *X* direction from the injection well at different time. It can be seen that the thermal front is ahead of the fracture front, which makes the fracture more likely to continue to propagate further. Figure 9.34 shows the fracture width profiles along *X* direction from the injection well at different time which indicates clearly the growth of the fracture.

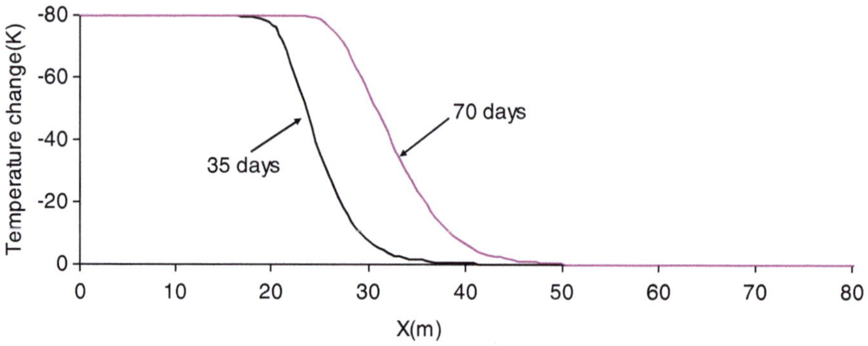

Fig. 9.33 Temperature change profiles at different time

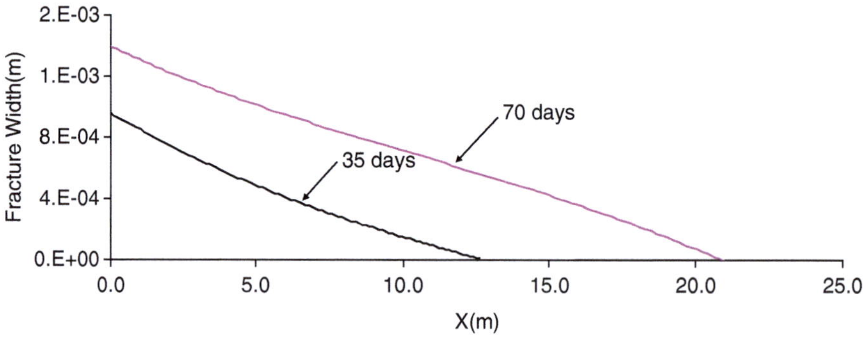

Fig. 9.34 Fracture width profiles at different time

9.4.2 Case 2: Impact of Injection Rate

To investigate the impact of injection rate, a comparison numerical experiment in which the injection rate is reduced from 240 to 220 m^3/day is performed to explore the effect of injection rate on pressure change and fracture development.

Figure 9.35 shows matrix pressure profile along *X* direction from the injection well at 35 days, compared with profile with higher injection rate. It can be seen that as the flow rate is reduced, the matrix pressure declines. Figure 9.36 shows the comparison of fracture width profiles along *X* direction from the injection well at 35 days. It can be seen that the fracture width is getting smaller as injection rate declines.

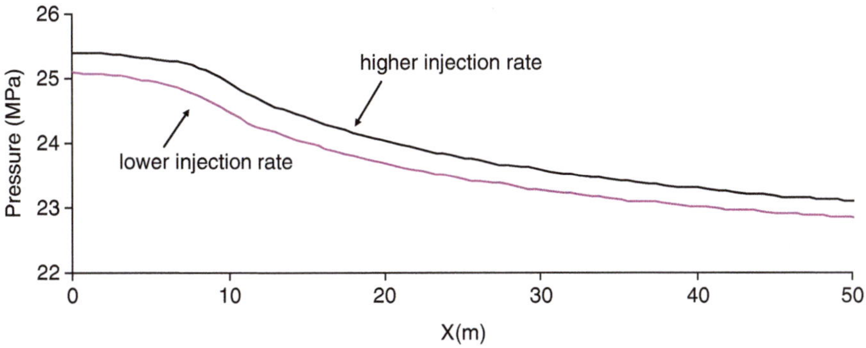

Fig. 9.35 Matrix pressure profiles at 35 days

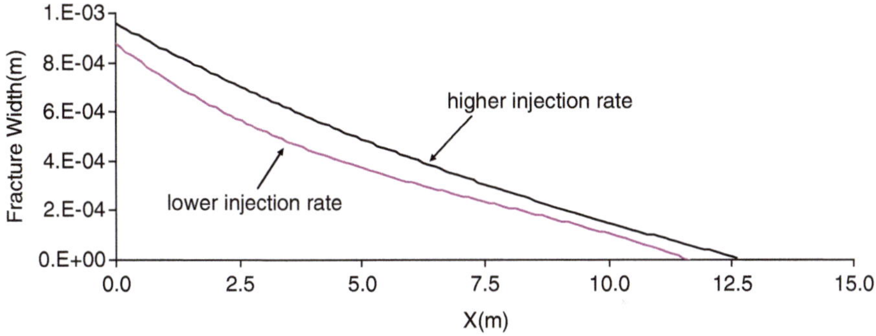

Fig. 9.36 Fracture width profiles at 35 days

9.4.3 Case 3: Impact of Injection Temperature

In this section, a comparison numerical experiment in which the injection temperature is increased from 20 to 30 °C is conducted to explore the effect of injection temperature on pressure change and fracture development.

Matrix pressure profile along *X* direction from the injection well at 35 days is shown in Fig. 9.37, compared with profile with lower injection temperature. It can be seen that as the injection temperature is increased, the matrix pressure increases. This can be explained by the narrower induced fracture with higher injection temperature which can be seen in Fig. 9.38. Figure 9.38 shows the comparison of fracture width profile along *X* direction from the injection well at 35 days. It can be seen that the fracture width is getting smaller as injection temperature increases because the temperature difference between the injection temperature and formation temperature is smaller.

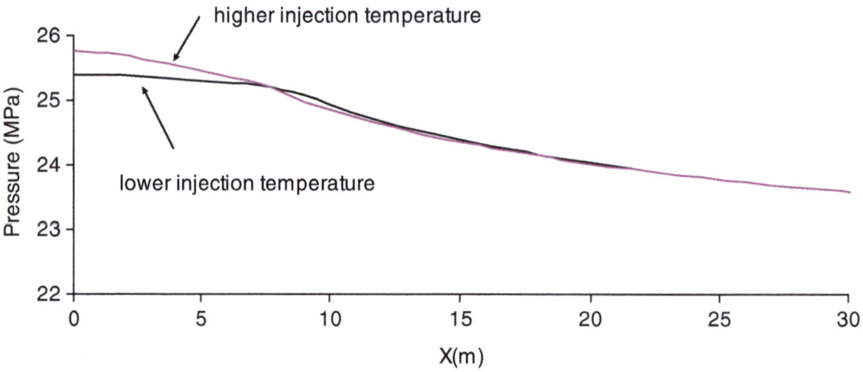

Fig. 9.37 Matrix pressure profiles at 35 days

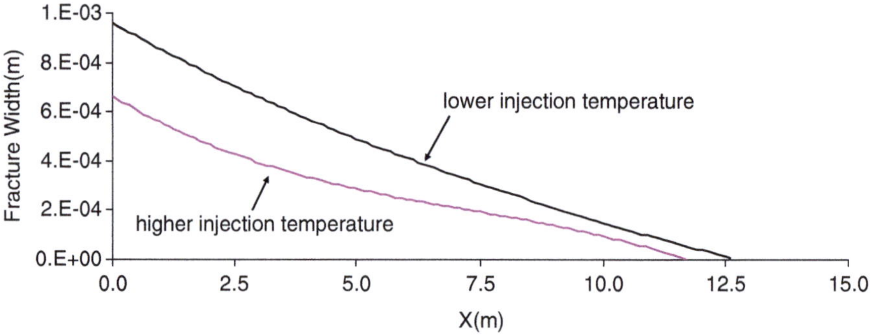

Fig. 9.38 Fracture width profiles at 35 days

9.4.4 Case 4: Impact of Aquifer Stiffness

In this section, a comparison numerical experiment in which the Young's modulus of the aquifer formation is reduced from 1.5×10^7 to 1.4×10^7 kPa is conducted to explore the impact of aquifer stiffness on pressure change and fracture development.

Figure 9.39 shows matrix pressure profile along *X* direction from the injection well at 35 days, compared with profile of stiffer aquifer. It can be seen that as the aquifer stiffness is reduced, the matrix pressure may increase due to the narrower induced fracture as can be seen in Fig. 9.40, which shows the comparison of fracture width profiles along *X* direction from the injection well at 35 days.

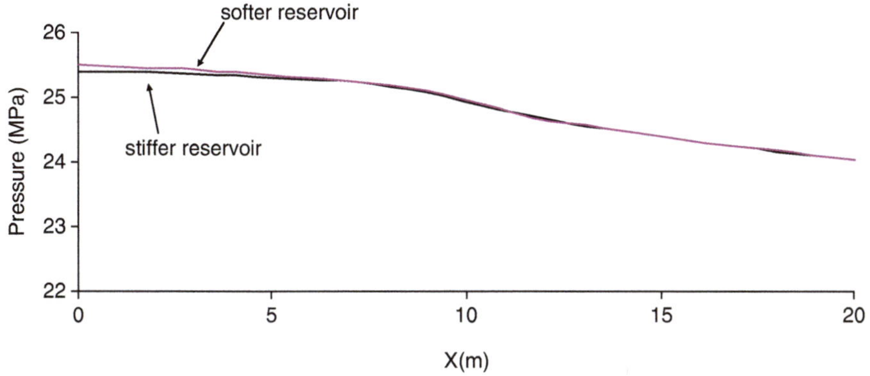

Fig. 9.39 Matrix pressure profiles at 35 days

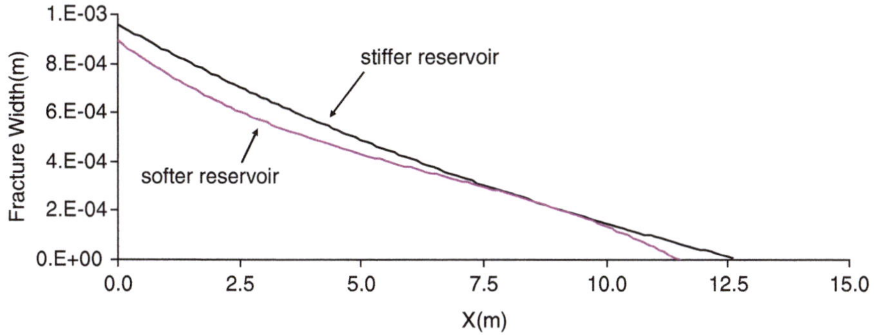

Fig. 9.40 Fracture width profiles at 35 days

9.4.5 Case 5: Impact of Aquifer Permeability

In this section, a comparison numerical experiment in which the aquifer permeability is reduced from 100 to 80 mD is conducted to explore the effect of aquifer permeability on pressure change and fracture development.

Figure 9.41 shows matrix pressure profile along fracture in *X* direction at 35 days, compared with profile with higher aquifer permeability. It can be seen that as the aquifer permeability is reduced, the matrix pressure is higher. Figure 9.42 shows the comparison of fracture width profiles along *X* direction from the injection well at 35 days. It can be seen that the permeability has little impact on the fracture width at this particular boundary condition.

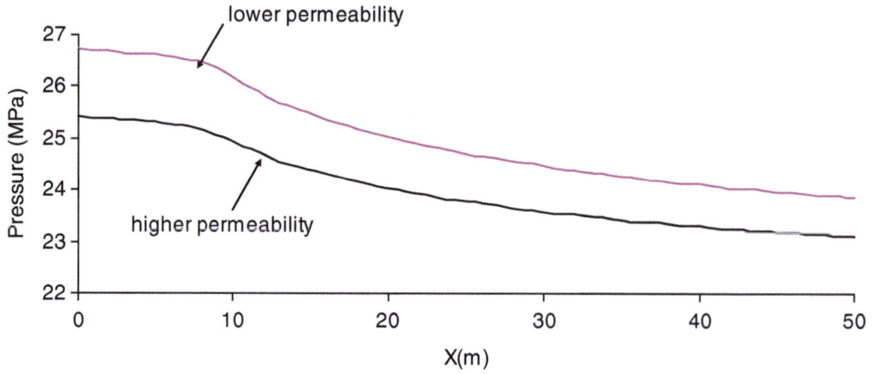

Fig. 9.41 Matrix pressure profiles at 35 days

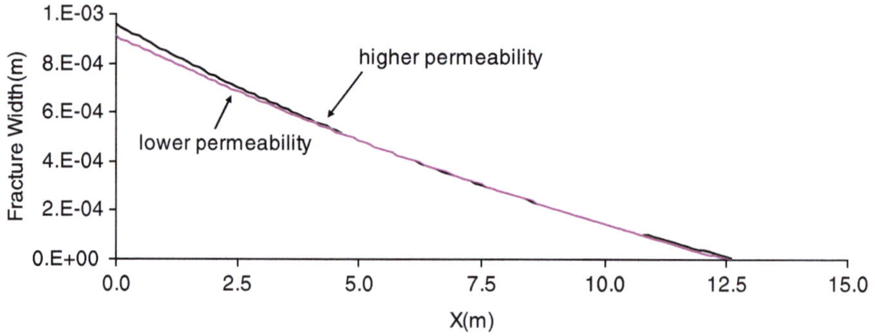

Fig. 9.42 Fracture width profiles at 35 days

9.4.6 Case 6: Impact of the Stabilized FEM Scheme

The impact of the stabilized FEM scheme on the results is explored with the initial data set (keeping the value of all parameters unchanged).

Figure 9.43 shows the temperature change profiles at 0.001 days computed by the stabilized SGS/GSGS FEM scheme [2] and the traditional Galerkin FEM scheme. It can be seen that the Galerkin FEM scheme leads to numerical oscillations whereas the stabilized FEM scheme avoids such oscillations. These numerical oscillations also result in subsequent instability of the solution in terms of pressures and displacements. This demonstrates the effectiveness of the stabilized finite element model in reducing the numerical oscillations in transient advection-diffusion problems.

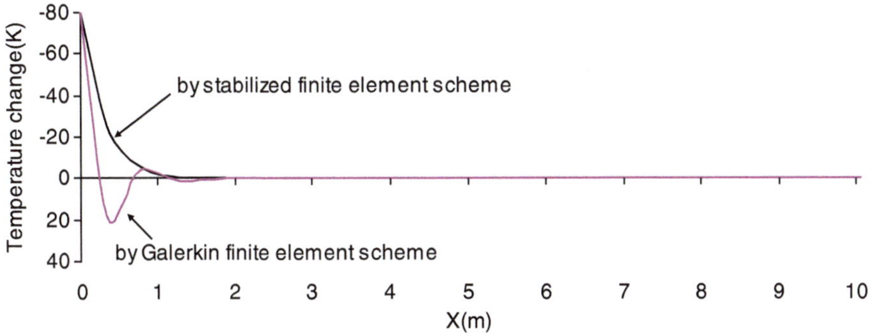

Fig. 9.43 Temperature change profiles generated by Galerkin FEM and stabilized FEM

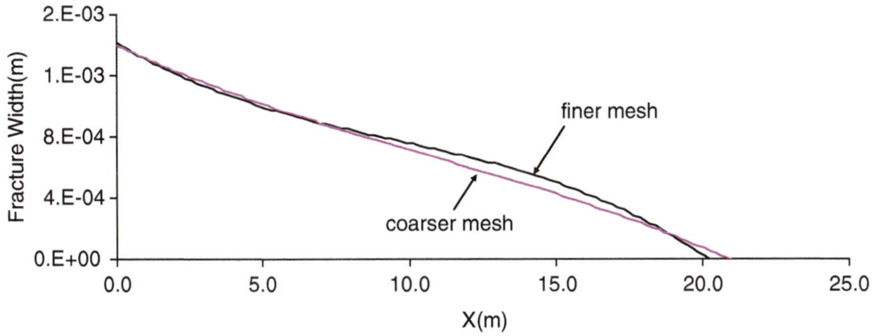

Fig. 9.44 Fracture width profiles at 70 days

9.4.7 Case 7: Impact of the FEM Mesh Size

The impact of the mesh size on the results is lastly explored with the initial data set.

Figure 9.44 shows the fracture width at 70 days computed on the initial mesh (100 elements in *X* direction in Fig. 9.30) and on a finer mesh with 150 elements in *X* direction. It can be seen that the result based on the initial finite element mesh is very close to the result based on the finer finite element mesh.

9.5 Conclusions

Petroleum geomechanics mostly deals with coupled THMC problems in subsurface energy resources extraction and waste disposal activities. A computational perspective is provided and proves effective and efficient in solving problems: a combination of boundary element and finite element methods is used to predict the subsidence in half-space; stabilized finite element and traditional finite element methods were employed to allow for more general cases of borehole stability analysis; extended finite element and stabilized finite element methods were used to simulate hydraulic fracture propagation with nonnegligible thermal contribution considered.

Petroleum geomechanics is far younger than its geoengineering sisters, and because of its nature, has followed a direction based far more on physics, analysis and field observations than on laboratory testing and empirical models. Other than the shared computational analysis, the greatest challenges are:

- Full integration of deformation and microseismic monitoring into numerical modeling
- Improving mathematical models based on correct physics to achieve better THMC coupling so that various processes can be more robustly simulated
- Achieving numerical simulation that can secure zero waste discharge through geomechanics principles and surveillance
- Improving predictability in cases of shale instability, sand influx, reservoir stress path, subsidence, casing shear, amongst others

References

1. OGJ. (2000–2010). Various news articles on subjects such as Ekofisk subsidence and redevelopment, casing shear, etc. *Oil and Gas Journal.*
2. Geertsma, J. (1966). Problems of rock mechanics in petroleum production engineering. *Proceedings of the First Congress of International Society of Rock Mechanics* (Vol. I, pp. 585–594). Lisbon.
3. Rothenburg, L., Bratli, R. K., & Dusseault, M. B. (1994). *A poroelastic solution for transient fluid flow into a well.* PMRI Publications, University of Waterloo, Canada, available upon request.
4. Hettema, M., Papamichos, E., & Schutjens, P. (2002). Subsidence delay: Field observations and analysis. *Oil and Gas Science and Technology., 57*, 443–458.
5. Salamon, M. D. G. (1963). Elastic analysis of displacements and stresses induced by the mining of seam or reef deposits. Part I. *Journal of the South African Institute of Mining and Metallurgy, 64*, 128–149.
6. Rothenburg, L., Carvalho Jr., A. L. P., & Dusseault, M. B. (2007). Performance of a mining panel over tachyhydrite in Taquari-Vassouras potash mine. *Proceedings of 6th Conference on the Mechanical Behavior of Salt*, Hannover, 9 p.
7. Crouch, S. L., & Starfield, A. M. (1983). *Boundary element methods in solid mechanics.* London: Allen and Unwin.
8. Detournay, E., & Cheng, A. H.-D. (1993). Fundamentals of poroelasticity. In J. A. Hudson (Ed.), *Comprehensive rock engineering: Principles, practice and projects* (Vol. 2, pp. 113–171). Oxford: Pergamon.
9. Zimmerman, R. W. (1991). *Compressibility of sandstones.* Amsterdam: Elsevier.
10. Biot, M. A. (1941). General theory of three-dimensional consolidation. *Journal of Applied Physics, 12*, 155–164.
11. Sandhu, R. S., & Wilson, E. L. (1969). Finite element analysis of seepage in elastic medium. *Journal of the Engineering Mechanics Division, ASCE, 95*, 641–651.
12. Gambolati, G., & Freeze, R. A. (1973). Mathematical simulation of the subsidence of Venice. I. Theory. *Water Resources Research, 9*, 721–733.

13. Lewis, R. W., & Schrefler, B. A. (1998). *The finite element method in the static and dynamic deformation and consolidation of porous media*. New York: John Wiley.
14. Zienkiewicz, O. C., & Taylor, R. L. (1991). *The finite element method, solid and fluid mechanics, dynamics and non-linearity* (Vol. 2). London: McGraw-Hill.
15. Dusseault, M. B. (1995). Radioactive waste disposal. *Nature, 375*, 625.
16. Dusseault, M. B., Rothenburg, L., & Bachu, S. (2002). Sequestration of CO_2 in salt caverns. In: *Proceedings of the Canadian international Petrol Conference*. Calgary, Alberta. Paper 2002-237.
17. Dusseault, M. B. (2003a). Coupled processes and petroleum geomechanics. *Proceedings of Geoproc 2003, International Conference on Coupled THMC Processes in Porous Media* (pp. 44–57). Stockholm, Sweden.
18. Fjær, E., Holt, R. M., Horsrud, P., Raaen, A. M., & Risnes, R. (1992). *Petroleum related rock mechanics*. Amsterdam: Elsevier.
19. Franklin, J. A., & Dusseault, M. B. (1989). *Rock engineering*. New York: McGraw-Hill Publishing Company.
20. Dusseault, M. B. (2003b). Coupled thermo-mechano-chemical processes in shales: The petroleum borehole. *Proceedings of Geoproc 2003, International Conference on Coupled THMC Processes in Porous Media* (pp. 571–578). Stockholm, Sweden.
21. Abousleiman, Y., Cui, L., Ekbote, S., Zaman, M., Roegiers, J.-C., & Cheng, A. H.-D. (1997). Applications of time-dependent pseudo-3D stress analysis in evaluating wellbore stability. *International Journal of Rock Mechanics and Mining Sciences, 34*, 421–422.
22. Dusseault, M. B. (1994). Analysis of borehole stability. *Proceedings of 8th IACMAB Conference* (Vol. 1, pp. 125–137) Morgantown, WV. Invited paper.
23. Dusseault, M. B. (1999). Petroleum geomechanics: Excursions into coupled behavior. *Journal of Canadian Petroleum Technology, 38*, 10–14.
24. Dusseault, M. B., Bruno, M. S., & Barrera, J. (2001). Casing shear: Causes, cases, cures. *SPE Drilling & Completion Journal, 16*, 98–107.
25. Fam, M. A., Dusseault, M. B., & Fooks, J. C. (2003). Drilling in mudrocks: Rock behavior issues. *Journal of Petroleum Science and Engineering, 38*, 155–166.
26. Rothenburg, L., & Bruno, M. S. (1997). Micromechanical modeling of sand production and arching effects around a cavity. *International Journal of Rock Mechanics and Mining Sciences. 34*(3–4), Paper #068.
27. Wang, Y., & Dusseault, M. B. (2003). A coupled conductive—convective thermoporoelastic solution and implications for wellbore stability. *Journal of Petroleum Science and Engineering, 38*, 187–198.
28. Yuan, Y. -G., Abousleiman, Y., & Roegiers, J. -C. (1995). Fluid penetration around a borehole under coupled hydro-electro-chemico-thermal potentials. *The 46th Annual Technical Meeting of the Petroleum Society of CIM*, Alberta, Canada.
29. Dusseault, M. B., & Hojka, K. (1992). Conduction and convection in thermoelastic porous solids as a tool in prediction of enhanced rock permeability. In M. Quintard & M. Todorovic (Eds.), *Heat and mass transfer in porous media* (pp. 203–219). Amsterdam: Elsevier.
30. Coussy, O. (1995). *Mechanics of porous continua*. New York: Wiley.
31. Sherwood, J. D. (1993). Biot poroelasticity of a chemically active shale. *Proceedings of the Royal Society of London A, 440*, 365–377.
32. Dusseault, M. B., Wang, Y., & Simmons, J. V. (1998). Induced stresses near a fire flood front. *AOSTRA Journal of Research., 4*, 153–170.
33. Yin, S., Dusseault, M. B., & Rothenburg, L. (2009). Thermal reservoir modelling in petroleum geomechanics. *International Journal for Numerical and Analytical Methods in Geomechanics, 33*, 449–485.
34. Yin, S., Towler, B. F., Dusseault, M. B., & Rothenburg, L. (2010). Fully coupled THMC modeling of wellbore stability with thermal and solute convection considered. *Transport in Porous Media, 84*, 773–798.

35. Strang, G., & Fix, G. (1973). *An analysis of the finite element method*. New Jersey: Prentice-Hall.
36. Donea, J., & Huerta, A. (2003). *Finite element methods for flow problems*. Chichester, England: John Wiley & Sons.
37. Hughes, T. J. R., Franca, L. P., & Hulbert, G. (1989). A new finite element formulation for computational fluid dynamics: VIII. The Galerkin/least-squares method for advective-diffusive equations. *Computational Methods in Applied Mechanics and Engineering, 73*, 173–189.
38. Gresho, P. M., Lee, R. L., & Sani, R. (1978). Advection-dominated flows with emphasis on the consequences of mass lumping. In R. H. Gallager, O. C. Zienkiewicz, J. T. Oden, M. Morandi Cecchi, & C. Taylor (Eds.), *Finite elements in fluids* (Vol. 3, pp. 335–351). Chichester: Wiley.
39. Harari, I. (2005). Stability of semidiscrete formulations for parabolic problems at small time steps. *Computer Methods in Applied Mechanics and Engineering, 193*, 1491–1516.
40. Hauke, G., & Doweidar, M. H. (2006). Fourier analysis of semi-discrete and space-time stabilized methods for the advective-diffusive-reactive equation: III. SGS/GSGS. *Computer Methods in Applied Mechanics and Engineering, 195*, 6158–6176.
41. Carslaw, H. S., & Jaeger, J. C. (1959). *Conduction of heat in solids* (2nd ed.). Oxford: Clarendon.
42. Perkins, T. K., & Gonzalez, J. A. (1985). The effect of thermoelastic stresses on injection well fracturing. *Society of Petroleum Engineers Journal, 25*, 78–88.
43. Settari, A., & Warren, G. M. (1994). Simulation and field analysis of waterflood induced fracturing. *Proceedings of SPE/ISRM Meeting Rock Mechanics in Petroleum Engineering* (pp. 435–445).
44. Svendsen, A. P., Wright, M. S., Clifford, P. J., & Berry, P. J. (1991). Thermally induced fracturing of Ula water injectors. *SPE Production Engineering, 6*, 384–393.
45. Yin, S. (2013). Numerical analysis of thermal fracturing in subsurface cold water injection by finite element methods. *International Journal for Numerical and Analytical Methods in Geomechanics, 37*, 2523–2538.
46. Belytschko, T., Gracie, R., & Ventura, G. (2009). A review of extended/generalized finite element methods for material modeling. *Modelling and Simulation in Materials Science and Engineering, 17*, 1–24.
47. Kraaijeveld, F. (2009). Propagating discontinuities in ionized porous media. PhD thesis, University of Technology of Eindhoven.
48. Larsson, R., Runesson, K., & Sture, S. (1996). Embedded localization band in undrained soil based on regularized strong discontinuity theory and FE-analysis. *International Journal of Solids and Structures, 33*, 3081–3101.
49. Roels, S., Moonen, P., de Proft, K., & Carmeliet, J. (2006). A coupled discrete-continuum approach to simulate moisture effects on damage processes in porous materials. *Computer Methods in Applied Mechanics and Engineering, 195*, 7139–7153.
50. Rethore, J., de Borst, R., & Abellan, M. A. (2007). A two-scale approach for fluid flow in fractured porous media. *International Journal for Numerical Methods in Engineering, 71*, 780–800.
51. Al-Khoury, R., & Sluys, L. J. (2007). A computational model for fracturing porous media. *International Journal for Numerical Methods in Engineering, 70*, 423–444.
52. Lamb, A. R., Gorman, G. J., Gosselin, O. R., & Onaisi, A. (2010). Finite element coupled deformation and fluid flow in fracture porous media. SPE 131725, *SPE EUROPEC/EAGE Annual Conference and Exhibition*, Barcelona, Spain.
53. Masters, I., Pao, W. K. S., & Lewis, R. W. (2000). Coupling temperature to a double-porosity model of deformable porous media. *International Journal for Numerical Methods in Engineering, 49*, 400–421.
54. Wells, G. N., & Sluys, L. J. (2001). A new method for modelling cohesive cracks using finite elements. *International Journal for Numerical Methods in Engineering, 50*, 2667–2682.

55. Gasser, T. C., & Holzapfel, G. A. (2005). Modeling 3D crack propagation in unreinforced concrete using PUFEM. *Computer Methods in Applied Mechanics and Engineering, 194*, 2859–2896.
56. Belytschko, T., Parimi, C., Moes, N., Sukumar, N., & Usui, S. (2003). Structured extended finite element methods for solids defined by implicit surfaces. *International Journal for Numerical Methods in Engineering, 56*, 609–635.
57. Zhang, J., Standifird, W. B., Roegiers, J.-C., & Zhang, Y. (2007). Stress-dependent fluid flow and permeability in fractured media: From lab experiments to engineering applications. *Rock Mechanics and Rock Engineering, 40*, 3–21.

Chapter 10
Insights on the REV of Source Shale from Nano- and Micromechanics

Katherine L. Hull and Younane N. Abousleiman

Abstract *Nano*. In the past decade, chemical, physical, and mechanical characterization of source rock reservoirs has moved towards micro- and nano-scale analyses. This is primarily driven by the fact that the representative elementary volume (REV) for characterizing shales is at the nanometer scale. Nanoindentation is now used in many industrial and university laboratories to measure both stiffness and strength and other mechanical properties of materials, such as anisotropic Young's moduli and plastic yielding parameters. However, standardized methods of testing and analysis are yet to be developed.

Micro. The shale matrix, composed of nano-granular clay and microscale non-clay minerals, also includes the hydrocarbon source material kerogen. This biopolymer is interbedded and intertwined with the clay and non-clay minerals at almost all scales. Kerogen not only has a Young's modulus in compression but also has a substantial Young's modulus value in tension and much higher tensile strength than rocks in general. This fact has now been observed at the micro- and nanoscale during nanoindentation while monitoring in situ via scanning electron microscopy (SEM). Load and unload experiments with micro-Newton forces (μN) and nanometer (nm) displacements have clearly shown the elastic nature of kerogen in the shale gas matrix.

Macro. Given that the organic matter has an elastic Young's moduli in tension, and viscoelastic characteristics, it is therefore capable of re-healing the hydraulic fracture. This is a major reason for our more or less unsuccessful gas shale stimulations. Keeping the fracture open even after proppant placement has proven to not be enough for gas and oil shale optimal well productivity. New macro-scale testing techniques are needed to evaluate the mechanical properties of shales that have not been possible to imagine outside of recent advances in nano- and micro-scale analyses.

K.L. Hull
Aramco Services Company, Houston, TX, USA

Y.N. Abousleiman (✉)
PoroMechanics Institute, University of Oklahoma, Norman, OK, USA
e-mail: yabousle@ou.edu

C. Jin, G. Cusatis (eds.), *New Frontiers in Oil and Gas Exploration*,
DOI 10.1007/978-3-319-40124-9_10

10.1 Introduction

Granular cohesionless earth materials with compaction histories, "memory" and compacted densities, have undergone complex geological processes and are still poorly understood when it comes to their mechanical properties [1, 2]. Clay-bearing sedimentary rocks, such as shale, formed under even more complex geological processes, are even less understood mechanistically. For example, the role of mineral composition in the overall mechanical property characterization has been the subject of many studies [3, 4]. For more than a decade, shale has been the subject of micro- and nano-mechanical characterization using the advances of nanotechnology [5, 6]. Kerogen-free shale (KFS) was found to be strongly transversely isotropic at the nano- and micro-scales while estimating the normal and transverse Young's moduli [7]. These early nanoindentation studies [7, 8] were attempts to measure shale mechanics at the smallest possible "porous unit" of a mudstone rock, i.e., attempting to identify the scale of the representative elementary volume (REV) [9], of fluid-saturated porous composites.

Shale anisotropy has been well known and documented in our mechanistic approaches, in late 1960s [10], as a dry or fluid-saturated porous media exhibiting transverse isotropy. These TI characteristics are likely due to mode of deposition, bedding planes, micro-fractures, and/or micro- and nano-clay grain shape and packing porosity. The findings of the mechanistic TI nature of shale were also modeled analytically and implemented for many field applications [11–13]. Experimental results, particularly acoustic measurements, provided the early evidence of organic shale transverse anisotropy [10]. However, later studies on source rock or kerogen-rich shale, KRS, with the acoustic measurements, have attributed shale anisotropy not only to fractures and bedding planes but also to the presence of kerogen interlayered with illite clay minerals [14]. Podio et al. [10] and Vernik and Nur [14], among many others, have paved the way for geomechanics anisotropy modeling of shale in wellbore stability analysis [13, 15], reservoir compaction simulation [11], hydraulic fracturing [12], and tiny shale sample laboratory testing characterization [16]. However, kerogen could not be definitively pinned as the culprit for anisotropy at all these scales. KFS has shown intrinsic anisotropy, and in many instances even higher than KRS anisotropy, at micro- and macro-scales [17, 18] and re-confirmed recently [19].

The shale samples used in early nano- and micro-scale experiments contained only "trace" levels of organic material [7], which means the organic matter had little effect on the overall mechanical response (the total clay content was more than 75 wt%). Therefore, the tested KFS, with in-bedding and perpendicular to bedding indentations, is granular in nature with anisotropic nano- and micro-mechanical properties which depend upon porosity, packing density, and the stiffnesses of its corresponding minerals. Not surprisingly, the KFS properties varied from one sample to the next, and the clay and other mineral composition varied along with their respective porosities. These micromechanical responses confirmed the

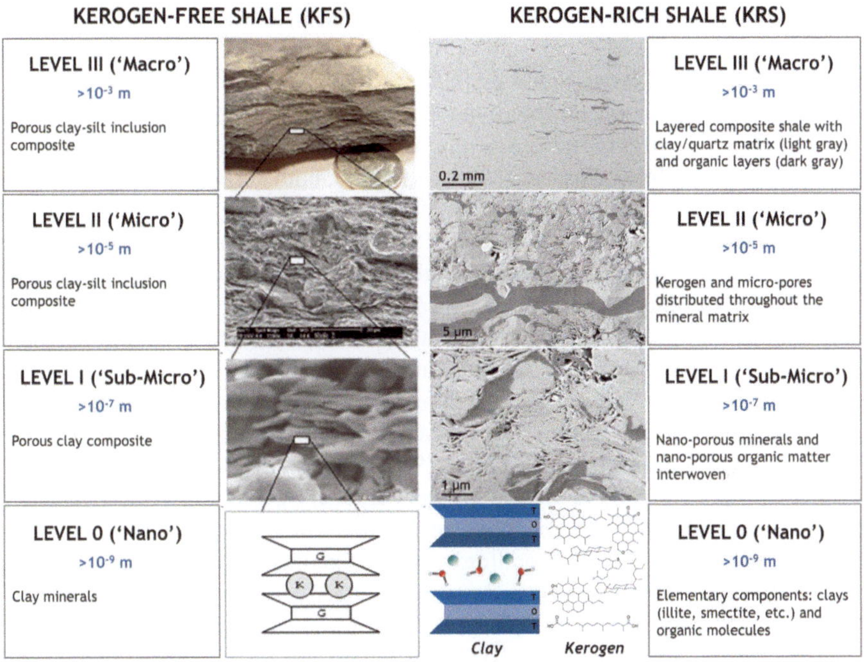

Fig. 10.1 After [21]; Level III—the lenticular dark colors kerogen; Level II—kerogen occupy 15–20 % of the surface area; Level I—kerogen and matrix intertwined, quartz grains, pyrites, micro fractures network; Level 0—atomistic scale

anisotropic nature of KFS at the micro- and macro-scales [3, 7]. A multiscale model of KRS [21] is shown in Fig. 10.1.

All shale source rocks have the major components of non-clay minerals like quartz, feldspar and pyrite (QFP), clays such as illite, mica, and smectite, and finally organic matter such as kerogen and bitumen where the generated oil and gas reside [22]. An unconventional reservoir with 5–6 wt% kerogen (~10–12 vol.%) is considered to be kerogen-rich shale (KRS) [23]. In this discussion, we will not differentiate between the various types of organic matter and rather group them with kerogen. The complexity of KRS, such as for Woodford shale, is strongly enhanced by the presence of the string-like kerogen intertwined with the rock matrix. In compiling this micro- to macro-structure with micro-bedding planes and micro-fractures shown under the SEM from level I to Level III, in the above figure, you could imagine that the failure mechanisms of such composite could be very complex [21].

Recently, an extensive campaign of nanoindentation was conducted on organic-rich Woodford shale ($\leq$30 % clay; 10–18 % kerogen) demonstrating the effects that the kerogen matrix has on the overall mechanical properties of KRS, including the elastic and plastic behavior [24–26]. The upscaling of the poromechanical anisotropic parameters of KRS from nanoindenter testing and characterization, to macro-scale rock mechanics laboratory measurements, all the way to field-scale logging tools has also been the subject of further studies [18, 25, 27]. However, it was found that the kerogen stiffness and the higher percent volume of kerogen, reduced the degree of anisotropy of the mechanical parameters [18]. The isotropic stiffness characteristics of the kerogen in the Woodford shale was confirmed using the nanoindenter while measuring Young's moduli, both normal and perpendicular to bedding [6].

In all these previous studies on the KRS, very little light has been shed on how the KRS fails in tension (such as in hydraulic fracturing) or in compression (such as in drilling) at the micro- and nano-scales. Also, missing in our literature, is how the polymer nature of kerogen and its spatial intertwined structure within the shale mineral matrix affects the overall mechanical failure. Classical macro-scale rock mechanics testing on KRS in both tension and compression has been performed with respect to deposition modes both parallel and perpendicular to the bedding planes of the Woodford shale [25, 28, 29]. However, the American Society for Testing and Materials (ASTM) and the International Society for Rock Mechanics (ISRM) standard test methods practiced did not reveal any novelties about the failure mechanisms of the Woodford KRS. Moreover, recently designed rock testing instrumentation for tiny shale samples under confining pressures (inclined direct shear testing device, IDSTD) [16] also showed results for compression and tension tests for the Woodford shale, that turned out to be very similar to the classical ISRM and ASTM methods.

10.2 Sample Preparation for Nano- and Micro-Scale Shale Characterization

Preparing shale samples for nano- and micro-scale mechanical characterization is tedious, challenging and requires multistage methods, since the surface roughness and the multi-phase components play very critical role in the measurement interpretation. Small samples (dimensions in the 1″ range) were first cut from larger shale preserved Woodford core [24]. These samples were then subdivided into smaller specimen to be used for various analyses including compositional methods and mechanical measurements, from the same geological stratigraphy [29]. Specimen used for averaged compositional studies such as X-ray diffraction (XRD), total organic carbon (TOC) analysis, and X-ray fluorescence (XRF) were milled with a Retsch Mixer Mill MM400 to obtain a powder. Other specimen used for mechanical measurements such as nano-indentation and energy-dispersive X-ray spectroscopy (EDS) characterization required surface preparation in order to achieve highly polished, smooth surfaces. These specimen were cut with precision saws and/or Dremel tools to obtain samples <1 cm in all dimensions. They were then polished either parallel or perpendicular to their bedding planes. These samples were mechanically polished first using 600 and 1200 grit silicon carbide paper then

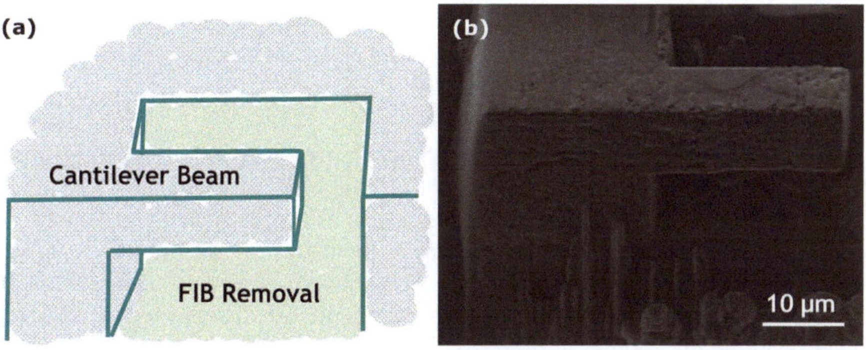

Fig. 10.2 The micro-cantilever beam is (**a**) represented as a schematic showing the regions of shale material which were removed during the milling process with the FIB. (**b**) Micro-beam Test A is also shown in the SEM image

with progressively finer diamond suspensions beginning at 3 μm and continuing until reaching 0.05 μm. The polished samples were subsequently imaged via SEM in pre-indentation analysis, to estimate topography accepted roughness. In an effort to isolate the effects of the KRS samples surface roughness, Argon ion-milled samples were also prepared and tested.

Samples for in situ indentation/loading, as micro-beams and micro-pillars, where prepared in slightly different procedures. After the small shale specimen (<1 cm in all dimensions) was obtained, it was mechanically polished to obtain a sharp 90° edge using progressively finer standard silicon carbide paper until reaching 4000 grit then polished with a 1 μm diamond suspension. The sample was subsequently placed inside an SEM and milled with a FIB in order to prepare micro-cantilever beam and micro-pillar geometries. Micro-beam sample preparation was performed with a Quanta 3D field emission gun (FEG) FIB-SEM. First, FIB surface milling was used to clean the surface for better sample imaging as well as to prepare the desired microgeometries. Four micro-cantilevers (Tests A–D) were manufactured using the following procedures utilized in previously tested materials [20, 30]. Each bend bar was shaped by cutting trenches on all three sides with widths of 20 μm and depths of 10 μm using a 15 nA beam current, resulting in a U-shaped trench. The geometry was then refined by applying a 1 nA beam current. Afterwards, the sample was tilted to 45° along the length axis to shape the cantilever. The base of the cantilever was undercut from both sides using a 3 nA beam current. The resulting cantilever geometry is shown schematically in Fig. 10.2a, with the corresponding SEM image of one of the micro-beams shown in Fig. 10.2b. EDS was performed on the front surfaces of the beams using an Oxford EDS attached to the FIB-SEM.

10.3 Test Methods

10.3.1 Compositional Analysis

Powder samples obtained from crushing the shale were analyzed via X-ray diffraction (XRD) with a Bruker D8 Advance Eco powder diffractometer and then analyzed using Rietveld refinement to determine the mineral constituents. A portion of the powder sample was also tested by a Shimadzu SSM-5000 total organic carbon (TOC) analyzer to determine the percent weight of organic matter. Furthermore, polished samples prepared for nanoindentation were first imaged via SEM to understand the distribution of minerals in the matrix and identify areas for phase mechanical analysis. Post-indentation, the indented areas were located under the SEM and high-resolution images were obtained. Similarly, cantilever beams prepared via FIB-SEM were imaged before testing and were also scanned with EDS to determine the distribution of components. The post-testing procedure utilized the same methods, with additional analysis of the broken areas.

10.3.2 Nanoindentation

Instrumented indentation methods are well established by now for determining mechanical properties of materials such as hardness and elastic moduli. The experiment is performed by pressing a tip with known geometry and mechanical properties into the surface of a material. The applied load P and the depth of the tip h are recorded while the indenter loads, holds for a period of time, then unloads. The resulting load-displacement curve, as shown in Fig. 10.3, can be analyzed to determine the unloading stiffness S from the initial elastic unloading curve after reaching the maximum load $P_{\max}$ [32].

The projected contact area A_c can also be estimated based on $h_{\max}$, the maximum indentation depth [31]. From these values, the indentation hardness H and indentation modulus M are determined [32]:

$$H = \frac{P_{\max}}{A_c} \tag{10.1}$$

$$M = \frac{\sqrt{\pi}}{2} \frac{S}{\sqrt{A_c}} \tag{10.2}$$

For materials such as shale which are characterized by transverse isotropic elasticity, the indentation moduli M_1 and M_3 measured normal and parallel to bedding, respectively, can be related to the five elastic stiffness constants C_{ij} of the shale [7, 33]:

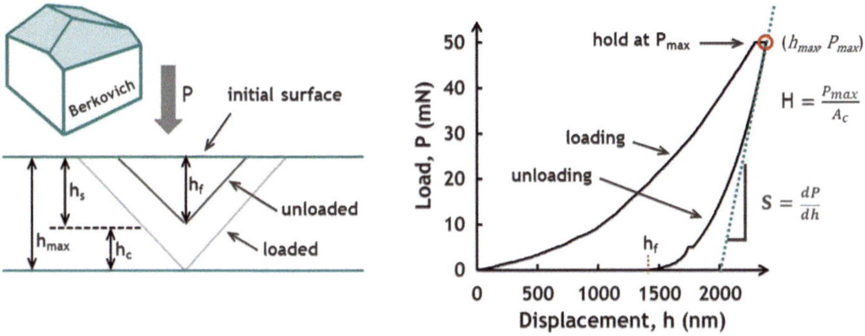

Fig. 10.3 Typical load-displacement curve generated by indenting into shale

$$M_3 = 2\sqrt{\frac{C_{11}C_{33} - (C_{13})^2}{C_{11}}\left(\frac{1}{C_{44}} + \frac{2}{\sqrt{C_{11}C_{33}} + C_{13}}\right)^{-1}} \quad (10.3)$$

$$M_1 = \sqrt{\sqrt{\frac{C_{11}}{C_{33}}}\frac{(C_{11})^2 - (C_{12})^2}{C_{11}}M_3} \quad (10.4)$$

Nanoindentation was performed on the polished shale samples using a Micro Materials NanoTest Vantage system. Samples were indented with a Berkovich indenter whose tip contact area had been calibrated by indenting into fused silica. Each shale specimen was indented under load control over a wide range of loads with the resulting depths ranging from 200 to 4000 nm. A typical series of indentations performed on the shale samples is shown in Fig. 10.4. Indentations were typically performed in sets of 10 at loads of 100, 50, 20, 10, 5, and 1 mN. A few experiments were also performed at 200 and 400 mN, although the hardness and moduli results were not substantively different from 100 mN experiments.

Following the indentation experiments, pyramidal analysis was performed on the load-displacement curves and the hardness H and indentation moduli M_1 and M_3 were determined as previously described. SEM imaging was conducted on the indented shale samples, and the resulting images were analyzed in detail for features such as minerals indented as well as surface delamination and breakage.

10.3.3 Micro-Cantilever Beams

A Hysitron Pi-85 Picoindenter was used to load the micro-cantilever beams under displacement control mode at 10 nm/s using a 5 μm diamond flat punch indenter tip geometry. All experiments were performed in situ under the SEM, and loading of the micro-beam shale samples continued until failure was observed. For the loading experiment, the indenter tip was placed at the end of the micro-cantilever beam,

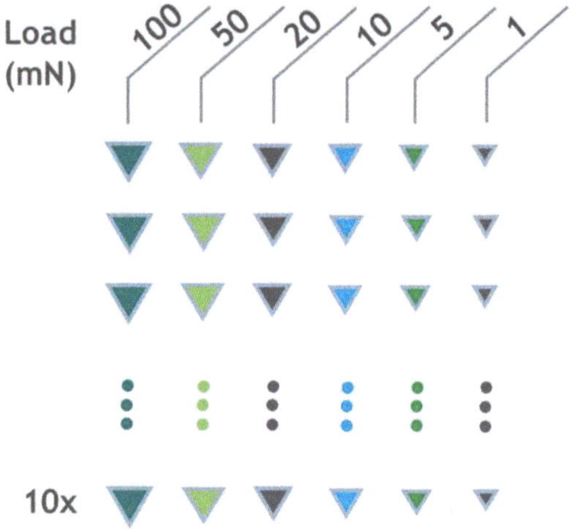

Fig. 10.4 Typical array of indentations performed on normal and parallel oriented shale samples

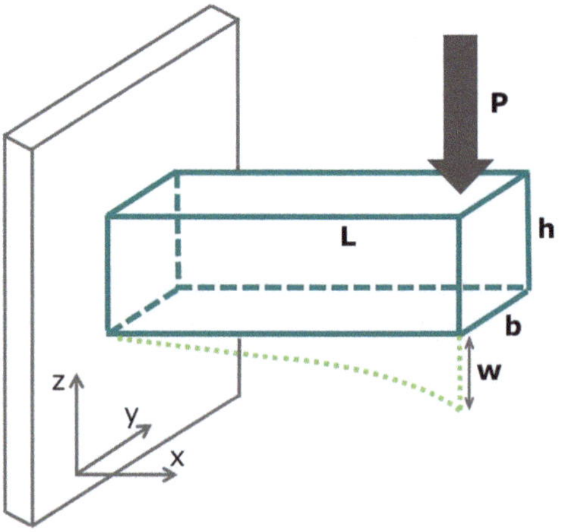

Fig. 10.5 The micro-cantilever beam is represented as a schematic showing the labeled dimensions *L*, *b*, and *h* as well as the deformation, *w*(*x*), and applied force, *P*

centered along the *y*-axis. Figure 10.5 provides a schematic of the micro-beam geometry with dimensions of length *L*, width, *b*, and height, *h*. When a load *P* is applied to the end of the beam, the cantilever tip is displaced by distance *w*(*x*).

When a solid metallic beam with dimensions in the micron range is subjected to loading, there is a strong evidence of size effects [34]. The effects of an intrinsic length l_{FE} on the overall deflection of a solid micro-cantilever beam has been formulated. Although expressions for micro-cantilever beams with various loading conditions have been derived, the expression for concentrated load is relevant to our experiments [34]:

$$w = \frac{Px^2(3L - x)}{6E\left(I + bhl_{\text{FE}}^2\right)} \tag{10.5}$$

However, for a granular material with polymer-like strands connected and interspersed throughout the structure, it is not clear what Eq. (10.5) should look like for our KRS micro-cantilever experiments. However, when we assumed $l_{\text{FE}} = 0$, the above expression turns into the classical theory of beams equation, and the expression for Young's modulus, E, in Eq. (10.5) becomes

$$E = \frac{PL^3}{3wI} \tag{10.6}$$

where I is the prismatic beam moment of inertia.

10.4 Nano- and Micro-Measurements

10.4.1 Compositional Analysis

The Woodford shale formation, deposited during the lower Mississippian and upper Devonian period in an anaerobic marine environment, is found throughout the central part of the US Midwest. The formation has long been known to be one of the major source rocks of the region, and for the past decade, it has been a great source of energy in gas and oil. The Woodford shale has high quartz content as revealed by X-ray diffraction (XRD) analysis, greater than 20 % in total porosity, and permeability ranging from 40 to 80 nano-Darcy. The composition of the Woodford shale sample used in the current study was determined by a number of methods including XRD, EDS, and TOC analyses. A full suite of Woodford shale characterization has been provided and can be used as a point of comparison [24]. The mineralogical matrix of the shale horizon of the present study, as determined by XRD, is composed of 64 % quartz, 4 % feldspar, 12 % illite, 12 % smectite, 3 % carbonate (primarily dolomite), 2 % plagioclase, 1 % mica, 1 % pyrite, and <1 % trace minerals such as chlorite and anatase. The material also includes within its matrix approximately 9.9 wt% total organic carbon, which is nearly 20 vol% organic matter.

The composition and distribution of minerals and organic matters in these samples are highly complex and vary widely from one shale source to the next. Surface preparation of these heterogeneous samples can be difficult due to large differences in hardness between non-clay and clay minerals, in addition to the soft kerogen, for example. Polishing these materials can result in various issues such as grooves in the surface due to hardness differences as well as grains pop-outs or smearing. For the Woodford shale of the present study, SEM images of polished surfaces are shown in Fig. 10.6a, b. In the left image (a), the specimen has been

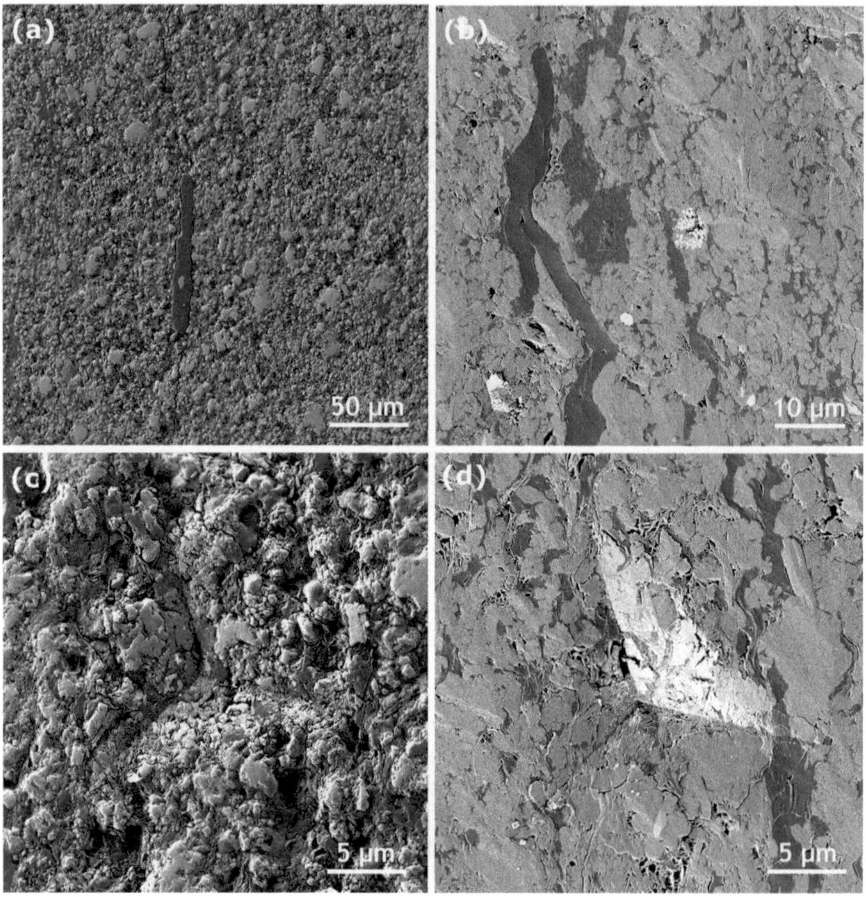

Fig. 10.6 SEM images of Woodford shale, viewed parallel to bedding of (**a**) mechanically polished and (**b**) argon ion-milled specimen, their nanoindentation footprints for 100 mN are captured in (**c**) and (**d**), respectively

ground and polished mechanically according to the procedure described earlier. In the right image (b), a specimen from the same geological horizon and in the same orientation (viewing parallel to bedding), the surface was polished identically but then further polished via argon ion milling. The two surfaces are clearly quite different, with many uneven surface features in (a). The argon ion-milled surface is much smoother, and the various components are much more evident including pores, minerals, and organics. However, the very near surface, 10 of nano-meters could be mechanically altered due to the heat level generated in the argon ion milling process. The gray matrix contains both the non-clay and clay minerals in addition to the kerogen (dark gray), with higher resolution images providing clear distinctions. The bright white inclusions are iron sulfide (pyrite) including both framboids which are clusters of small crystallites as well as larger euhedral pyrite minerals. Dark gray regions dispersed among the minerals and within the pyrite framboids are composed of kerogen which is the organic hydrocarbon source material. While it is typical of source rock shale to have kerogen dispersed in its structures, the Woodford shows pronounced intertwined kerogen strings shown in two-dimensions when compared to the overall granular mineral matrix [23].

10.4.2 Nanoindentation

The micron-sized quartz and pyrite inclusions within the nanogranular clay matrix interwoven with kerogen polymer-like strings necessitate an approach to mechanical characterization which incorporates micro- and nanoscale measurements. However, interpretation of such tests requires careful analysis due to the sample complexity. Examples of micro- and nanoindentation performed on the Woodford shale specimen are provided here with the indenter oriented parallel to bedding. These are provided in order to illustrate some of the challenges as well as some of the key findings of shale indentation. A series of maximum loads including 100, 50, 20, 10, 5, and 1 mN were tested in multiples of 10 each. Indentations were performed on the two polished samples—mechanically polished and argon ion milled. Two sets of ten indentations collected during experiments performed at maximum loads of 100 and 50 mN into the mechanically polished specimen are shown in Fig. 10.7.

Furthermore, six sets of indentations collected during experiments performed at maximum loads of 100 mN down to 1 mN into the argon ion milled specimen are shown in Fig. 10.8. Direct comparison of Fig. 10.7a, b with Fig. 10.8a, b demonstrates that indentation into two specimen of the same shale from the same horizon produce different results. Average displacements per maximum load are similar across the two specimen, but the statistical spread is quite different. The difference likely stems from the differences in surface preparation, as the mechanically polished specimen with the rougher surface results in indentation data with a larger spread.

The load-displacement curves were analyzed post-indentation to determine the hardness H and moduli M. Figure 10.9 provides plots of the hardness versus the maximum depth for each experiment. At low indentation depths, the scatter in the hardness values is high while the hardness is much more uniform at high indentation depths. The average hardness values also decrease mildly as the maximum depth increases, as previously observed in other materials. Similarly, indentation moduli were determined from each of the indentations. Figure 10.9 shows the plots of moduli versus the maximum depth in the two orientations. As in the hardness values, the moduli values are spread over a wide distribution at low indentation depth, while they converge to the same value when the indentations are deep. For both hardness and moduli, it is expected that sampling homogeneity improves with deeper indentations whereas shallow indentations are likely to come in contact with fewer minerals/components. In the latter case, the response of the material will be much different than in the former (bulk) type of measurement.

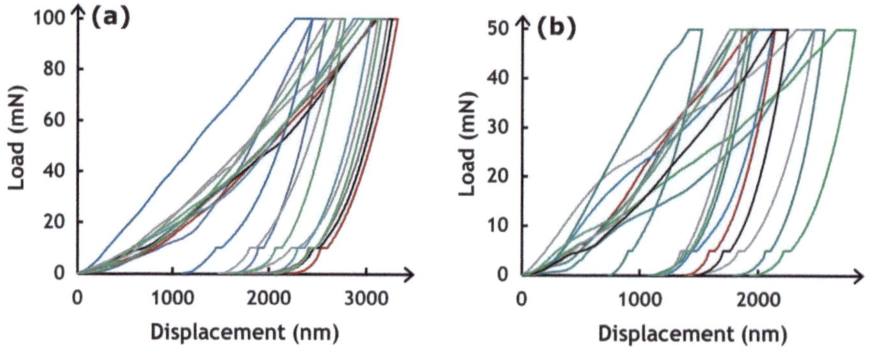

Fig. 10.7 Load-displacement curves obtained from indentation into a mechanically polished Woodford shale sample. The indenter was oriented parallel to bedding and brought to maximum loads of (**a**) 100 mN and (**b**) 50 mN

10.4.3 Micro-Cantilever Beams loading

EDS was performed on the front face of the micro-beams prior to the loading tests in order to approximate the local concentrations of each mineral. Figure 10.10 shows the EDS data collected on the front faces of Test A and Test C. The combined map containing each of the elements analyzed is superimposed on an SEM image of the micro-beam. To the right of the image is a set of 2-D maps of the elements, which allows the reader to more easily distinguish between the various minerals included in the structure. Both beams contain calcium/magnesium features far back in the support region corresponding to dolomite, but only micro-beam Test B shows a small amount of dolomite within the beam itself. Also, both beams contain iron sulfide framboids close to the middle of the analyzed area, which fall near the interface between the beam and the support. Finally, when comparing the two beams, it is apparent that there is more carbon in micro-beam A than in B. These features will again appear and be explored after the loading experiment.

Following the compositional analyses, shale micro-beam Tests A–D were loaded and failed. As demonstrated above, this composite geomaterial is highly heterogeneous in terms of type and distribution of minerals and organic matter. The loading tests of the shale samples were performed at the micro-scale, which is still larger than the representative elementary volume (REV). The four tests were performed inside the SEM with the small-scale nanoindenter, and movies of the loading and failure were captured in real time. This unique experimental setup provides not only the ability to load and fail micro- and nanoscale shale structures but also the advantage of visualizing the crack initiation, propagation, and ultimate failure of the beams. Subsequent high-resolution imaging of the support and beam fracture faces as well as complementary EDS allows us to analyze the minerals and organic matter that are associated with the fracture.

Four frames captured while Test A was in the loading configuration are shown in Fig. 10.11. Approximately 2200 μN of applied force was required to fail the beam.

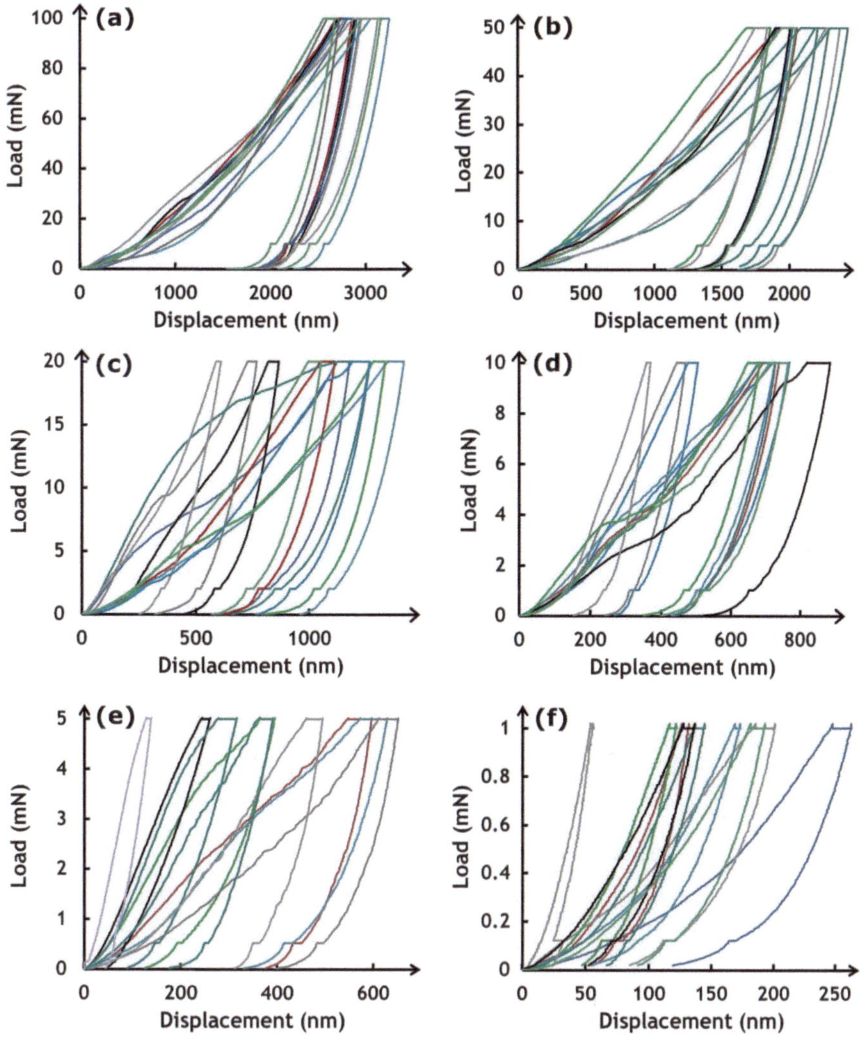

Fig. 10.8 Load-displacement curves obtained from indentation into an argon ion-milled Woodford shale sample. The indenter was oriented parallel to bedding and brought to maximum loads of (**a**) 100 mN, (**b**) 50 mN (**c**) 20 mN, (**d**) 10 mN, (**e**) 5 mN, and (**f**) 1 mN

Frame (a) was collected during linear elastic loading. After reaching an applied load of 2000 μN (corresponding to 650 nm of deflection), the cantilever beam undergoes a small amount of plastic deformation before ultimate failure at $w = 900$ nm shown in Fig. 10.11d. During the entire course of the experiment, it is very difficult to visually detect the changes in this micro-beam because the total deflection pre-failure is very small—on the order of hundreds of nanometers. It is interesting to note that the failure mode exhibited is largely brittle, whereas subsequent beams will exhibit more ductility.

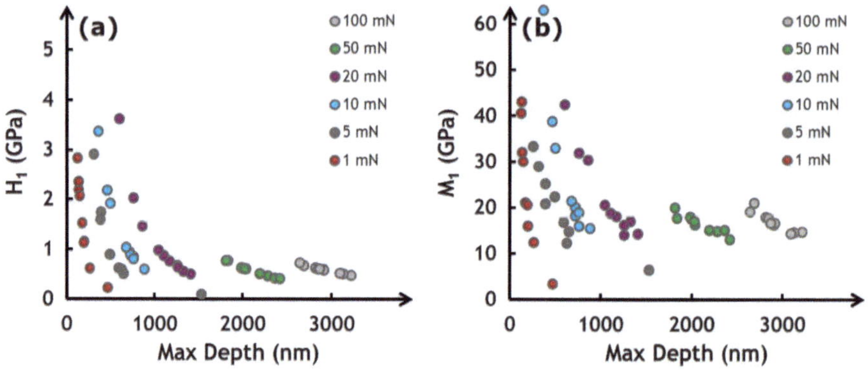

Fig. 10.9 (**a**) Hardness and (**b**) indentation moduli for indentations performed on the argon ion-milled Woodford shale with the indenter oriented parallel to the bedding planes

SEM images of micro-cantilever beam Test A are shown in Fig. 10.12a–c. Panel (a) shows Test A immediately after fracturing the micro-beam, as a transversal crack can be seen near the support region to the left. The face of the broken micro-beam is shown in panel (b) where an iron sulfide inclusion can be seen. The fracture face of the support for Test B was also analyzed by EDS as shown in panel (c), and the iron sulfide cluster is not observed there. This indicates that the fracture propagated entirely around the inclusion and not through it. The EDS of the broken support face furthermore shows the presence of only a small amount of carbon, indicative of a small quantity of organic matter.

Four frames captured while Test B was in the loading configuration are shown in Fig. 10.13. Approximately 1200 μN of applied force were required to fail the beam, which is significantly less than required to fail Test A. Figure 10.13a was collected during linear elastic loading. In this test, the cantilever beams reaches an applied load of ~1200 μN (corresponding to ~800 nm of deflection) then fails abruptly with little or no plastic deformation. During the entire course of the experiment, it is very difficult to visually detect the changes in this micro-beam because the total deflection pre-failure is very small—on the order of hundreds of nanometers. It is interesting to note that the failure mode exhibited is largely brittle, whereas subsequent beams will exhibit more ductility.

Figure 10.14 provides load-displacement curves collected during beam Test C with frame (a) captured at a force of 2500 μN in the linear elastic region of the loading curve. As the loading continues, a dip is observed at 3800 μN while a small crack appears on the beam. Then the micro-beam recovered shortly to a load value close to 4050 μN before the noticeable failure shown where the load decreased to almost 3000 μN. As the loading continues beyond this point, clear strain hardening behavior is observed, while a major fracture develops. However, the micro-beam continues to gain energy before total failure is observed in frame (d). The maximum force achieved corresponds to the ultimate tensile stress, UTS, while the area under the force-displacement curve corresponds to the amount of energy required to fail this micro-beam.

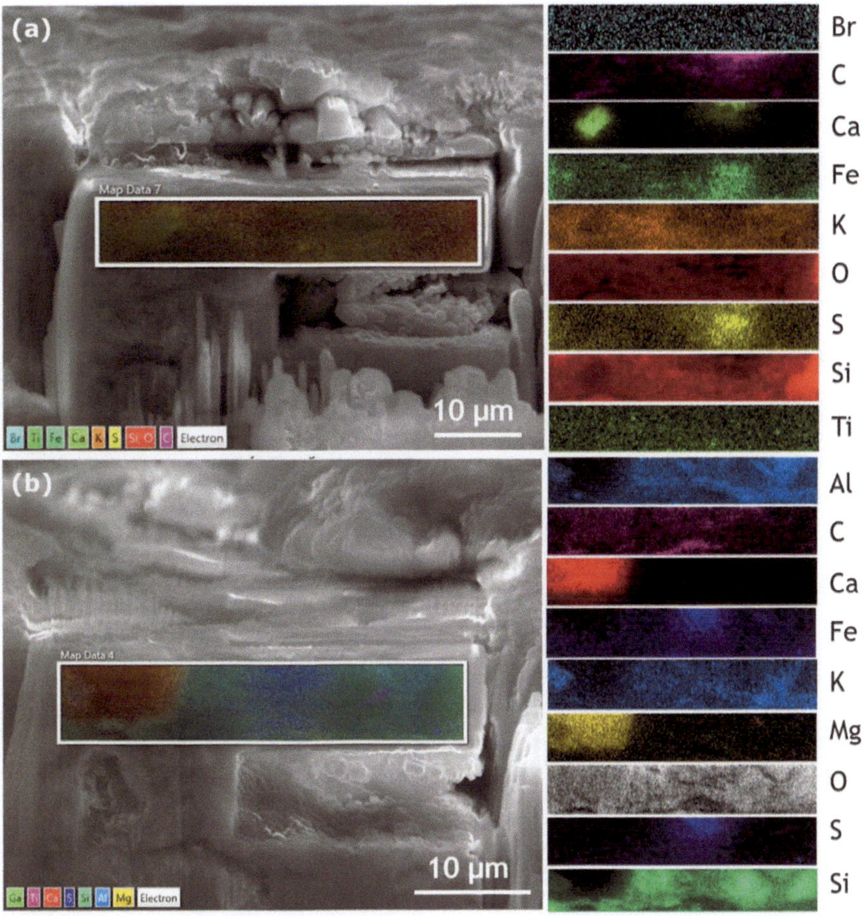

Fig. 10.10 EDS data is shown for micro-beam (**a**) Test A and (**b**) Test C. Maps with all of the elements are overlaid with the SEM images of each of the micro-beams. The individual elemental maps are also shown to the right

The chemical composition of the micro-beam in Test B was also examined using SEM and EDS. A number of observations can be made from the SEM images of Test B shown in Fig. 10.15a–c. First, panel (a) shows the beam pre-failure from the top view. Panels (b) and (c) show the broken sample from different orientations. In (b), only the beam is shown, whereas (c) shows only the support region where the beam originated. The green rectangles in panels (a) and (b) highlight an inclusion through which the fracture propagated. The identity of this inclusion was determined by EDS as shown in panel (c). The SEM image in (c) is looking directly at the fracture face of the broken beam and has been rotated such that the inclusion is in the upper right hand corner, and the image is furthermore overlaid with EDS data.

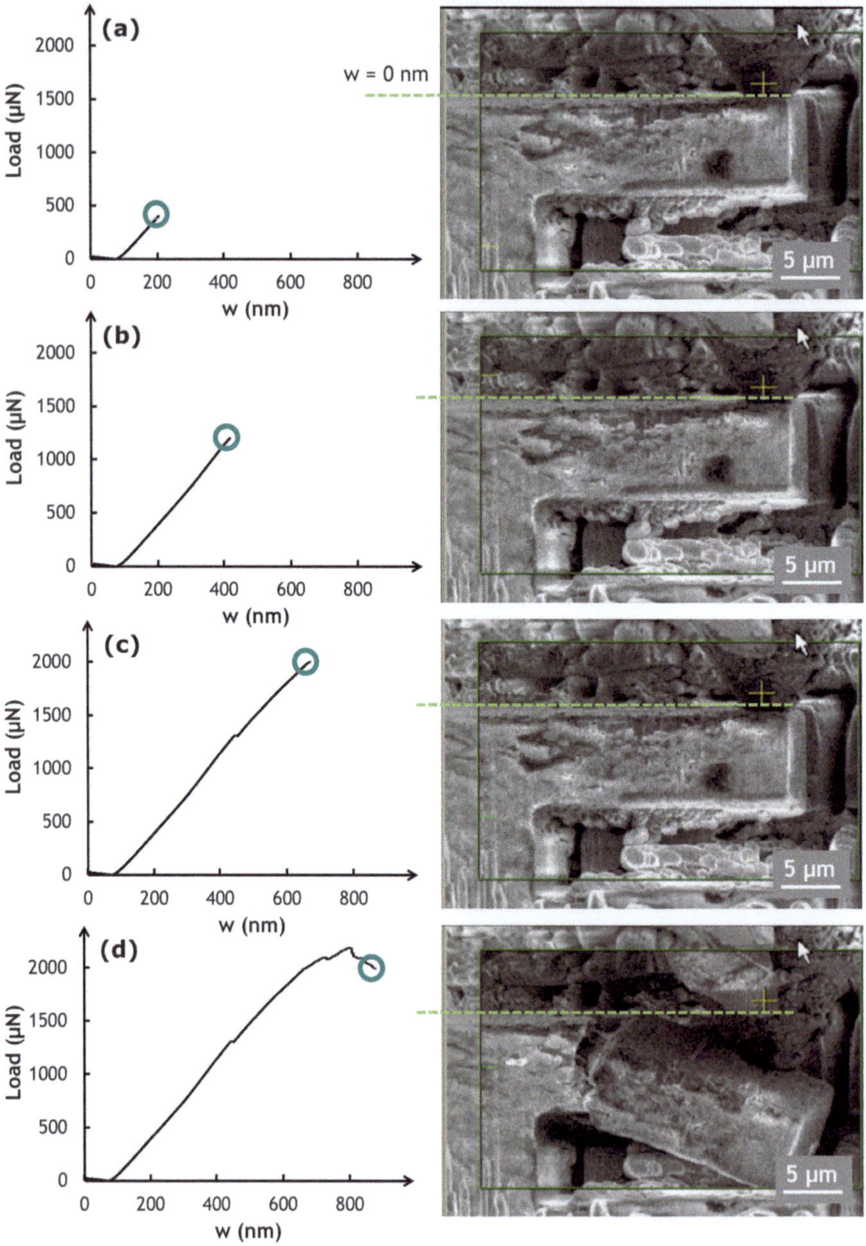

Fig. 10.11 The plots of load versus displacement for the micro-cantilever testing are shown for Test A with frames (**a**)–(**d**) provided at different stages of testing. Frame (**d**) was collected just before beam failure

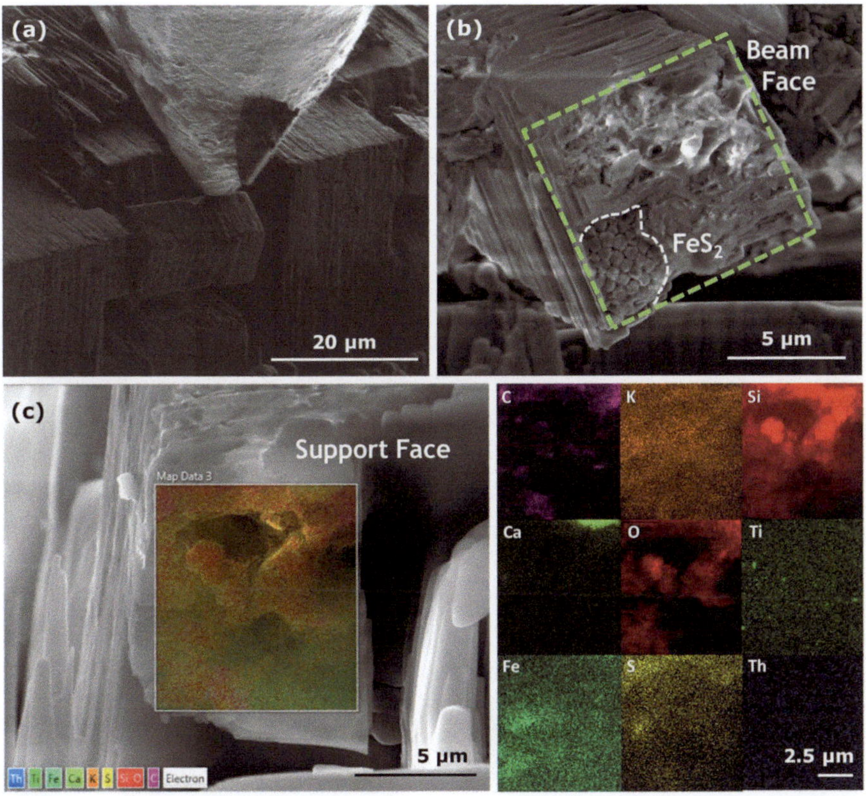

Fig. 10.12 Several views of Test A post-failure are shown including (**a**) the beam and the support together just after failure; (**b**) the failure face of the broken beam highlighting an iron sulfide inclusion; and (**c**) the broken face of the support where the beam was previously attached

To the right, the EDS maps for various elements analyzed are shown and clearly demonstrate that the inclusion is iron sulfide.

A second important observation from Test C is the manner in which the micro-beam crack forms and propagates through the structure. First, as the fracture initiates as seen in Fig. 10.14c, the crack moves transversally from the top of the cantilever near the support towards the bottom of the cantilever. However, as the load continues to increase, the crack eventually takes a turn and moves along the length of the cantilever, propagating horizontally. After the beam has completely failed, the failure pathway becomes quite transparent. In Fig. 10.16a, there is a large piece of shale sticking out from the bottom of the support region, and a large piece is missing in the broken micro-beam as seen in Fig. 10.16b. The broken piece of the support was further analyzed by EDS from the front side as shown in Fig. 10.16c. From the individual element maps, it can be seen that there is a prominent vane of carbon (kerogen) running through the broken piece of shale all the way out to the tip of the broken structure. Similarly, the presence of a large

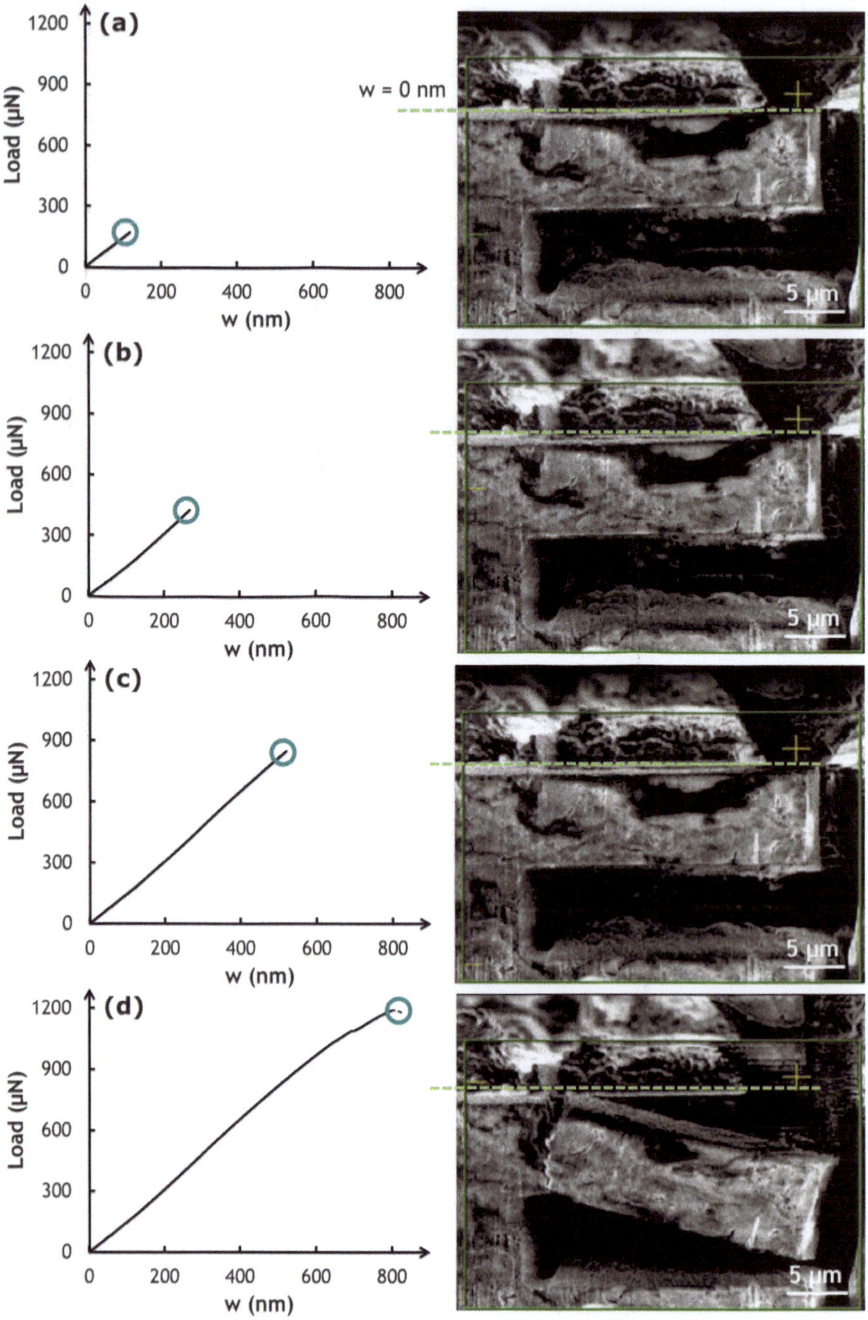

Fig. 10.13 The plots of load versus displacement for the micro-cantilever testing are shown for Test B with frames (**a**)–(**d**) provided at different stages of testing. Frame (**d**) was collected just before beam failure [21]

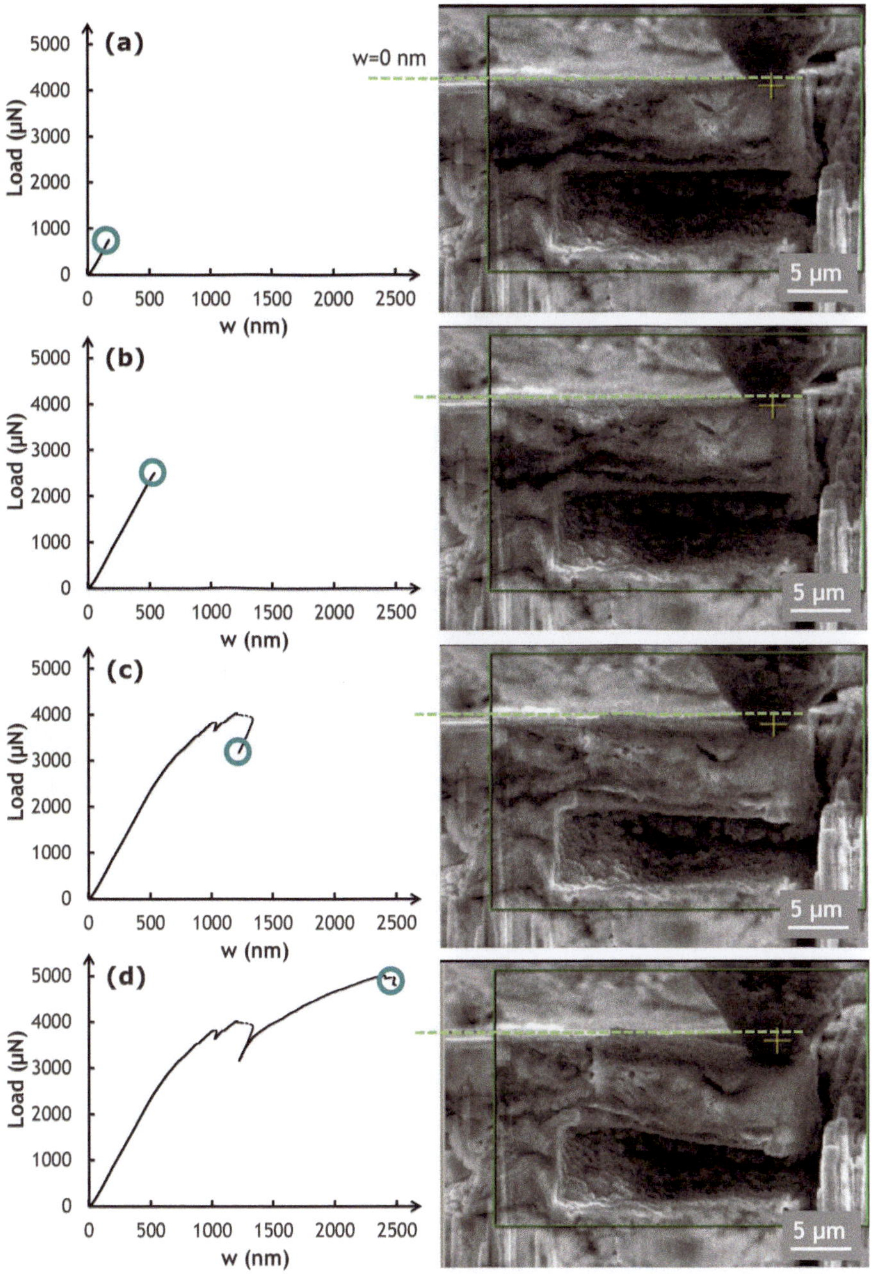

Fig. 10.14 The plots of load versus displacement for the micro-cantilever testing are shown for Test B with frames (**a**)–(**d**) provided at different stages of testing. Frame (**d**) was collected just before beam failure [21]

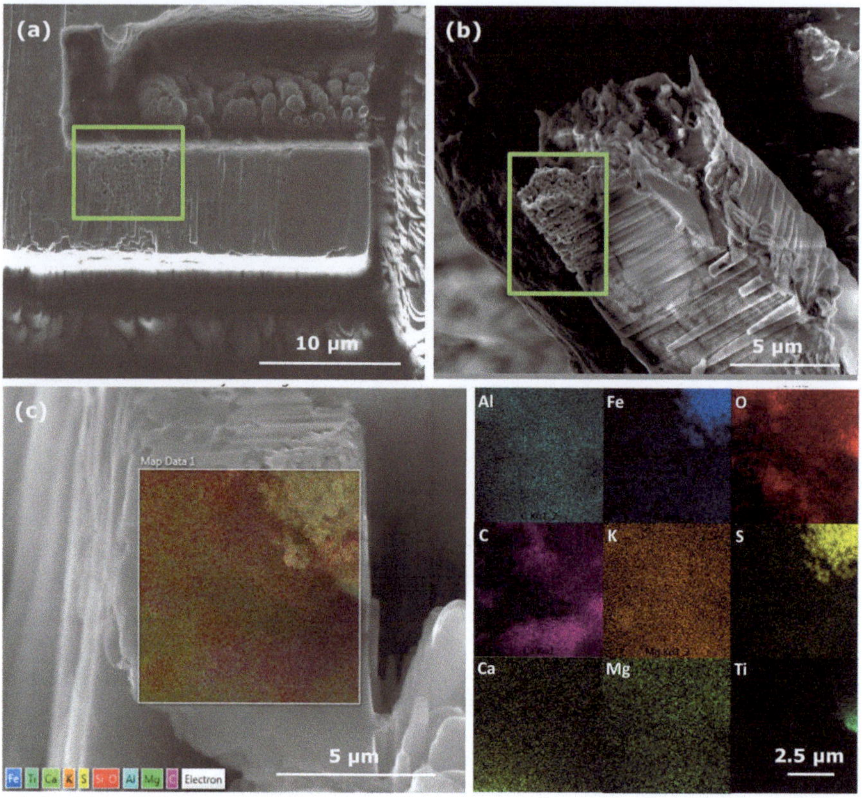

Fig. 10.15 SEM images are shown of Test C from various orientations: (**a**) the top of the beam pre-failure; (**b**) the side of the broken beam post-failure; and (**c**) the face of the broken support post-failure

amount of organic matter in the failure region is supported by the EDS of the failure face in Fig. 10.15c, where a large amount of carbon is present. As will be described in the proceeding section, the organic matter that was near the support region for this particular cantilever test plays a significant role in influencing the tensile strength characteristics of composite shales such as the Woodford.

Figure 10.17a–d shows four stages of Test D, with the load-displacement correlated to the in situ real-time SEM pictures of the micro-cantilever beam progressive loading to failure. In Fig. 10.17a the beam has been continuously loaded up to $P = 809$ μN with a displacement $w_1 = 697$ nm, in a linear elastic load deformation curve. A sudden drop in stress occurs, as shown in frame (b) just after point 1, and a crack is observed close to the top of the beam. However, the beam continues to deflect and soften as the indenter continuously loads the tip of the micro-beam to point 2 in frame (b). A complex strain softening post yield is developing, and the fracture continues to propagate towards the bottom of the micro-beam as shown in frame (c) at 3. In this frame, observe that the micro-

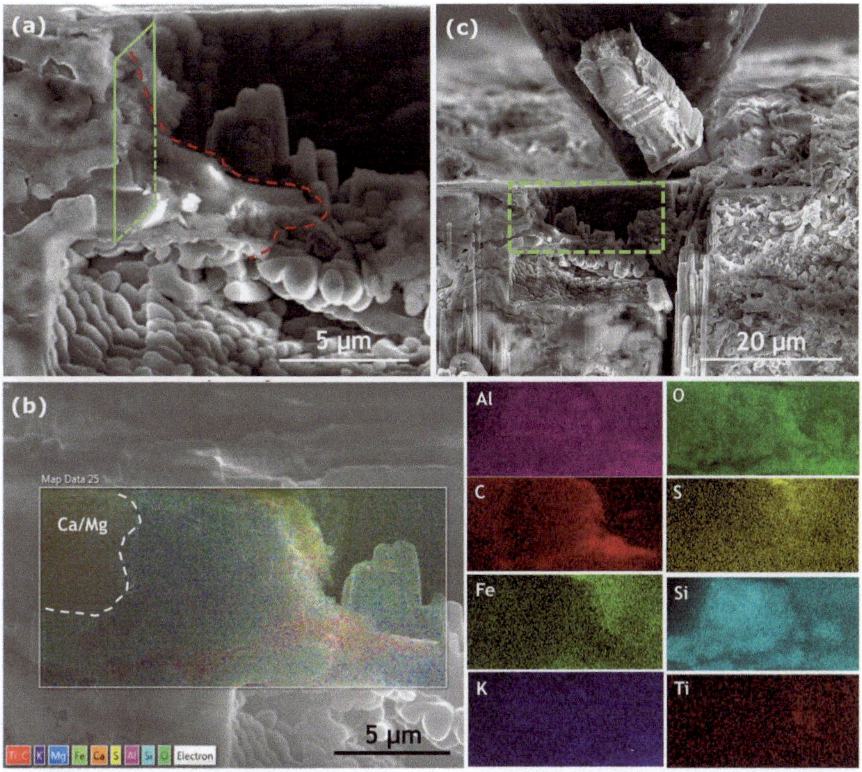

Fig. 10.16 SEM images are shown of Test C from various orientations: (**a**) the broken support with a large piece protruding to the right as outlined by the *red line*; (**b**) both the broken support and failed beam; and (**c**) the front of the broken support overlaid with 2-D maps of EDS data

cantilever beam has totally failed, and it is "almost" detached from its support with a maximum deflection, $w_3 = 4499$ nm.

Finally, in frame (d), an elastic rebound is observed between point 3 and 4 and the final deformation when the indenter is lifted, and the deflection $w_3 > w_4$ is shown in dotted lines, evidenced by the displacement elastic recovery from frame (c) to frame (d). Interpreting this behavior is extremely interesting. A polymer-based rod linking the failed beam to the support can be seen moving from frame (b) to (c). SEM images showing the string-like kerogen are displayed in Figs. 10.1 and 10.6 of the Woodford under the same horizon of the micro-beam. Notice the beam in Fig. 10.17c, d with total length of 22 µm and the white kerogen string holding the beam to the support. It is easily imagined that this polymer-like kerogen can be embedded in the total length of the beam and way further into the micro-cantilever beam fixed support.

In contrast to granular material failure, obviously we can notice that the polymer-like string keeps the beam attached to the support after a total tensile

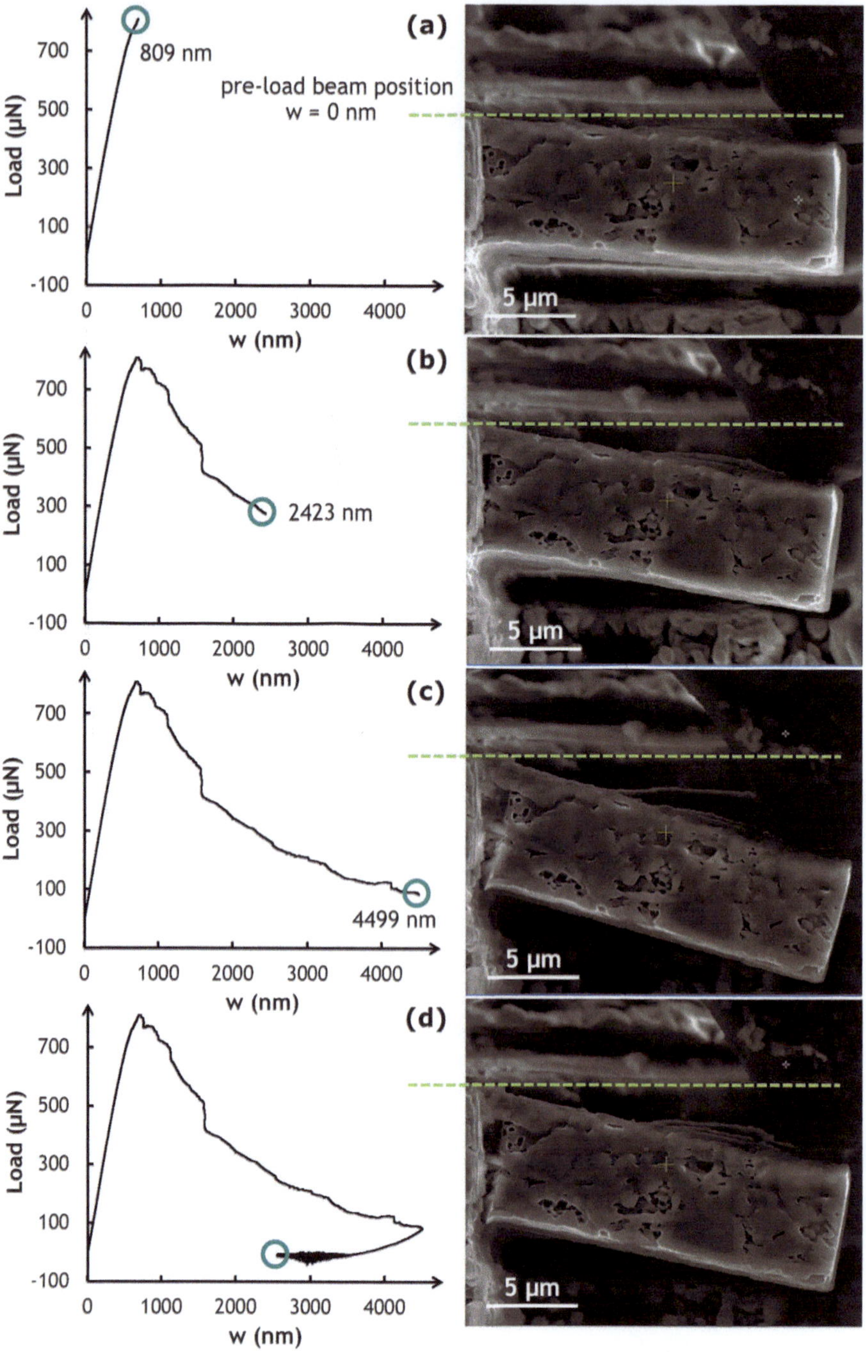

Fig. 10.17 The plots of load versus displacement for the Test D micro-cantilever testing are shown for with frames (**a**) through (**d**) given in chronological order [21]

failure of the micro-beam. The shale matrix granular failure is clearly broken as shown below in Fig. 10.17 yet the micro-cantilever beam is still hanging on after the nanoindenter load was released! This behavior is typical of composite beams such as reinforced concrete beam [35] or in geogrid reinforced site constructions [36]. Post failure analysis shows strain softening behavior that can be easily reproduced using numerical simulation. Since the organic content in these shales, that is kerogen, was never observed in tensile loading or tensile failure to have any effects, the constitutive model for mechanical behavior of the kerogen matrix intertwined with other shale minerals is nonexistent. To feel comfortable with our explanation of the observed strain softening behavior of the micro-beam, we built a simple two dimensional numerical model mimicking the micro-beam response in this one test. This is an attempt to explore the potential constitutive model for the micro-cantilever beam mechanical behavior, through matching the force-displacement curve by placing a percent of volume of organic matter with sustainable tensile strength characteristics at the support.

10.5 Micro-Measurement Cantilever-Beam Overview

Tests A–D are not only from the same horizon of the Woodford shale formation, but were also milled within ~100 μm of each other. Their different mechanical responses to loading require a deeper understanding of how the heterogeneous shale matrix is influencing the mechanical characteristics. Before complete failure, beam Tests C and D exhibited significant ductile behavior while Tests A and B underwent only a small degree of plastic deformation. Furthermore, the amount of force and the resulting displacement required to fail each micro-beam varied significantly from the most brittle case of Test B to the toughest case of Test C. Determining the reason for these differences is important for us to be able to upscale and convert this information into reservoir scale models.

The dimensions of each of the micro-beams that were tested are summarized in Table 10.1. The beams are similar in size and the corresponding moments of inertia (I) were calculated. For each beam, the applied force P was captured at the halfway point in the linear elastic region. These values were used to calculate Young's modulus (E, in GPa). Tests A and B have an E that is ~14 GPa, whereas Test A resulted in a much higher value of E at ~30 GPa. The former value is within 10 % of the Young's moduli determined in other studies of Woodford shale [24, 26]. The ultimate tensile load is proportional to the ultimate tensile stress, UTS, which is a value that carries much significance in fracturing source shales. In fact, the UTS is much more important than the unconfined compressive strength (UCS) since hydraulic fracturing is Mode I tensile crack failure rather than a compressive or shear failure.

The differences observed when failing the two beams are best captured by viewing them plotted together. Figure 10.18 shows Tests A–D plotted as load versus displacement with the areas under each curve corresponding to the energy

Table 10.1 The dimensions of each of the cantilever beams are given along with their respective moments of inertia

Symbol	Unit	Description	Test A	Test B	Test C	Test D
L	μm	Beam length	23.18	24.12	21.37	21.69
b	μm	Beam width	7.95	7.25	7.36	9.49
h	μm	Beam height	9.94	9.47	9.23	8.80
I	μm^4	Moment of inertia	651	514	483	539
P	μN	Force (elastic regime)	478	353	1016	290
E	FPa	Young's modulus	13.5	14.2	30.4	9.1

The forces were obtained from the linear loading curve at the halfway point [21]

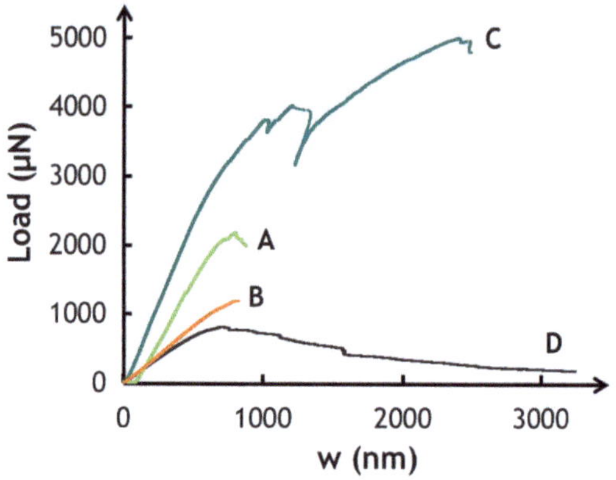

Fig. 10.18 Plots of load versus displacement for micro-beam Tests A–D are superimposed on each other [21]

require to fail the beams [21]. These plots plainly demonstrate that approximately ten times more energy is needed to fail beam C than beam B. The ability to identify and capture the true tensile mechanical characteristics and failure strength of unconventional source rocks on the nano- and micro-scales is expected to have a profound effect upon the way we understand the macro-mechanics of kerogen-rich shales.

The reason for the large difference in tensile strength between two specimens from the same sample is easily described by Figs. 10.1 and 10.6. The organic matter (kerogen, bitumen, etc.) is interwoven with the mineral matrix in a complex manner that spans across nano-, micro-, and macro-scales. The Woodford shale used in the present study contains almost 10 wt% organic matter which corresponds to roughly 20 vol.%. This organic matter has rubber- or polymer-like elastomeric qualities, including low compressive strength but high tensile strength. Because the kerogen is interwoven with the clay and non-clay minerals at nano- and micro-scales, determining the tensile strength of the composite material requires testing on the nano- and micro-scales.

Beams A and B exhibited largely brittle failure, characteristic of their mineral constituents, and contained very little kerogen in the support/fracture region. In Test C the amount of kerogen was significantly higher and even overwhelming at the support with little volume of the clay or non-clay granular material. The volume of the organic matter that stayed behind at the support is evident by the large cavity left on the micro-beam after total collapse shown in Fig. 10.16a, b as well as by the corresponding EDS data shown in Figs. 10.15c and 10.16c. The volume of the kerogen was large enough and strong enough to carry the micro-beam into a strain hardening behavior at the post-yield stage, with a large modulus of toughness contributing to the overall shale micro-beam behavior at failure. The progress of failure in Test C reinforced our early hypothesis that the cross-linked polymer nature of kerogen and its intertwined structure with the non-clay and clay mineral matrix is the one holding the granular shale matrix together, resulting in large and unexpected values for granular material in tensile failures. Also from Fig. 10.14c, the fracture has developed almost entirely across the depth of the beam, yet the beam continued taking more load and exhibiting strain hardening until complete failure.

10.5.1 Macro-Measurements of Kerogen-Rich Shale Following ASTM and ISRM Methods

In this section we address the macro-scale classical measurement of tensile strength and tensile resistance of rocks [24, 28, 29]. However, efforts were made to minimize the size of the Woodford shale used in these tests [24, 37], given the prohibitive cost to retrieve as much as we need of core preserved shale samples. Shale rock preservations, at the rig, were very costly in drilling and non-productive time, NPT, and in many instances impossible.

10.5.1.1 Brazilian Tensile Test

The Woodford shale samples were sectioned into halves for the Brazilian and Chevron notch semicircular bending, CNSCB, tensile strength tests as shown in Figs. 10.19 and 10.21. Only, 2 cm diameter cores with longitudinal axis parallel to sample bedding planes were extracted from the 5.0 cm diameter cores as demonstrated in Fig. 10.19, with 1 cm in thickness. The two samples obtained from the 2 cm diameter core and polished to 1 cm thickness to achieve a length to diameter ratio of 0.5 according to the ASTM specifications [37].

An Axial-Torsion MTS 319 loading frame was used to compress the samples diametrically. The schematic of the experiment setup and a Woodford sample under the Brazilian test are shown in Fig. 10.20. For each depth, a sample was loaded with the loading line parallel to the bedding plane ($\theta_b = 90°$) while the other sample was loaded with the loading line orthogonal to the bedding plane ($\theta_b = 0°$). The loading rate was set at 8.9 N/s (2 lb/s) so that the total testing time for each sample was less than 1 min according to ASTM standards [37].

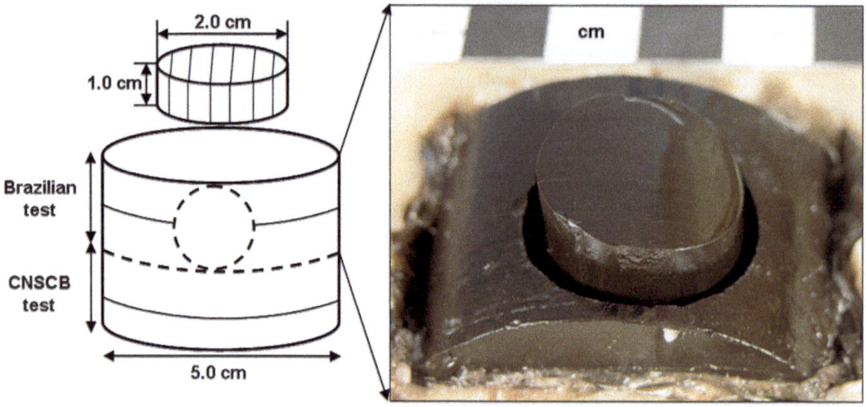

Fig. 10.19 Schematic for sample preparation for the Brazilian test [28]

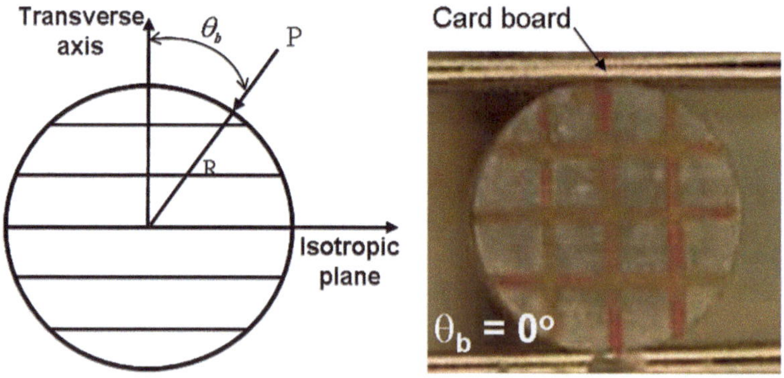

Fig. 10.20 Schematic and picture of Brazilian test setup

10.5.2 Three-Point Chevron Notch Semicircular Bending Shale Sample (CNSCB)

Fracture toughness describes the resistance of the material to the propagation of a preexisting crack. Also, mode-I fracture was chosen to simulate wellbore hydraulic fracturing failure and propagation [21, 38]. Table 10.2 gives the dimensions of the prepared CNSCB samples.

The Chevron notch was chosen as it can produce a stable crack growth and guarantee self-pre-cracking during the test [39]. The notch was formed by two cuts, as illustrated in Fig. 10.21, using a Diamond-Laser saw with thickness of 0.3 mm, outer diameter of 10.2 cm, and inner diameter of 1.3 cm. The initial crack length a_0 was designed to be approximately 6 mm.

The schematic of the experimental setup and a prepared Woodford sample is shown in Fig. 10.22. The notch was cut at 90° with respect to the base of the

Table 10.2 CNSCB test samples' dimensions

Depth (m)	Radius (cm)	Thickness (cm)
33.81	2.31	2.01
36.85	2.56	2.36
41.36	2.46	2.67
44.28	2.46	2.49
50.59	2.41	2.49

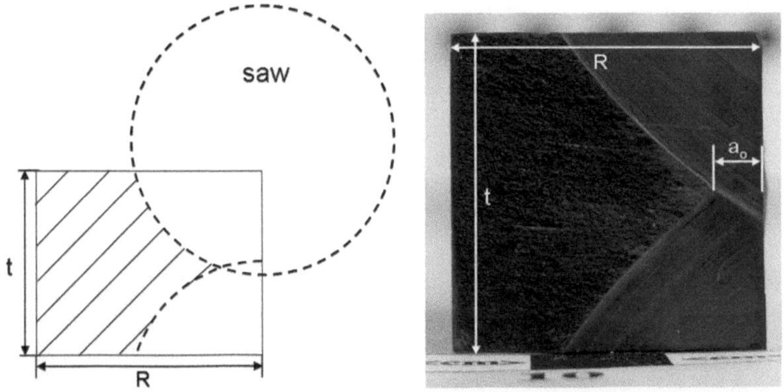

Fig. 10.21 Schematic of Chevron-notch preparation and picture of a Chevron-notch on a tested Woodford sample

specimen for mode-I fracture toughness test. Following Chang et al. [38], the loading span (2S) was fixed at 4.06 cm which results in a span-to-diameter ratio (S/D) of 0.8. During the test, the lateral base displacement of the crack was recorded using an MTS clip gage. The gage was mounted to the Plexiglas pads glued on the base of the sample as shown in Fig. 10.22.

Axial load was applied to the CNSCB samples by Axial-Torsion MTS 319 loading frame. Displacement controlled instead of load control was chosen to provide a better control of fracture growth. Displacement rate was set constant at 0.0005 mm/s so that total test time is within 5 min. Figure 10.22 shows a CNSCB sample ready for testing. The values of the fracture toughness varied 100 % between the upper Woodford to the Lower Woodford shale [28].

10.5.2.1 Anisotropic Tensile Strength

Pictures of tested Woodford samples with loading directions normal and parallel to bedding planes are shown in Fig. 10.23.

Summarized in Table 10.3 is the maximum applied load at which the tensile strength Brazilian test samples failed and from which the sample tensile strength (value of tensile stress at center of the core) can be obtained following [39, 40]:

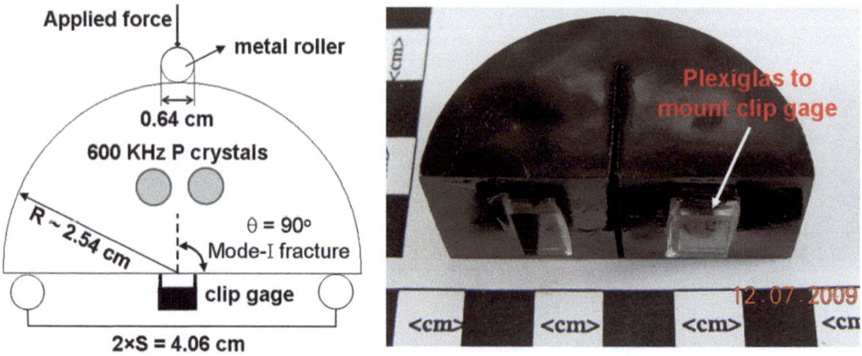

Fig. 10.22 CNSCB test and Chevron notch specifications

Fig. 10.23 Tested Woodford samples with the transverse isotropy assumption

Table 10.3 Brazilian tensile strength [28]

Depth (m)	Load direction (rel. to bedding)	Max. Load (N)	Tensile strength (MPa)
33.81	⊥	3747	12.8
	//	2351	6.2
36.85	⊥	3556	12.7
	//	1989	7.4
41.36	⊥	3347	11.4
	//	1956	5.0
44.28	⊥	3140	11.2
	//	1920	5.1
50.59	⊥	3276	11.7
	//	1609	4.4

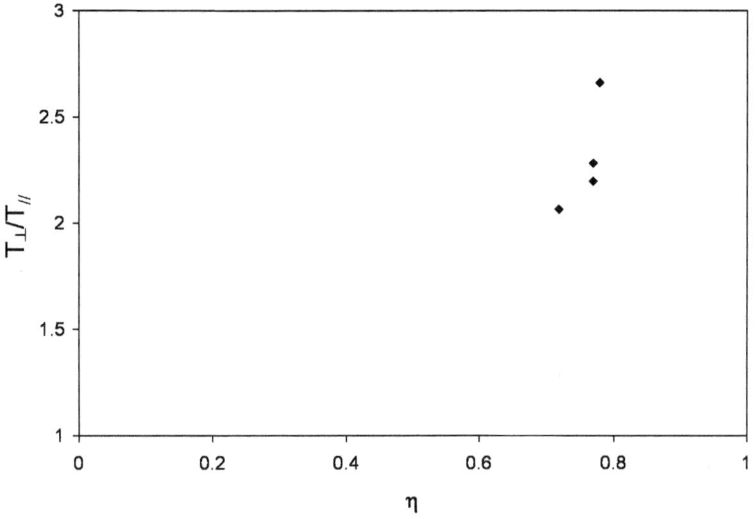

Fig. 10.24 Variation of tensile strength ratio, $T\perp/T_{//}$, with clay packing density [28]

$$T = \frac{\sqrt[4]{E_1/E_3}^{\cos(2\theta_b)} P}{\pi RL} - \frac{P\cos(4\theta_b)}{4\pi RL}\left(\frac{\sqrt{E_1E_3}}{2}\left(\frac{1}{G_3} - \frac{2\nu_3}{E_3}\right) - 1\right) \tag{10.7}$$

where T is sample tensile strength. E_1 and ν_1 are Young's modulus and Poisson's ratio, respectively, in the plane of isotropy; E_3, ν_3, and G_3 are Young's modulus, Poisson's ratio, and shear modulus along the transverse axis. R and L represent the sample radius and thickness respectively. The calculated Woodford tensile strengths in directions perpendicular to bedding planes, denoted as $T\perp$, and parallel to bedding planes, denoted as $T_{//}$, are summarized in Table 10.3.

The tensile strength anisotropy ratio ($T\perp/T_{//}$) exhibits an increasing trend with increase clay packing density as shown in Fig. 10.24 [7, 28]. The effects of carbonate on tensile strength were also investigated for tested samples (samples 36.85–50.59 m) with relatively constant quartz content (36–39 %). An increase of tensile strength in both loading direction with carbonate content is observed as shown in Fig. 10.25, and that may be a pure coincidence. The proportional behavior of tensile strengths with carbonate content suggests that the Woodford Shale carbonate minerals can be precipitations in addition to detrital shell fragments as observed with the thin sections [28].

10.6 Summary and Future Direction (Macro-Scale)

The macro-scale rock mechanics testing methods were used to characterize source rocks such as shales as discussed in the previous section. These existing mechanical testing methods cannot account for the properties of the polymer-like kerogen

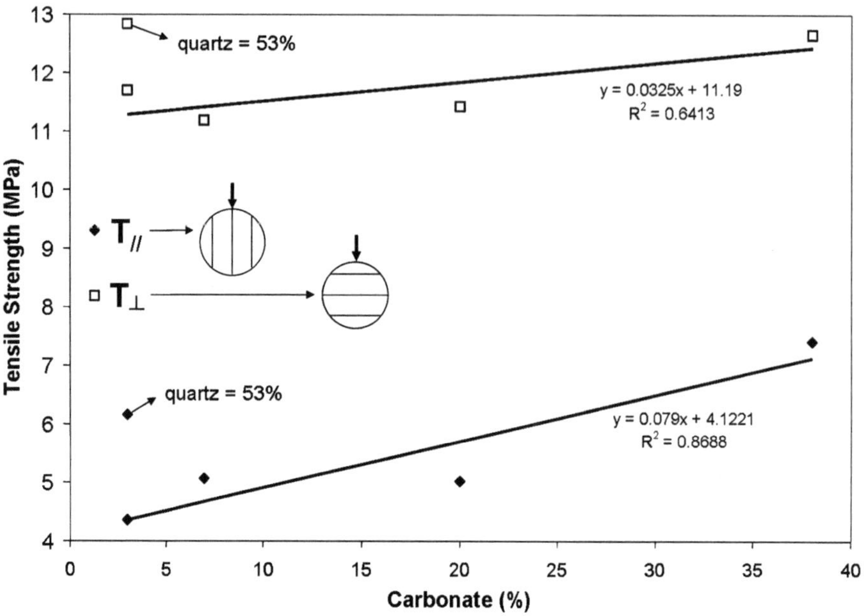

Fig. 10.25 Variation of tensile strengths, $T\perp$ and $T_{//}$, with carbonate content [28]

material which is interbedded and interwoven within the mineral matrix. While these ASTM and ISRM methods failed to pick up the tensile attributes of kerogen on the Woodford shale the nano- and micro-cantilever beams testing methods were very successful. The reason is that the tensile characteristics of polymers are easily masked in the ISRM and ASTM standard testing methods such as the Brazilian test and other approved tensile strength measurements for rocks, as well as the fracture toughness. These tests were never designed to isolate or measure the tensile strength of polymer-like composite porous materials. This natural cross-linked polymer component, kerogen, with its tensile characteristics, was not known previously to contribute to the tensile strength of any known rock loaded in tension. Now that the organic rich source shale formations are loaded under tensile forces at the micro- and nano-scales, the implications for this discovery are far-reaching. Kerogen, once thought a weak and compliant material under compression loads, is now understood as very strong in tensile loading configurations. In fact, the tensile strength of the organic matter far exceeds that of the surrounding rock matrix. It remains to be seen how this challenge will be addressed in the laboratory in terms of revising our current tensile testing methods as well as in hydraulic fracturing applications.

Acknowledgements The authors would like to thank Aramco for allowing the publication of this work. The second author would like to specially acknowledge the GeoMechanics Gas Shale Consortium at the integrated Poromechanics Institute, University of Oklahoma.

References

1. Richard, P., Nicodemi, M., Delannay, R., Ribière, P., & Bideau, D. (2005). Slow relaxation and compaction of granular systems. *Nature Materials, 4*, 121–128.
2. Han, Y., & Cundall, P. A. (2013). LBM-DEM modeling of fluid-solid interaction in porous media. *International Journal for Numerical and Analytical Methods, 37*(10), 1391–1407.
3. Hornby, B. E., Schwartz, L. M., & Hudson, J. A. (1994). Anisotropic effective-medium modeling of the elastic properties of shales. *Geophysics, 59*(10), 1570–1583.
4. Wenk, H.-R., Lonardelli, I., Franz, H., Nihei, K., & Nakagawa, S. (2007). Preferred orientation and elastic anisotropy of illite-rich shale. *Geophysics, 72*(2), E69–E75.
5. Abousleiman, Y., Ulm, F. -J. (2003). "TheGeoGenome™ Industry Consortium" JIP internal report, The Poromechanics Institute, Oklahoma University and Massachusetts Institute of Technology.
6. Zeszotarski, J. C., Chromik, R. R., Vinci, R. P., Messmer, M. C., Michels, R., & Larsen, J. W. (2004). Imaging and mechanical property measurements of kerogen via nanoindentation. *Geochimica et Cosmochimica Acta, 68*(20), 4113–4119.
7. Ulm, F.-J., & Abousleiman, Y. (2006). The nano granular nature of shale. *Acta Geotechnica, 1* (2), 77–88.
8. Ulm, F.-J., Constantinides, G., Delafargue, A., Abousleiman, Y., Ewy, R., Duranti, L., et al. (2005). Material invariant poromechanics properties of shales. In Y. Abousleiman, A. H.-D. Cheng, & F.-J. Ulm (Eds.), *Poromechanics III. Biot centennial (1905-2005)* (pp. 637–644). London: A.A. Balkema Publishers.
9. Biot, M. A. (1941). General theory of three-dimensional consolidation. *Journal of Applied Physics, 12*, 155–164.
10. Podio, A. L., Gregory, A. R., & Gray, K. E. (1968). Dynamic properties of dry- and water-saturated green river shale under stress. *Society of Petroleum Engineers Journal, 8*(4).
11. Abousleiman, Y., Cheng, A. H.-D., Cui, L., Detournay, E., & Roegiers, J.-C. (1996). Mandel's problem revisited: Consolidation of a porous anisotropic rock. *Geotechnique, 46*(2), 187–195.
12. Abousleiman, Y., Hoang, S., & Liu, C. (2014). Anisotropic porothermoelastic solution and hydro-thermal effects on fracture width in hydraulic fracturing. *International Journal for Numerical and Analytical Methods, 38*(5), 493–517.
13. Ekbote, S., & Abousleiman, Y. (2006). Porochemoelastic solution for an inclined borehole in a transversely isotropic formation. *Journal of Engineering Mechanics ASCE, 132*(7), 754–763.
14. Vernik, L., & Nur, A. (1992). Ultrasonic velocity and anisotropy of hydrocarbon source rocks. *Geophysics, 57*(5), 727–735.
15. Abousleiman, Y., & Cui, L. (1998). Poroelastic solutions in transversely isotropic media for wellbore and cylinders. *International Journal of Solids and Structures, 35*(34-35), 4905–4930.
16. Abousleiman, Y., Hoang, S., & Tran, M. (2010). Mechanical characterization of small shale samples subjected to fluid exposure using the inclined direct shear testing device. *International Journal of Rock Mechanics and Mining Sciences, 47*(3), 355–367.
17. Ortega, J. A., Ulm, F.-J., & Abousleiman, Y. (2007). The effect of the nanogranular nature of shale on their poroelastic behavior. *Acta Geotechnica, 2*(3), 155–182.
18. Ortega, A., Ulm, F.-J., & Abousleiman, Y. (2009). The nanogranular acoustic signature of shale. *Geophysics, 74*(3), 65–84.
19. Allan, M., Kanitpanyacharoen, W., & Vanorio, T. (2015). A multiscale methodology for the analysis of velocity anisotropy in organic-rich shale. *Geophysics, 80*(4), C73–C88.
20. Frazer, D., Abad, M. D., Krumwiede, D., Back, C. A., Khalifa, H. E., Deck, C. P., et al. (2015). Localized mechanical property assessment of SiC/SiC composite materials. *Composites Part A: Applied Science & Manufacturing, 70*, 93–101.
21. Abousleiman, Y. N., Hull, K. L., Han, Y., Al-Muntasheri, G., Hosemann, P., Parker, S., et al. (2016). The granular and polymer composite nature of kerogen-rich shale. *Acta Geotechnica*. doi:10.1007/s11440-016-0435-y.

22. Kelemen, S. R., Walters, C. C., Ertas, D., Kwiatek, L. M., & Curry, D. J. (2006). Petroleum expulsion part 2. Organic matter type and maturity effects on kerogen swelling by solvents and thermodynamic parameters for kerogen from regular solution theory. *Energy & Fuels, 20*(1), 301–308.
23. Passey, Q. R., Bohacs, K. M., Esch, W. L., Klimentidis, R., & Sinha, S. (2010). From oil-prone source rock to gas-producing shale reservoir—geologic and petrophysical characterization of unconventional shale gas reservoirs. CPS/SPE International Oil & Gas Conference & Exhibition in China, Beijing, 8–10 June.
24. Abousleiman, Y., Tran, M., Hoang, S., Bobko, C., Ortega, J. A., Ulm, F. -J. (2007). Geomechanics field and lab characterization of Woodford shale: The next gas play. *SPE Annual Technical Conference, Society of Petroleum Engineers*, Anaheim, CA, 11–14 November.
25. Abousleiman, Y., Tran, M., Hoang, S., Ortega, J. A., & Ulm, F. -J. (2009) GeoMechanics field characterization of the two prolific U.S. mid-west gas plays with advanced wire-line logging tools. *SPE Annual Technical Conference*, New Orleans, Louisiana, 4–7 October.
26. Bennett, K. C., Berla, L. A., Nix, W. D., & Borja, R. I. (2015). Instrumented nanoindentation and 3D mechanistic modeling of a shale at multiple scales. *Acta Geotechnica, 10*, 1–14.
27. Abousleiman, Y., Tran, M., Hoang, S., Ulm, F. -J., Ortega, J. A., & Bobko, C. (2013). Method of predicting mechanical properties of rocks using mineral compositions provided by in-situ logging tools. U.S. Patent: 8,380,437.
28. Sierra, R., Tran, M. H., Abousleiman, Y. N., & Slatt, R. M. (2011). Woodford shale mechanical properties and the impacts of lithofacies. 44th U.S. Rock Mechanics Symposium and 5th U.S. Canada Rock Mechanics Symposium, Salt Lake City, Utah, 27–30 June.
29. Slatt, R., & Abousleiman, Y. (2011). Merging sequence stratigraphy and geomechanics for unconventional gas shales. *The Leading Edge, 30*(3), 274–282.
30. Maio, D. D., & Roberts, S. G. (2005). Measuring fracture toughness of coatings using focused-ion-beam-machined microbeams. *Journal of Materials Research, 20*, 299–302.
31. Oliver, W. C., & Pharr, G. M. (1992). An improved technique for determining hardness and elastic modulus using load and displacement sensing indentation experiments. *Journal of Materials Research, 7*(6), 1564–1583.
32. Oliver, W. C., & Pharr, G. M. (2004). Measurement of hardness and elastic modulus by instrumented indentation: Advances in understanding and refinements to methodology. *Journal of Materials Research, 19*(1), 3–20.
33. Delafargue, A., & Ulm, F.-J. (2004). Explicit approximations of the indentation modulus of elastically orthotropic solids for conical indenters. *Journal of Solids and Structures, 41*(26), 7351–7360.
34. Chen, S. H., & Feng, B. (2011). Size effect in micro-scale cantilever beam bending. *Acta Mechanica, 219*, 291–307.
35. Bazant, Z. P., & Oh, B. H. (1984). Deformation of progressively cracking reinforced concrete beams. *ACI Journal, 81*(3), 268–278.
36. Bhandari, A., Han, J., & Parsons, R. L. (2015). Two-dimensional DEM analysis of behavior of geogrid-reinforced uniform granular bases under a vertical cyclic load. *Acta Geotechnica, 10*, 469–480.
37. Newman, D. A., & Bennett, D. G. (1990). The effect of specimen size and stress rate for the Brazilian test—a statistical analysis. *Rock Mechanics and Rock Engineering, 23*, 123–134.
38. Chang, S. H., Lee, C. I., & Jeon, S. (2002). Measurement of rock fracture toughness under modes I and II and mixed-mode conditions by using disc-type specimens. *Engineering Geology, 66*, 79–97.
39. Kuruppu, M. D. (1997). Fracture toughness measurement using chevron-notched semi-circular bend specimen. *International Journal of Fracture, 86*, L33–L38.
40. Claesson, J., & Bohloli, B. (2002). Brazilian test: Stress field and tensile strength of anisotropic rocks using analytical solution. *International Journal of Rock Mechanics and Mining Sciences, 39*, 991–1004.

Chapter 11
Experimental and Numerical Investigation of Mechanical Interactions of Proppant and Hydraulic Fractures

Congrui Jin

Abstract Hydraulic fracturing has recently received a great amount of attention not only for its economic importance but also for its potential environmental impact. The basic intention of the hydraulic fracturing process is to increase the productivity of the stimulated well by maximizing the reservoir's permeability, but the permeability of the fractured reservoir is strongly affected by the apertures of the fractures. Proppants are often utilized during hydraulic fracturing to aid the retention of the fracture aperture, and laboratory experiments and field observations have shown a strong correlation between the volume of proppant deployed in hydraulic fracturing operations and reservoir productivity. However, the factors controlling proppant performance in real rock fractures are still poorly understood. Considering the high cost of a hydraulic fracturing treatment, a more informed selection of design parameters, such as proppant size, shape, concentration and properties, fracture fluid viscosity, and pumping schedule is needed. A better understanding of the behavior of fluid and proppant within a fracture and their relationship to fracture conductivity is of great practical interest. The goal of this chapter is to provide a summary of recent experimental and numerical investigations on the interactions of proppant and hydraulic fractures.

11.1 Introduction

A well's ability to produce hydrocarbons or receive injection fluids is limited by the reservoir's natural permeability and near-wellbore changes resulting from drilling or other operations. Hydraulic fracturing, or fracking, is the process of initiation, propagation, and branching of cracks by pumping fluids at relatively high flow rate

---

C. Jin (✉)
Department of Mechanical Engineering, Binghamton University, State University of New York, Binghamton, NY 13902, USA
e-mail: cjin@binghamton.edu

C. Jin, G. Cusatis (eds.), *New Frontiers in Oil and Gas Exploration*,
DOI 10.1007/978-3-319-40124-9_11

and pressure. It is a widespread stimulation technique extensively used since 1947 for oil and natural gas production [39, 41–43, 52, 54, 65]. About 40 % of today's wells are hydraulically fractured, which has made 25–30 % of US oil reserves economically producible. The role of this technique is expected only to become more important in the years to come as the formations with increasingly lower permeability will be involved in oil and gas production.

The basic intention of the hydraulic fracturing is to increase the productivity of the stimulated well by maximizing the reservoir's permeability, but the permeability of a fractured formation is strongly affected by the aperture of the tensile fractures. However, the created fractures tend to close and rapidly heal due to overburden pressure on the reservoir once the well is depressurized, and therefore the fracture usually needs to be propped by a suitable agent [42, 78]. This agent, called proppant, which ranges from natural sands to synthetic materials, blended in a certain ratio with the fracturing fluid, prevents the fracture from closing after the injection has stopped by counteracting the reservoir's compressive stresses, and creates a filling with relatively high permeability inside the fracture to sustain the flow long after the treatment [41, 114]. The residual fracture openings are the main outcome of such hydraulic stimulations as these openings significantly affect the permeability of the reservoirs and, subsequently, well productivity [11, 111]. The stimulation treatment ends when the planned pumping schedule has been completed or when a sudden rise in pressure indicates that a screenout has taken place. A screenout is a blockage caused by bridging of the proppant across the fracture width that restricts fluid flow into the hydraulic fracture. The deposited proppant must have sufficiently high permeability to provide low resistance to flow and also satisfy certain requirements regarding its compressive strength that would allow it to withstand the in situ stresses after the fracture has closed. Furthermore, the proppant must sit firmly enough to remain in place during flowback.

As new oil and gas extraction techniques continue developing, proppant fracturing technique has evolved greatly over the past 60 years and continue to evolve to provide the solution to increasing the effectiveness of hydraulic fracturing. There are three types of proppant fracturing: hydraulic proppant fracturing, water fracturing, and hybrid fracturing. Each type has unique characteristics, and each possesses its own positive and negative performance traits [98, 101].

Hydraulic proppant fracturing, preferentially applied in medium to high permeable formations, uses highly viscous gels with high proppant concentrations, creating highly conductive, but relatively short fractures in a permeable reservoir with a porous matrix. The fracture connects the well and the reservoir and reduces permeability impairments near the well, resulting in an increase in productivity. However, this treatment also induces serious problems in well performance. For example, high proppant concentration can screenout in the near wellbore region due to pressure losses. Although gels are very efficient for transporting proppant, these gels often damage the fracture, and are difficult to clean-up. Under these conditions, minimal effective stimulation was achieved, sometimes resulting in sub-economic or even uneconomic wells [98, 101, 102].

Water fracturing, as defined by Schein, is a fracture treatment that utilizes a large volume of water to create an adequate fracture geometry and conductivity to obtain commercial production from low permeability, large net pay reservoirs [103]. It was developed as an alternative to conventional gel treatments mainly because it reduced the potential for gel damage, lowered costs, and provided more complex fracture geometry which was evident from microseismic data [92]. Water containing friction-reducing chemicals (slickwater) with low or zero proppant concentration is used as a pumping fluid to create long and skinny fractures. Smaller proppants are often employed since these smaller diameter proppants are easier to place in the narrower fractures, as well as transport in the fracture [20]. This treatment aims at connecting reservoir parts at some distance from the borehole, and inflow area is maximized by connecting the well to a network of natural joints. Previous studies have suggested that, when compared to conventional gel treatments, water fracturing can generate similar or sometimes better production responses [28, 74, 77, 83, 84, 113]. Moreover, even when conventional gel treatments generate longer propped fracture lengths, the presence of damaging gels may adversely affect well performance [83, 84, 113]. Water fracturing has been successfully applied to thousands of wells in shale formation development. This treatment offers a lower cost and greater safety profile, but slickwater's low viscosity limits its ability to transport proppant, usually resulting in an accumulation of proppant at the fracture bottom, which is called proppant bank, leaving the top portion of the fracture unpropped and affecting the fracture conductivity in the vicinity of wellbore area.

Hybrid fracturing is a fracturing treatment that relies on a fluid system in which some combination of slickwater fluids, linear gels, and/or cross-linked gels may be used as part of the fluid formulation. Such a treatment begins with a slickwater formulation to create fracture geometry, which is called pad, and later is shifted to a cross-linked gel treatment to transport proppant down the fracture more effectively.

During the 1950s, oil-based fracturing fluids were replaced by water-based slickwater fluids composed of low concentrations of guar and sand, but during the late 1960s, the industry determined that greater long-term production could be achieved by pumping much larger volumes of fluid and sand, and that resulted in a widespread replacement of slickwater treatments with massive cross-linked gel treatments. However, in the late 1980s, cross-linked gel treatments were identified to cause greater than 80 % damage to fracture permeability due to the gel residue retained in the proppant pack after fracturing treatments. This recognition made many operators return to slickwater fluids during the 1990s [15, 53]. Later, slickwater treatments with little or no proppant were successfully applied [84]. The stimulated volume was gradually increased until the fracture conductivity became the productivity-limiting factor, leading to the invention of the hybrid fracturing treatment. Hydraulic operations composed of intervals of gradually increasing proppant concentrations are common practices today [53].

Design of an appropriate fracture geometry as well as placement of a sufficient proppant pack of the right proppant type is a key parameter to maintain long-term productivity. Historically sand has been the most commonly used proppant material, but as the well depths increase, the pressures increase, and thus sand is unable to resist the high pressures and crushes, generating free fines, which reduce the permeability by occupying the pore space in the pack. The urgent need to improve proppant performance has recently led to the development of new-generation proppants and propping strategies, such as the development of ultra-light-weight materials as proppant candidates, either as a single material type or as a mixture of hard and soft materials [73], to reduce settling rates and improve transport in low viscosity fluids [1, 15–17, 21, 47, 51, 73, 99], high-aspect-ratio proppants to control proppant flowback [81], and the development of strategies employing multiple modal proppant size distributions to improve propping at the fracture tip. A recent trend in the oil industry is to use cyclic pumping and variable proppant concentrations to create heterogeneous proppant deposition [87, 100].

However, the factors controlling performance of even traditional proppants in real rock fractures are still not well understood due to the complex range of behaviors. Very often, proppant selection is simplified to single parameters such as depth, stress, and crush, or the notion that in extremely tight rock any proppant is good enough. In some cases, proppant selection is just limited to what the last engineer did, or what the operator next door is using [20]. A more informed selection of design parameters, such as proppant size, shape, concentration and properties, fracture fluid viscosity, and pumping schedule is needed for the optimization of fracturing treatment to maximize hydrocarbon production and subsequent profits. Recognizing the significance of a better understanding of the behavior of fluids and proppants within a fracture and their relationship to fracture conductivity, many recent studies have been devoted to the experimental and numerical investigation on the interactions of proppant and hydraulic fractures. The goal of this chapter is to provide a summary of the investigations.

11.2 Experimental Investigation

Experimental investigations have shown that when proppant is present in the fracture, factors such as proppant concentration, size, and properties, and closure stresses have a significant influence on fracture conductivity. By the mid-1950s, “everyone” knew that a proppant pack’s flow capacity increased with fracture width, e.g., increasing numbers of proppant layers. In a 1959 turnabout, however, Darin and Huitt showed that partial monolayer proppant packs, i.e., a single layer of sparsely spread proppant particles, could achieve superior fracture conductivity in comparison with a full monolayer, i.e., a complete coverage of the fracture surface with one proppant grain thick, or multi-layer packs due to the open fracture spaces available around and between proppant particles. To measure the fracture conductivity in a laboratory setting, they flowed light hydrocarbon oil through a proppant pack of steel balls placed between two steel plates and calculated the fracture permeability by using a modified form of the Kozeny–Carman equation. The results showed that one half of a monolayer achieved similar conductivity values to a multi-layer proppant pack with a concentration higher than 4.0 lb/ft^2 [36].

On the other hand, with few exceptions field treatments designed on the basis of monolayer or partial monolayer concepts have not responded any better than conventionally designed treatments [117]. This was explained by the fact that fractures with a partial monolayer proppant were found to be more susceptible to loss in conductivity under high confining stresses [55]: in relatively soft formations, the loss in conductivity was attributed to proppant embedment in the fractured rock [57, 108]; while in hard formations, the loss in conductivity was explained in terms of crushing or excessive deformation of proppant particles under high confining stresses [58, 64]. The partial monolayer concept is therefore still just a concept. Because of the complexity of fracture, fluids, pressures, proppant, and a host of other factors that come into play, most industry experts believe that it is not possible to achieve a uniform partial monolayer throughout a propped fracture [15, 42, 49]. However, Chambers and Meise argued that it is not impossible to create partial monolayers with the use of ultra-light-weight proppants in fluids of similar specific gravity [19].

Cooke conducted laboratory experiments using a brine solution to study the permeability of a proppant pack squeezed between two steel pistons at different stress levels using Brady sand with different sizes [26]. He concluded that fracture conductivity had an inverse relationship with closure stress and also showed that gel residue could significantly reduce in situ conductivity [27]. van der Vlis et al. carried out an experimental investigation on a similar conductivity cell in an attempt to establish theoretical relationships describing fracture conductivity in terms of proppant type, size and concentration, closing pressure, and formation hardness [110]. In 1989, the America Petroleum Institute (API) standardized the procedures for measuring the conductivity of proppant in the laboratory using a Cooke conductivity cell to provide a means to compare the performance of proppants in a way that was reasonable and repeatable with in all proppant experiments, commonly known as API-RP-61 [3]. The experimental procedures in API-RP-61 include placing proppants of known concentration uniformly between two steel sheets at ambient temperature, applying closure stress at a stress rate of 100 psi per min and maintaining each closure stress for about 15 min, then pumping deionized or distilled water, measuring the differential pressure across the proppant pack and recording flow rates and calculating proppant conductivity. API-RP-61 is considered a short-term conductivity measurement procedure [3]. However, the API-RP-61 procedure does not take many realistic downhole conditions into account, and as a result, elevated temperatures, fracture fluid residues, embedment, lower proppant concentration, flow convergence in transverse fractures, and fines migration could reduce conductivity by 90 % or more [6, 91, 104, 116]. Moreover, during the production stage, the width of the propped fracture also diminishes due to the compaction of the proppant pack and the embedment of proppant particles into the fracture surface [6]. In fact, Vincent demonstrated that a 50-h test, non-Darcy effects, lower achieved fracture widths, multiphase flow, gel damage, fines migration, and cyclical stress could reduce effective conductivity by

99.9 % with low-quality sand, 99.7 % with best quality Ottawa sand, and 98.6 % with premium light-weight ceramic proppants, compared with the API-RP-61 test [112]. Since each proppant type and size is affected differently by these conditions, a more realistic fracture model is required to determine a proper estimate of realistic conductivity and subsequent proppant selection.

For long-term conductivity testing, Penny developed experimental procedures and equipment to test proppants placed between two metal shims or two Ohio sandstones [94]. The measurement conditions ranged from 3000 psi and 150°F to 10,000 psi and 300°F and common proppant concentrations were used (2 lb/ft^2). To understand the effects of fracture properties on conductivity, Fredd et al. performed a series of long-term conductivity measurements on fractured cores from the east Texas Cotton Valley sandstone formation. The results demonstrated that fracture displacement is required for surface asperities to provide residual fracture width and sufficient conductivity in the absence of proppants. However, the conductivity may vary by at least two orders of magnitude, depending on formation properties such as the degree of fracture displacement, the size and distribution of asperities, and rock mechanical properties [44]. In 2007, the International Organization of Standardization (ISO) publication ISO 13503-5 "Procedures for measuring the long-term conductivity of proppants" was elaborated as an update of API-RP-61 to measure the long-term conductivity of proppants [66]. This standard established a testing time limit of 50 h as the time when the conductivity reaches a semi-steady state value, replaced the stainless steel shims with sandstone cores, increased the testing temperature to a value more representative of the reservoir formation, and recommended an oxygen-free silica-saturated testing fluid.

With the standard testing procedures, modifications of the testing apparatus, experiment conditions, and operation procedures have been made by various investigators. The most common modification is to use a modified API conductivity cell to accommodate thicker samples to account for fluid leakoff through the sample during the experiment, or instead of flowing 2 % KCl through the fracture recommended by ISO standards, the modified tests can flow dry gas, wet gas, fresh water, brine of various concentrations, and multi-phase flow for different purposes. A modified API conductivity cell of 10 in. long, 3.25 in. wide and 8 in. in height has been used in fracture conductivity studies of gel damage problems in tight sand gas wells [90]. The cell body, designed to accommodate core samples 7 in. long, 1.65 in. wide and 3 in. in height, allows leakoff through the samples to simulate real reservoir conditions. They examined the filter cake formation by pumping the fracture fluid through the conductivity cell, allowing leakoff to build the filter cake, measuring the cake thickness, and flowing gas through the cell to simulate the cleanup process. The results showed that the yield stress of the residual gel plays a critical role in gel cleanup. Very often, in conductivity tests with shale, cylindrical samples such as 1–1/2 in. or 1 in. diameter drilled cores are used with fracture surfaces either saw cut or Brazil fractured [96]. Morales et al. performed fracture conductivity measurements on Barnett shale samples with fractured surfaces by using tap water in propped and unpropped fractures and under varying conditions of stresses and temperatures. They studied the effects of asperity, creep, and variations in proppant mesh sizes and temperature [86]. Guzek performed fracture conductivity measurements on Eagle Ford shale samples with fractured surfaces by applying dry nitrogen gas as a working fluid instead of water or brine to simulate natural gas. Proppant concentrations of 0.1 lb/ft^2 and 0.2 lb/ft^2 were used instead of the standard 2 lb/ft2 as specified by API-RP-61 [53].

The partial monolayer technique was examined by Densirimongkol with particular emphasis on the impact of acid in possibly improving fracture conductivity of carbonate rocks, which is often referred to as "closed fracture acidizing" [38]. After obtaining a partial monolayer distribution on the fracture face, gelled acid was injected through the fracture face. Fracture conductivity before and after acid injection was evaluated. The results clearly demonstrated that acid injection does not enhance fracture conductivity of partial monolayer proppant fracturing, and the more the acid is injected, the more rapidly fracture conductivity decreases.

Most of the experimental studies on fracture conductivity were carried out using a static conductivity cell. During the experiment, the proppant was first placed into the conductivity cell and then the fracture fluid was pumped through the cell. This procedure underestimates the ability of the fluid to carry the proppant and merely evaluates the damage caused by the gel in the proppant pack. Dynamic fracture conductivity tests have been successfully designed to simulate realistic proppant placement [4, 18, 80]. In dynamic proppant placement, slurry is pumped by a multi-stage centrifugal pump and flows through a created fracture with fixed width in the conductivity cell [80]. After the pump is shut down, the fracturing fluid trapped in the fracture leaks off through the rocks leaving the proppants inside the fracture. This new approach simulates the process of the fracturing fluid and proppant mixing and pumping, and examines the combination of parameters such as polymer and breaker concentrations, among others, affecting the fracture conductivity.

A new formation damage mechanism, called mechanically induced fracture face skin, was proposed and investigated by Reinicke et al. [97, 98]. They proposed that the mechanically induced fracture face skin results from the interaction between proppants and rock under non-isostatic stress conditions: differential stress may cause crushing of grains and fines production at the fracture face and the embedment of proppants into the rock matrix, and the produced fine particles block the pores at the fracture face, resulting in a reduced permeability and an additional pressure drop normal to the fracture face. To prove the mechanical destruction at the rock-proppant interface and associated permeability reduction due to fines generation, they developed two different flow cells to separate the permeability evolution with increasing stress of the reservoir rock, the fracture face skin, and the propped fracture: (1) the acoustic emission flow cell to analyze the crushing and damage at the rock-proppant interface; and (2) the bi-directional flow cell to simulate the geometric flow conditions in reservoirs intersected by a proppant filled fracture. A schematic view of both flow cells is given in Fig. 11.1. They performed permeability measurements on tight Flechtingen sandstones with propped fractures under stress. The results revealed a permeability reduction down to 77 % of initial rock permeability at 50 MPa differential stress leading to a permeability reduction

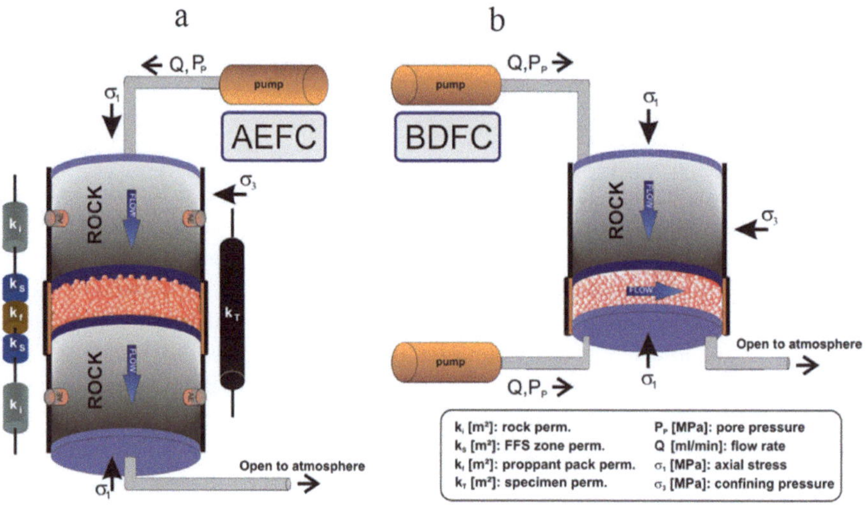

Fig. 11.1 Two different flow cells were developed by Reinicke et al. to separate the permeability evolution with increasing stress of the reservoir rock, the fracture face skin, and the propped fracture: (**a**) the acoustic emission flow cell (AEFC); and (**b**) the bi-directional flow cell (BDFC) [98]

in the fracture face skin zone up to a factor of 6. Serious mechanical damage at the rock-proppant interface was observed under stress as low as 5 MPa. Microstructure analysis identified quartz grain crushing, fines production, and pore space blocking at the fracture face leading to the observed mechanically induced fracture face skin [97].

11.3 Theoretical Study and Numerical Modeling

Over the past 60 years, many theoretical and numerical studies have been devoted to estimating the conditions for fracture initiation and propagation in rocks [2, 42, 79, 95, 115, 118]. These models are used to predict the geometry of the hydraulic fracture for particular fracturing treatment conditions or to identify the fracturing conditions under which an optimum fracture geometry could be achieved. However, the increase in reservoir production rate due to hydraulic fracture ultimately depends on the performance of the proppant pack, which controls the length, opening, and conductivity of a fracture during the production stage. Past research in this aspect has been largely focused on proppant transport modeling and investigations of the effects of proppant settling [5, 22–24, 35, 40, 46, 67, 85, 89, 105, 109]. Surprisingly, numerical simulation models to study the behavior of fluid and proppant within a fracture, which are directly linked to fracture conductivity, have received much less attention.

The shape and structure of hydraulically driven fractures are very complex [95], and thus any approach to the description and evaluation of the hydraulically driven fractures will require some radical simplifications. Papanastasiou and Dam et al. utilized oversimplified models to assess the residual fracture profiles, merely neglecting the proppant compressibility and its distribution inside the fracture [34, 93]. A recent study by Kotousov deals with the final opening of two semi-infinite planes compressing a rigid circular inclusion of linear elastic behavior [71]. This work is of great practical interest in situations where the hydraulic fracture length is not known a priori and it is supported by individual particles, but not an ideal solution to simulate the residual opening of a hydraulic fracture supported by a pack of proppant with nonlinear behavior. Bortolan Neto et al. then developed a nonlinear mathematical model capable of predicting the residual opening of a hydraulic fracture taking into account both the proppant distribution and its compressibility [11]. The developed model can also be used to understand, investigate, and describe the stress state around the fracture due to the residual opening. The approach is based on the Distribution Dislocation technique [8–11] and the Gauss–Chebyshev quadrature. The fracture is considered to be a two-dimensional centered straight crack subjected to plane strain conditions, and the surrounding medium is assumed to be impermeable, isotropic, homogeneous, and linearly elastic, as illustrated in Fig. 11.2. The combined behavior of non-consolidated proppant particles is described by the model developed by Bortolan Neto et al. [12, 13], in which the particles are assumed to be of spherical shape, having a small contact area and the deformation of the particles follows the classical Hertz contact theory [56]. Bortolan Neto et al. later considered a more general problem of a fracture partially filled with a loose granular assembly of proppant particles [14]. The fracture is subjected to confining stresses, which result in the closure of the unpropped fracture segments as well as rearrangement of proppant particles in the pack, leading to a reduction in fracture opening and conductivity. A simple fluid flow model is utilized to quantify the effect of confining stress and partial filling of the fracture on the productivity index of the well.

Khanna et al. developed a simplified approach for calculating the conductivity of narrow fractures filled with a sparse monolayer of proppant particles [68]. The proppant particles are modeled as rigid spheres and the deformation of the fracture faces is assumed to be purely elastic. Hertz contact theory and the principle of superposition are utilized to obtain the fracture opening profile as a function of the proppant concentration and the value of confining stress. The conductivity of the deformed fracture channel is determined by using computational fluid dynamics.

Compared with continuum methods, which require complex constitutive models with many variables to describe the behavior of quasi-brittle materials and do not address successfully the local damage modes such as shear planes and particle fragmentation [119], discrete element method (DEM) has more advantages in modeling rock-proppant system [37]. The DEM model was originally introduced by Cundall and Strack about 30 years ago [29–33], and since then it has become a promising approach to study rock mechanics tests and fracture processes. It has been widely used by the geotechnical engineering community to model the

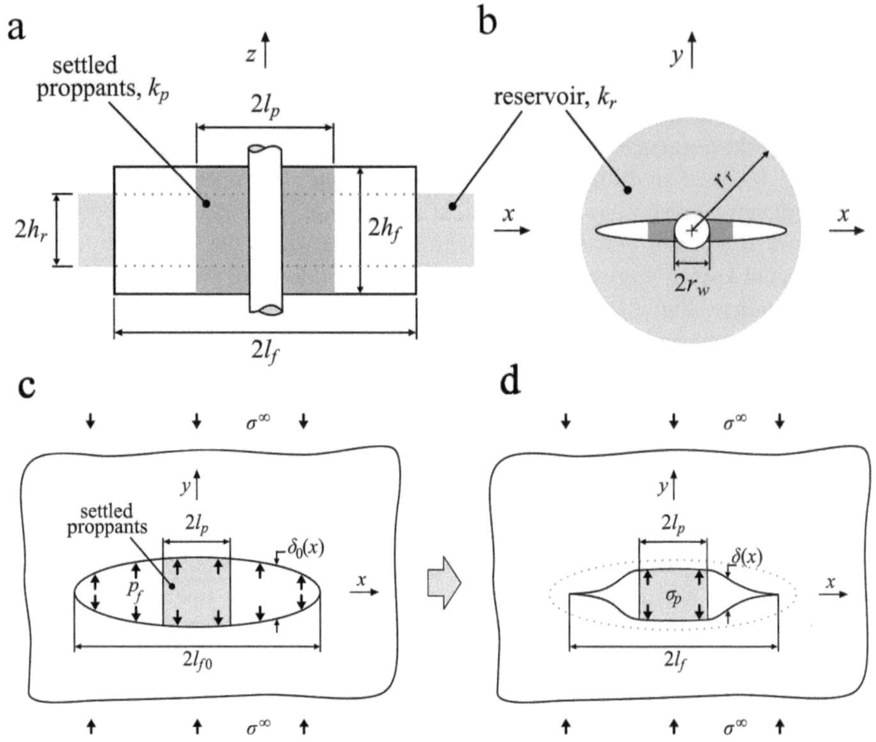

Fig. 11.2 Schematic diagram of a partially propped hydraulic fracture considered in the problem formulation: (**a**) cross-sectional view; and (**b**) plan view. The 2D approximation of a hydraulic fracture partially filled with proppant and subject to remote confining stress: (**c**) The initial opening and length of the fracture due to a uniform internal pressure; and (**d**) the residual opening and length of the fracture once the stimulating fluid pressure is removed [14]

mechanical deformation and fracturing of polycrystalline rocks at various scales, ranging from grain-scale microcracks to large-scale faults associated with earthquakes [59–62]. DEM treats the material as an assembly of separate blocks or particles. According to Cundall and Hart, DEM is any modeling technique that (1) allows finite displacements and rotations of discrete bodies, including complete detachment; and (2) recognizes new contacts automatically as the simulation progresses [31]. DEM was originally developed to efficiently treat solids having pre-existing discontinuities with spacing comparable to the scale of interest of the problem, such as blocky rock masses, ice plates, masonry structures, and flow of granular materials. DEM can be classified according to several different criteria, such as the type of contact between bodies, the representation of deformability of solid bodies, the methodology for detection and revision of contacts, and the adopted solution algorithm [63]. Main advantages of DEM include the simple mathematical treatment of the problem, whereby complex constitutive relationships are replaced by simple particle contact logic, and the natural predisposition of the approach to account for material heterogeneity [70]. The major disadvantage of

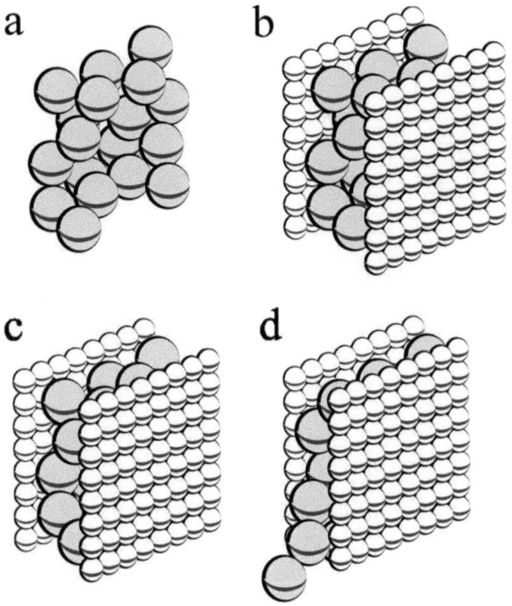

Fig. 11.3 The simulation procedure begins with (**a**) the generation of a proppant pack, (**b**) addition of fracture walls, (**c**) compaction, and (**d**) application of fluid flow and proppant production [106]

DEM is the computation intensity of the simulation, since computation time grows as a function of particle count, which can be quite large for fracture–proppant models. Comprehensive reviews of DEM have recently been provided by Bobet et al. [7] and Lisjak and Grasselli [76].

To study proppant flowback under confining pressures, Shor and Sharma developed a simple, scalable, 3D model using DEM particle simulator to simulate representative cubic volumes consisting of fracture openings, fracture walls, and the confining formation, as illustrated in Fig. 11.3 [106, 107]. Fluid flow exerts a point drag force on proppant grains and an open boundary allows proppant to flow out of the fracture and be produced. The base model is built upon an open source granular simulator called LAMMPS Improved for General Granular and Granular Heat Transfer Simulations, and described in detail by Kloss and Goniva [69]. The effects of fracture width, confining stress, fluid flow velocity, and proppant cohesion are studied for a variety of conditions. Fracture width is found to be dependent on confining stress and fluid flow velocity while proppant production is also dependent on cohesion. Three regimes are observed, with complete fracture evacuation occurring at high flow rates and low confining stresses, fully packed fractures occurring at high confining stresses, and open but mostly evacuated fractures occurring in between. This study is important for the evaluation of proppant back stability and flowback.

To evaluate the fracture aperture under different proppant sizes, Young's moduli and pressure levels, Deng et al. proposed and deployed a 3D DEM model for the simulation of shale–proppant interaction in hydraulic fracturing [37]. The shale matrix is represented by bonded discrete spherical particles with diameters following a uniform distribution within the range of 0.16–0.24 mm. A pre-existing

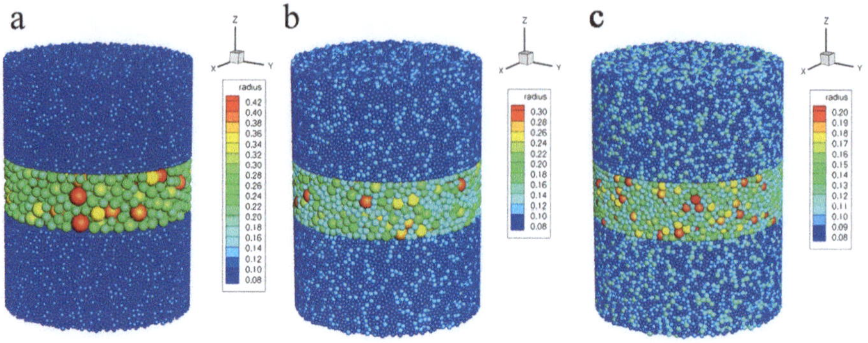

Fig. 11.4 Three particle packages for different proppant sizes [37]. (**a**) 20/40, (**b**) 30/60, and (**c**) 40/70

hydraulic fracture with a given average aperture size of 2.54 mm is used with proppant particles filled in, corresponding to a 1 lbs/ft^2 proppant load, as shown in Fig. 11.4. The proppants are represented by the particles with relatively larger size. The shale particle is treated as a particle with a cement layer, whereas proppant particle is assumed to be a particle without cement. Furthermore, they extended grain's interaction and cement bond from elasticity to elasto-plasticity, and thus the model can directly be applied to high stress and high temperature cases. The velocity Verlet method [50] is implemented to substitute the traditional central time integration scheme [59, 60]. The results reveal that, the more soft shale particle, the higher pressure and the larger proppant size imply smaller crack aperture and larger plastic zone.

Recent studies by the StimLab proppant consortium have indicated that the flow field across a Cooke proppant conductivity testing cell may not be uniform as initially believed, resulting in significantly different conductivity results [25]. To evaluate these experimental findings, a physics-based 2D DEM model was developed and applied to simulate the stress distribution in the Cooke cell and proppant rearrangement during conductivity testing as a function of stress [82]. The model was constructed to represent a realistic cross-section of the Cooke cell with a distribution of four material properties, three that represent the Cooke cell (steel, sandstone, square rings), and one representing the proppant. The model domain with approximately 15,000 particles was used to simulate tests in the Cooke cell. Only the top half of the Cooke cell was simulated since there would be stress symmetry along the midpoint of the proppant pack, as illustrated in Fig. 11.5. The experimental cell is represented as a cohesive body composed of a large number of discrete elements, and proppants can be modeled as the individual discrete particles with various sizes (following the proppant size distribution density function used in the test) that exhibit no cohesive strength between the particles. The results suggest that proppant rearrangement and nonuniform stress distribution can develop across the proppant pack due to square ring modifications. Compaction along the edge of the proppant pack beneath the square ring seal results in a disproportionate lower flow field along these edges as compared to the middle of the proppant pack.

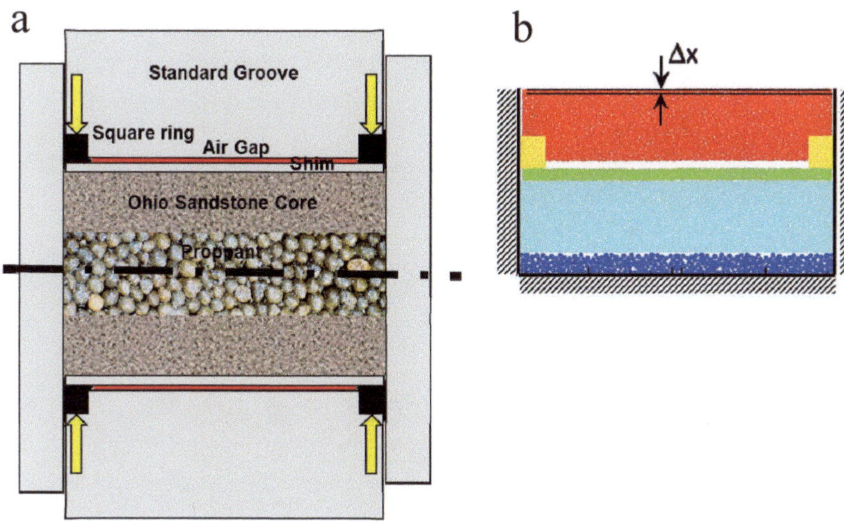

Fig. 11.5 Conceptual cross-section of (**a**) standard Cooke cell and (**b**) DEM model cross-section used in simulations. *Red* is stainless steel, *yellow* is Viton rubber, *white* is the air gap, *green* is stainless steel, *light blue* is sandstone, and *dark blue* is the proppant [82]

For granules made of highly deformable material, having a complex nonlinear relationship, the overall response of the particle bed will significantly differ if the particles are treated as linearly elastic as in the case of DEM approach. Moreover, particle angularity also gives significantly different results compared with the simulations in which the particles are modeled as spheres. Munjiza proposed the combined FEM/DEM approach where the particles are modeled as a continuum while the contact detections are carried out based on the DEM code [88]. In this approach like the discrete approach each particle is modeled individually, but unlike the DEM approach particle deformation is captured by meshing it with finite elements. Because the particles are modeled as a continuum it is possible to introduce large deformation, and fracture in the particles. This method has been applied extensively to model powder compaction [45, 48, 75].

A 2D FEM/DEM model has been recently developed to model hard and soft proppant mixtures by Kulkarni and Ochoa to better understand the mechanics governing a proppant pack and help develop a virtual parametric test bed capability to screen proppant samples [72, 73]. The model has been developed in the software ABAQUS using the explicit time integration scheme. Various mixtures of hard and soft particles are investigated as a function of shape, size, and inter-particle friction, as shown in Fig. 11.6. Single particle compression tests on individual particles have been carried out to obtain mechanical properties, and compression tests on proppant mixtures have been performed according to industry standards. The load vs displacement response of the computational model has been compared to the experimental results. The results clearly demonstrated that the rock material description, proppant mixture composition, and inter-particle friction have significant impact on both the flowback and compressive response.

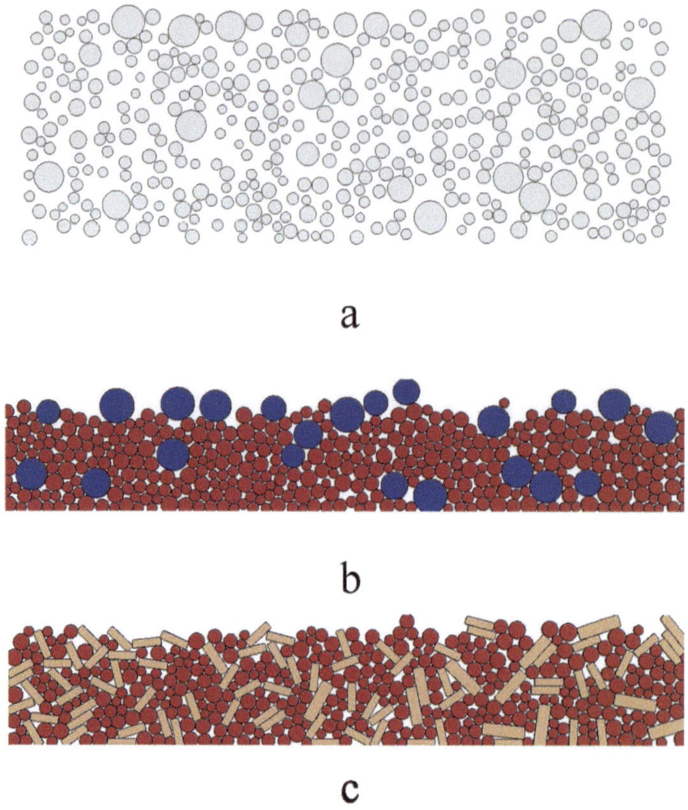

Fig. 11.6 Type-A mixture consists of 10 % walnut shell by weight and ceramic particles: (**a**) Randomly generated polydisperse particles; and (**b**) particles at the end of free fall. Type-B mixture consists of prismatic 99 % pure aluminum and ceramic particles: (**c**) Particle configuration after free fall for 25 % pure Al by weight [72]

11.4 Discussion

Currently, according to industrial practices, proppant characterization involves expensive long-term conductivity tests, and the tests usually do not yield better understanding of the mechanics governing the proppant pack response. Theoretical and numerical investigations are urgently needed to successfully predict the response of proppant packs, facilitate the screening of proppant samples, and reduce the current dependency on conductivity tests.

However, based on the literature review, it can be seen that there has been a lack of comprehensive and robust formulation equipped with realistic description of shale's heterogeneity and anisotropy; complex fracture geometries; physically sound constitutive equations; fully coupled fluid-granular approach; as well as extensive calibration and validation against a large variety of experimental data.

To formulate a comprehensive and robust computational framework to establish the fundamental physical principles that govern the shale–proppant interactions, the following aspects should be taken into account.

- It is hypothesized that properly accounting for the heterogeneity and anisotropy of shale is critical for the investigation of mechanical interactions of shale–proppant system. In particular, the laminations and bedding planes will significantly affect shale's elastic moduli, shear strength, permeability, and crack network creation, since cracks always initiate and propagate along possible planes of weakness. Therefore, the shale model should include realistic description of internal structures and mechanical properties. The effect of different bedding plane directions on proppant embedment should be investigated.
- Although traditional ceramic proppants are spherical, it is hypothesized that the shape of the proppant will have a significant impact on how it packs with other proppant particles and the surrounding area, and thus the shape of the proppant can significantly alter the permeability and conductivity of a proppant pack in a fracture. Moreover, different shapes of the same material offer different strengths and resistance to closure stress. It is desirable to engineer the shape of the proppant to provide high strength and a packing tendency that will increase oil/gas flow. The optimum shape may differ for different depths, closure stresses, and materials to be extracted. Therefore, the proppant model should be easily extended to include the particle polydispersity, i.e., size, shape, and material type variation in the pack.
- In most of the existing models, the rock and the proppants are simply considered as elastic-plastic. It is hypothesized that nonlinear and inelastic behavior, such as fracture and cohesion due to tension and tension-shear, friction due to compression-shear, and compaction and pore collapse from compression, is of critical importance for an accurate prediction of the pack response. Therefore, these physical mechanisms should be incorporated in the constitutive equations for both the shale and the ceramic proppants.
- In almost all the existing models, the proppant pack is squeezed between two rigid or flat surfaces. It is hypothesized that shale's natural fracture topology and surface deformation due to particle indentation plays a significant role in fracture conductivity measurements. Therefore, the numerical simulation system should include realistic material properties and a reconstruction of natural fracture surfaces preserving the fractures' key topological features.
- Fluid behavior abstracted as point drag forces has been applied successfully in the past [107]. However, it is hypothesized that as particle concentration increases, as is the case in proppant packs, the hydraulic interaction in the shale–proppant system will be significantly different if the point drag force model is replaced with a fully coupled fluid–granular model. Therefore, a fully coupled fluid–granular approach should be applied to accurately model the interaction of the proppants and fracture fluid within a fracture.

According to the hypotheses, existing modeling approaches do not have the predictive capabilities required to determine a proper estimate of realistic fracture conductivity and subsequent proppant selection. An improved simulation package based on those hypotheses will significantly reduce the uncertainties and risks related to the design and operation of hydraulic fracturing by providing a fundamental understanding of the hydraulic and mechanical interactions in the shale–proppant system. Such a computational framework will also enable the development of new experimental methods as well as improve current laboratory tests by providing a priori insight.

11.5 Concluding Remarks

Hydraulic fracturing is an essential stimulation method. The operation cost for a hydraulic fractured well can reach millions of dollars and the benefits from better understanding and controlling this technology are obvious. It is of utmost importance to understand the mechanisms of hydraulic fracturing with complex geological structures and stress conditions. A key design parameter in well stimulation is fracture conductivity, which relates the capacity of the fracture to transmit fluids into the wellbore to the ability of the formation to deliver fluid into the fracture [42]. Selecting the appropriate size, shape, concentration, and properties of proppant for the fracture is critical for the success of a hydraulic fracturing treatment. The limitation in the knowledge on the behavior of fluid and proppant within a fracture and their relationship to fracture conductivity in complex geological setting has restricted the invention and application of innovative proppants and propping strategies. Based on the literature review as above, some problems that still need to be further studied. For example, a fully coupled fluid–granular approach should be applied to more accurately model the interaction of the proppant grains and fracture fluid within a fracture. Fluid abstracted as point drag forces has been applied successfully in the past [107], but faces limitations as particle concentration increase, as is the case in proppant packs. Moreover, with the standard testing procedures API-RP-61 and ISO 13503-5, modifications of the testing apparatus, experiment conditions, and operation procedures have been made by various investigators. Numerical simulations of the testing apparatus are critical to understanding the impact of modification to the testing cell as well as understanding the key proppant conductivity issues. Computational models also offer us the opportunity to develop new experimental methods as well as improve the current laboratory tests by providing a priori insights.

Acknowledgements This work is supported by start-up funds provided by the Department of Mechanical Engineering at State University of New York at Binghamton.

References

1. Aboud, R. S., & Melo, R. C. B. (2007). Past technologies emerge due to lightweight proppant technology: Case histories applied on mature fields. SPE 107184.
2. Adachi, J. I., Siebrits, E., Peirce, A., & Desroches, J. (2007). Computer simulation of hydraulic fractures. *International Journal of Rock Mechanics and Mining Sciences, 44*, 739–757.
3. American Petroleum Institute. (1989). Recommended practices for evaluating short term proppant pack conductivity. API-RP-61.
4. Awoleke, O., Romero, J., Zhu, D., & Hill, A. D. (2012). Experimental investigation of propped fracture conductivity in tight gas reservoirs using factorial design. SPE 151963.
5. Barree, R. D., & Conway, M. W. (1994). Experimental and numerical modeling of convective proppant transport. SPE 28564.
6. Barree, R. D., Cox, S. A., Barree, V. L., & Conway, M. W. (2003). Realistic assessment of proppant pack conductivity for material selection. SPE 84306.
7. Bobet, A., Fakhimi, A., Johnson, S., Morris, J., Tonon, F., & Yeung, M. R. (2009). Numerical models in discontinuous media: Review of advances for rock mechanics applications. *Journal of Geotechnical and Geoenvironmental Engineering ASCE, 135*(11), 1547–1561.
8. Bortolan, N. L., & Khanna, A. (2013). The performance of hydraulic fractures partially filled with compressible proppant. *Australian Journal of Structural Engineering, 10*, 185–197.
9. Bortolan, N. L., & Kotousov, A. (2012). Residual opening of hydraulically stimulated fractures filled with granular particles. *Journal of Petroleum Science and Engineering, 100*, 24–29.
10. Bortolan, N. L., & Kotousov, A. (2013). On the residual opening of hydraulic fractures. *International Journal of Fracture, 181*, 127–137.
11. Bortolan, N. L., & Kotousov, A. (2013). Residual opening of hydraulic fractures filled with compressible proppant. *International Journal of Rock Mechanics and Mining Sciences, 61*, 223–230.
12. Bortolan, N. L., Kotousov, A., & Bedrikovetsky, P. (2011). Application of contact theory to evaluation of elastic properties of low consolidated porous media. *International Journal of Fracture, 168*(2), 267–276.
13. Bortolan, N. L, Kotousov, A., & Bedrikovetsky, P. (2011). Elastic properties of porous media in the vicinity of the percolation limit. *Journal of Petroleum Science and Engineering, 78*(2), 328–333.
14. Bortolan, N. L., Khanna, A., & Kotousov, A. (2015). Conductivity and performance of hydraulic fractures partially filled with compressible proppant packs. *International Journal of Rock Mechanics and Mining Sciences, 74*, 1–9.
15. Brannon, H. D., Malone, M. R., Richards, A. R., Wood, W. D., Edgeman, J. R., & Bryant, J. L. (2004) Maximizing fracture conductivity with proppant partial monolayers: Theoretical curiosity or highly productive reality? SPE 90698.
16. Brannon, H. D., & Starks, T.R. (2009). Maximizing return-on-fracturing-investment by using ultra-lightweight proppants to optimize effective fracture area: Can less be more? SPE 119385.
17. Card, R. J., Howard, P. R., & Feraud, J. P. (1995). A novel technology to control proppant back production. SPE 31007.
18. Castro, J. C. C. (2011). *Evaluation and effect of fracturing fluids on fracture conductivity in tight gas reservoirs using dynamic fracture conductivity test*. Master Thesis, Department of Petroleum Engineering, Texas A&M University.
19. Chambers, R. W., & Meise, K. D. (2005). Comparison of fracture geometries utilizing ultralightweight proppants provide evidence that partial monolayer can be created: A case history. SPE 96818.

20. Chapman, M., & Palisch, T. (2014). Fracture conductivity–Design considerations and benefits in unconventional reservoirs. *Journal of Petroleum Science and Engineering, 124*, 407–415.
21. Chen, T., Wang, Y., Yan, C., Wang, H., Xu, Y., & Ma, R. (2015). Preparation of heat resisting poly(methyl methacrylate)/graphite composite microspheres used as ultra-lightweight proppants. *Journal of Applied Polymer Science, 132*, 41924.
22. Clark, P. E., & Guler, N. (1983). Prop transport in vertical fractures: Settling velocity correlations. SPE 11636.
23. Clark, P. E., & Quadir, J. A. (1981). Prop transport in hydraulic fractures: A critical review of particle settling velocity equations. SPE 9866.
24. Clifton, R. J., & Wang, J.-J. (1988). Multiple fluids, proppant transport, and thermal effects in three-dimensional simulation of hydraulic fracturing. SPE 18198.
25. Conway, M. W., & O'Connell, L. (2013). Conductivity variations between piston designs. *Stim-Lab Proppant Consortium Meeting*, Rohnert Park, CA, July 18–19.
26. Cooke, Jr. C. E. (1973). Conductivity of fracture proppants in multiple layers. *Journal of Petroleum Technology, 25*, 1101–1107.
27. Cooke, Jr. C. E. (1975). Effect of fracturing fluids on fracture conductivity. *Journal of Petroleum Technology, 27*, 1273–1282.
28. Craddock, D. L., Goza, B. T., & Bishop, J. C. (1983). A case history–fracturing the morrow in southern Blaine and western Canadian counties, Oklahoma. SPE 11567.
29. Cundall, P. A. (1988). Formulation of a 3-dimensional distinct element model. *International Journal of Rock Mechanics and Mining Sciences, 25*, 107–116.
30. Cundall, P. A. (1990). Numerical modeling of jointed and faulted rock. In *Proceedings of the international conference on mechanics of jointed and faulted rock* (pp. 11–18).
31. Cundall, P.A., & Hart, D. H. (1992). Numerical modeling of discontinua. *Engineering with Computers, 9*, 13.
32. Cundall, P. A., & Strack, O. D. L. (1979). Discrete numerical model for granular assemblies. *Geotechnique, 29*, 47–65.
33. Cundall, P. A., & Strack, O. D. L. (1980). A discrete numerical model for granular assemblies—reply. *Geotechnique, 30*, 335–336.
34. Dam, D. B., Pater, C. J., & Romijn, R. (2000). Analysis of hydraulic fracture closure in laboratory experiments. SPE 65066.
35. Daneshy, A. A. (1978). Numerical solution of sand transport in hydraulic fracturing. SPE 5636.
36. Darin, S. R., & Huitt, J. L. (1959). Effect of a partial monolayer of propping agent on fracture flow capacity. SPE 1291-G.
37. Deng, S, Li, H., Ma, G., Huang, H., & Li, X. (2014). Simulation of shale–proppant interaction in hydraulic fracturing by the discrete element method. *International Journal of Rock Mechanics and Mining Sciences, 70*, 219–228.
38. Densirimongkol, J. (2009). *The role of acidizing in proppant fracturing in carbonate reservoirs*. Master Thesis, Department of Petroleum Engineering, Texas A&M University.
39. Donaldson, E. C., Alam, W., & Begum, N. (2013). *Hydraulic fracturing explained: Evaluation, implementation, and challenges*. Houston: Gulf Publishing.
40. Dontsov, E. V., & Peirce, A. P. (2015). Proppant transport in hydraulic fracturing: Crack tip screen-out in KGD and P3D models. *International Journal of Solids and Structures, 63*, 206–218.
41. Economides, M. J., & Martin, T. (2007). *Modern fracturing—enhancing natural gas production*. Houston: BJ Services Company.
42. Economides, M. J., & Nolte, K. G. (2000). *Reservoir stimulation* (3rd ed.). New York: Wiley.
43. Fjaer, E., Holt, R. M., Horsrud, P., Raaen, A. M., & Risnes, R. (2008). *Petroleum related rock mechanics* (2nd ed.). Amsterdam: Elsevier.

44. Fredd, C. N., McConnell, S. B., Boney, C. L., & England, K. W. (2001). Experimental study of fracture conductivity for water-fracturing and conventional fracturing applications. SPE 74138.
45. Freening, G. (2007). Analysis of pharmaceutical powder compaction using multiplicative hyperelasto-plastic theory. *Powder Technology, 172*, 103–112.
46. Gadde, P. B., Liu, Y., Norman, J., Bonnecaze, R., & Sharma, M.M. (2004). Modeling proppant settling in water-fracs. SPE 89875.
47. Gaurav, A., Dao, E. K., & Mohanty, K. K. (2012). Evaluation of ultra-light-weight proppants for shale fracturing. *Journal of Petroleum Science and Engineering, 92–93*, 82–88.
48. Gethin, D. T., Ransing, R. S., Lewis, R. W., Dutko, M., & Crook, A. J. L. (2001). Numerical comparison of a deformable discrete element model and an equivalent continuum analysis for the compaction of ductile porous material. *Computers & Structures, 79*, 1287–1294.
49. Gidley, J. L., Holditch, S. A., Nierode, D. E., & Veatch, R. W. (1989). Recent advances in hydraulic fracturing. SPE Monograph (Vol. 12).
50. Griebel, M., Knapek, S., & Zumbusch, G. (2007). *Numerical simulation of molecular dynamics*. Berlin: Springer.
51. Gu, M., Dao, E., & Mohanty, K. K. (2015). Investigation of ultra-light weight proppant application in shale fracturing. *Fuel, 150*, 191–201.
52. Guo, B. Y., Lyons, W. C., & Ghalambor, A. (2007). *Hydraulic fracturing. In Petroleum production engineering: A computer-assisted approach*. Houston: Gulf Publishing.
53. Guzek, J. J. (2014). *Fracture conductivity of the Eagle Ford shale*. Master Thesis, Department of Petroleum Engineering, Texas A&M University.
54. Haimson, B. (2004). Hydraulic fracturing and rock characterization. *International Journal of Rock Mechanics and Mining Sciences, 41*, 188–194.
55. Harrington, L., & Hannah, R. R. (1975). Fracturing design using perfect support fluids for selected fracture proppant concentrations in vertical fractures. SPE 5642.
56. Hertz, H. (1896). In D. E. Jones & G. A. Schott (Eds.), *Miscellaneous papers*. New York: MacMillan.
57. Huitt, J. L., & McGlothlin, B. B. (1958). The propping of fractures in formations susceptible to propping-sand embedment. API-58-115.
58. Huitt, J. L., McGlothlin, B. B., & McDonald, J. F. (1959). The propping of fractures in formations in which propping sand crushes. API-59-120.
59. Itasca Consulting Group Inc. (2008). *PFC2D particle flow code in 2 dimensions user's guide*, Minneapolis, MS, USA.
60. Itasca Consulting Group Inc. (2008). *PFC3D particle flow code in 3 dimensions user's guide*, Minneapolis, MS, USA.
61. Itasca Consulting Group Inc. (2011). *UDEC universal distinct element code user's guide*, Minneapolis, MS, USA.
62. Itasca Consulting Group Inc. (2013). *3DEC 3 dimensional distinct element code user's guide*, Minneapolis, MS, USA.
63. Jing, L., & Stephansson, O. (2007). *Fundamentals of discrete element methods for rock engineering: Theory and applications*. Amsterdam/Oxford: Elsevier.
64. Kamenov, A., Zhu, D., Hill, A. D., & Zhang, J. (2013). Laboratory measurement of hydraulic fracture conductivities in the Barnett shale. SPE 163839.
65. Kasza, P., & Wilk, K. (2012). Completion of shale gas formations by hydraulic fracturing. *Przemysl Chemiczny, 91*, 608–612.
66. Kaufman, P. B., Anderson, R. W., Ziegler, M., Neves, A. R., Parker, M. A., Abney, K., et al. (2007). Introducing new API/ISO procedures for proppant testing. SPE 110697.
67. Kern, L. R., Perkins, T. K., & Wyant, R. E. (1959). The mechanics of sand movement in fracturing. *Journal of Petroleum Technology, 11*, 55–57.
68. Khanna, A., Kotousov, A., Sobey, J., & Weller, P. (2012). Conductivity of narrow fractures filled with a proppant monolayer. *Journal of Petroleum Science and Engineering, 100*, 9–13.

69. Kloss, C., & Goniva, C. (2011). LIGGGHTS–Open source discrete element simulations of granular materials based on LAMMPS. *Annual Meeting and Exhibition* (Vol. 2); Minerals, Metals and Materials Society.
70. Komodromos, P., & Williams, J. R. (2001). *Development and implementation of a combined discrete and finite element multibody dynamics simulation environment*. Ph.D Dissertation, Massachusetts Institute of Technology.
71. Kotousov, A., Bortolan, N. L., & Rahman, S. S. (2011). Theoretical model for roughness induced opening of cracks subjected to compression and shear loading. *International Journal of Fracture, 172*(1), 9–18.
72. Kulkarni, M. C. (2012). *Mechanics of light weight proppants: A discrete approach*. Ph.D. Dissertation, Department of Mechanical Engineering, Texas A&M University.
73. Kulkarni, M. C., & Ochoa, O. O. (2012). Mechanics of light weight proppants: A discrete approach. *Composites Science and Technology, 72*, 879–885.
74. Kundert, D. P., & Smink, D. E. (1979). Improved stimulation of the Escondido sandstone. SPE 7912.
75. Lewis, R.W., Gethin, D. T., Yang, X. S., & Rowe, R. C. (2005). A combined finite-discrete element method for simulating pharmaceutical powder tableting. *International Journal for Numerical Methods in Engineering, 62*, 853–869.
76. Lisjak, A., & Grasselli, G. (2014). A review of discrete modeling techniques for fracturing processes in discontinuous rock masses. *Journal of Rock Mechanics and Geotechnical Engineering, 6*, 301–314.
77. Mack, D. J., & Myers, R. R. (2001). Proppants: Is bigger better or is placement the key? SPE 72381.
78. Mader, D. (1989). *Hydraulic proppant fracturing and gravel packing*. New York: Elsevier Science Publishing Company Inc.
79. Mahrer, K. D. (1999). A review and perspective on far-field hydraulic fracture geometry studies. *Journal of Petroleum Science and Engineering, 24*, 13–28.
80. Marpaung, F., Chen, F., Pongthunya, P., Zhu, D., & Hill, A. D. (2008). Measurement of gel cleanup in a propped fracture with dynamic fracture conductivity experiments. SPE 115653.
81. Mathur, A. K., Semary, M., Edelman, J., Maghrabia, K., Rojas, J. M. B., & Zaki, A. S. (2013). Rod-shaped proppant provides superior proppant flowback control in the Egyptian Eastern Desert. SPE 164014.
82. Mattson, E. D., Huang, H., Conway, M., & O'Connell, L. (2014). Discrete element modeling results of proppant rearrangement in the Cooke conductivity cell. SPE 168604.
83. Mayerhofer, M. J., & Meehan, D. N. (1998). Waterfracs–Results from 50 Cotton Valley wells. SPE 49104.
84. Mayerhofer, M. J., Richardson, M. F., Walker, R. N., Meehan, D. N., Oehler, M.W., & Browning, Jr. R. R. (1997). Proppants? We don't need No proppants. SPE 38611.
85. Mobbs, A. T., & Hammond, P. S. (2001). Computer simulations of proppant transport in a hydraulic fracture. SPE 69212.
86. Morales, R. H., Suarez-Rivera, R., & Edelman, E. (2011). Experimental evaluation of hydraulic fracture impairment in shale reservoirs. ARMA-11-380.
87. Morris, J. P., Chugunov, N., & Meouchy, G. (2014) Understanding heterogeneously propped hydraulic fractures through combined fluid mechanics, geomechanics, and statistical analysis. ARMA 14-7408.
88. Munjiza, A. (2004). *The combined finite-discrete element method*. West Sussex: Wiley.
89. Novotny, E. J. (1977). Proppant transport. SPE 6813.
90. Ouyang, L., Yango, T., Zhu, D., & Hill, A. D. (2011). Theoretical and experimental modeling of residual gel filter cake displacement in propped fractures. SPE 147692.
91. Palisch, T. T., Duenckel, R., Bazan, L., Heidt, H., & Turk, G. (2007). Determining realistic fracture conductivity and its impact on well performance–theory and field examples. SPE 106301.

92. Palisch, T. T., Vincent, M. C., & Handren, P. J. (2010). Slickwater fracturing: Food for thought. *SPE Production & Operations, 25*(3), 327–344.
93. Papanastasiou, P. (2010) Hydraulic fracture closure in a pressure-sensitive elastoplastic medium. *International Journal of Fracture, 103*(2), 149–161.
94. Penny, G. S. (1987). An evaluation of the effects of environmental conditions and fracturing fluids upon the long-term conductivity of proppants. SPE 16900.
95. Rahman, M. M., & Rahman, M. K. (2010). A review of hydraulic fracture models and development of an improved pseudo-3D model for stimulating tight oil/gas sand. *Energy Sources Part A, 93*, 1416–1436.
96. Ramurthy, M., Barree, R. D., Kundert, D. P., Petre, E., & Mullen, M. (2011). Surface area vs conductivity type fracture treatments in shale reservoirs. SPE 140169.
97. Reinicke, A. (2010). *Mechanical and hydraulic aspects of rockproppant systems–laboratory experiments and modelling approaches*. Doctoral Dissertation, Universität Potsdam, Germany.
98. Reinicke, A., Rybacki, E., Stanchits, S., Huenges, E., & Dresen, G. (2010). Hydraulic fracturing stimulation techniques and formation damage mechanisms–Implications from laboratory testing of tight sandstone-proppant systems. *Chemie der Erde, 70*, S3, 107–117
99. Rickards, A. R., Brannon, H. D., Wood, W. D., & Stephenson, C. J. (2006). High strength ultra lightweight proppant lends new dimensions to hydraulic fracturing applications. SPE 84308.
100. Roy, P., Du Frane, W. L., & Walsh, S. D. C. (2015). Proppant transport at the fracture scale: Simulation and experiment. ARMA 15–0662.
101. Rushing, J. A., & Sullivan, R. B. (2003). Evaluation of hybrid water-frac stimulation technology in the bossier tight gas sand play. SPE 84394.
102. Rushing, J. A., & Sullivan, R. B. (2007). Evaluating the impact of waterfrac technologies on gas recovery efficiency: Case studies using elliptical flow production data analysis. SPE 110187.
103. Schein, G. (2005). The application and technology of slickwater fracturing. SPE 108807.
104. Schubarth, S. K., Spivey, J. P., & Huckabee, P. T. (2006) Using reservoir modeling to evaluate stimulation effectiveness in multilayered "tight" gas reservoirs: A case history in the Pinedale Anticline area. SPE 100574.
105. Shokir, E. M., & Al-Quraishi, A. A. (2007). Experimental and numerical investigation of proppant placement in hydraulic fractures. SPE 107927.
106. Shor, R. J. (2014). *Modeling proppant flow in fractures using LIGGGHTS, a scalable granular simulator*. Master Thesis, Department of Petroleum and Geosystems Engineering, University of Texas at Austin.
107. Shor, R. J., & Sharma, M. M. (2014). Reducing proppant flowback from fractures: Factors affecting the maximum flowback rate. SPE 168649.
108. Terracina, J. M., Turner, J. M., Collins, D. H., & Spillars, S. E. (2010). Proppant selection and its effect on the results of fracturing treatments performed in shale formations. SPE 135502.
109. Unwin, A. T., & Hammond, P. S. (1995). Computer simulations of proppant transport in a hydraulic fracture. SPE 29649.
110. van der Vlis, A. C., Haafkens, R., Schipper, B. A., & Visser, W. (1975). Criteria for proppant placement and fracture conductivity. SPE 5637.
111. Vincent, M. C. (2002). Proving it–A review of 80 published field studies demonstrating the importance of increased fracture conductivity. SPE 77675.
112. Vincent, M. C. (2009). Examining our assumptions–Have oversimplifications jeopardized our ability to design optimal fracture treatments? SPE 119143.
113. Walker, R. N., Hunter, J. L., Brake, A. C., Fagin, P. A., & Steinsberger, N. (1998). Proppants, We still don't need no proppants–A perspective of several operators. SPE 49106.
114. Walsh, J. B. (1981). Effect of pore pressure and confining pressure on fracture permeability. *International Journal of Rock Mechanics and Mining Sciences & Geomechanics Abstracts, 18*, 429–435.

115. Warpinski, N. R., Moschovidis, Z. A., Parker, C. D., & Abou-Sayed, I. S. (1994). Comparison study of hydraulic fracturing models–test case: GRI staged field experiment no.3. *SPE Production & Facilities, 9*, 7–16.
116. Weaver, J., Rickman, R., & Luo, H. (2008). Fracture-conductivity loss due to geochemical interactions between manmade proppants and formations. SPE 118174.
117. Wendorff, C. L., & Alderman, E. N. (1969). Prop-packed fractures–A reality on which productivity increase can be predicted. SPE-2452-MS.
118. Yew, C. H. (1997). *Mechanics of hydraulic fracturing*. Houston: Gulf Professional Publishing.
119. Zienkiewics, O. C., Chan, A. H. C., Pastor, M., Schrefler, B. A., & Shiomi, T. (1999). *Computational geomechanics with special reference to earthquake engineering*. West Sussex: Wiley.

Chapter 12
Integrated Experimental and Computational Characterization of Shale at Multiple Length Scales

Weixin Li, Congrui Jin, and Gianluca Cusatis

Abstract Shale is known as a hydrocarbon source rock and is emerging as a potentially key component of the worldwide energy landscape via the recent development of hydraulic fracturing technique. A fundamental understanding of the fracturing behaviors of intact shale is the basis of any technological innovation aiming at increasing extraction efficiency. Considering the highly heterogeneous and fine-grained nature of shale, investigation and characterization should be conducted at multiple length scales. Through a brief overview of the experimental and computational studies for mechanical characterization of shale at different scales, this study investigates the possibility of integrating experimental and computational characterization research into a unified multiscale framework. Such multiscale framework is needed to predict macroscale shale fracturing behavior from the microscopic events. As preliminary results, an experimental characterization campaign of Marcellus shale at the macroscopic level and a micromechanical discrete model for predicting the mechanical behaviors of anisotropic shale are also presented.

12.1 Introduction

With the recent growth of the shale gas/oil industry, especially the development of hydraulic fracturing techniques, the study to promote deep understanding of the mechanical properties of shale-like rocks is becoming more important. Shale, made of highly compacted clay particles of submicrometer size, nanometric porosity, and diverse mineralogy, is probably one of the most complicated and intriguing natural materials present on earth [1]. The multiphase composition is permanently evolving

W. Li • G. Cusatis (✉)
Department of Civil and Environmental Engineering, Northwestern University, 2145 N Sheridan Rd, Evanston, IL 60208, USA
e-mail: g-cusatis@northwestern.edu

C. Jin
Department of Mechanical Engineering, Binghamton University, State University of New York, Binghamton, NY 13902, USA

C. Jin, G. Cusatis (eds.), *New Frontiers in Oil and Gas Exploration*,
DOI 10.1007/978-3-319-40124-9_12

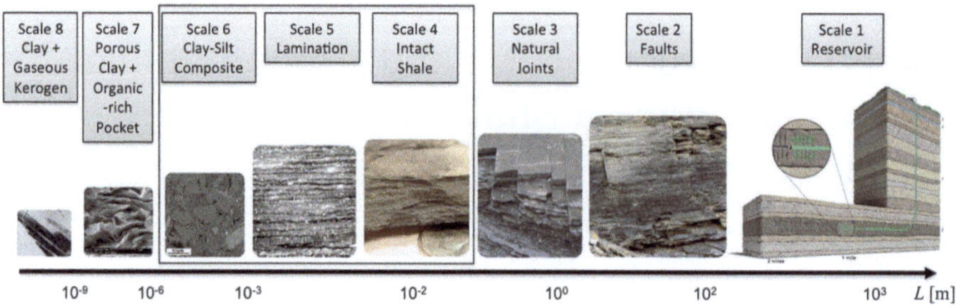

Fig. 12.1 Shale multiple length scales

over various scales of length and time, creating in the course of this process the most heterogeneous class of materials in existence. The heterogeneities manifest themselves from the nanoscale to the macroscopic level, which all contribute to a pronounced anisotropy and large variety of macroscopic behavior.

Shale exhibiting substantial heterogeneities can be characterized by several levels of hierarchy as illustrated in Fig. 12.1. Scale 1 represents the reservoir level characterized by porous or fracture rock formations where hydrocarbon exploration events occur. At scale 2, the detail of medium size faults with spacing in the order of tens of meters are fully resolved while other small features are accounted for only in the average sense in the behavior of material bulk outside the faults. Scale 3 is relevant to the scale of pre-existing joints and fractures naturally occurring in the shale formation and characterized by a spacing on the order of centimeters. At scale 3 and scale 2, the mechanical and hydraulic properties of many rock masses are affected significantly by the additional mechanical compliance and fluid conductivity that result from joints and faults. Depending on structure, porosity, and geometry, existing and ensuing faults may act either as seals or preferential pathways for fluid flow. For example, faults can affect hydraulic stimulations channelized to propagate along faults. A fault can effectively dominate the fracture growth and redirect all the energy of the treatment into the fault system and out of the target zone. The injection-induced seismicity can occasionally travel upwards thousands of feet, in most cases caused by inducing fracturing along faults.

Length scale 4 corresponds to that of a typical laboratory specimens used in standard mechanical characterization experiments such as acoustic measurements for elasticity and triaxial testing for strength properties. At this scale, shale is considered a transversely isotropic continuum. Down to scale 5, shale is characterized by the presence of weak planes, a consequence of the sedimentation process; thus, the continuum assumption breaks down and a certain set of discontinuity needs to be considered. Hence, at this scale, many shale specimens exhibit mm- and micron-scale grain-size variability, lamination/bedding planes and stratification, scour surfaces, and burrows, which play a significant role in shale geomechanical characteristics, such as the degree of elastic anisotropy, tensile strength, and fracture toughness. At the microscopic level (length scale 6), shale is a composite material made of porous clay, silt inclusion, and organic matter and at a lower scale

(scale 7) it exists as a porous clay/organic matter composite. Especially, the nanometer length level (scale 8) is the fundamental scale where elementary clay minerals are bounded to kerogen.

The adequate knowledge and prediction of the structure and mechanical properties at all several length scales identified above are needed for the analysis of reservoir stimulation and for the precise interpretation of microseismic and other measurements. To date, no predictive model of fracture nucleation/propagation in shale has been developed based on an integrated experimental and computational framework spanning all multiple scales. The formulation of such a multiscale predictive model requires three fundamental ingredients: (a) mathematical formulations describing the behavior of interest at a certain given scale; (b) experimental and field observations for calibration—identification of model parameters—and validation—proof of predictive capability; and (c) multiscale theories and technologies allowing the upscaling of the selected model to scale 1. Herein a study into the status quo and problems of the mechanical characterization of shale at multiple length scales is presented with the main focus of the integrated experimental and computational studies at the mesolevel (length scales 4, 5, and 6).

12.2 Experimental Studies

A thorough experimental campaign of shale is needed to interpret its material properties governed by the interplay of mechanisms and material textures featured at different length scales, and thus to develop fundamental new insight in the phenomena of hydraulic fracturing. Furthermore, a comprehensive database of shale characteristics and its mechanical properties at various length scales is necessary to calibrate any analytical theories and simulation tools. The database can also be used to prove that the modeling approaches have reliable predictive capabilities.

12.2.1 Overview of Experimental Studies for the Mechanical Characterization of Shale

12.2.1.1 Field Scale

In field conditions, geophysical well logging and seismic profiling are commonly used. Depending on the tools used, the measurement scale ranges from the meter scale at which a typical logging tool senses, to the scale of hundreds of meters at which a seismic wavelength senses.

Reflection seismic techniques have been applied widely to fracture detection in shale plays, specially detection of faults or fault zones [2]. Seismic data can be used for lithofacies classification, and to calculate the elastic attributes for shale-gas reservoir characterization through acoustic/elastic inversion [3, 4]. Seismic petrophysics also provides tools to detect the orientation and intensity of anisotropy due to stress or fractures, offering the opportunity to identify fracture-prone zones. Such identification is essential to optimize the development of unconventional gas plays [5]. Through a comprehensive understanding of the geological framework coupled with a focused processing, inversion, and interpretation of the seismic data [6], the seismic methods can be a powerful tool in reservoir characterization. For instance, 3D seismic data were used to help characterize reservoir quality variations, avoid drilling hazards, and help predict sweet spots in the Eagle Ford shale formation [6].

Well logging tools, on the contrary, are used to measure both borehole and in situ formation properties, whether thermal, magnetic, electric, radioactive, or acoustic at the entire length of drilled hole [7]. Common logging data include Gamma Ray, Resistivity, Neutron Porosity, Bulk Density, and Sonic logs [8]. The elastic properties are usually obtained from sonic logs. Most in situ data from wireline log measurements confirmed that shale behaves as a transversely isotropic medium in terms of dynamic elastic moduli [9–13]. However, Wong et al. [13] reported that the dynamic elastic moduli determined from the ultrasonic method are higher than those measured in the acoustic logs, which could have resulted from the drainage condition or dispersion effect. Besides elastic properties, the unconfined compressive strength (UCS) can be estimated from log porosity [14, 15] or sonic log and neutron log data [16, 17]. Chang and Zoback et al. [18] related UCS and internal friction angle of sedimentary rocks to physical properties, like velocity, modulus, and porosity, measurable with well logs through empirical equations. With these advanced logging tools, Abousleiman and coworkers [9, 10] estimated the anisotropic elastic and poroelastic properties of Woodford and Barnett shales.

Despite the great convenience of using logging tools for formation characterization, Abousleiman [19] pointed out that obtaining essential parameters for drilling and hydraulic fracturing activity still has to rely on laboratory rock mechanics testing and results. Unfortunately, the log obtained mechanical properties not only have large uncertainty on their accuracy, but also are insufficient for the comprehensive characterization of shale behaviors. Especially, there are no correlations available for estimating the tensile strength and fracture toughness, which are vital parameters for hydraulic fracturing design.

12.2.1.2 Macroscopic Scale

At the macroscopic scale, the elastic–plastic and fracture properties of gas shale are typically assessed in the laboratory using conventional rock mechanics testing methods such as Ultrasonic Pulse Velocity (UPV) measurements, splitting (Brazilian) tensile tests, uniaxial compression tests, triaxial tests, and three-point-bending tests. These conventional testing methods allow obtaining poromechanical parameters such as Young's modulus, Poisson's ratio, compressive strength (UCS), tensile strength, and pore pressure coefficient, in addition to measurements of shale failure parameters such as cohesion and friction angle [19].

Many reported results show that the elastic behavior is highly anisotropic with the elastic modulus being greater in the direction parallel to the bedding planes (E_1) than perpendicular to the bedding planes (E_3). For example, Abousleiman, Ulm, and coworkers [19] reported that the dynamic modulus $E_1 = 17.91$ GPa, $E_3 = 10.46$ GPa for Woodford Shale at a depth of 131 ft below ground through UPV measurements. Similar results can be found from UPV measurements or static moduli measurements on other types of shale [20–24]. Most of reported results confirm that shale can be approximated by a transversely isotropic elasticity model at small strain levels. Sone and Zoback [24, 25] carried out an extensive campaign encompassing Barnett, Haynesville, Eagle Ford, and Fort St John samples. The static and dynamic Young's modulus was found to decrease with the clay and Kerogen content.

Similarly, anisotropy of strength followed by the direction dependency of failure pattern in standard rock mechanics testing of shale can be also found in the literature. For example, Fjær and Nes [26] conducted a set of uniaxial compression tests and consolidated drained triaxial tests on Mancos outcrop shale samples. They reported that the observed strength anisotropy was related to a set of heterogeneous weakness planes. Simpson et al. [27] performed Brazilian tensile tests with Mancos shale. Although no apparent anisotropy of Brazilian tensile strength was observed, the dependency of the mode of failure on sample orientation was apparent. Abousleiman et al. [28] investigated the Woodford Shale fracturing properties including anisotropic tensile strength through a suite of Brazilian Tensile and Chevron Notched Semicircular Bend (CNSCB) tests. An increasing trend of Woodford Shale tensile strength with clay packing density and with carbonate content was observed. Hiandou et al. [29] explored the anisotropic elastic response, plastic deformation, and failure behavior of Tournemire shale through hydrostatic compressibility tests and triaxial compression tests. It was observed that the failure mode depends on confining pressures and loading orientation, and the main mechanisms of failure are the extension and sliding of the bedding planes, the splitting and the shear bands in shale matrix. Finally, Sone and Zoback [25] have shown through laboratory experiments that the ductile creep property and brittle strengths of shale-gas reservoir rocks are dependent on material composition and sample anisotropy.

Researchers [30–33] concluded that the anisotropic nature of shale has an impact on breakdown pressure, fracture initiation, and fracture containment during hydraulic fracturing process. For example, at great depths, where the magnitude of the stress anisotropy exceeds the strength anisotropy, fractures will therefore tend to propagate vertically; conversely, at shallower depths where the strength anisotropy exceeds the stress anisotropy, fracture are more likely to propagate horizontally [34]. Thus shale anisotropy at the macroscopic level needs to be characterized and taken into account when making decisions on the design of hydraulic fracturing.

In practical field situations, brittleness is a common term used to describe how rocks fail, and has been used as a descriptor in evaluating the potential effectiveness of hydraulic fracture initiation and propagation in low permeability shale formations [30, 35–38]. There is, however, no unique definition of a brittleness parameter. Various types of brittleness indices with varying degree of scientific basis are summarized in [39–42]. Yang and Zoback [42] investigated four brittleness indices based on triaxial test data on shale plugs from four different reservoirs. They reported that different brittleness indices are not consistent even using the data from the same samples, and do not correlate well with rock strength or elastic properties. Holt et al. [40] calculated different brittleness indices based on the experimental results of unconfined and trial compression plus Brazilian tensile tests on Mancos shale and Pierre shale, and showed that brittleness in anisotropic shale depends on loading direction, character of the failure process, and confining pressure. Although it is often assumed that formation with high brittleness is easy to fracture [37], Jin et al. [43] claimed that this viewpoint is not reasonable because brittleness does not indicate rock strength. Instead, they introduced a new fracability index by integrating both brittleness and fracture energy (critical strain energy release rate).

Fracture toughness represents the ability of rock to resist fracturing and propagation of pre-existing cracks, and may be a more reliable means of assessing quantitative measure of shale fracability thanks to its solid physical basis. The fracture toughness of gas/oil shale, however, is less known than the elastic–plastic properties and brittleness indices. Through CNSCB tests on Woodford shale, Abousleiman et al. [28] reported a fracture toughness in the range 0.74–1.17 MPa $\sqrt{m}$. Chandler et al. [34] carried out short rod fracture testing experiments on Mancos shale and reported a fracture toughness higher in the divider orientation than in the short transverse and arrester ones, which is similar to the observation by Schmidt [44] through three-point-bending tests. Finally, Akono and Ulm [45] carried out macroscopic scratch tests on Niobrara shale, and reported also an anisotropic fracture response with the toughness being higher for the divider orientation. The reported data of fracture toughness were mostly calculated based on Linear Elastic Fracture Mechanics (LEFM). Shale is also often assumed to be a quasi-brittle material with a non-negligible fracture process zone ahead of crack tip in some numerical models for hydraulic fracturing [46–48]. In these cases, LEFM-based methods are claimed either to give conservative predictions or to be invalid. However, no fracture properties on ductile or quasi-brittle gas/oil shale are reported to the best of writers' knowledge.

Although the experimental studies of shale at the macroscopic level are abundant and of great importance to any drilling and fracking activities in shale formation, investigations of shale at lower scales are also necessary. The meso- and microstructure and numerous parameters of shale at lower scales can have significant effects on rock properties. Investigation into the correlation between meso- and microscale parameters and macroscale behavior is necessary to understand the sources of shale material anisotropy, heterogeneity, and failure mechanism, and thus pivotal to the development of a reliable and physical-based predictive model.

12.2.1.3 Mesoscopic Scale

The common features of shale internal structure at the mesoscopic scale are lamination or bedding and microcracks. They are important sources of anisotropy leading to a directional dependence for the measured elastic properties, shear and tensile failures, and the permeability at the macroscopic level. Eseme et al. [49] classified the thin laminations into two categories: the microlamination of organic matter and minerals (corresponding to microscale in Fig. 12.1); and the macrolamination with variation in properties laterally due to compositional differences (corresponding to mesoscale in Fig. 12.1). Sone and Zoback [24] suggested that due to the limited kerogen volume present in their samples, they did not observe clear lamination in which most organic matter appeared to be separated. However, they proposed that fabric anisotropy forming the bedding planes is defined by the combination of the following: preferred orientations of matrix clay, shape/orientation/distribution of organics, and alignment of elongated fossils. Sayers [50] mentioned that anisotropy in shale results from a partial alignment of anisotropic clay particles, kerogen inclusions, bedding-parallel microcracks, low-aspect ratio pores, and layering. Slatt et al. [22] suggested that the boundaries between strata due to lamination, which is the product of transport/depositional events, are planes of weakness rather than being physio-chemically bonded.

Characterization of lamination and microcracks strongly relies on advanced imaging techniques such as thin-section, computed tomography (CT), and scanning electron microscopy (SEM) analyses. Sampling cores of black shale coming from a variety of locations and depths in the Mississippian-Devonian Bakken formation, Vernik and Nur [51] showed clear or obscure bedding-parallel lamination that can be identified either in thin sections as alternating 0.2 to 3 mm thick laminae of shale and silty shale or in SEM backscatter images as up to 20 μm thick dark laminae apparently enriched in organic matter. In [52, 53], Mokhtari showed the clear lamination structure of Eagle Ford Shale characterized by calcite-filled discontinuities in a thin section view, a microfracture in Mancos shale detected in a CT scan, and microfractures in Eagle Ford shale characterized by SEM imaging. From backscattered SEM imaging, Eseme et al. [54] showed prominent lamina of organic matter for Posidonia shale from N. Germany and the Condor oil shale. Josh et al. [55] compared the SEM micrographs of weak and strong weak shales, and suggested that no laminations are evident for weak shale while apparent laminations can be observed for strong shale. Note that a weak shale of less than a certain stiffness is not considered the best material for hydraulic fracturing operations [30] (Fig. 12.2).

Although lamination or bedding of shale is considered to be an important geometrical factor that is of great influence on the mechanical anisotropy at the macroscopic level, research on the mechanical characterization of lamina/bedding planes, rock matrix, and the interface between them is limited. Given some failure criteria for anisotropic geomaterials [57], it is possible to determine the parameters describing lamination effect in failure tests. For instance, Aoki [58] determined the

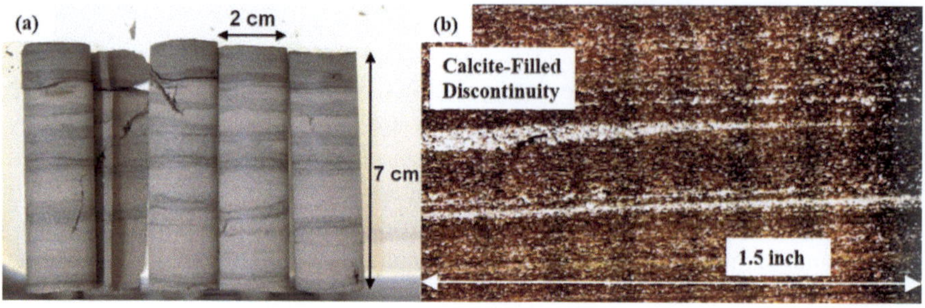

Fig. 12.2 Laminated structure of gas/oil shale: (**a**) Layered oil well shale stored in PG1 bath [56], (**b**) Thin section of the Eagle Ford shale with calcite-filled discontinuities [53]

strength parameters, effective cohesion and effective angle of internal friction, for rock matrix and bedding planes respectively from a series of CU triaxial tests based on the plane of weakness theory by Jaeger [59]. However, the direct measurement of lamina properties in laboratory is difficult. Advanced testing techniques, such as microindentation [60–62] and scratch tests [45, 63], are good candidates for conducting mechanical tests at the mesoscopic level.

12.2.1.4 Microscopic and Nanometer Scales

At the microscopic level, an organic shale sample is often considered as being comprised of essentially three types of constituents: organic matter, clay, and other minerals (quartz, feldspar, and pyrite carbonate). With the help of mineralogy quantification methods like X-Ray Diffraction (XRD) and Elemental Capture Spectroscopy (ECS), researchers are able to build the relationship between the volumetric composition of shale samples and the macroscopic mechanical properties. The organics in the form of kerogen have significant influence on the shale properties as they lower density, increase porosity, impart anisotropy, alter wettability, and introduce absorption [64]. Combining results from powder XRD analysis and pyrolysis, Sone and Zoback [24] determined the volumetric composition of tested samples from four different shale-gas reservoirs, and concluded that the deformation anisotropy of the gas shale samples, described by Thomsen parameters, increases with clay and organic content. A similar conclusion was obtained by Rybacki et al. [65] based on the mechanical tests performed on European black shales with different mineralogical composition, porosity, and maturity.

In addition to volumetric composition, the orientation and distribution of constituents has also significant effects on the mechanical behavior of shale. Sayers [50] investigated the effect of partial orientation of clay particles and the presence of silt particles, kerogen inclusions, microcracks, and low-aspect-ratio pores on the Young's moduli and Poisson's ratios of shales, and claimed that the Young's moduli and Poisson's ratios are sensitive to the degree of alignment of the clay particles. Sayers [66] also suggested that a model of discontinuous shale inclusions in kerogen, with which kerogen is assumed to form discrete inclusions in shale matrix, gives a better description of the measured elastic stiffness. With the help of SEM techniques, many studies [64, 67, 68] show that patches of kerogen occupying discrete spaces within the inorganic-rich matrix of shale. Ulm and Abousleiman [62] related the degree of anisotropy to the clay packing density which describes how tightly the clay particles are packed together. Abousleiman et al. [28] investigated the effects of micromorphology and mineralogy on Woodford shale anisotropic mechanical properties, and showed correlation between the degrees of anisotropy with clay packing density and carbonate content. By atomic force microscopy (AFM) technique, Eliyahu et al. [69] created quantitative moduli maps of the organic and inorganic components in selected shale samples with nanoscale resolution, and concluded that the elastic modulus of organic matter is heterogeneous at the nanometer scale.

Coupled with advanced observational methods such as SEM, transmission electron microscopy (TEM), and AFM, nanoindentation technique provided direct measurements of mechanical properties for the constituents of gas shale at the microscopic and lower levels. Ulm and Abousleiman [62] adapted the nanoindentation technique for highly heterogeneous and anisotropic shale. In light of the statistical analysis of massive nanoindentation tests (indentation depths of 200–300 nm), microindentation (maximum indentation depths of 1500–2500 nm) results, and UPV measurements, they concluded that shale behaves mechanically like a nanogranular material, whose behavior is driven by particle-to-particle contact and by characteristic packing densities, and that the much stiffer mineral properties play a secondary role. By means of an extensive nanoindentation campaign on a large range of shale materials, Bobko and Ulm [70] revealed that the nanomechanical elementary building block of shales is anisotropic in stiffness, and isotropic and frictionless in strength. However, they did not include the effect of organic matter in the nanoindentation analysis. Later, Ulm and coworkers extended their model to treat kerogen-rich shales [71], and used nanoindentation technique to extract properties of Woodford Shale, which is organic rich [10]. Kumar et al. [72] conduct a nanoindentation study on shales from the Barnett, Woodford, Ordovician, Eagle Ford, and Haynesville plays. They reported an excellent match between the average indentation Young's modulus and the dynamic Young's modulus measured from various samples, and that Young's modulus anisotropy increases with an increase in Total Organic Carbon (TOC) and clay content. Via nanoindentation and microindentation tests, Borja and coworkers [73] showed that a penetration depth less than 200 nm was enough to provide in situ mechanical property measurements of the constituents comprising Woodford shale samples, whereas indentation depths higher than 3 μm yielded the composite response (Fig. 12.3).

Although nanoindentation technique provides a powerful tool to extract mechanical properties of shale constituents like elastic modulus and hardness at the micro- and nanoscales, it does not provide information on fracture properties. Scratch test developed by Akono and Ulm [45, 63] provides a method for the multiscale assessment of intrinsic fracture properties. They carried out microscopic scratch tests on Niobrara shale, and reported a strong directionality of the fracture behavior.

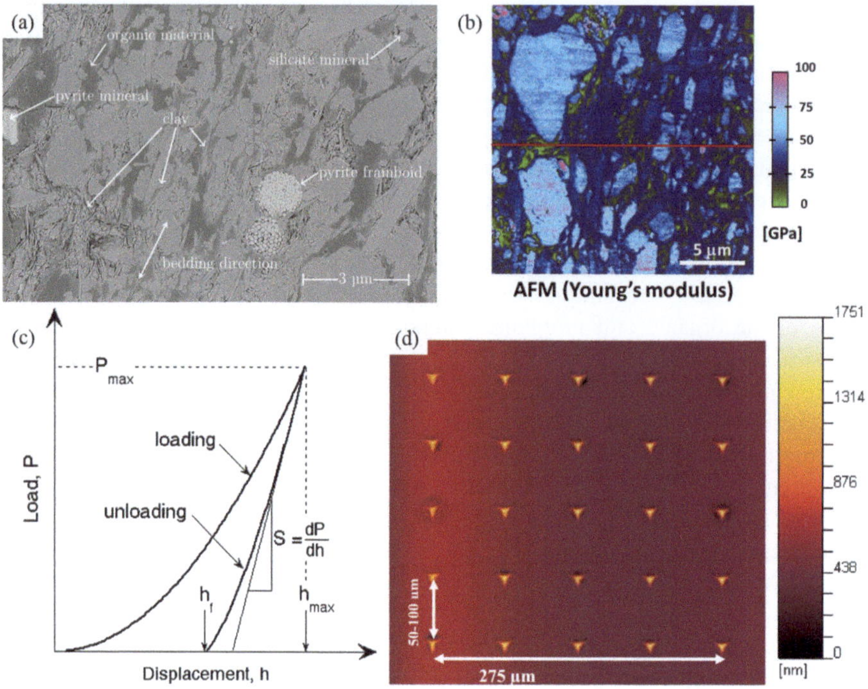

Fig. 12.3 (**a**) SEM image of polished Woodford shale sample surface [73]. (**b**) Young's modulus map for an organic-rich shale sample through AFM [69]. (**c**) Schematic illustration of indentation load–displacement data showing important measured parameters [74]. (**d**) 2D survey scanning map obtained from nanoindenter [72]

A toughening behavior was also observed in the scratch tests at the microscopic level. Compared to the fracture toughness measured via the macro scratch tests, the one measured via micro scratch tests was reported to increase by 80 %. It was claimed that this microscopic toughening phenomenon may be attributed to the presence of inhomogeneities and inclusions.

12.2.2 Experimental Characterization of Marcellus Shale at the Macroscopic Scale

Recently, at Northwestern University, a systematic analysis was performed on the macroscopic anisotropic behaviors in strength and fracture patterns observed during Brazilian tests and compression tests on Marcellus black shale samples. This includes comprehensive data on the mechanical behavior of Marcellus shale samples, including the dynamic elastic properties, and the variations of compressive and tensile strength as functions of the anisotropy angle measured as angle with respect to the lamination. Three-point-bending tests were also performed to study the failure mechanisms of shale. For the purpose of comparison, some tests were also conducted on grey shale specimens, taken from Thornton Quarry, located in Thornton, Illinois, just south of Chicago.

12.2.2.1 Sample Preparation

The tested Marcellus black shale samples have an average mass density of 2558 kg/m^3. The specimens were free of surface cracks and voids. Prior to testing, all the specimens were ground and polished by hand. The hand grinding was done to avoid disturbances from machining during the sample preparation. Three types of specimens were prepared for further tests.

The first type consisted of cube specimens with a nominal dimension of 1 in. × 1 in. × 1 in. Cube specimens were cut by a band saw from a large shale block in such a way that the bedding plane direction and the cube edge have an angle, called anisotropy angle, denoted by the angle θ, which is measured clockwise from the loading direction relative to the bedding plane, as shown in Fig. 12.4. The anisotropy angle θ varied between 0° (perpendicular to the loading direction) and 90° (parallel to the loading direction). Bedding direction was marked by solid lines after cutting. Specimens with 5 different anisotropy angles of 0°, 30°, 45°, 60°, and 90° were prepared. These cube specimens were used for UPV measurement, Brazilian tensile, and uniaxial compression tests.

The second type consisted of disc specimens with a nominal diameter, $D = 38$ mm, and a nominal height, $t = 19$ mm. The shale block was properly clamped to prevent any unwanted movement during the coring process. The speed of coring was constant to avoid any irregularities or defects on the cutting surface. The specimens were cored in the direction parallel to the bedding planes.

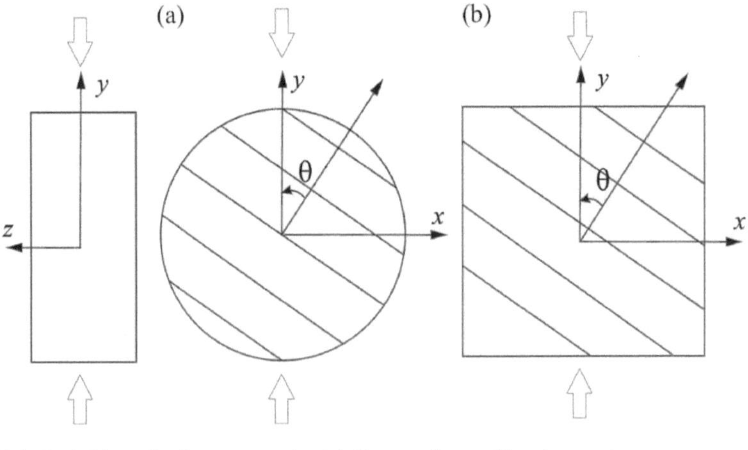

Fig. 12.4 Definition of anisotropy angle: (**a**) disc specimen; (**b**) cube specimen

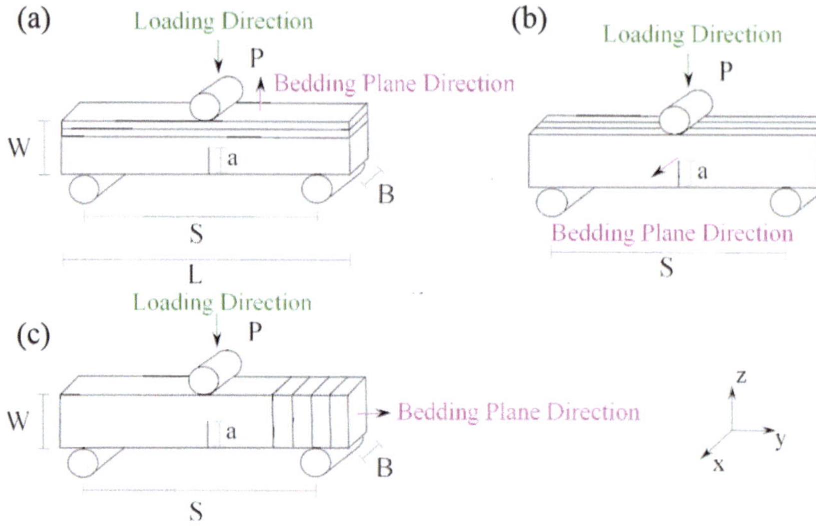

Fig. 12.5 Three types of notched beam specimens for different bedding plane orientations: (**a**) specimens with bedding plane parallel to x-y plane (arrester); (**b**) specimens with bedding plane parallel to x-z plane (divider); and (**c**) specimens with bedding plane parallel to y-z plane (short transverse)

These specimens were used for Brazilian tensile tests. Different anisotropy angles were achieved by simply rotating the disc during the tests.

The third type was notched beam specimens. The nominal dimensions of the length L, width W, and thickness B of the specimens were 100 mm, 25 mm, and 12.5 mm, respectively. The dimensions of the specimens were defined in compliance with ASTM requirements. A band saw was used to cut a notch of length $a = 6$ mm for each specimen. Three types of notched beam specimens for different bedding plane orientations were prepared, as shown in Fig. 12.5.

12.2.2.2 Ultrasonic Pulse Velocity

A UPV measurement system was used to determine the longitudinal elastic wave (P-wave) velocity of Marcellus shale samples, as shown in Fig. 12.6. The ultrasonic system consists of several functional units, such as the pulser/receiver, transducer, and a display device. A pulser/receiver is an electronic device that can produce high voltage electrical pulses. Driven by the pulser, the transducer generates high frequency ultrasonic energy. This energy propagates through the materials in the form of stress waves. The wave signal is transformed into an electrical signal by the transducer and is displayed on a screen. The velocity of the wave is calculated as the distance that the signal traveled divided by the travel time.

Figure 12.7 shows the variation of the P-wave velocity with respect to the anisotropy angle for Marcellus shale. The maximum values occurred in the

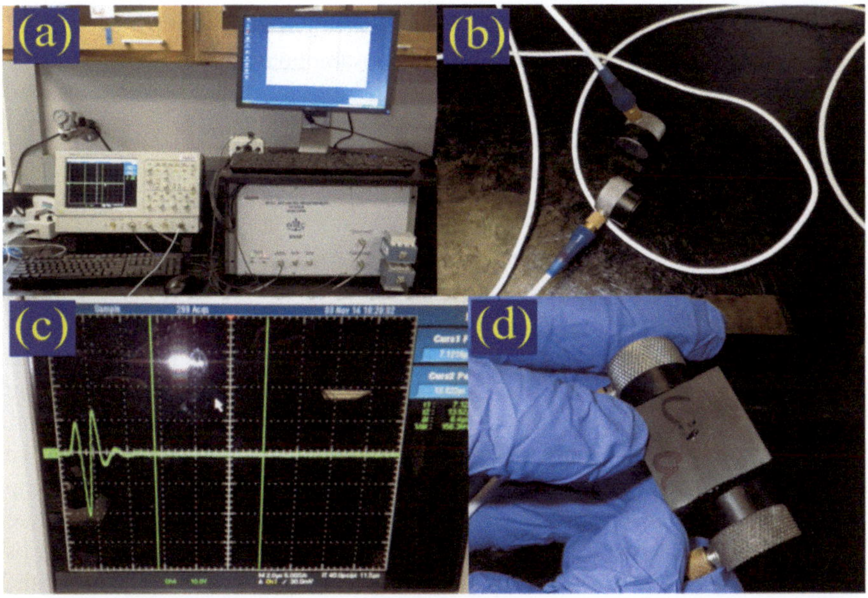

Fig. 12.6 The measurement system used to determine the P-wave velocity of each sample. (**a**) RAM-5000 computer controlled ultrasonic system; (**b**) Transmitter and receiver; (**c**) Oscilloscope; and (**d**) Experimental sample

Fig. 12.7 Variation of measured P-wave velocity with anisotropy angle

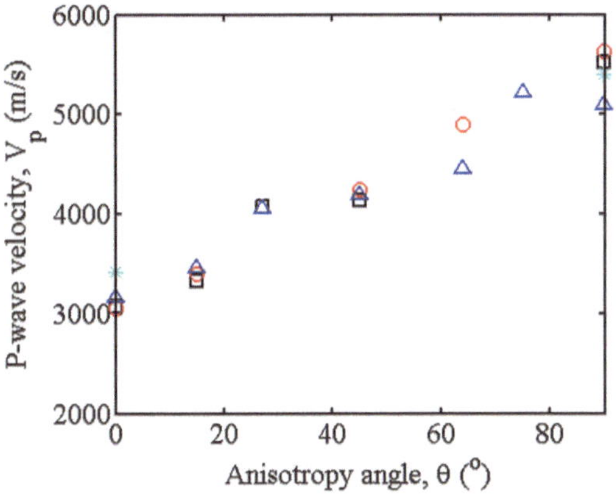

direction parallel to the isotropic plane, and the minimum value occurred in the direction perpendicular to the isotropic plane. The measured P-wave velocity for Marcellus shale varies from 3104 m/s to 5481 m/s. The maximum and the minimum P-wave velocities occurred at the anisotropy angles of 90° and 0° respectively. The anisotropy ratios of P-wave velocity, defined as V_P^{max}/V_P^{min}, is about 1.8.

To estimate the elastic moduli, the P-wave modulus M is approximated by $M = \rho V_P^2$ based on isotropy and homogeneity assumptions in the theory of linear elasticity, although an assumption of transverse isotropy should be rather adopted for an accurate estimation. It is shown that the estimated P-wave modulus varies from 24.6 GPa to 76.8 GPa.

12.2.2.3 Brazilian Tensile Tests

The Brazilian test, often called splitting tensile test, is a simple indirect testing method to obtain the tensile strength of brittle and quasi-brittle material such as concrete, rock, and rock-like materials. In this study, the Brazilian tests were conducted on both Marcellus black shale specimens and grey shale specimens. The tests were conducted with a 1000 kips MTS loading system operated in stroke control mode. The MTS loading system consists of a main frame, a hydraulic pump unit, a controller, a load cell, and a computer. Specimens were loaded up to failure at a constant displacement rate of 0.003 mm/s. For cube specimens, wood strips were used to limit stress concentration at points of application of the load. After the failure of specimens, pieces of failed rock specimens were collected to investigate the relationship between the bedding plane and the failure plane.

The indirect tensile strength, perpendicular to the loaded diameter, based on linear elastic calculations for homogeneous and isotropic rock is written as $\sigma_t = 2kP/(Dt)$, where P is the load at failure, D is the diameter of the test specimen (or the edge length of the cube specimen), t is the thickness of the test specimen measured at the center, and k is a coefficient depending on the specimen geometry and experimental conditions. For disc tests, $k = 1/\pi$, and for cube tests $k \approx 0.3$. Of course this formula is considered for the loaded diameter in the condition of a typical vertical splitting, which is often not the case for transversely isotropic rock material. Therefore, this formula is used just for comparison purposes and the results do not necessarily represent the true tensile strength of the samples.

The results of Brazilian tests on Marcellus shale are presented in Table 12.1 and Fig. 12.8b. In Table 12.1, for disc tests $k = 0.318$, and $k = 0.307$ for cube tests. From

Table 12.1 Brazilian test results on Marcellus shale specimens

θ[°]	Specimen type	σ_t[MPa]				Mean [MPa]
0	Disc	6.18	4.60	5.42	4.03	5.06
10	Disc	5.22	5.28	4.60	5.28	5.10
25	Disc	7.25	5.93	5.10	7.11	6.35
45	Disc	4.66	5.26	6.31	4.70	5.23
60	Disc	5.53	5.74	5.48	6.41	5.79
80	Disc	5.26	4.29	6.10	4.98	5.16
90	Disc	3.83	3.75	3.66	3.77	3.750
0	Cube	10.22	10.30	9.37	9.70	9.90
90	Cube	4.31	3.95	5.13	4.74	4.53

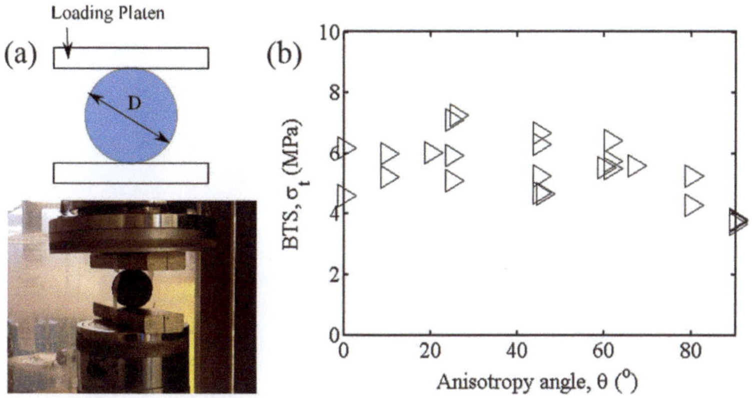

Fig. 12.8 (**a**) Loading configuration used in splitting tensile tests. (**b**) Variation of Brazilian tensile strength with anisotropy angle

Fig. 12.8b, it can be seen that the maximum tensile strength occurred at or near $\theta = 0°$, while the minimum value occurred at $\theta = 90°$. Figure 12.8b implies that Marcellus shale shows a trend of the variation of tensile strength with anisotropy angle: the tensile strength increases slightly from 0° to 45°, followed by a decrease from 45° to 90°. The anisotropy ratio calculated by $\sigma_t^{\max}/\sigma_t^{\min}$ is about 1.6. The shale specimens are observed to fail suddenly during the tests, demonstrating that the material is brittle. By studying the shale samples after failure, we observed that, in almost all the samples, the crack propagates along the loaded diameter, as shown in Fig. 12.9.

For the purpose of comparison, some tests were also conducted on grey shale core specimens. For the experiments conducted on disc specimens, the loading direction is always perpendicular to the bedding; and for cube specimens, the loading direction is either perpendicular or parallel to the bedding, as shown in Fig. 12.10a. In sharp contrast, for grey shale specimens, two different types of fractures are observed, as shown in Fig. 12.10b: (a) one propagates along the loaded diameter; and (b) the other fails along the bedding layers. The results of Brazilian tests on grey shale are presented in Table 12.2. The grey shale specimens after failure are shown in Fig. 12.11. It can be seen that the specimens show a failure by either Mode A only or a combination of both Mode A and Mode B. The rupture of a chain occurs at the weakest link. It is a law of nature that a fracture grows and propagates along a path and direction that needs less energy to dissipate. In a layered rock, the boundaries of the bedding layers are considered as the planes of weakness. When mentioned planes are frequent, it is logical that fractures use one or more of them to propagate. It has been observed that the bedding layer boundaries' frequency and weak mineral (mica and carbonate) percentage are directly proportional to each other [75]. More weak minerals cause layer activation, becoming more dominant in the fracture pattern. For larger weak minerals percentage, a decrease of the tensile strength is reported [75]. This is consistent with the results

Fig. 12.9 Marcellus shale specimens after failure in Brazilian tests

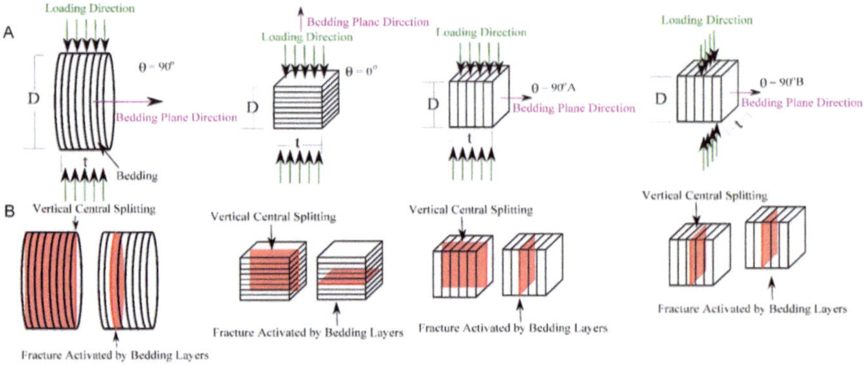

Fig. 12.10 Brazilian tests on grey shale: (**a**) Loading configuration; (**b**) Two failure modes

presented in Table 12.2, which shows that when Mode A and Mode B happen simultaneously, the values of tensile strength are considerably lower. In the case of $\theta = 90°$B, if a straight fracture parallel to the bedding layers between both loading platens is induced, it is classified as layer activation. In that case, the weak layer is in the same direction of central splitting mode. Without a microscope, we do not

Table 12.2 Brazilian test results on grey shale specimens

θ[°]	Specimen type	Failure mode	σ_t[MPa]				Mean [MPa]
90	Disc	A and B	9.91	6.32	7.51	7.83	7.89
90	Disc	A	9.20	13.71	12.22	12.54	11.92
0	Cube	A and B	8.25	6.59	9.69	7.01	7.89
0	Cube	A	11.66	10.16	12.17	11.39	11.35
90A	Cube	A and B	6.53	4.22	3.14	3.41	4.33
90A	Cube	A	9.86	13.82	12.61	10.98	11.82
90B	Cube	A or B	7.28	5.12	7.91	6.07	6.60

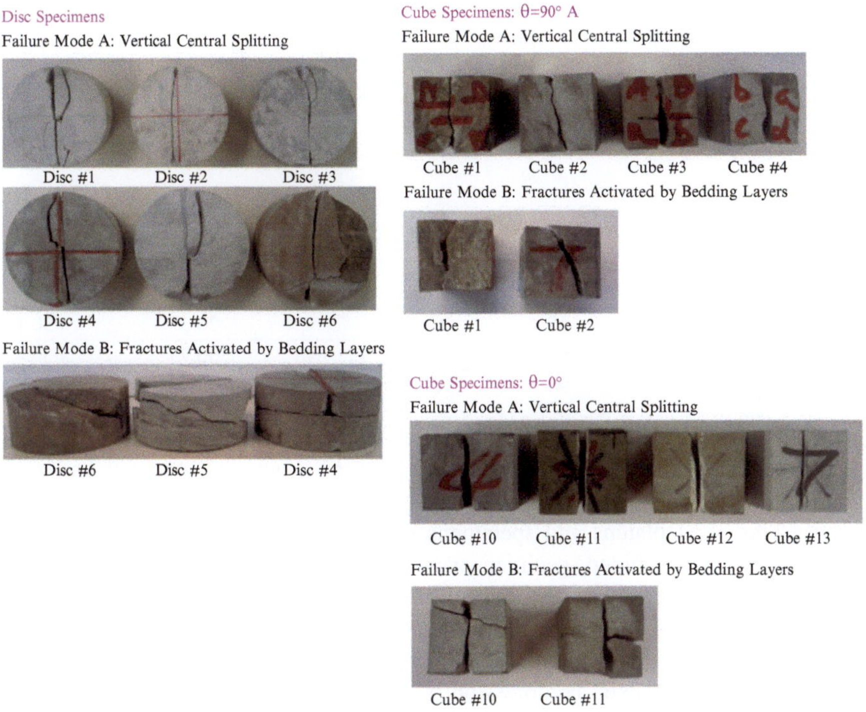

Fig. 12.11 Grey shale specimens after failure in Brazilian tests

know it cracks along weak layer or not. It may crack just along the central line, which is not a weak layer, although it is possible that there is a weak layer along the central line

12.2.2.4 Uniaxial Compression Tests

Cube specimens were loaded in compression up to failure at a constant displacement rate of 0.001 mm/s using the same 1000 kips MTS loading system as the one used in Brazilian tensile strength test. Specimens were placed between two loading

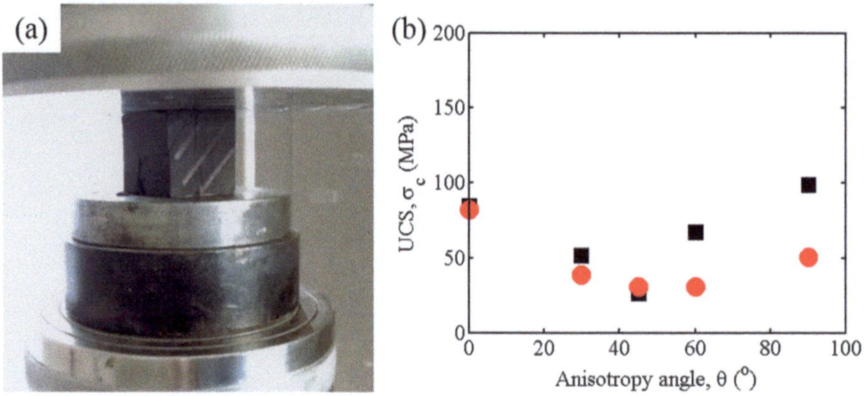

Fig. 12.12 (**a**) Loading configuration used in uniaxial compression tests. (**b**) Variation of uniaxial compressive strength with anisotropy angle

Table 12.3 Uniaxial compression test results on Marcellus shale specimens

θ[°]	Specimen type	σ_c[MPa]	
0	Cube	84.31	81.90
30	Cube	51.91	38.59
45	Cube	26.37	30.72
60	Cube	66.95	30.42
90	Cube	98.99	50.53

platens, as shown in Fig. 12.12a. The tests were carried out with specimen and steel loading platens in direct contact with each other, and Dry Moly Lube was sprayed between the steel platens and specimens to reduce friction. After the failure of specimens, pieces of failed specimens were collected to investigate the relationship between the bedding plane and the failure plane. The Unconfined Compressive Strength (UCS) was approximated by the maximum nominal stress, i.e., $\sigma_c = F/A$, where F is the peak load, and A is the initial area of the specimen cross section.

The results of uniaxial compression tests on Marcellus shale are presented in Table 12.3 and Fig. 12.12b. The compression strength varies from 26 to 99 MPa. The maximum uniaxial compressive strength occurs at $\theta = 0°$ or $\theta = 90°$, and they exhibit minimum strength at approximately $45° < \theta < 75°$. The anisotropy ratio of uniaxial compressive strength, calculated by $\sigma_c^{max}/\sigma_c^{min}$, is about 3.2.

The variation of uniaxial compressive strength can be explained by the effect of weakness plane, i.e., when the failure plane coincides with the weakness plane, failure occurs at a lower stress level. This is due to the fact that shear cracks develop more easily along the weakness plane. The bedding layers can be assumed as the weakness planes. Figure 12.13 shows the specimens after uniaxial compression failure. The specimens mostly failed along the bedding layers when $45° < \theta < 90°$.

Fig. 12.13 Marcellus shale specimens after failure in compression tests

12.2.2.5 Three-Point-Bending Tests

This section covers the determination of the fracture toughness of shale using a three-point-bending test on a single-edge notched beam specimen. These fracture toughness results are applicable whenever there is a fracture involved in operation, e.g., in the resource recovery process such as stimulation of oil and gas wells by hydraulic fracturing, fragmentation of oil shale beds for in situ retorting, and stress-induced fracturing of geothermal sources.

Notched beams were placed on the three-point-bending fixture with a nominal span $S = 75$ mm, as shown in Fig. 12.14a. The experimental setup consists of two parallel supports for the sample and a single loading pin in the middle. The support pins must be mounted in such a way that they can rotate freely on their axes in order to minimize the influence of friction on the measurement. One of the supports must also be able to rotate about an axis perpendicular to this and parallel to the axes of the sample so that the test piece can align itself when under stress. The loading pin must also possess similar rotational axes in order to ensure the uniform application of force on the test piece. The crack growth is monitored by measuring crack opening displacement with an extensometer mounted. The samples are loaded using a 20 kips MTS loading system. The rate of loading is controlled by the notch opening displacement, and the crack mouth opening rate for all the tests is 0.0005 mm/s. The specimens are tested for three different orientations.

Assuming that Linear Elastic Fracture Mechanics (LEFM) applies, the critical stress intensity factor, K_{IC}, can be calculated using Griffith's relationship as: $K_{\mathrm{IC}} = 6Ma^{1/2}F(a/W)/(BW^2)$, where $F(a/W)$ is a correction factor, $M = P_c S/4$ is the applied bending moment, a is notch length, B is thickness of the specimen, W is depth of the specimen, S is span length, and P_c is peak load. This equation is valid for a/W between zero and 0.6.

The results for three-point-bending tests for Marcellus shale are tabulated in Table 12.4. The load–displacement curves are linear up to peak. In the experiment, it was not possible to capture the post-peak behavior for Marcellus shale specimens, probably because Marcellus shale is very brittle. Table 12.4 shows that the specimen with the bedding layer parallel to the loading direction exhibits lower K_{IC} than the other two specimens with the bedding plane perpendicular to the loading direction. The anisotropy ratio of K_{IC} calculated by $K_{\mathrm{IC}}^{\max}/K_{\mathrm{IC}}^{\min}$ is about 1.3. The fracture surface appeared quite smooth consistently with the brittle character of the response at failure, as shown in Fig. 12.14b.

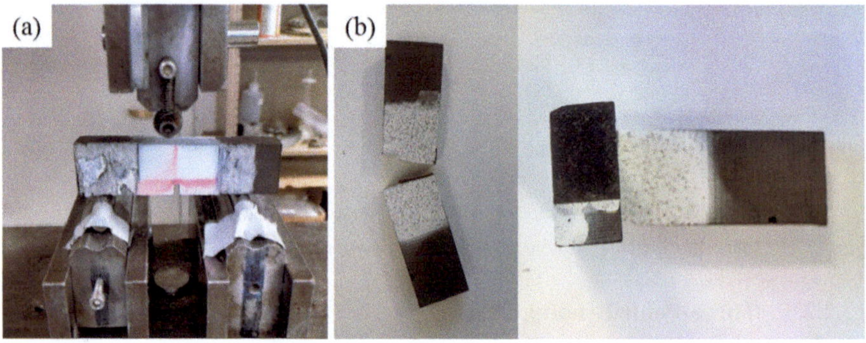

Fig. 12.14 (**a**) Loading configuration used in three-point-bending tests. (**b**) Marcellus shale specimens after failure in three-point-bending tests

Table 12.4 Three-point-bending test results on Marcellus shale specimens

Bedding layer orientation	Peak load [N]	$K_{IC}(\text{MPa}\sqrt{m})$
x–y	678	1.34
x–z	533	1.06
y–z	622	1.23

12.2.3 Discussion

It has previously been assumed that shale was not a reservoir rock, and thus not interesting in terms of hydrocarbon production, which has led to a lack of published experimental data on shale samples. Especially, there has been little research on the experimental characterization of mechanical behaviors of shale at the mesoscopic level (millimeter and micrometer scales), which are indeed of vital importance for understanding the influence of mesostructure features, such as lamination and microfractures, on the shale macroscopic behaviors and the mechanism of fracture initiation, nucleation, and propagation. Thanks to the development of advanced testing methodologies at the micro- and nanoscales, such as nanoindentation and micro scratch tests, it is nowadays possible to measure mechanical properties of each constituent directly. However, a micromechanics-based framework is necessary to predict the composite responses at higher levels by incorporating the macro-, micro-, and nano-measurements, which is important for explaining the link between micro- and macro-behaviors. In addition, although the properties and structure of inorganic minerals have been extensively examined, little is known about their coupling with organic matter. A better understanding of the organic–inorganic interfaces and the porous and anisotropic microstructure is needed to fully characterize mechanical behaviors of shale. Finally, due to the extreme mineralogical, morphological, and topographical variability of shale, the experimental characterization should be on a case-by-case basis, and a comprehensive data set is needed.

12.3 Computational Studies

A predictive macroscale model for fracture nucleation/propagation in gas/oil shale is needed to better understand the influence of material properties on induced fracture initiation, propagation, and fracturing treatment, and to provide a tool to optimize fracturing process in hydraulic fracturing design. To capture the micromechanics of the formation and propagation of fractures at the grain scale, which are important to explain the link between micro- and macro-behavior, the adopted model should be developed based on an integrated experimental and computational framework spanning several length scales identified previously in this chapter.

12.3.1 Overview of Modeling Techniques for the Mechanical Characterization of Shale

Numerical models of hydraulic fracturing at field scale have undergone a development from 2D simulators, such as the KGD [76, 77] and PKN [78, 79] models, to pseudo-3D [80, 81], planar 3D [82–84], and fully 3D [85, 86] simulators. In these models, the rock deformation is modeled using linear elasticity, which is used to determine the relationship between the fracture width and the fluid pressure. As a result, they generally bear little or no relation with the underlying microscopic processes and can be overly complex when their mathematical structure is adapted to cope with a large range of phenomenological responses. In recent years, a number of new numerical hydraulic fracture propagation models, largely based on the variations of finite element models (FEM), such as extended FEM (XFEM) [87, 88] and cohesive zone FEM methods [89–91], have been developed to handle mechanical deformations and fracture propagation, and coupled with various continuum flow models. However, those models still handle only simple fracture geometries, or assume unrealistic boundary conditions and material properties, or require using empirical failure criteria and post-failure mechanical/hydraulic constitutive laws, and simplified fluid-solid coupling methods. As a result, these models do not capture random initiation of fractures, or strongly nonlinear coupling among deformation, fracturing, and fluid flow in fracture apertures and leakage into porous rock matrix.

A reliable hydraulic fracture model requires a comprehensive and robust three-dimensional formulation equipped with correct fracture mechanics; realistic description of shale internal geometry; physically sound constitutive equations; strongly nonlinear coupling between fluid flow and heterogeneities at various scales; as well as an extensive calibration/validation against a large variety of experimental data. Especially, a predictive macroscale model of fracture nucleation/propagation in shale is needed. The predictive ability critically depends on the constitutive assumption at the micro- and mesoscales. In addition, a multiscale framework is required to link the shale behaviors at different levels and to reduce computational costs.

12.3.1.1 Macroscopic Scale

Considering shale as general sedimentary rock at the macroscopic level, a brief overview of the modeling techniques of rock mechanical behaviors is presented herein. Mechanical breakdown and fracture of rock at laboratory scale are due to various meso- and microscopic processes, including the formation of microcracks at multiple grain or defect boundaries, and the growth and coalescence of microcracks [92]. Rock initial defects such as grain boundaries, microcracks, and pores can be regarded as stress concentrators, leading to progressive breakdown of intact rock samples under external loads and the initiation and growth of cracks. For example, the formation and extension of cracks for rock samples under compression can be explained by two representative models, i.e., the open crack model [93, 94] and the sliding crack model [95–97], as shown in Fig. 12.15a, b respectively. Both models regard the initiation of cracks as being due to the intensified tensile stress near the tips of the initial defects. Although fracture mechanics allows the prediction of nonlinear mechanical behaviors of rock, it requires the information of pre-existing cracks within the material, and cannot predict the localization and coalescence of microcracks during rock failure [92]. Due to the limitations of the fracture mechanics model, macroscopic numerical models characterizing fracture and failure behaviors of rock are needed.

The numerical models can be usually classified into continuum-based and discrete models [98]. The continuum-based model treats the material domain as a continuous body and capture material failure behaviors through proper damage models. Typical damage models adopt internal state variables to capture damage evolution at the microstructural level, and to relate strains to stresses through appropriate constitutive laws [99]. The numerical implementation of continuum-based models includes FEM, finite difference method (FDM), and boundary element method (BEM). Although capable of simulating material deterioration by averaging the mechanical response corresponding to damage evolution, classic continuum damage models are unable to reproduce the localization and coalescence of microcracks. This is because the models do not consider microscopic events or assume a uniform distribution of microdefects [92]. Therefore they exhibit the lack of an internal length scale. As a result, localization occurs in a region of zero

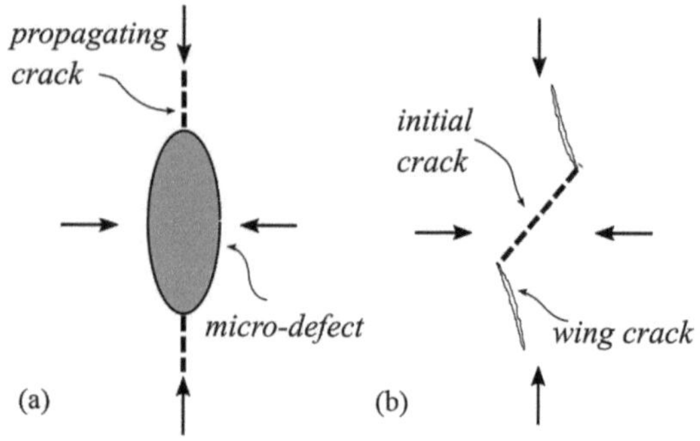

Fig. 12.15 Schematic of (**a**) open crack model and (**b**) sliding crack model

thickness and a mesh sensitivity arises [100, 101]. A concept of fictitious crack [102, 103] has been proposed to represent the fracture process zone (FPZ) where phenomena such as growth and coalescence of microcracks take place. Adapting this concept, Bažant and Oh [102] introduced the crack band model to ensure mesh independence of global properties and that the correct amount of energy is dissipated. By enriching the classic continuum formulations, regularization techniques, such as gradient damage model [104, 105], nonlocal damage model [106], and other higher order continuum models [107, 108], offer more complicated localization limiters to avoid the spurious mesh sensitivity problem. Statistical damage model [94, 109–111] with nonuniform distributed defects thought the material provides another solution to the localization problem. One drawback of the statistical approach is the need to determine an appropriate distribution function for the heterogeneity [92].

Discrete approaches, based on the pioneering work of Cundall and Strack [112, 113], treat the material domain as a system of separate blocks or particles. The approach applied to rock engineering problems has undergone a development from the early stage of rigid-body movements to motion and deformation of deformable block system [114]. Early models of this type were referred to as the discrete element method (DEM). Discrete models are capable of incorporating the length scale automatically. For instance, the introduction of particle size in the distinct element model provides boundaries for shear band formation, and guarantees the thickness of the shear band [115]. A particular group of DEMs referred to as particle-type model, with the most notable implementations represented by the universal distinct element code (UDEC) [116] and the particle flow code (PFC) [117], mimics the grain-cement system in real rock [118] with rigid particles representing grains bounded or unbounded at their contacts. For example, PFC uses a so-called soft contact approach: the particles are permitted numerically to overlap each other, and the degree of overlap controls the contact forces through a contact law; the bonded-particle model (BPM) [118] based on PFC introduces bonds between the particles in the assembly corresponding with the addition of cement between the grains of a sedimentary rock. In the particle model, the breaking or weakening of bonds simulates damage evolution within rock, which finally leads to rock failure at the macroscopic level through propagation, nucleation, and coalescence of the microscopic events.

Lattice models, often called dynamic lattice network models [92, 98], are closely related to the DEM models and have been used to study fracture initiation and propagation in solid materials and conductivity behavior of porous media. This kind of lattice models was proposed by Schlanger and Mier [119–123], Schlangen et al. [124, 125], Bažant [126], Song and Kim [127], Place and Mora [128], Bolander et al. [129, 130], and Katsman et al. [131] to analyze concrete and rock fracture. In this approach, the material domain is modeled by a lattice of basic shapes, such as triangles, squares, or hexagons, with particles of lumped masses located at the vertices of the lattice. A site in the lattice represents a grain in the rock; a bond, usually simulated by a massless spring or beam element along the edges of the lattice, represents a potential crack in the medium. Damage evolution is

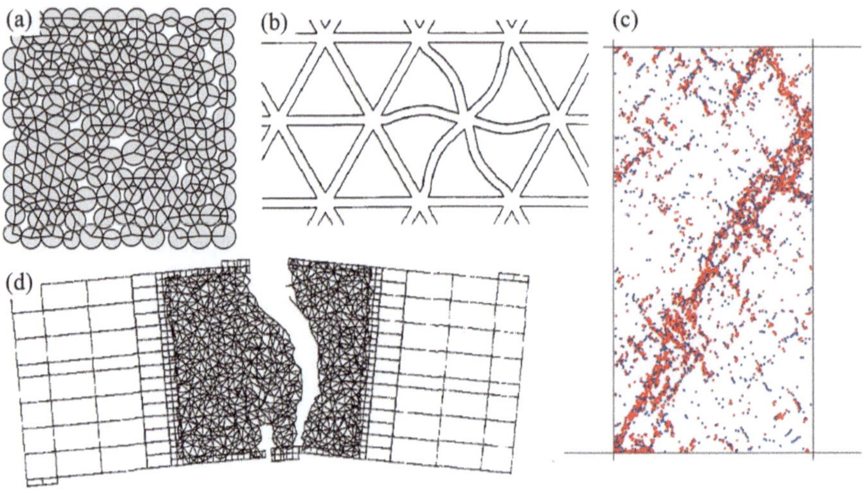

Fig. 12.16 (**a**) Parallel-bond network in a 2D specimen created by BPM [118]. (**b**) Triangular lattice of beam elements in a lattice model [120]. (**c**) Damage patterns formed during biaxial tests for a specimen of the PFC2D granite model [118]. (**d**) Simulated crack patters in DEN beams with fixed supports using the random lattice [120]

simulated by removal of a bond once the maximum stress locally exceeds a certain threshold. Original proposed lattice configurations were regular, but Jirásek and Bažant [132], Schlangen [124], and Bolander et al. [130] suggested that a random or irregular lattice configuration is necessary to avoid a bias for the overall crack propagation direction. Recently, 3D versions of lattice model are proposed to simulate 3D fracture processes [125, 133–135]. Although similar to the particle model, lattice models can represent a continuous medium composed of heterogeneous elements, with which the heterogeneity of the material is taken into account by assigning different properties to the lattice elements according to the real material structure.

Both particle and lattice models require numerical trials to determine small-scale parameters, such as particle bond strength, contact conditions, and grain size distribution, which are difficult to be extracted from experimental techniques. Although current advanced testing methods at micro- and nanoscales, such as nanoindentation, provide opportunities for accessing material properties at small scales, most models are phenomenological to some extent. Therefore, not all the model parameters can be determined through experiments, and model calibration through numerical trials is necessary. In addition, the particle and lattice models are based on micromechanics; it may be difficult to apply to large-scale analyses of rock engineering problems because of high computational costs [92] (Fig. 12.16).

Hybrid continuum/discontinuum approaches regard rock as a material mostly dominated by continuum region while inserted by a set of discontinuity. Typical examples include the combined finite-discrete element method (FDEM) [136, 137] and the Adaptive Continuum/Discontinuum Code (AC/DC) [138]. Some other numerical schemes, either continuum-based or discontinuum-based, such as XFEM [139, 140], smooth-particle hydrodynamics (SPH) [141], and Peridynamics method [142], have also been used to characterize fracture and failure behaviors of rock.

12.3.1.2 Mesoscopic Scale

At the mesoscopic level, the laminated material microstructure along with the presence of schistosity, foliation, or bedding planes can be regarded as physical discontinuities, which also exist at a larger length scale as joints, natural fractures, faults etc. From the mechanical point of view, discontinuities serve as planes of weakness or weakness zones with less cohesion which subdivide the medium of interest into small number of continuous regions. Inside each region the displacement field will be continuous while may be discontinuous across regions. Therefore, the mathematical representations and the corresponding numerical modeling techniques of discontinuous are similar at either field, laboratory, or mesoscopic scales. As discussed in the previous sections, particularly at the mesoscopic level, the laminated structure of sedimentary rock is an important source of mechanical anisotropy at the laboratory level. A good numerical model should be able to capture the mechanical anisotropy induced by rock discontinuities.

Computational techniques that are used for modeling discontinuities are often classified into smeared and discrete approaches [143]. The smeared approach utilizes an implicit representation of discontinuities to produce a fictitious continuous material which exhibits mechanical behaviors similar to the original discontinuous medium. The continuum damage model can be considered as a smeared approach with the effect of microdefects, which lead to discontinuous displacement field, smeared over the material volume, and represented by degrading the material properties according to a constitutive law. In terms of simulating mechanical anisotropy, the smeared approach introduces the effect of lamination, bedding-parallel microcracks, and other sources of anisotropy at the mesoscopic and microscopic levels via proper constitutive laws. The directional dependence of deformation properties for sedimentary rock during loading and unloading processes, i.e., deformation anisotropy, is commonly captured using the theory of elasticity for transversely isotropic medium [144]. Strength anisotropy, as well as anisotropic features induced during progressive damage evolution, is usually captured by various anisotropic failure criteria [57, 145–147] along with specifying the directional dependence of material properties, or the continuum formulation of plasticity and damage for anisotropic solids [145, 148–150]. A representative failure criterion is the well-known plane of weakness model [59], which assumes that the failure in the rock matrix and along weakness planes is described by the Mohr-Coulomb type criterion with different values of cohesion and friction separately.

The smeared approach is a phenomenological approach without the need to simulate the real structure and parameters at the mesoscopic and microscopic levels, and may be thus computationally more effective. It has been applied to the numerical characterization of mechanical anisotropic behaviors of shale and other types of sedimentary rock. For instance, Mahjoub et al. [145] proposed a failure criterion based on the Drucker-Prager yield function and an anisotropic damage behavior law for transversely isotropic solids. They simulated 3D Brazilian tests for Boryeong shale, Tournemire shale, and two other types of rocks with different orientation angles and showed satisfactory results compared with experimental data. Lisjak et al. [143] developed a smeared approach via the FEM/DEM framework by imposing the orientation-dependent cohesive strength at the crack element level. Comparing the experimental results on clay shales, they calibrated the model and demonstrated that macroscopically observed strength anisotropy can be captured correctly through the proposed smeared approach (Fig. 12.17).

The discrete approach, on the contrary, employs an explicit representation of discontinuities with a distribution of cohesion-less fractures, weakness planes, etc. embedded in a homogeneous medium. In this approach, strength and damage anisotropy can be captured naturally through the initiation of shear deformation along weakness surfaces induced by discontinuous, which causes a stress concentration, and finally leads to macroscopic failure. With the discrete modeling technique, mechanical anisotropy emerges as a natural property of the discrete system and is introduced by a geometric description of the internal structure, which is usually approximated by a discrete particle or lattice system inserted by weakness zones, representing plane of weakness, pre-existing cracks [153], parallel continuous weak layers [154], continuous smooth joints [151], and discontinuous joints [155].

The discrete approach is a physical-based approach without complicated mathematical formulation at the constitutive level. However, the computational cost of the discrete approach is quite large for simulating field scale problem since the mesh resolution is constrained by the spacing of discontinuities, typically millimeter scale for bedding or lamination of shale. In addition, the discrete approach requires the geometrical and material parameters at the mesoscopic and microscopic levels, which are usually difficult to obtain from experimental observation and testing. Despite their limitations, discrete approaches have been widely used in simulating material anisotropy. For example, Park et al. [152, 156, 157] used the smooth joint model along with BPM to simulate BTS and UCS tests on Boryeong shale, and captured correctly the dependency of failure mode on loading orientation. Lisjak et al. [153] introduced the discrete approach based on the FEM/DEM method by imposing the distribution of pre-existing cracks aligned with the bedding plane orientation and several broken crack elements with the simulation domain. By simulating the uniaxial compression, indirect tension, and biaxial compression tests on Opalinus Clay, they confirmed that a distinctive variation of specimen compressive strength, crack pattern, and damage mechanism can be captured as a function of the orientation of specimen anisotropy.

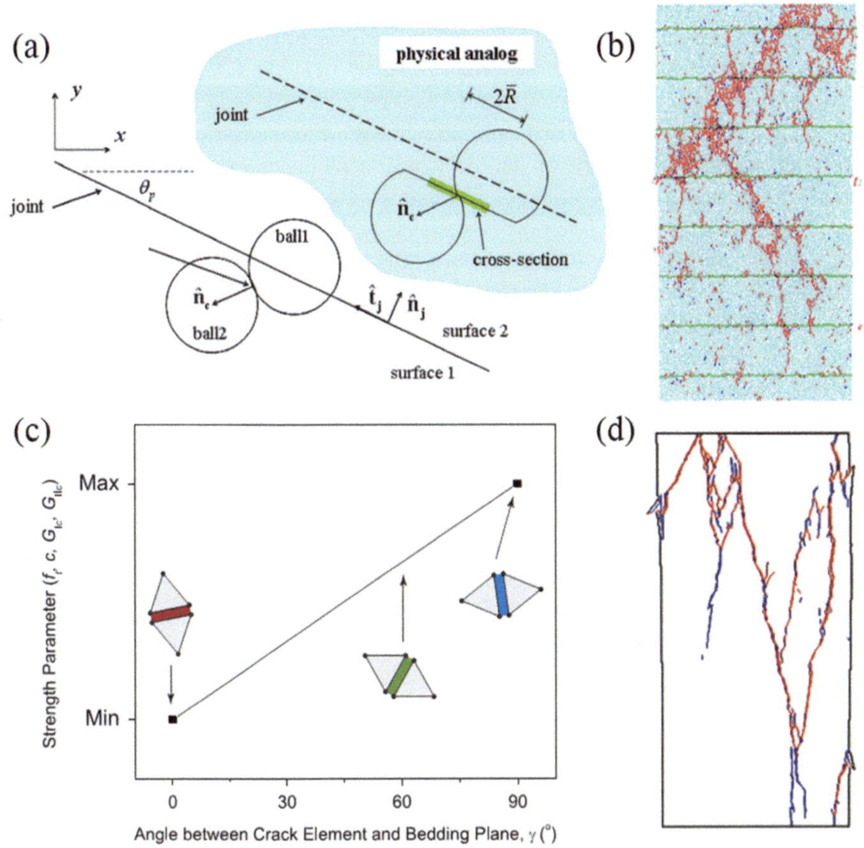

Fig. 12.17 (**a**) Smooth joint creation [151]. (**b**) Modeling of Boryeong shale specimen with 0° anisotropy angle by smooth joint model after uniaxial compressive strength test [152]. (**c**) Smeared approach for modeling strength anisotropy with bedding plane orientation in FEM/DEM [143]. (**d**) Simulated fracture patters for Opalinus Clay sample with 0° loading angle orientation by smeared approach and FEM/DEM [143]

12.3.1.3 Microscopic and Nanometer Scales

Models at the microscopic level and nanoscale provide a mean of interpreting experimental results and extracting material properties from tests at the corresponding length scales, such as nanoindentation. For example, Borja et al. [73] developed a FE model of the nanoindentation experiments on organic-rich Woodford shale, and implemented with Drucker-Prager/Cap plasticity and consolidation creep. Through the calibrated simulations, plastic material parameters were extracted from nanoindentation experiments and were used to simulate triaxial compression tests, highlighting the anisotropic behavior of shale. In addition, simulations at the atomic and molecular structure levels, such as Molecular Dynamics (MD) simulations and Monte Carlo simulations, have been carried out to provide the precise characterization of intact shale and shale constituents. Especially, close attention has been paid to the role of organic matter [158–160] and clay minerals [161, 162] in material properties of shale. For instance, Brochard et al. [158] performed molecular simulations on a silica-kerogen nano-composite, and suggested that the ductility and toughness of organic-rich shale could be attributed to the pulling out of ductile kerogen inclusions out of a brittle silica matrix. From the results of molecular simulations on a typical clay mineral, illite, Hantal et al. [162] reported that the fracture resistance under tensile loading is low, and the mechanism of yield and fracture failures is decohesion in the interlayer space, while the nanoscale failure mechanism under shear loading is a stick–slip between clay layers. They suggested that the low fracture resistance in mode I and the stick–slip failure in mode II are the consequence of the lack of chemical bonds between clay layers.

12.3.1.4 Multiscale Algorithm

From the computational point of view, a multiscale algorithm is necessary because "full" microscopic simulations are too expensive to perform in order to predict the macroscopic behavior of very fine-grained shale, even with the computational power currently available. A multiscale algorithm is needed to link the shale micromechanics behaviors all the way up to the fracture and failure behaviors at the field level. In addition, the multiscale framework should integrate experimental measurement and observation with current numerical simulation techniques, with a proper explanation of the micromechanics of the formation and propagation of fractures at lower levels.

Ulm and coworkers [62, 71] have developed a theoretical model, the GeoGenome model, for estimating shale anisotropic elastic and poroelastic properties based mainly on nanoindentation results and UPV measurement. In this approach, macroscopic elastic and poroelastic properties of shale depend on the clay packing density, the volume fraction of the nonclay phase, and the organic volume fraction [71]. Abousleiman et al. [9] used the GeoGenome upscaling model to estimate the Young's modulus and shear modulus of Woodford and Barnett shale, and achieved good agreement with sonic log data. Later, Bobko et al. [163] developed a strength homogenization approach to study scaling relationships for indentation hardness with clay packing density, providing an interpretation of the nanogranular origin of friction and cohesion in the porous clay binder phase in shale.

Homogenization techniques provide an opportunity for upscaling micro- and mesoscale computational models. In this approach, micro- and meso-material levels can be replaced by a homogenized poroelastic material in the higher level problem domain, based on two assumptions: the existence of a Representative Volume Element (RVE) and the scale separation assumption. Although no literature can be found to apply homogenization approach to shale modeling, homogenization theory has been widely used in many braches of engineering. For instance, Hassani [164] investigated formulation of homogenization theory and topology optimization and its numerical application to materials with periodic microstructure like laminate composite. Chung [165] presented detailed derivation of multiple scale formulation for elastic solids through Asymptotic Expansion Homogenization (AEH). Fish employed this approach to study elastic as well as elasto-plastic composites [166]. Ghosh [167] adopted Mathematical Homogenization (MH) along with Voronoi Cell Finite Element Method (VCFEM) to study the behavior of composites with random mesostructure [168]. More recently, Fish [169] introduced the Generalized Mathematical Homogenization (GMH) to derive thermo-mechanical continuum equations starting from MD. Feyel [170] built a homogenization scheme to couple a Cauchy continuum formulation at the micro-scale giving rise to a Cosserat continuum formulation at the macroscale. Asymptotic homogenization technique was employed by Forest [171] for upscaling elastic Cosserat solids. Rezakhani and Cusatis [172] derived a general multiscale homogenization scheme suitable for upscaling materials whose fine-scale behavior can be successfully approximated through the use of discrete models featuring both translational and rotational degrees of freedom. Developed homogenization theory has been used to evaluate macroscopic properties of concrete through RVE simulations, and to homogenize concrete elastic and nonlinear behavior under different types of loading conditions (see [172, 173]).

12.3.2 A Micromechanical Discrete Approach

A discrete approach based on a micromechanical framework [174] is presented herein, which is built upon the mesoscopic level of shale (length scale 5 as illustrated in Fig. 12.1) with an explicit representation of the laminated internal structure. Features of smaller length scales are introduced by constitutive laws, while simulations of the macroscopic behavior of shale are allowed by introducing a proper multiscale algorithm. The proposed micromechanical approach is formulated within the Lattice Discrete Particle Model (LDPM) framework, developed by Cusatis and coworkers [175, 176], which offers the following unique advantages:

1. LDPM has been extensively proven to be a powerful tool to accurately predict mechanical behavior of quasi-brittle materials under various loading conditions in both tension and unconfined, confined as well as hydrostatic compression.
2. LDPM is formulated within the framework of discrete models, which provides an inherent potential to take the heterogeneous nature of shale into account.
3. The model is able to explicitly reproduce the material internal structure, which provides a potential to capture the fabric anisotropy of shale.

Since LDPM was first proposed to simulate the failure behavior of concrete with a statistically isotropic random mesostructure, it needs to be extended to capture the aspects of material anisotropy based on the composition and internal texture of shale. Hereinafter, the LDPM constitutive equations are introduced, and modified to accommodate material anisotropy.

12.3.2.1 Geometrical Characterization of Shale Internal Structure

Based on the composition and texture of shale described above, a shale sample at the mesoscopic level can be represented by a laminated structure model with multiple layers of weakness embedded in a block matrix, shown in Fig. 12.18b, which is an analogy for the typical laminated shale as shown in Fig. 12.18a. The weak layers represent stratification or layering of rock due to alternately operating depositional processes, which are results of macrolamination as Eseme et al. mentioned in [49]. The layers or beds are weaker and are characterized by lower moduli and strengths in constitutive laws. A block matrix is dominated by the constituent components of shale including organic matter, clay, quartz, and other minerals. Apart from the discrete representation, a smeared representation of anisotropy with transversely isotropy assumption is also introduced. The smeared approach accounts for the sources of mechanical anisotropy that cannot be represented by the discrete approach and that has length scales lower than the scale of grains (mostly microscopic level). The transverse isotropy assumption is used because these sources of shale anisotropy at the microscopic level, such as microlamination of organic matter and minerals, partial alignment of clay minerals, microcracks, and low-aspect ratio pores, are mostly bedding/lamination-parallel. Thus, the plane of isotropy is consistent with the plane of bedding/lamination. The proposed model considers both the deformation anisotropy and the failure anisotropy. The deformation anisotropy at the mesoscopic level is simulated by the explicit representation of layering while the deformation anisotropy at the microscopic level is tackled by a proper constitutive formulation. The failure anisotropy is simulated mostly by layering at the mesoscopic level as the weak layers can work as a weak zone where fracture can initiate and propagate. The failure modes could be different for specimens with different bedding/lamination inclination with respect to loading directions.

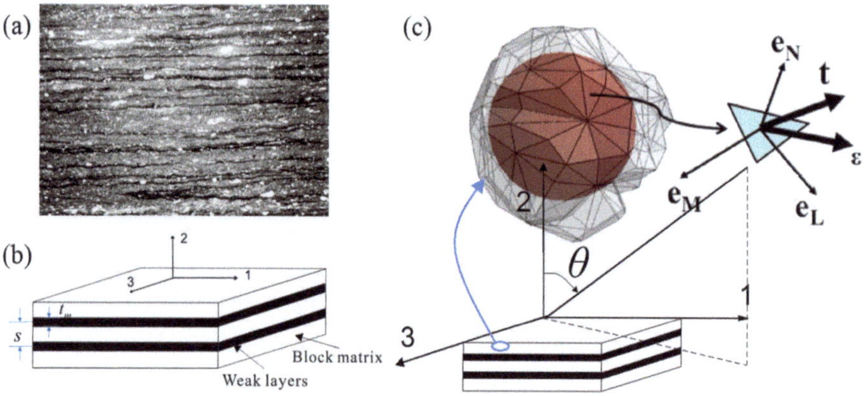

Fig. 12.18 (**a**) Fine lamination in cherty shale [177]. (**b**) Schematic diagram of the laminated structure model with tm representing the thickness of weak layers and S representing the layer spacing. (**c**) LDPM cell and facets

The LDPM simulates the shale internal structure by considering only the coarse grains. The model generation procedure is briefly introduced here but details can be found in [176]. Particles with assumed spherical shape and in accordance with granulometric distributions (the Fuller curve with Fuller coefficient n_F) are first introduced randomly into the volume through the algorithm that avoids particle overlapping. A Delaunay tetrahedralization of the particle centers is used to define the lattice system that represents the mesostructure topology. A 3D domain tessellation, anchored to the Delaunay tetrahedralization, creates a system of polyhedral cells. These LDPM cells are assumed to represent the coarse grains of materials and thus material heterogeneity. Note that particles are generated such that the maximum cell size matches typical heterogeneity size $d_{max} \approx 50$ μm. Adjacent cells interact through the triangular facets where they are in contact (see Fig. 12.18c). In the LDPM formulation, the interface facets are interpreted as potential crack surfaces. Rigid-body kinematics is adopted to describe the mesostructure deformation with which the strain vector and the stress vector at the facet level can be derived. The LDPM constitutive law, which provides the relationship between the strain vector and the stress vector, is presented in the next section. The laminated structure can be represented by assigning different material properties for the layers of weakness and the block matrix. The constitutive laws for the weak layers or beds represent weak bonds between different layers of block matrix and are also presented in the next section. Together with the grain distribution parameters including the maximum grain size d_{max}, the minimum grain size d_{min} and the Fuller coefficient n_F, the lamination parameters including the layer spacing s and the thickness of weak layers t_m geometrically characterize the internal structure of shale at the mesoscale.

12.3.2.2 Constitutive Equations

Mesostructure facet strains, one normal strain ε_N, and two shear strains ε_M and ε_L, are defined through the relative displacement at the centroid of a given facet [176] which is computed based on the displacements and rotations of the particles associated with the facet and the assumption of particles rigid-body motion. The strains then can be used to compute LDPM facet stresses, σ_N, σ_M and σ_L through LDPM constitutive law.

Elastic Behavior

At the framework of LDPM, the elastic behavior is formulated by the linear relations between normal/shear stresses and the corresponding strains, i.e.

$$\sigma_N = E_N \varepsilon_N; \quad \sigma_M = E_T \varepsilon_M; \quad \sigma_L = E_T \varepsilon_L \tag{12.1}$$

where E_N is the effective normal modulus; E_T is the effective shear modulus, and α is the shear-normal coupling parameter. Through a smear representation of deformation anisotropy and the transversely isotropy assumption of shale, E_N and E_T are assumed to be the function of the observed orientation, i.e., $E_N = E_N(\theta)$, $E_T = E_T(\theta)$, where θ is defined as angle between the normal vector of LDPM facets and the normal vector of lamination plane, as shown in Fig. 12.18c. Therefore, $E_N(\theta)$ and $E_T(\theta)$ read

$$E_T(\theta) = \left(\frac{\sin^2\theta}{E_{T1}} + \frac{\cos^2\theta}{E_{T0}}\right)^{-1} = E_{T0}(\beta_T \sin^2\theta + \cos^2\theta)^{-1} \quad (12.2)$$

$$E_T(\theta) = \left(\frac{\sin^2\theta}{E_{T1}} + \frac{\cos^2\theta}{E_{T0}}\right)^{-1} = E_{T0}(\beta_T \sin^2\theta + \cos^2\theta)^{-1} \quad (12.3)$$

where $E_{N0} = E_N(0°)$, $E_{N1} = E_N(90°)$, $E_{T0} = E_T(0°)$, $E_{T1} = E_T(90°)$ respectively; $\beta_N = E_{N0}/E_{N1}$ and $\beta_T = E_{T0}/E_{T1}$ are the ratios of elastic moduli at 0° to moduli at 90°. The relationship between the mesoscale LDPM parameters and the macroscopic elastic parameters can be obtained through the kinematically constrained formulation of the microplane model of the normal component, which reads

$$\mathbf{E}^* = \frac{2}{3\pi}\int_\Omega \mathbf{P}^T \mathbf{E}_P \mathbf{P} d\Omega \quad (12.4)$$

The integration is done over a unit hemisphere with the surface Ω. Matrix $\boldsymbol{P}$ collects the information on the microplane orientation, while the matrix $\boldsymbol{E}_p$ depends on LDPM effective moduli, i.e., $\mathbf{E}_P = \mathrm{diag}(E_N, E_T, E_T)$.

Through the smeared approach described above, the deformation anisotropy of shale at the microscopic level is properly addressed. At the mesoscopic level, the effect of weak layers, which separate the matrix layers above and below through weak bonds, on the elastic coefficients is considered by introducing the reduction factors C_N and C_T for effective normal and shear moduli associated with weak layers/beds respectively, as follows

$$E_N^l = C_N E_N; \quad E_T^l = C_T E_T \quad (12.5)$$

The reduction factors typically satisfy $0 < C_N, C_T \leq 1$. To simplify our model, we assume that the reduction factors for the normal and shear modulus are the same, i.e., $C_N = C_T$. As a result, the macroscopic stiffness matrix of the weaker layers is proportional to that of the block matrix.

Fracturing Behavior

In the LDPM formulation, inelastic fracturing and cohesive behavior due to tension and tension/shear given $\varepsilon_N > 0$ are formulated through an effective strain $\varepsilon = \sqrt{\varepsilon_N^2 + \alpha(\varepsilon_M^2 + \varepsilon_L^2)}$, and stress $\sigma = \sqrt{\sigma_N^2 + (\sigma_M^2 + \sigma_L^2)/\alpha}$, which define the normal and shear stresses as $\sigma_N = \varepsilon_N\ (\sigma/\varepsilon)$; $\sigma_M = \varepsilon_M\ (\sigma/\varepsilon)$; $\sigma_L = \varepsilon_L\ (\sigma/\varepsilon)$. The effective stress σ is incrementally elastic ($\dot{\sigma} = E_N\dot{\varepsilon}$), and must satisfy the inequality $0 \leq \sigma \leq \sigma_{bt}(\varepsilon, \omega)$, where $\sigma_{bt}(\varepsilon, \omega) = \sigma_0(\omega)\exp[-H_0(\omega)\langle\varepsilon_{max} - \varepsilon_0(\omega)\rangle/\sigma_0(\omega)]$, $\langle x \rangle = \max\{x, 0\}$, and $\tan(\omega) = \varepsilon_N/\sqrt{\alpha}\varepsilon_T$. The post peak softening modulus is defined as $H_0\ (\omega) = H_t\ (2\omega/\pi)^{nt}$, where n_t is the softening exponent; H_t is the softening modulus in pure tension ($\omega = \pi/2$) expressed as $H_t = 2E_N/(l_t/l_e - 1)$; $l_t = 2E_N G_t/\sigma_t^2$; l_e is the length of the tetrahedron edge; and G_t is the mesoscale fracture energy. LDPM provides a smooth transition between pure tension and pure shear ($\omega = 0$) with parabolic variation for strength given by

$$\sigma_0(\omega) = \sigma_t \frac{-\sin\omega + \sqrt{\sin^2(\omega) + 4\alpha\cos^2(\omega)/r_{st}^2}}{2\alpha\cos^2(\omega)/r_{st}^2} \tag{12.6}$$

where $r_{st} = \sigma_s/\sigma_t$ is the ratio between the shear strength and the tensile.

The smeared approach of modeling anisotropy is applied to the shear and tensile strength respectively with σ_t and σ_s expressed as

$$\sigma_t(\theta) = \sigma_{t0}\left(\beta_t \sin^2\theta + \cos^2\theta\right)^{-1} \tag{12.7}$$

$$\sigma_s(\theta) = \sigma_{s0}\left(\beta_s \sin^2\theta + \cos^2\theta\right)^{-1} \tag{12.8}$$

where σ_{t0} and σ_{s0} are the tensile strength and the shear strength at $\theta = 0°$ respectively; $\beta_t = \sigma_{t0}/\sigma_{t1}$ and $\beta_s = \sigma_{s0}/\sigma_{s1}$ are the ratios of the tensile strength and the shear strength at 0° to those at 90°.

Similar to modeling elastic behavior, the effect of lamination can be considered by introducing the reduction factors C_{Nt} and C_{Ts}. The tensile strength and shear strength associated with weak layers can then be written as

$$\sigma_t^l = C_{Nt}\sigma_t; \quad \sigma_s^l = C_{Ts}\sigma_s \tag{12.9}$$

in which $0 < C_{Nt}, C_{Ts} \leq 1$.

Therefore, the strength limit for the effective stress $\sigma_0(\omega)$ is a function of the facet orientation θ and location in addition to ω. In the stress space σ_N–σ_T, Eq. (12.6) is a parabola with the axis of symmetry coinciding with the σ_N axis. The parabola represents the envelop of the elastic limit of stress status (σ_N, σ_T), varies with facet orientation, and shrinks for facets associated with weak layers.

Frictional Behavior

Due to frictional effects, the shear strength increases in the presence of compressive stresses. This phenomenon can be simulated effectively through classical incremental plasticity. The plastic potential can be expressed as $\varphi = |\sigma_t| - \sigma_{bs}(\sigma_N)$ in which the shear boundary is formulated with the following frictional law:

$$\sigma_{bs}(\sigma_N) = \sigma_s + (\mu_0 - \mu_\infty)\sigma_{N0} - \mu_\infty\sigma_N - (\mu_0 - \mu_\infty)\sigma_{N0} \times \exp(\sigma_N/\sigma_{N0}) \quad (12.10)$$

where μ_0 and μ_∞ are the initial and final internal friction coefficients; σ_{N0} is the normal stress at which the internal friction coefficient transitions from μ_0 to μ_∞. Since there are no available experimental data, we simply adopt the classical Coulomb linear frictional law with slop μ_0 by setting $\sigma_{N0} = \infty$ or $\mu_\infty = \mu_0$.

The shear boundary governs the unconfined and low confinement macroscopic compression for low values of σ_N and high confinement macroscopic compression for high values of σ_N. It is practical to make the internal friction coefficient vary in a way similar to the elastic moduli, i.e.,

$$\mu_0 = \mu_{00}\left(\beta_\mu \sin^2\theta + \cos^2\theta\right)^{-1} \quad (12.11)$$

$$\mu_0^l = C_\mu \mu_0 \quad (12.12)$$

where μ_{00} is the initial internal friction coefficient at $\theta = 0°$; β_μ is the ratio of the friction coefficient at 0° to the one at 90°; the reduction factor C_μ considers the effect of lamination on internal friction. In addition, LDPM has the capability to simulate the triaxial compressive behavior at the macroscopic scale with a compressive boundary capturing the material compaction and rehardening phenomenon. Since this is outside our current research scope, an elastic behavior is assumed for compression, i.e., $\sigma_N = E_N \varepsilon_N$.

To summarize, the LDPM material parameters for shale governing the elastic behavior, the fracturing behavior, and the frictional behavior are shown in Table 12.5.

12.3.2.3 Preliminary Results

The proposed micromechanical discrete approach has been implemented into the MARS software [178], which is a structural analysis computer code with an object-oriented architecture. The model was calibrated against experimental results for Boryeong shale provided by J. Chao et al. [20, 179] through a trial-and-error calibration procedure. The simulated experiments include static elastic analysis, Brazilian tensile, and uniaxial compression tests. As the preliminary results reported in [174] show, the dependence of the simulated elastic modulus, BTS, UCS, and failure modes, on sample orientation can be captured

Table 12.5 LDPM material parameters for anisotropic shale

Parameters	Units
Normal modulus E_{N0}	GPa
Ratio of normal modulus β_N	
Lamination reduction factor for normal modulus C_N	
Shear modulus E_{T0}	GPa
Ratio of shear modulus β_T	
Lamination reduction factor for shear modulus C_T	
Tensile strength σ_{t0}	MPa
Ratio of tensile strength β_t	
Lamination reduction factor for tensile strength C_{Nt}	
Shear strength σ_{s0}	MPa
Ratio of shear strength β_s	
Lamination reduction factor for shear strength C_{Ts}	
Characteristic length l_t	mm
Softening exponent n_t	
Initial frictional coefficient μ_{00}	
Ratio of initial frictional coefficient β_u	
Lamination reduction factor for initial friction C_μ	

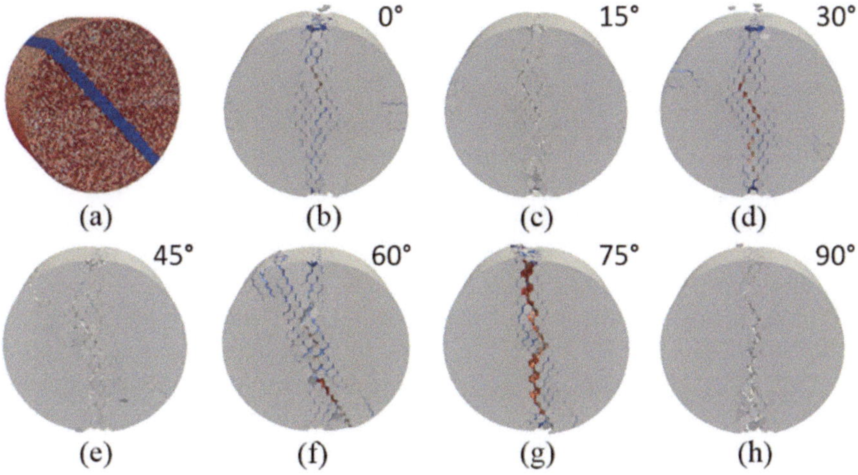

Fig. 12.19 Cylinder specimens for Brazilian tensile test. (**a**) Snapshot of mesostructure; (**b**)–(**h**) simulated specimens with anisotropy angles of 0°, 15°, 30°, 45°, 60°, 75°, and 90° after failure in the test [174]

successfully. Especially, simulated specimens after failure in both Brazilian and uniaxial compression tests illustrated different failure mechanisms at different anisotropy angles or loading orientations, as shown in Figs. 12.19 and 12.20 respectively, which are in close agreement with the corresponding experimental data.

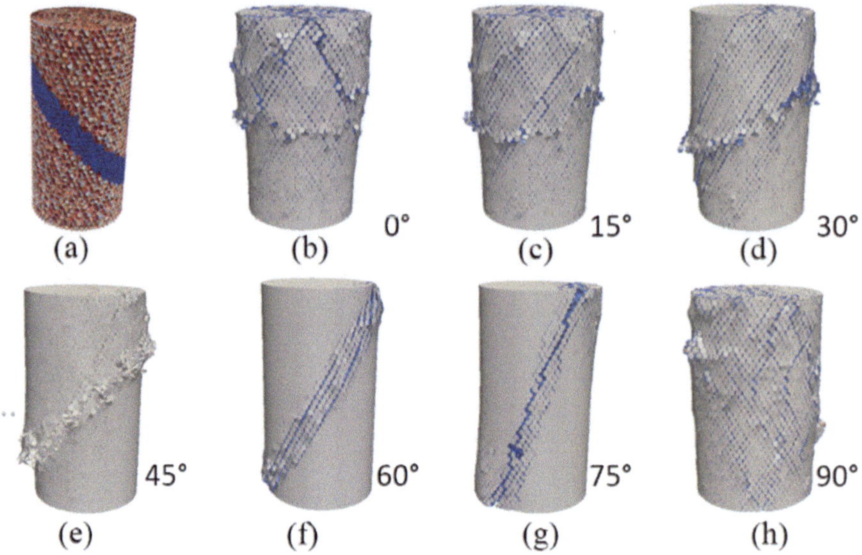

Fig. 12.20 Cylinder specimens for unconfined compressive strength test. (**a**) Snapshot of mesostructure; (**b**)–(**h**) simulated specimens with anisotropy angles of 0°, 15°, 30°, 45°, 60°, 75°, and 90° after failure in the test [174]

12.3.3 Discussion

In order to capture the micromechanics of fracture initiation, propagation, and nucleation, which finally lead to the failure and fracture of shale rock at the macroscopic and field levels, a numerical model needs to incorporate the microscopic events, especially the growth and coalescence of microdefects at the grain scale. Both advanced continuum and discontinuum modeling techniques nowadays provide features to simulate damage evolution at the microstructural level, and are thus able to capture localization events. Although continuum models usually offer more economical solutions in terms of computational efficiency, discontinuum modeling techniques provide straightforward and physical-based solutions when simulating intrinsic discontinuities at multiple length scales, especially the weakness zones often observed at the mesoscopic level. Therefore, the micromechanical discrete based model presented here yields a unique advantage in characterizing mechanical behaviors of shale by taking into account its heterogeneous nature. However, the predictive capability critically relies on the constitutive assumption at the meso- and microscales, which are powered by the knowledge of shale microstructure and microscopic failure mechanism, and the experimental data used for model calibration at different scales. Unfortunately, a comprehensive set of experimental measurements spanning all multiple scales concerned is currently not available. In addition, a multiscale scheme is needed to predict shale macroscopic behavior starting from high-fidelity fine-scale models with the limited computational power currently available.

References

1. Dormieux, L., & Ulm, F.-J. (2005). *Applied micromechanics of porous materials.* Vienna: Springer.
2. Gale, J. F. W., Laubach, S. E., Olson, J. E., Eichhubl, P., & Fall, A. (2014). Natural fractures in shale: A review and new observations. *AAPG Bulletin, 98*, 2165–2216. doi:10.1306/08121413151.
3. Koesoemadinata, A., El-Kaseeh, G., Banik, N., Dai, J., Egan, M., & Gonzalez, A., et al. (2011). *Seismic reservoir characterization in Marcellus shale.* 2011 SEG Annual Meeting. Society of Exploration Geophysicists.
4. Zhu, Y., Liu, E., Martinez, A., Payne, M. A., & Harris, C. E. (2011). Understanding geophysical responses of shale-gas plays. *The Leading Edge, 30*, 332–338. doi:10.1190/1.3567265.
5. Goodway, B., Perez, M., Varsek, J., & Abaco, C. (2010). Seismic petrophysics and isotropic-anisotropic AVO methods for unconventional gas exploration. *The Leading Edge, 29*, 1500–1508. doi:10.1190/1.3525367.
6. Treadgold, G., Campbell, B., McLain, B., Sinclair, S., & Nicklin, D. (2011). Eagle Ford shale prospecting with 3D seismic data within a tectonic and depositional system framework. *The Leading Edge, 30*, 48–53. doi:10.1190/1.3535432.
7. Fertl, W. H. (1976). Evaluation of oil shales using geophysical well-logging techniques. In G. V. Chilingarian & T. F. Yen (Eds.), *Developments in petroleum science* (pp. 199–213). Amsterdam: Elsevier.
8. Sondergeld, C. H., Newsham, K. E., Comisky, J. T., Rice, M. C., & Rai, C. S. (2010). *Petrophysical considerations in evaluating and producing shale gas resources.* SPE Unconventional Gas Conference, Society of Petroleum Engineers. doi:10.2118/131768-MS.
9. Abousleiman, Y., Tran, M. H., Hoang, S., Ortega, A., & Ulm, F.-J. (2009). *GeoMechanics field characterization of the two prolific U.S. Mid-west gas plays with advanced wire-line logging tools.* SPE Annual Technical Conference and Exhibition, Society of Petroleum Engineers. doi:10.2118/124428-MS.
10. Abousleiman, Y., Tran, M., Hoang, S., Ortega, J. A., & Ulm, F.-J. (2010). Geomechanics field characterization of Woodford shale and Barnett shale with advanced logging tools and nano-indentation on drill cuttings. *The Leading Edge, 29*, 730–736. doi:10.1190/1.3447787.
11. Brevik, I., Ahmadi, G. R., Hatteland, T., & Rojas, M. A. (2007). Documentation and quantification of velocity anisotropy in shales using wireline log measurements. *The Leading Edge, 26*, 272–277. doi:10.1190/1.2715048.
12. Kebaili, A., & Schmitt, D. (1996). Velocity anisotropy observed in wellbore seismic arrivals: Combined effects of intrinsic properties and layering. *Geophysics, 61*, 12–20. doi:10.1190/1.1443932.
13. Wong, R. C. K., Schmitt, D. R., Collis, D., & Gautam, R. (2008). Inherent transversely isotropic elastic parameters of over-consolidated shale measured by ultrasonic waves and their comparison with static and acoustic in situ log measurements. *Journal of Geophysics and Engineering, 5*, 103. doi:10.1088/1742-2132/5/1/011.
14. Horsrud, P. (2001). Estimating mechanical properties of shale from empirical correlations. *SPE Drilling and Completion, 16*, 68–73. doi:10.2118/56017-PA.
15. Lashkaripour, G. R. (1993). A statistical study on shale properties: Relationship among principal shale properties. PMGE Aust. Probabilistic Method Geotech. Eng.
16. McNally, G. H. (1987). Estimation of coal measures rock strength using sonic and neutron logs. *Geoexploration, 24*, 381–395. doi:10.1016/0016-7142(87)90008-1.
17. Onyia, E. C. (1988). Relationships between formation strength. *Drilling Strength, and Electric Log Properties.* doi:10.2118/18166-MS.
18. Chang, C., Zoback, M. D., & Khaksar, A. (2006). Empirical relations between rock strength and physical properties in sedimentary rocks. *Journal of Petroleum Science and Engineering, 51*, 223–237. doi:10.1016/j.petrol.2006.01.003.

19. Abousleiman, Y. N., Tran, M. H., Hoang, S., Bobko C. P., Ortega, A., & Ulm, F.-J. (2007). *Geomechanics field and laboratory characterization of the woodford shale: The next gas play*. SPE Annual Technical Conference and Exhibition, Society of Petroleum Engineers. doi:10.2118/110120-MS.
20. Kim, H., Cho, J.-W., Song, I., & Min, K.-B. (2012). Anisotropy of elastic moduli, P-wave velocities, and thermal conductivities of Asan Gneiss, Boryeong Shale, and Yeoncheon Schist in Korea. *Engineering Geology, 147–148*, 68–77. doi:10.1016/j.enggeo.2012.07.015.
21. Nes, O.-M., Sonstebo, E. F., Horsrud, P., & Holt, R. M. (1998). *Dynamic and static measurements on mm-size shale samples*. SPE/ISRM Rock Mechanics in Petroleum Engineering, Society of Petroleum Engineers. doi:10.2118/47200-MS.
22. Slatt, R., & Abousleiman, Y. (2011). Merging sequence stratigraphy and geomechanics for unconventional gas shales. *The Leading Edge, 30*, 274–282. doi:10.1190/1.3567258.
23. Sondergeld, C. H., & Rai, C. S. (2011). Elastic anisotropy of shales. *The Leading Edge, 30*, 324–331. doi:10.1190/1.3567264.
24. Sone, H., & Zoback, M. D. (2013). Mechanical properties of shale-gas reservoir rocks—Part 1: Static and dynamic elastic properties and anisotropy. *Geophysics, 78*, D381–D392. doi:10.1190/geo2013-0050.1.
25. Sone, H., & Zoback, M. D. (2013). Mechanical properties of shale-gas reservoir rocks—Part 2: Ductile creep, brittle strength, and their relation to the elastic modulus. *Geophysics, 78*, D393–D402. doi:10.1190/geo2013-0051.1.
26. Fjær, E., & Nes, O.-M. (2014). The impact of heterogeneity on the anisotropic strength of an outcrop shale. *Rock Mechanics and Rock Engineering, 47*, 1603–1611. doi:10.1007/s00603-014-0598-5.
27. Simpson, N. D. J., Stroisz, A., Bauer, A., Vervoort, A., & Holt, R. M. (2014). *Failure mechanics of anisotropic shale during Brazilian tests*. 48th U.S. Rock Mechanics/Geomechanics Symposium.
28. Sierra, R., Tran, M. H., Abousleiman, Y. N., & Slatt, R. M., et al. (2010). *Woodford Shale mechanical properties and the impacts of lithofacies*. 44th US Rock Mech. Symp. 5th US-Can. Rock Mech. Symp.
29. Niandou, H., Shao, J. F., Henry, J. P., & Fourmaintraux, D. (1997). Laboratory investigation of the mechanical behaviour of Tournemire shale. *International Journal of Rock Mechanics and Mining Sciences, 34*, 3–16. doi:10.1016/S1365-1609(97)80029-9.
30. Britt, L. K., & Schoeffler, J. (2009). *The geomechanics of a shale play: What makes a shale prospective*. SPE Eastern Regional Meeting, Society of Petroleum Engineers. doi:10.2118/125525-MS.
31. Higgins, S. M., Goodwin, S. A., Donald, A., Bratton, T. R., & Tracy G. W. (2008) *Anisotropic stress models improve completion design in the Baxter shale*. SPE Annual Technical Conference and Exhibition, Society of Petroleum Engineers. doi:10.2118/115736-MS.
32. Khan, S., Williams, R. E., Ansari, S., & Khosravi, N. (2012). *Impact of mechanical anisotropy on design of hydraulic fracturing in shales*. Abu Dhabi International Petroleum Conference and Exhibition, Society of Petroleum Engineers. doi:10.2118/162138-MS.
33. Suarez-Rivera, R., Green, S. J., McLennan, J., & Bai, M. (2006). *Effect of layered heterogeneity on fracture initiation in tight gas shales*. SPE Annual Technical Conference and Exhibition, Society of Petroleum Engineers. doi:10.2118/103327-MS.
34. Chandler, M., Meredith, P., & Crawford, B. (2013). *Experimental determination of the fracture toughness and brittleness of the mancos shale, Utah*. p EGU2013.
35. Jarvie, D. M., Hill, R. J., Ruble, T. E., & Pollastro, R. M. (2015). Unconventional shale-gas systems: The Mississippian Barnett Shale of north-central Texas as one model for thermogenic shale-gas assessment. *AAPG Bulletin, 91*(4), 475–499.
36. Li, Q., Chen, M., Zhou, Y., Jin, Y., Wang, F. P., & Zhang, R. (2013). *Rock mechanical properties of shale gas reservoir and their influences on hydraulic fracture*. IPTC 2013: International Petroleum Technology Conference. doi:10.2523/16580-MS.

37. Rickman, R., Mullen, M. J., Petre, J. E., Grieser, W. V., & Kundert, D. (2008). *A practical use of shale petrophysics for stimulation design optimization: All shale plays are not clones of the Barnett shale*. SPE Annual Technical Conference and Exhibition, Society of Petroleum Engineers. doi:10.2118/115258-MS.
38. Wang, F. P., & Gale, J. F. W. (2009). Screening criteria for shale-Gas systems. *Gulf Coast Association of Geological Societies Trans, 59*, 779–793.
39. Holt, R. M., Fjaer, E., Nes, O. M., & Alassi, H. T. (2011). *A shaly look at brittleness*. 45th U.S. Rock Mechanics/Geomechanics Symposium.
40. Holt, R. M., Fjær, E., Stenebråten, J. F., & Nes, O.-M. (2015). Brittleness of shales: Relevance to borehole collapse and hydraulic fracturing. *Journal of Petroleum Science and Engineering, 131*, 200–209. doi:10.1016/j.petrol.2015.04.006.
41. Hucka, V., & Das, B. (1974). Brittleness determination of rocks by different methods. *International Journal of Rock Mechanics and Mining Science and Geomechanics Abstracts, 11*, 389–392. doi:10.1016/0148-9062(74)91109-7.
42. Yang, Y., Sone, H., Hows, A., & Zoback, M. D. (2013). Comparison of brittleness indices in organic-rich shale formations. 47th U.S. Rock Mechanics/Geomechanics Symposium.
43. Jin, X., Shah, S. N., Roegiers, J.-C., & Zhang, B. (2014). *Fracability evaluation in shale reservoirs—an integrated petrophysics and geomechanics approach*. SPE Hydraulic Fracturing Technology Conference, Society of Petroleum Engineers. doi:10.2118/168589-MS.
44. Schmidt, R. A. (1977). Fracture mechanics of oil shale—unconfined fracture toughness, stress corrosion cracking, and tension test results. The 18th U.S. Symposium on Rock Mechanics (USRMS).
45. Akono, A.-T. (2013). *Assessment of fracture properties and rate effects on fracture of materials by micro scratching: Application to gas shale*. Thesis, Massachusetts Institute of Technology.
46. Haddad, M., & Sepehrnoori, K. (2015). Simulation of hydraulic fracturing in quasi-brittle shale formations using characterized cohesive layer: Stimulation controlling factors. *Journal of Unconventional Oil and Gas Resources, 9*, 65–83. doi:10.1016/j.juogr.2014.10.001.
47. Parisio, F., Samat, S., & Laloui, L. (2014). An elasto-plastic-damage model for quasi-brittle shales. Proceedings of 48th US Rock Mechanics Symposium.
48. Yao, Y. (2011). Linear elastic and cohesive fracture analysis to model hydraulic fracture in brittle and ductile rocks. *Rock Mechanics and Rock Engineering, 45*, 375–387. doi:10.1007/s00603-011-0211-0.
49. Eseme, E., Urai, J. L., Krooss, B. M., & Littke, R. (2007). Review of the mechanical properties of oil shales: Implications for exploitation and basin modelling. *Oil Shale, 24*(2), 159–175.
50. Sayers, C. M. (2013). The effect of anisotropy on the Young's moduli and Poisson's ratios of shales. *Geophysical Prospecting, 61*, 416–426. doi:10.1111/j.1365-2478.2012.01130.x.
51. Vernik, L., & Nur, A. (1992). Ultrasonic velocity and anisotropy of hydrocarbon source rocks. *Geophysics, 57*, 727–735. doi:10.1190/1.1443286.
52. Mokhtari M (2015). *Characterization of anisotropy in organic-rich shales: Shear and tensile failure, wave velocity, matrix and fracture permeability*. Thesis Colorado School of Mines.
53. Mokhtari, M., Bui, B. T., & Tutuncu, A. N. (2014). *Tensile failure of shales: Impacts of layering and natural fractures*. SPE Western North American and Rocky Mountain Joint Meeting, Society of Petroleum Engineers. doi:10.2118/169520-MS.
54. Eseme, E., Littke, R., & Krooss, B. M. (2006). Factors controlling the thermo-mechanical deformation of oil shales: Implications for compaction of mudstones and exploitation. *Marine and Petroleum Geology, 23*, 715–734. doi:10.1016/j.marpetgeo.2006.02.007.
55. Josh, M., Esteban, L., Delle Piane, C., Sarout, J., Dewhurst, D. N., & Clennell, M. B. (2012). Laboratory characterisation of shale properties. *Journal of Petroleum Science and Engineering, 88–89*, 107–124. doi:10.1016/j.petrol.2012.01.023.
56. Abousleiman, Y. N., Hoang, S. K., & Tran, M. H. (2010). Mechanical characterization of small shale samples subjected to fluid exposure using the inclined direct shear testing device.

International Journal of Rock Mechanics and Mining Sciences, 47, 355–367. doi:10.1016/j.ijrmms.2009.12.014.
57. Duveau, G., Shao, J. F., & Henry, J. P. (1998). Assessment of some failure criteria for strongly anisotropic geomaterials. *Mech Cohesive-Frict Mater, 3*, 1–26.
58. Aoki, T., Tan, C. P., & Bamford, W. E. (1993). Effects of deformation and strength anisotropy on borehole failures in saturated shales. *International Journal of Rock Mechanics and Mining Science and Geomechanics Abstracts, 30*, 1031–1034. doi:10.1016/0148-9062(93)90067-N.
59. Jaeger, J. C. (1960). Shear failure of anisotropic rocks. *Geological Magazine, 97*, 65–72. doi:10.1017/S0016756800061100.
60. CHEN, P., HAN, Q., Ma, T., & Lin, D. (2015). The mechanical properties of shale based on micro-indentation test. *Petroleum Exploration and Development, 42*, 723–732. doi:10.1016/S1876-3804(15)30069-0.
61. Ringstad, C., Lofthus, E. B., Sonstebo, E. F., Fjaer, E., Zausa, F., & Fuh, G.-F. (1998). *Prediction of rock parameters from micro-indentation measurements: The effect of sample size*. SPE/ISRM Rock Mechanics in Petroleum Engineering, Society of Petroleum Engineers. doi:10.2118/47313-MS.
62. Ulm, F.-J., & Abousleiman, Y. (2006). The nanogranular nature of shale. *Acta Geotechnica, 1*, 77–88. doi:10.1007/s11440-006-0009-5.
63. Akono, A.-T., & Ulm, F.-J. (2011). Scratch test model for the determination of fracture toughness. *Engineering Fracture Mechanics, 78*, 334–342. doi:10.1016/j.engfracmech.2010.09.017.
64. Sondergeld, C. H., Ambrose, R. J., Rai, C. S., & Moncrieff, J. (2010). *Micro-structural studies of gas shales*. SPE Unconventional Gas Conference, Society of Petroleum Engineers. doi:10.2118/131771-MS.
65. Rybacki, E., Reinicke, A., Meier, T., Makasi, M., & Dresen, G. (2015). What controls the mechanical properties of shale rocks?—Part I: Strength and Young's modulus. *Journal of Petroleum Science and Engineering, 135*, 702–722. doi:10.1016/j.petrol.2015.10.028.
66. Sayers, C. M. (2013). The effect of kerogen on the elastic anisotropy of organic-rich shales. *Geophysics, 78*, D65–D74. doi:10.1190/geo2012-0309.1.
67. Curtis, M. E., Ambrose, R. J., & Sondergeld, C. H. (2010). *Structural characterization of gas shales on the micro- and nano-scales*. Canadian unconventional resources and international petroleum conference, Society of Petroleum Engineers. doi:10.2118/137693-MS.
68. Curtis, M. E., Sondergeld, C. H., Ambrose, R. J., & Rai, C. S. (2012). Microstructural investigation of gas shales in two and three dimensions using nanometer-scale resolution imaging. *AAPG Bulletin, 96*, 665–677. doi:10.1306/08151110188.
69. Eliyahu, M., Emmanuel, S., Day-Stirrat, R. J., & Macaulay, C. I. (2015). Mechanical properties of organic matter in shales mapped at the nanometer scale. *Marine and Petroleum Geology, 59*, 294–304. doi:10.1016/j.marpetgeo.2014.09.007.
70. Bobko, C., & Ulm, F.-J. (2008). The nano-mechanical morphology of shale. *Mechanics of Materials, 40*, 318–337. doi:10.1016/j.mechmat.2007.09.006.
71. Ortega, J. A., Ulm, F.-J., & Abousleiman, Y. (2009). The nanogranular acoustic signature of shale. *Geophysics, 74*, D65–D84. doi:10.1190/1.3097887.
72. Kumar, V., Sondergeld, C. H., & Rai, C. S. (2012). *Nano to macro mechanical characterization of shale*. SPE Annual Technical Conference and Exhibition, Society of Petroleum Engineers. doi:10.2118/159804-MS.
73. Bennett, K. C., Berla, L. A., Nix, W. D., & Borja, R. I. (2015). Instrumented nanoindentation and 3D mechanistic modeling of a shale at multiple scales. *Acta Geotechnica, 10*, 1–14. doi:10.1007/s11440-014-0363-7.
74. Oliver, W. C., & Pharr, G. M. (2004). Measurement of hardness and elastic modulus by instrumented indentation: Advances in understanding and refinements to methodology. *Journal of Materials Research, 19*, 3–20.

75. Tavallali, A., & Vervoort, A. (2010). Failure of layered sandstone under Brazilian test conditions: Effect of micro-scale parameters on macro-scale behaviour. *Rock Mechanics and Rock Engineering, 43*, 641–653. doi:10.1007/s00603-010-0084-7.
76. Geertsma, J., & De Klerk, F. (1969). A rapid method of predicting width and extent of hydraulically induced fractures. *Journal of Petroleum Technology, 21*, 1571–1581. doi:10.2118/2458-PA.
77. Khristianovic, S., & Zheltov, Y. (1955). *Formation of vertical fractures by means of highly viscous fluids*. In Proceedings 4th World Pet Congress Rome. pp 579–586.
78. Nordgren, R. P. (1972). Propagation of a vertical hydraulic fracture. *Society of Petroleum Engineers Journal, 12*, 306–314. doi:10.2118/3009-PA.
79. Perkins, T. K., & Kern, L. R. (1961). Widths of hydraulic fractures. *Journal of Petroleum Technology, 13*, 937–949. doi:10.2118/89-PA.
80. Morales, R. H., & Abou-Sayed, A. S. (1989). Microcomputer analysis of hydraulic fracture behavior with a pseudo-three-dimensional simulator. *SPE Production Engineering, 4*, 69–74. doi:10.2118/15305-PA.
81. Settari, A., & Cleary, M. P. (1986). Development and testing of a pseudo-three-dimensional model of hydraulic fracture geometry. *SPE Production Engineering, 1*, 449–466. doi:10.2118/10505-PA.
82. Advani, S. H., Lee, T. S., & Lee, J. K. (1990). Three-dimensional modeling of hydraulic fractures in layered media: Part I—finite element formulations. *Journal of Energy Resources Technology, 112*, 1–9. doi:10.1115/1.2905706.
83. Gu, H., & Leung, K. H. (1993). 3D numerical simulation of hydraulic fracture closure with application to minifracture analysis. *Journal of Petroleum Technology, 45*, 206–255. doi:10.2118/20657-PA.
84. Morita, N., Whitfill, D. L., & Wahl, H. A. (1988). Stress-intensity factor and fracture cross-sectional shape predictions from a three-dimensional model for hydraulically induced fractures. *Journal of Petroleum Technology, 40*, 1329–1342. doi:10.2118/14262-PA.
85. Carter, B., Desroches, J., Ingraffea, A., & Wawrzynek, P. (2000). Simulating fully 3D hydraulic fracturing. *Model Geomechanics, 200*, 525–557.
86. Li, L. C., Tang, C. A., Li, G., Wang, S. Y., Liang, Z. Z., & Zhang, Y. B. (2012). Numerical simulation of 3D hydraulic fracturing based on an improved flow-stress-damage model and a parallel FEM technique. *Rock Mechanics and Rock Engineering, 45*, 801–818. doi:10.1007/s00603-012-0252-z.
87. Dahi-Taleghani, A., & Olson, J. E. (2011). Numerical modeling of multistranded-hydraulic-fracture propagation: Accounting for the interaction between induced and natural fractures. *SPE Journal, 16*, 575–581. doi:10.2118/124884-PA.
88. Lecampion, B. (2009). An extended finite element method for hydraulic fracture problems. *Communications in Numerical Methods in Engineering, 25*, 121–133. doi:10.1002/cnm.1111.
89. Carrier, B., & Granet, S. (2012). Numerical modeling of hydraulic fracture problem in permeable medium using cohesive zone model. *Engineering Fracture Mechanics, 79*, 312–328. doi:10.1016/j.engfracmech.2011.11.012.
90. Chen, Z., Bunger, A. P., Zhang, X., & Jeffrey, R. G. (2009). Cohesive zone finite element-based modeling of hydraulic fractures. *Acta Mechanica Solida Sinica, 22*, 443–452. doi:10.1016/S0894-9166(09)60295-0.
91. Sarris, E., & Papanastasiou, P. (2011). Modeling of hydraulic fracturing in a poroelastic cohesive formation. *International Journal of Geomechanics, 12*, 160–167. doi:10.1061/(ASCE)GM.1943-5622.0000121.
92. Yuan, S. C., & Harrison, J. P. (2006). A review of the state of the art in modelling progressive mechanical breakdown and associated fluid flow in intact heterogeneous rocks. *International Journal of Rock Mechanics and Mining Sciences, 43*, 1001–1022. doi:10.1016/j.ijrmms.2006.03.004.
93. Sammis, C. G., & Ashby, M. F. (1986). The failure of brittle porous solids under compressive stress states. *Acta Metallurgica, 34*, 511–526. doi:10.1016/0001-6160(86)90087-8.

94. Zhang, J., Wong, T.-F., & Davis, D. M. (1990). Micromechanics of pressure-induced grain crushing in porous rocks. *Journal of Geophysical Research - Solid Earth, 95*, 341–352. doi:10.1029/JB095iB01p00341.
95. Ashby, M. F., & Hallam (Née Cooksley), S. D. (1986). The failure of brittle solids containing small cracks under compressive stress states. *Acta Metallurgica, 34*, 497–510. doi:10.1016/0001-6160(86)90086-6.
96. Brace, W. F., & Bombolakis, E. G. (1963). A note on brittle crack growth in compression. *Journal of Geophysical Research, 68*, 3709–3713. doi:10.1029/JZ068i012p03709.
97. Horii, H., & Nemat-Nasser, S. (1985). Compression-induced microcrack growth in brittle solids: Axial splitting and shear failure. *Journal of Geophysical Research - Solid Earth, 90*, 3105–3125. doi:10.1029/JB090iB04p03105.
98. Jing, L. (2003). A review of techniques, advances and outstanding issues in numerical modelling for rock mechanics and rock engineering. *International Journal of Rock Mechanics and Mining Sciences, 40*, 283–353. doi:10.1016/S1365-1609(03)00013-3.
99. Borst, R. D., Crisfield, M. A., Remmers, J. J. C., & Verhoosel, C. V. (2012). *Nonlinear finite element analysis of solids and structures*. Hoboken: John Wiley & Sons.
100. Bobet, A., Fakhimi, A., Johnson, S., Morris, J., Tonon, F., & Yeung, M. (2009). Numerical models in discontinuous media: Review of advances for rock mechanics applications. *Journal of Geotechnical and Geoenvironmental Engineering, 135*, 1547–1561. doi:10.1061/(ASCE)GT.1943-5606.0000133.
101. Lisjak, A., & Grasselli, G. (2014). A review of discrete modeling techniques for fracturing processes in discontinuous rock masses. *Journal of Rock Mechanics and Geotechnical Engineering, 6*, 301–314. doi:10.1016/j.jrmge.2013.12.007.
102. Bažant, Z. P., & Oh, B. H. (1983). Crack band theory for fracture of concrete. *Materiales de Construcción, 16*, 155–177. doi:10.1007/BF02486267.
103. Hillerborg, A., Modéer, M., & Petersson, P.-E. (1976). Analysis of crack formation and crack growth in concrete by means of fracture mechanics and finite elements. *Cement and Concrete Research, 6*, 773–781. doi:10.1016/0008-8846(76)90007-7.
104. Collin, F., Chambon, R., & Charlier, R. (2006). A finite element method for poro mechanical modelling of geotechnical problems using local second gradient models. *International Journal for Numerical Methods in Engineering, 65*, 1749–1772. doi:10.1002/nme.1515.
105. Mühlhaus, H.-B., & Alfantis, E. C. (1991). A variational principle for gradient plasticity. *International Journal of Solids and Structures, 28*, 845–857. doi:10.1016/0020-7683(91)90004-Y.
106. Bažant, Z. P., & Pijaudier-Cabot, G. (1988). Nonlocal continuum damage, localization instability and convergence. *Journal of Applied Mechanics, 55*, 287–293. doi:10.1115/1.3173674.
107. Adhikary, D. P., & Dyskin, A. V. (1997). A Cosserat continuum model for layered materials. *Computers and Geotechnics, 20*, 15–45. doi:10.1016/S0266-352X(96)00011-0.
108. Riahi, A., & Curran, J. H. (2009). Full 3D finite element Cosserat formulation with application in layered structures. *Applied Mathematical Modelling, 33*, 3450–3464. doi:10.1016/j.apm.2008.11.022.
109. Fang, Z., & Harrison, J. P. (2002). Application of a local degradation model to the analysis of brittle fracture of laboratory scale rock specimens under triaxial conditions. *International Journal of Rock Mechanics and Mining Sciences, 39*, 459–476. doi:10.1016/S1365-1609(02)00036-9.
110. Fang, Z., & Harrison, J. P. (2002). Development of a local degradation approach to the modelling of brittle fracture in heterogeneous rocks. *International Journal of Rock Mechanics and Mining Sciences, 39*, 443–457. doi:10.1016/S1365-1609(02)00035-7.
111. Tang, C. A., Liu, H., Lee, P. K. K., Tsui, Y., & Tham, L. G. (2000). Numerical studies of the influence of microstructure on rock failure in uniaxial compression—Part I: Effect of heterogeneity. *International Journal of Rock Mechanics and Mining Sciences, 37*, 555–569. doi:10.1016/S1365-1609(99)00121-5.

112. Cundall, P. A. (1971). A computer model for simulating progressive large scale movements in blocky rock systems. Proceedings symposium on rock fracture. ISRM Nancy 1
113. Cundall, P. A., & Strack, O. D. (1979). A discrete numerical model for granular assemblies. *Geotechnique, 29*, 47–65.
114. Jing, L., & Stephansson, O. (2007). *Fundamentals of discrete element methods for rock engineering: Theory and applications: Theory and applications*. Amsterdam: Elsevier.
115. Cundall, D. P. A. (1989). Numerical experiments on localization in frictional materials. *Ing-Arch, 59*, 148–159. doi:10.1007/BF00538368.
116. Itasca Consulting Group Inc. (2011). *Universal Distinct Element Code (UDEC), Version 5.0*. Minneapolis: Itasca Consulting Group, Inc.
117. Itasca Consulting Group Inc. (1999). *PFC2D/3D (Particle Flow Code in 2/3 Dimensions), Version 2.0*. Minneapolis, MN: Itasca Consulting Group, Inc.
118. Potyondy, D. O., & Cundall, P. A. (2004). A bonded-particle model for rock. *International Journal of Rock Mechanics and Mining Sciences, 41*, 1329–1364. doi:10.1016/j.ijrmms.2004.09.011.
119. Schlangen, E., & Mier, J. G. M. V. (1992). Micromechanical analysis of fracture of concrete. *International Journal of Damage Mechanics, 1*, 435–454. doi:10.1177/105678959200100404.
120. Schlangen, E., & Mier, J. G. M. V. (1995). Crack propagation in sandstone: Combined experimental and numerical approach. *Rock Mechanics and Rock Engineering, 28*, 93–110. doi:10.1007/BF01020063.
121. Schlangen, E., & van Mier, J. G. M. (1992). Experimental and numerical analysis of micromechanisms of fracture of cement-based composites. *Cement and Concrete Composites, 14*, 105–118. doi:10.1016/0958-9465(92)90004-F.
122. Schlangen, E., & van Mier, J. G. M. (1992). Simple lattice model for numerical simulation of fracture of concrete materials and structures. *Materials and Structures, 25*, 534–542. doi:10.1007/BF02472449.
123. Schlangen, E., & Van Mier, J. (1994). Fracture simulations in concrete and rock using a random lattice. *Computer Methods Advance Geomechanics, 2*, 1641–1646.
124. Schlangen, E., & Garboczi, E. J. (1997). Fracture simulations of concrete using lattice models: Computational aspects. *Engineering Fracture Mechanics, 57*, 319–332. doi:10.1016/S0013-7944(97)00010-6.
125. Schlangen, E., & Qian, Z. (2009). 3d modeling of fracture in cement-based materials. *Journal of Multiscale Modeling, 01*, 245–261. doi:10.1142/S1756973709000116.
126. Bazant, Z. P., Tabbara, M., Kazemi, M., & Pijaudier-Cabot, G. (1990). Random particle model for fracture of aggregate or fiber composites. *Journal of Engineering Mechanics, 116*, 1686–1705. doi:10.1061/(ASCE)0733-9399(1990)116:8(1686).
127. Song, J. S., & Kim, K. S. (1995). *Blasting induced fracturing and stress field evolution at fracture tips*. The 35th U.S. Symposium on Rock Mechanics (USRMS).
128. Place, D., & Mora, P. (2000). Numerical simulation of localisation phenomena in a fault zone. In P. Mora, M. Matsu'ura R. Madariaga, J.-B. Minster (Eds.), Microsc. Macrosc. Simul. Predict. Model. Earthq. Process. (pp. 1821–1845) Birkhäuser Basel.
129. Bolander, J. E., & Saito, S. (1998). Fracture analyses using spring networks with random geometry. *Engineering Fracture Mechanics, 61*, 569–591.
130. Bolander, J. E., & Sukumar, N. (2005). Irregular lattice model for quasistatic crack propagation. *Physical Review B, 71*, 094106. doi:10.1103/PhysRevB.71.094106.
131. Katsman, R., Aharonov, E., & Scher, H. (2005). Numerical simulation of compaction bands in high-porosity sedimentary rock. *Mechanics of Materials, 37*, 143–162. doi:10.1016/j.mechmat.2004.01.004.
132. Jirásek, M., & Bažant, Z. P. (1994). Macroscopic fracture characteristics of random particle systems. *International Journal of Fracture, 69*, 201–228. doi:10.1007/BF00034763.

133. Cusatis, G., Bažant, Z. P., & Luigi, C. (2003). Confinement-shear lattice model for concrete damage in tension and compression: I. Theory. *The Journal of Engineering, 129*, 1439–1448. doi:10.1061/(ASCE)0733-9399(2003)129:12(1439).
134. Lilliu, G., & van Mier, J. G. M. (2003). 3D lattice type fracture model for concrete. *Engineering Fracture Mechanics, 70*, 927–941. doi:10.1016/S0013-7944(02)00158-3.
135. Yip, M., Mohle, J., & Bolander, J. E. (2005). Automated modeling of three-dimensional structural components using irregular lattices. *Computer-Aided Civil and Infrastructure Engineering, 20*, 393–407. doi:10.1111/j.1467-8667.2005.00407.x.
136. Munjiza, A. A. (2004). *The combined finite-discrete element method.* Hoboken: John Wiley & Sons.
137. Munjiza, A., Owen, D. R. J., & Bicanic, N. (1995). A combined finite-discrete element method in transient dynamics of fracturing solids. *Engineering Computations, 12*, 145–174. doi:10.1108/02644409510799532.
138. Fabian, D., Peter, C., Daniel, B., & Torsten, G. (2007). Evaluation of damage-induced permeability using a three-dimensional Adaptive Continuum/Discontinuum Code (AC/DC). *Physics and Chemistry of the Earth, Parts ABC, 32*, 681–690. doi:10.1016/j.pce.2006.01.006.
139. Deb, D., & Das, K. C. (2010). Extended finite element method for the analysis of discontinuities in rock masses. *Geotechnical and Geological Engineering, 28*, 643–659. doi:10.1007/s10706-010-9323-7.
140. Fries, T.-P., & Belytschko, T. (2006). The intrinsic XFEM: A method for arbitrary discontinuities without additional unknowns. *International Journal for Numerical Methods in Engineering, 68*, 1358–1385. doi:10.1002/nme.1761.
141. Ma, G. W., Wang, X. J., & Ren, F. (2011). Numerical simulation of compressive failure of heterogeneous rock-like materials using SPH method. *International Journal of Rock Mechanics and Mining Sciences, 48*, 353–363. doi:10.1016/j.ijrmms.2011.02.001.
142. Ha, Y. D., & Bobaru, F. (2010). Studies of dynamic crack propagation and crack branching with peridynamics. *International Journal of Fracture, 162*, 229–244. doi:10.1007/s10704-010-9442-4.
143. Lisjak, A., Grasselli, G., & Vietor, T. (2014). Continuum–discontinuum analysis of failure mechanisms around unsupported circular excavations in anisotropic clay shales. *International Journal of Rock Mechanics and Mining Sciences, 65*, 96–115. doi:10.1016/j.ijrmms.2013.10.006.
144. Goodman, R. E. (1989). *Introduction to rock mechanics* (2nd ed.). New York: Wiley.
145. Mahjoub, M., Rouabhi, A., Tijani, M., & Granet, S. (2015). An approach to model the mechanical behavior of transversely isotropic materials. *International Journal for Numerical and Analytical Methods in Geomechanics n/a–n/a.* doi:10.1002/nag.2469.
146. Rouabhi, A., Tijani, M., & Rejeb, A. (2007). Triaxial behaviour of transversely isotropic materials: Application to sedimentary rocks. *International Journal for Numerical and Analytical Methods in Geomechanics, 31*, 1517–1535. doi:10.1002/nag.605.
147. Tien, Y. M., & Kuo, M. C. (2001). A failure criterion for transversely isotropic rocks. *International Journal of Rock Mechanics and Mining Sciences, 38*, 399–412.
148. Adhikary, D. P., & Guo, H. (2014). An orthotropic Cosserat elasto-plastic model for layered rocks. *Rock Mechanics and Rock Engineering, 35*, 161–170. doi:10.1007/s00603-001-0020-y.
149. Pietruszczak, S., Lydzba, D., & Shao, J. F. (2002). Modelling of inherent anisotropy in sedimentary rocks. *International Journal of Solids and Structures, 39*, 637–648. doi:10.1016/S0020-7683(01)00110-X.
150. Salager, S., François, B., Nuth, M., & Laloui, L. (2012). Constitutive analysis of the mechanical anisotropy of Opalinus Clay. *Acta Geotechnica, 8*, 137–154. doi:10.1007/s11440-012-0187-2.
151. Mas Ivars, D., Pierce, M. E., Darcel, C., Reyes-Montes, J., Potyondy, D. O., Paul Young, R., et al. (2011). The synthetic rock mass approach for jointed rock mass modelling.

International Journal of Rock Mechanics and Mining Sciences, 48, 219–244. doi:10.1016/j.ijrmms.2010.11.014.

152. Park, B., & Min, K.-B. (2015). Bonded-particle discrete element modeling of mechanical behavior of transversely isotropic rock. *International Journal of Rock Mechanics and Mining Sciences, 76*, 243–255. doi:10.1016/j.ijrmms.2015.03.014.
153. Lisjak, A., Tatone, B. S. A., Grasselli, G., & Vietor, T. (2014). Numerical modelling of the anisotropic mechanical behaviour of Opalinus clay at the laboratory-scale using FEM/DEM. *Rock Mechanics and Rock Engineering, 47*, 187–206. doi:10.1007/s00603-012-0354-7.
154. Debecker, B., & Vervoort, A. (2013). Two-dimensional discrete element simulations of the fracture behaviour of slate. *International Journal of Rock Mechanics and Mining Sciences, 61*, 161–170. doi:10.1016/j.ijrmms.2013.02.004.
155. Duan, K., & Kwok, C. Y. (2015). Discrete element modeling of anisotropic rock under Brazilian test conditions. *International Journal of Rock Mechanics and Mining Sciences, 78*, 46–56. doi:10.1016/j.ijrmms.2015.04.023.
156. Park, B., & Min, K.-B. (2013). Discrete element modeling of transversely isotropic rock. 47th U.S. Rock Mechanics/Geomechanics Symposium.
157. Park, B., Park, B., & Min, K.-B. (2012). Discrete element modelling of shale as a transversely isotropic rock. ISRM Regional Symposium—7th Asian Rock Mechanics Symposium.
158. Brochard, L., Hantal, G., Laubie, H., Ulm, F., & Pellenq, R. (2013). Fracture mechanisms in organic-rich shales: Role of Kerogen. *Fifth Biot Conference on Poromechanics* (pp. 2471–2480).
159. Katti, D. R., Katti, K. S., & Alstadt, K. An insight into molecular scale interactions and in-situ nanomechanical properties of kerogen in green river oil shale. In V. S. Poromechanics (ed.), Proceedings Fifth Biot Conf. Poromechanics. ASCE, pp 2510–2516.
160. Zhang, Z., & Jamili, A. (2015). Modeling the Kerogen 3D molecular structure. doi:10.2118/175991-MS
161. Cosenza, P., Prêt, D., Giraud, A., & Hedan, S. (2015). Effect of the local clay distribution on the effective elastic properties of shales. *Mechanics of Materials, 84*, 55–74. doi:10.1016/j.mechmat.2015.01.016.
162. Hantal, G., Brochard, L., Laubie, H., Ebrahimi, D., Pellenq, R. J.-M., Ulm, F.-J., et al. (2014). Atomic-scale modelling of elastic and failure properties of clays. *Molecular Physics, 112*, 1294–1305. doi:10.1080/00268976.2014.897393.
163. Bobko, C. P., Gathier, B., Ortega, J. A., Ulm, F.-J., Borges, L., & Abousleiman, Y. N. (2011). The nanogranular origin of friction and cohesion in shale-A strength homogenization approach to interpretation of nanoindentation results. *International Journal for Numerical and Analytical Methods in Geomechanics, 35*, 1854–1876. doi:10.1002/nag.984.
164. Hassani, B., & Hinton, E. (1998). A review of homogenization and topology optimization II—analytical and numerical solution of homogenization equations. *Computers and Structures, 69*, 719–738. doi:10.1016/S0045-7949(98)00132-1.
165. Chung, P. W., Tamma, K. K., & Namburu, R. R. (2001). Asymptotic expansion homogenization for heterogeneous media: Computational issues and applications. *Composites Part Applied Science and Manufacturing, 32*, 1291–1301. doi:10.1016/S1359-835X(01)00100-2.
166. Fish, J., Shek, K., Pandheeradi, M., & Shephard, M. S. (1997). Computational plasticity for composite structures based on mathematical homogenization: Theory and practice. *Computer Methods in Applied Mechanics and Engineering, 148*, 53–73. doi:10.1016/S0045-7825(97)00030-3.
167. Ghosh, S., Lee, K., & Moorthy, S. (1995). Multiple scale analysis of heterogeneous elastic structures using homogenization theory and Voronoi cell finite element method. *International Journal of Solids and Structures, 32*, 27–62. doi:10.1016/0020-7683(94)00097-G.
168. Ghosh, S., Lee, K., & Moorthy, S. (1996). Two scale analysis of heterogeneous elastic-plastic materials with asymptotic homogenization and Voronoi cell finite element model. *Computer Methods in Applied Mechanics and Engineering, 132*, 63–116. doi:10.1016/0045-7825(95)00974-4.

169. Fish, J., Chen, W., & Li, R. (2007). Generalized mathematical homogenization of atomistic media at finite temperatures in three dimensions. *Computer Methods in Applied Mechanics and Engineering, 196*, 908–922. doi:10.1016/j.cma.2006.08.001.
170. Feyel, F. (2003). A multilevel finite element method (FE2) to describe the response of highly non-linear structures using generalized continua. *Computer Methods in Applied Mechanics and Engineering, 192*, 3233–3244. doi:10.1016/S0045-7825(03)00348-7.
171. Forest, S., Pradel, F., & Sab, K. (2001). Asymptotic analysis of heterogeneous Cosserat media. *International Journal of Solids and Structures, 38*, 4585–4608. doi:10.1016/S0020-7683(00)00295-X.
172. Rezakhani, R., & Cusatis, G. (2015). Asymptotic expansion homogenization of discrete fine-scale models with rotational degrees of freedom for the simulation of quasi-brittle materials. *Journal of the Mechanics and Physics of Solids, 88*, 320–345.
173. Cusatis, G., Rezakhani, R., Alnaggar, M., Zhou, X., & Pelessone, D. (2014). Multiscale computational models for the simulation of concrete materials and structures. In N. Bícanić, H. Mang, G. Meschke, & R. de Borst (Eds.), *Computer modeling concrete structures* (pp. 23–38). Boca Raton: CRC Press.
174. Li, W., Jin, C., Salviato, M., & Cusatis, G. (2015). *Modeling of failure behavior of anisotropic shale using lattice discrete particle model*. 49th U.S. Rock Mechanics/Geomechanics Symposium.
175. Cusatis, G., Mencarelli, A., Pelessone, D., & Baylot, J. (2011). Lattice Discrete Particle Model (LDPM) for failure behavior of concrete. II: Calibration and validation. *Cement and Concrete Composites, 33*, 891–905. doi:10.1016/j.cemconcomp.2011.02.010.
176. Cusatis, G., Pelessone, D., & Mencarelli, A. (2011). Lattice Discrete Particle Model (LDPM) for failure behavior of concrete. I: Theory. *Cement and Concrete Composites, 33*, 881–890. doi:10.1016/j.cemconcomp.2011.02.011.
177. Bramlette, M. N. (1946). *The Monterey Formation of California and the origin of its siliceous rocks*. (Vol. 212). US Government Printing Office.
178. Pelessone, D. (2006). MARS, modeling and analysis of the response of structures. User's Man. ES3 Inc
179. Cho, J.-W., Kim, H., Jeon, S., & Min, K.-B. (2012). Deformation and strength anisotropy of Asan gneiss, Boryeong shale, and Yeoncheon schist. *International Journal of Rock Mechanics and Mining Sciences, 50*, 158–169. doi:10.1016/j.ijrmms.2011.12.004.

Chapter 13
Recent Advances in Global Fracture Mechanics of Growth of Large Hydraulic Crack Systems in Gas or Oil Shale: A Review

Zdeněk P. Bažant and Viet T. Chau

Abstract This chapter reviews the recent progress toward computer simulation of the growth of vast systems of branched hydraulic cracks needed for the efficient extraction of gas or oil from shale strata. It is emphasized that, to achieve significant gas extraction, the spacing of parallel hydraulic cracks must be on the order of 0.1 m, which means that the fracturing of the entire fracking stage would require creating about a million vertical cracks. Another emphasized feature is that the viscous flow of fracking water along the hydraulic cracks must be combined with Darcy diffusion of a large amount of water into the pores and flaws in shale. The fracture mechanics on the global scale is handled by the crack band model with gradual postpeak softening and a localization limiter in the form of a material characteristic length. Small scale computer simulations demonstrate that the computational approach produces realistically looking results.

13.1 Introduction

Extraction of gas or oil from shale strata by hydraulic fracturing of oil and gas bearing rocks, aka "fracking," is by now a well-established technology [2, 14, 29, 40, 45]. It has been developed gradually since 1947, under private funding except for some public funding during the last few years, granted after the viability had

Z.P. Bažant (✉)
McCormick Institute Professor and W.P. Murphy Professor of Civil and Mechanical Engineering and Materials Science, Northwestern University, 2145 Sheridan Road, CEE/A135, Evanston, IL 60208, USA
e-mail: z-bazant@northwestern.edu

V.T. Chau
Graduate Research Assistant, Northwestern University, Evanston, IL, USA

C. Jin, G. Cusatis (eds.), *New Frontiers in Oil and Gas Exploration*,
DOI 10.1007/978-3-319-40124-9_13

already been proven. Nevertheless, despite astonishing progress, the knowledge of the growth of the hydraulic crack system is still mostly empirical and no method exists for predicting the growth of an extensive system of branched hydraulic fractures which must develop in successful wells. The purpose of this chapter is to review the recent advances previously presented in full detail in [11] and [21].

13.2 Brief Overview of Fracking Technology

The gas bearing stratum of tight shale, typically about 3 km below the surface and 20–150 m in thickness [22, 44], is accessed by parallel horizontal boreholes, about 500 m apart (Fig. 13.1). They are drilled from a single drillpad in the direction of the minimum principal tectonic stress σ_h, whose magnitude is about 1/5 to 4/5 of the overburden stress [32] σ_g.

The horizontal boreholes used to be about 3 km long, but 11 km is now being attained. Each of them is subdivided into segments about 600 m long, each of which consists of about 5–10 fracturing stages. Each stage, typically 70 m long, is further subdivided into about 5–8 perforation clusters. In each cluster, about 14 m long, the steel casing (or pipe), of typical inner diameter 3.5 in. (77 mm) [28, 49], is perforated at 5–8 locations by detonating groups of shaped explosive charges (Fig. 13.1).

Powerful pumps on the surface drillpad inject the fracking fluid into the shale stratum. The fluid, about 99 % water, contains proppant (a fine sand needed to prevent crack closure), and contains various additives, particularly gellants, various acids or pH controlling ions. Each stage requires injection of several million gallons of water (which is equivalent to about 0.5–2 mm of rain over the area of the lease, 3×5 to $10\,\mathrm{km}^2$). The flowback of water, representing about 15 % of the injected total, is very salty and highly contaminated with dissolved minerals. Strict controls are required to prevent its accidental release to the environment. Often, the water

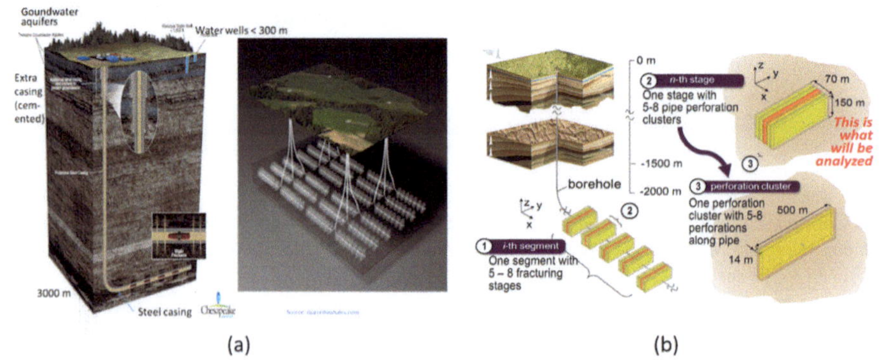

Fig. 13.1 Main features of frac operations. (**a**) Overview of Frac Operations. (**b**) Main Features of Frac Operation

Fig. 13.2 Vertical joints in Devonian Shale (Marcellus); (**a**) drilled cores, (**b**) surface outcrop

flowback is reinjected underground. Minimization and treatment of this flowback is an important goal of research.

Pumps, currently attaining at the surface level the pressure of about 25 MPa, push the fracking fluid through the perforations into the shale stratum. To begin with, the shale is intersected by a system of natural fractures or rock joints, nearly vertical, which are either tightly closed or filled by calcite or other minerals. They are typically 15–50 cm apart [29, 42] (Fig. 13.2). The shale is also intersected by numerous finer faults and inclined slip planes. The shale contains weak near-horizontal bedding planes with sub-millimeter spacing, which have higher permeability and higher concentration of kerogen filled nano-voids [36]. The first, large, hydraulically produced cracks must be roughly normal to the horizontal wellbore, since it is always drilled in the direction of the minimum tectonic stress. No horizontal cracks are expected to form since the overburden pressure exceeds the tectonic stresses.

Most of the gas in gas shale, mainly methane, is contained in kerogen-filled nanopores of diameters from 0.5 nm to about 10 nm [33, 34, 36]. Drilled cores brought to the surface (Fig. 13.2) reveal the total gas content of shale, and thus it is estimated that at most 15 %, and often as little as 5 %, of the gas content of the shale stratum gets extracted (percentages as high as 50 % have been rumored but probably are rare local aberrations).

13.3 Estimation of Hydraulic Crack Spacing from Gas Flow History Observed at Wellhead

According to reported measurement, the shale permeability, k, ranges from 10^{-9} darcy to 10^{-7} darcy, which is 10 to 1000 times lower than the typical permeability of concrete; e.g., [38]. The huge spread of the measured k-values reported in the literature [23, 36, 39, 46] is probably caused less by differences among locations and more by differences among the methods of measurement [3, 31]. These are transient methods based on the long-term decay of gas loss using: (1) a drilled core or (2) powderized shale, obtained by grinding the shale to particle size of about 0.5–0.85 mm.

The tests of the cores are dominated by diffusion along the kerogen-filled bedding layers in which the clay mineral platelets are oriented mainly horizontally, making the permeability higher than in the rest of shale. On the other hand, the powder test depends more on the compacted clay rock between the bedding layers in which the clay platelets do not have a preferential orientation. This feature reduces the permeability. Since the hydraulic cracks must be orthogonal to the bedding layers, the core test appears to be more relevant. The average permeability, 10^{-8} darcy, has been considered in calculations.

13.3.1 *Diffusion of Gas from Shale into Hydraulic Cracks*

To quantify the gas escape from the nano-voids into the nearest hydraulic cracks, we consider an idealized typical situation—one-dimensional gas diffusion in direction x into adjacent parallel cracks of spacing s, normal to x (Fig. 13.3). The Darcy law, given by the equation in Fig. 13.3b, is assumed to apply [1, 2]; and the condition of mass conservation appears below. $t =$ time, $v =$ velocity of gas, $p =$ gas pressure, $\phi =$ gas porosity, $\mu =$ dynamic viscosity of the gas, and $C_s =$ bulk compressibility of shale; and $\rho =$ mass density of the gas. The gas closely follows the ideal gas equation, i.e., $p/\rho = RT =$ const., which means that the gas compressibility is $C_g = 1/p$. Using $p = \rho\, R\, T$ in the previous two equations, and noting that constant RT cancels out, one obtains, after some rearrangements. Because of nonlinear compressibility of gas, the diffusion equation is strongly nonlinear [13].

For a shale layer between two parallel cracks at $x = 0$ and $x = s$, the flux of gas from the layer into the cracks at $x = 0$ is given by the equation in Fig. 13.3b (middle left). This represents an interface condition for the flow of gas into the hydraulic crack system. With this interface condition, the diffusion equation is solved numerically by central finite differences in the spatial direction and explicitly by forward finite differences, while the time step Δt is lengthened as the flux decays, to the extent allowed by the limit of numerical stability.

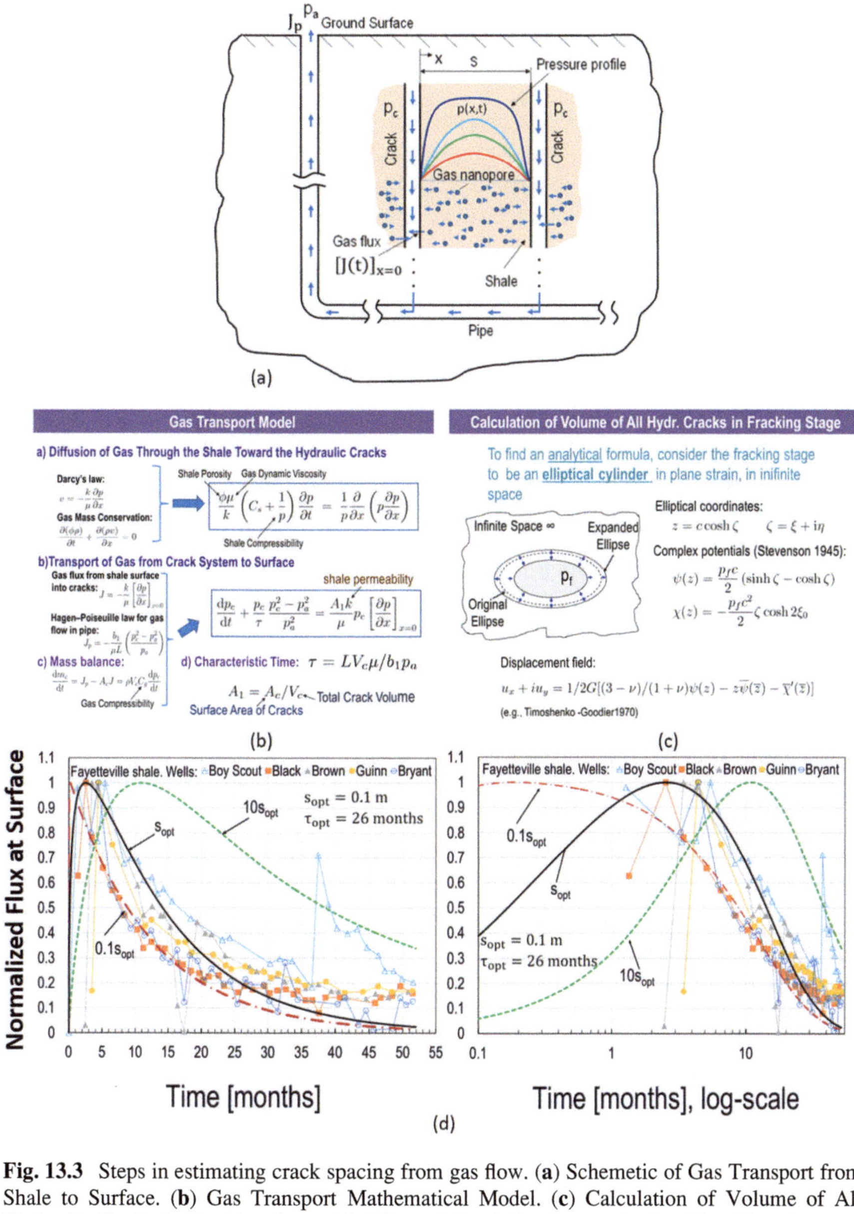

Fig. 13.3 Steps in estimating crack spacing from gas flow. (**a**) Schemetic of Gas Transport from Shale to Surface. (**b**) Gas Transport Mathematical Model. (**c**) Calculation of Volume of All Hydraulic Cracks in Fracking Stage. (**d**) Effect of Crack Spacing on the Gas Flux Histories

13.3.2 Total Volume and Surface Area of Hydraulic Crack System

To relate the J to the flux of gas observed at the wellhead in the initial transient period about 1 month long, we need to estimate the total volume V_c and total surface area S_c of all the hydraulic cracks in a fracturing stage. Since at most only 15 % of the gas is currently getting extracted, we optimistically assume that the fracked shale volume is 15 % of the volume of a typical fracking stage (70 m long, 150 m deep, and 500 m wide, Fig. 13.3c), which is $V_{\text{cyl}} = 495{,}000\,\text{m}^3$

The pressure of fracking water generally does not exceed the overburden pressure. Therefore, all the cracks must be approximately vertical. They are assumed to consist of two approximately orthogonal systems of parallel cracks. There are two reasons for the orthogonality: (1) The rock joints, which are the preferred crack locations, are roughly orthogonal, and (2) equilibrium of a horizontal element bounded by four non-orthogonal vertical cracks would require transmission of shear stresses across the cracks, which is impossible if the cracks are fully open.

To make analytical estimates possible, we idealize the fracked zone as an elliptical cylinder with a vertical axis, height h and a generating horizontal ellipse of axes a and b (Fig. 13.3c), such that the cylinder volume, $\pi\, a\, b\, h$, be equal to V_{cyl}. The fractured rock inside the cylinder is under uniform pressure p equal to the fracking pressure, from which one can calculate the combined volume decrease of these rocks, ΔV_{in} (see the inner ellipse in Fig. 13.3c). The total surface area of all cracks (each crack having two faces) is $S_c = 4\pi abh/s$.

Further one needs to calculate the enlargement of the area of the idealized ellipse considered as a pressurized hole in an infinite elastic plane (elastic modulus $E = 37.5\,\text{GPa}$ and Poisson ratio $\nu = 0.3$ in the planes of the bedding layers). The area increase ΔA of the elliptical hole in infinite plane due to internal fracking pressure p_f was calculated using Stevenson's solution [48, 50] in complex variables, which was extended to obtain also the displacements. The main equations are shown in Fig. 13.3c), and in detail see [11] (p_f is here considered to be in excess of the average remote tectonic pressure $p_{tectonic} = (\sigma_h + \sigma_H)/2 \approx 40$ MPa). The area increase multiplied by cylinder height gives a crude estimate of the volume increase ΔV_{out} due to compression of the surrounding shale mass. The total volume of the crack system of the fracking stage is $V_c = \Delta V_{in} + \Delta V_{out}$.

13.3.3 Flow of Gas from the Hydraulic Crack System to the Wellhead

To estimate the crack spacing, we need to figure out the rate of gas flow through the hydraulic cracks and through the horizontal and vertical pipes. Since the gas is mostly methane and methane does not dissolve in water, the gas must move as bubbles, which start as microscopic but soon coalesce into big bubbles filling the crack width and the pipe diameter. In the hydraulic crack system, as well as in the horizontal and vertical pipes, the bubbles are driven by gas pressure gradient and move in turbulent flow. In the vertical pipe, the movement of gas bubbles is, additionally, also propelled gravitationally, by buoyancy. Detailed modeling of the movement of gas bubbles is quite involved, but for a crude estimate it suffices to use the Hagen–Poiseuille law for the flow of a compressible fluid in pipes, shown in Fig. 13.3b (bottom left) where J_p = flux of gas out of the fracked zone into the hydraulic crack system; p_c = inlet pressure = gas pressure at, approximately, the center of the fracked zone; p_a = atmospheric pressure = gas pressure at exit on the surface; L = distance of flow; and b_1 = constant characterizing the resistance to gas flow through the hydraulic crack system and the horizontal and vertical sections of the pipe. The condition of mass balance is written for the entire volume of the hydraulic cracks (see Fig. 13.3b where $dm_c/dt = J_p$ = rate of change of gas mass contained in all the opened hydraulic cracks, where p_c = average gas pressure in the crack system, C_g = gas compressibility, V_c = total volume of the opened hydraulic cracks, A_c = area of the surfaces of all opened hydraulic cracks = double of the total crack area, S_c). Combining all the equations, one can obtain the final approximate equation for the decay of gas pressure p_c in the hydraulic crack system shown in Fig. 13.3b (bottom right), in which the most important parameter is the gas flux halftime τ, giving approximately the time at which the gas flux decays to 20 % of its peak.

Now we are ready to make comparisons with the observed gas flux. Figure 13.3d shows the gas flux histories observed at five wells in Fayetteville shale in Arkansas (freely downloaded from the internet) [37]. The histories are plotted in actual time scale on the left and the logarithmic time scale on the right. They have two key features that must be matched by computations:

1. The halftime of flux decay, τ; and
2. The time to reach the peak flow, τ_{peak}.

Matching these two times by numerical solution of the aforementioned equations has made it possible to estimate the crack spacing s [11]. The calculated history of gas flux giving the least-square optimum fit is shown in Fig. 13.3d by the solid curve. It corresponds to crack spacing

$$s_{opt} \approx 0.1 \text{ m} \tag{13.1}$$

and to halftime $\tau_{\text{opt}} = 26$ months. To realize how credible this estimate is, see the curve for $s = 1$ m and for $s = 0.001$ m. Obviously there is no match at all (Fig. 13.3d). How well the time to peak gas flow is matched can better be seen in the logarithmic time scale in Fig. 13.3d right.

13.3.4 Long-Term Gas Flow as the Main Indicator of Crack Spacing

How big is the error in crack spacing, s, due to simplifications of the analysis of the gas flow from the hydraulic cracks to the wellhead? In fact, small. These simplifications can affect significantly only the time to peak flow, τ_{peak}, but not the flux after a few years. In fact, even if the transmission of gas from the hydraulic cracks to the wellhead was considered immediate, the terminal part of the curves of the *relative* decay of the gas flow in Fig. 13.3d would hardly be affected.

The slope of the long-time descending curve in linear time, or the location of the steep descent in log-time (compare the descending solid and dashed curves in Fig. 13.3d) depends almost only on the crack spacing and thus on the permeability of shale, which can be directly measured (this is, of course, not true for the initial rising part of gas flow history which depends strongly on the delay of gas flow through the pipes; without this delay, the gas flow history would descend from the start). The reason is that, after approximately 1 year, the gas flow is declining very slowly and is close to a steady state. In a steady state, the rate of gas flow from the shale into the cracks would be exactly equal to the rate of gas flow at the wellhead.

Another reason why the crack spacing estimate of 0.1 m [11] ought to be quite reliable is that the halftime of the rate gas flow from the shale into the hydraulic cracks scales is approximately proportional to the *square* of crack spacing. If the crack spacing was 1 m, the half time, τ, would increase from 26 to 2600 months, or 217 years, and if the crack spacing was 10 m, to 21,700 years. Were the cracks spaced at 1 cm, the halftime would decrease to 0.26 month or 7.6 days. The fact that all of these halftime values grossly disagree with experience reinforces the spacing estimate of 0.1 m.

The consequence of the crack spacing estimate of 0.1 m is that complete fracturing of one typical fracking stage would require creating about one million vertical cracks, in two orthogonal crack systems. Yet most previous studies have dealt with one hydraulic crack [2, 16, 24–26, 30, 35], and lately with five parallel (i.e., nonintersecting) cracks [15, 17–19]. These studies are valuable for clarifying the micromechanics of fracking, especially the complicated interaction of crack tip singularity with viscous flow near the singularity, and with capillary meniscus and surface adsorption of water near the singular crack tip. But they have little relevance for the overall fracking operation.

There is, nevertheless, one phenomenon which could alter the crack spacing estimate appreciably—the creep, which is in fact quite pronounced in shale. Its main consequence is that proppant particles (sand grains) must be getting gradually embedded into the walls of hydraulic cracks, thus causing that, with the lapse of time, the proppant can prevent crack closure only partly. The creep embedment must lead to a decrease of crack width and perhaps even to crack closure, thus reducing or even stopping the flow of gas. This might be one reason why refracking, with more proppant, helps. However, data on the creep of shale are scant at present and more testing is needed.

13.4 Evolution of a System of Parallel Hydraulic Cracks

13.4.1 Hydrothermal Analogy

For the steady states of multiple hydraulic cracks, a previous study [11] pointed out a simple analogy with thermal cracks. Consider an elastic halfspace that is getting cooled from the surface (Fig. 13.4). If the temperature drop is high enough, parallel cooling cracks will form. The heat conduction produces a temperature field that decays from the surface as the error function.

First imagine that some external agency would apply on the crack surfaces a pressure that would close the cracks, and then glue them. Second, the agency would unglue the cracks. The ungluing is equivalent to applying opposite pressure on the crack surfaces; see Fig. 13.4 right. Since, for the glued cracks, the stress intensity factors K_c at the crack tips vanish, the solution of the crack system according to linear elastic fracture mechanics (LEFM) is, for the thermal cracks, the same as it is for the hydraulic cracks.

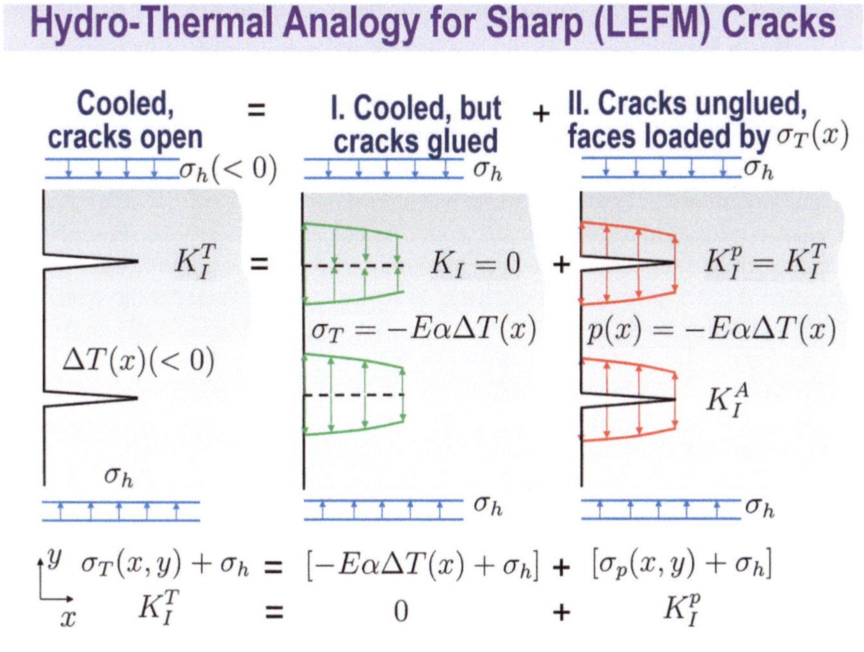

Fig. 13.4 Superposition leading to hydrothermal analogy

13.4.2 Review of Stability of Parallel Crack Systems

At the time of the first energy crisis in the mid 1970s, extraction of heat from hot dry rock located relatively close to the earth surface was studied. It was speculated that if a large vertical crack was created hydraulically from a borehole in hot granite (Fig. 13.5a), and if this crack was then intersected by another borehole, circulation of water could deliver enough steam to generate electricity. Success would have required the cooling to produce many closely spaced parallel cracks propagating laterally form the walls of the main crack. However, drilling into the ancient Valles Caldera in the Jemez Mountains of New Mexico gave a negative result, which was explained [7, 9] by analysis of the localization instability of parallel cooling cracks.

This negative result nevertheless provides today some useful information for the understanding of one aspect of fracking, particularly the stability of a parallel crack system (Fig. 13.5c) against localization. The stability can be assessed on the basis of the Helmholtz free energy, F, of a structure with interacting cracks of lengths a_1, $a_2, \ldots a_N$ and fracture energy Γ for each crack (LEFM being assumed); see the top equation in Fig. 13.5c, in which U = strain energy of the elastic solid; p = control parameter, such as the depth of penetration D of the fluid pressure front into the halfspace, analogous to the cooling front.

In fracking as well as cooling, there are many possible fracture equilibrium solutions but, according to thermodynamics, the stable one is that which minimizes F. The equilibrium and stability of the crack system are, respectively, decided by the first and second variations [7], written at the bottom of Fig. 13.5c ($i = 1, \ldots m$ are the cracks that are growing ($\delta a_i > 0$) and dissipating fracture energy Γ; $i = m + 1, \ldots n$ are the cracks that are shortening ($\delta a_i < 0$), for which the fracture energy is 0, and $i = n + 1, \ldots N$ are the cracks that are immobile ($\delta a_i = 0$), which occurs if the energy release rate $-\partial \mathcal{U}/\partial a_i$ is non-zero but less than critical. Equilibrium (or static) crack propagation requires vanishing of the first variation (which yield the standard crack propagation criterion. Fracture stability requires the matrix of the second variation $\delta^2 F$ (Fig. 13.5c bottom) to be positive for all admissible variations δa_i [5, 7, 9, 41].

Consider that cooling or hydraulic cracks may develop alternating lengths $a_i (i = 1, 2)$ although initially they are equal (Fig. 13.5b). The positive definiteness of the matrix of $\delta^2 F$ was shown to be lost first by the vanishing of its determinant. But this indicates neither instability nor bifurcation because the corresponding eigenvector implies every other crack to shorten ($\delta a_2 = -\delta a_1 \neq 0$), which is impossible since the energy release rates of all cracks are nonzero. After further crack advance, when the crack length becomes about $1.5s$ to $2s$, the positive definiteness of the matrix of $\delta^2 F$ is lost due to the vanishing of the first diagonal term while the corresponding eigenvector of δa_i is found to be admissible. This implies a stable bifurcation, in which the length of every other crack jumps ahead, while the remaining cracks unload and close. Thus the spacing of the leading cracks doubles, and their opening width w eventually does, too [7, 8, 41].

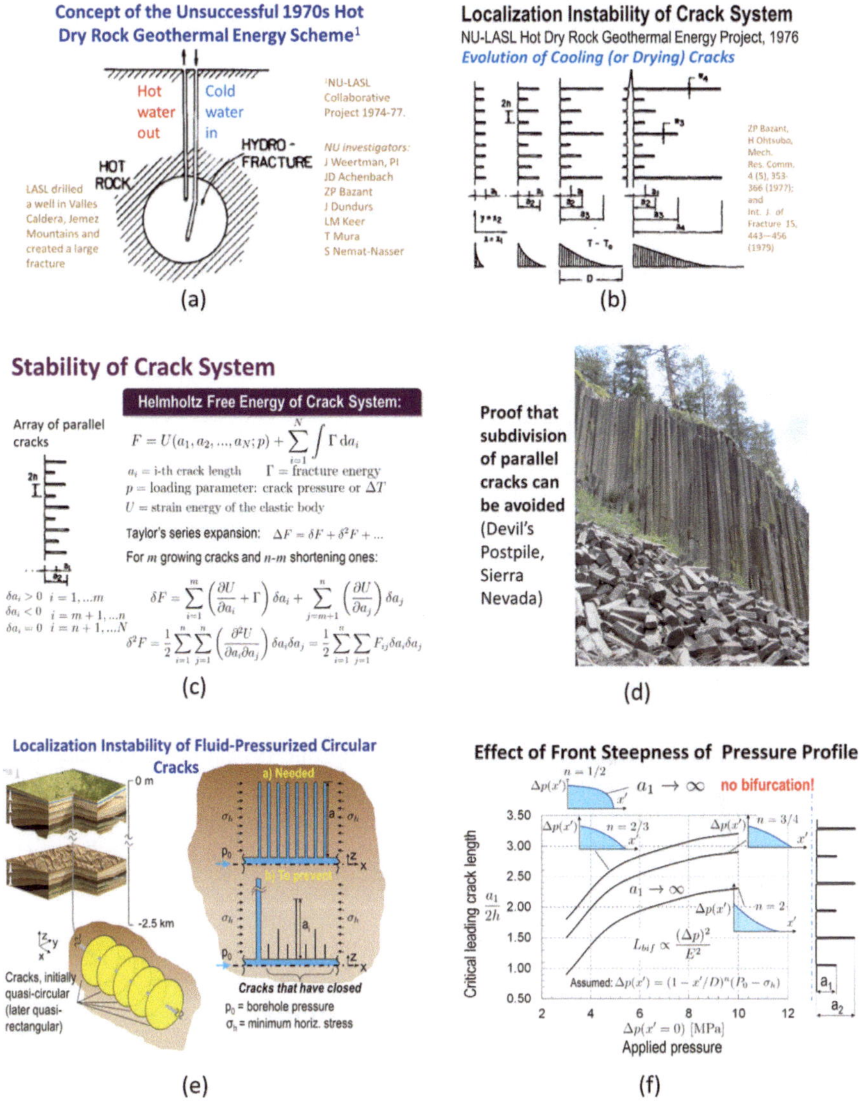

Fig. 13.5 Bifurcations and localization instabilities of parallel crack system. (**a**) Concept of the Unsuccessful Hot Dry Rock Geothermal Energy Scheme. (**b**) Localization Instability of Geothermal Crack System in Hot Dry Rock. (**c**) Stability Analysis of Crack System. (**d**) Evidence of Parallel Cracks. (**e**) Localization Instability of Fluid-Pressurized Circular Cracks. (**f**) Effect of Front Steepness of Pressure Profile

The remaining leading cracks of doubled spacing $2s$ eventually reach another bifurcation of the same kind, at which the spacing every other crack stops growing and gradually closes while the spacing of the open cracks doubles to $4s$ (Fig. 13.5b). This doubling of crack spacing, in which the crack system localizes into fewer and fewer cracks, is periodically repeated as the pressure front (or cooling front) advances (see also [5, Sect. 12.5]). The rate of gas flux into the leading cracks decreases approximately in inverse proportion to crack spacing s.

However, these bifurcation instabilities with cracking localization can be avoided. This is, for example, blatantly demonstrated by the photo of Devil's Postpile in Sierra Nevada (Fig. 13.5d) which shows the denuded side wall of a solidified ancient lava flow. Obviously, here the cracking localization was prevented by a complex thermal regime affected by latent heat of solidification and by circulation of water along the cracks.

For the present problem, it was found in 1979 [5, 12] that if the error-function profile is replaced by a profile with nearly uniform pressure and a steep drop at front, the parallel cracks can propagate at constant spacing, with no localizations into fewer cracks (Fig. 13.5f). The same behavior was recently demonstrated for parallel radially growing circular cracks (Fig. 13.5e [11]). Such behavior is, of course, desirable for the efficacy of fracking.

So we may conclude that the effect of temperature profile on fracture propagation is generally the same as that of a similar crack pressure profile. Thus the previous analysis of cooling cracks makes it possible to state, even without any calculations, that by achieving a sufficiently uniform crack pressure profile, with a sufficiently steep pressure front, the parallel cracks should not get localized and should propagate indefinitely, at constant spacing. This is what is needed to create densely spaced channels by which the shale gas could escape from the nanopores.

From the foregoing considerations it transpires that a nearly uniform pressure profile is one feature that should help the efficacy of fracking. To make the pressure along the cracks more uniform, a slower pumping rate is desirable. This observation, however, will have to be balanced against many other influencing factors. And, of course, for the 3D intersecting crack systems, and in presence of water diffusion into the pores, the problem of preventing localization of hydraulic cracks is much more complex and deserves further study.

13.5 Evolution of Two Orthogonal Systems of Hydraulic Cracks

Given that the spacing (Fig. 13.3) is about 0.1 m, the number of vertical near-orthogonal cracks in a typical fracking stage must be about a million. The hydraulic fracture system doubtless includes fractures running along preexisting rock joints [47, 52, 54]. Although the joints are usually filled by calcite, their tensile strength and fracture energy are certain to be smaller, probably much smaller, than those of shale. Hence, the hydraulic cracks must be expected to run along the rock joints whenever possible. That they do, is supported by the fact that the spacing of rock joints is on the order of 0.1 m, coinciding with our estimate of crack spacing.

An effective way to deal with a vast number of cracks is the crack band model [6, 20], in which the crack is smeared and represented as equivalent damage over the finite element of a width equal to the crack band width. Compared to the cohesive crack model, an additional advantage is that atensorially enriched cohesive softening law, sensitive to compression stress parallel to the crack plane, can be used. The finite element size is taken equal to the crack spacing, about 0.1 m. This way, of course, the local stress and flow fields at crack tip cannot be captured, but the global evolution of crack system can. Even with this element size, about a billion finite elements would be needed to simulate the whole fracking stage, unless a method of coarse graining taking into account the global cracking localizations is developed.

13.5.1 Cracked Finite Elements for Crack Band Model

The shale mass is subdivided into six-node cubic (hex) finite elements with linear shape functions (Fig. 13.7a). Each element contains two vertical, mutually orthogonal, cracks normal to the x_1 and x_2 axes, coinciding with the principal tectonic stresses T_1 and T_2 ($x_1 \equiv x, x_2 \equiv y$). The opening widths of the cracks normal to these stresses, denoted as h_1 and h_2, respectively, govern the components of the vectors (q_1, q_3) and (q_2, q_3) of the flux vectors of water along these cracks, q_3 being the vertical flux component. The water pressure, p, is considered to be uniform within each element, and since the cracks intersect, pressure p in each element must be the same in both intersecting cracks. Pressure p, of course, varies from one element to the next.

13.5.2 Secondary Lateral Crack Initiation and the Necessity to Include Diffusion

In the previous studies, the fracking water was assumed to flow only along the opened cracks, pressuring the crack walls. In this way, however, no secondary cracks would branch from the crack walls (V-shaped branching at the crack tip is possible only in dynamics, at crack speeds close to Rayleigh wave speed). The secondary cracks initiate if the horizontal normal stress, σ_{11}, parallel to the face of a vertical cracks is tensile and exceeds the strength limit, σ_0, of the cohesive crack or the crack band model [6, 10, 20]. However, according to LEFM, the stress field on the sides of a pressurized crack (Fig. 13.6a, b top left and bottom) is compressive. This can be shown not only by finite element analysis but also from Westergaard's complex variable solution (Fig. 13.6b bottom).

By contrast, if a spherical or wide ellipsoidal cavity is pressurized (Fig. 13.6b, c) [50, p. 395], a tensile tangential normal stress along the cavity wall is produced and can give rise to a crack. The same is true when the pores in a porous spherical or ellipsoidal cavity are pressurized, producing on the cavity walls an effective

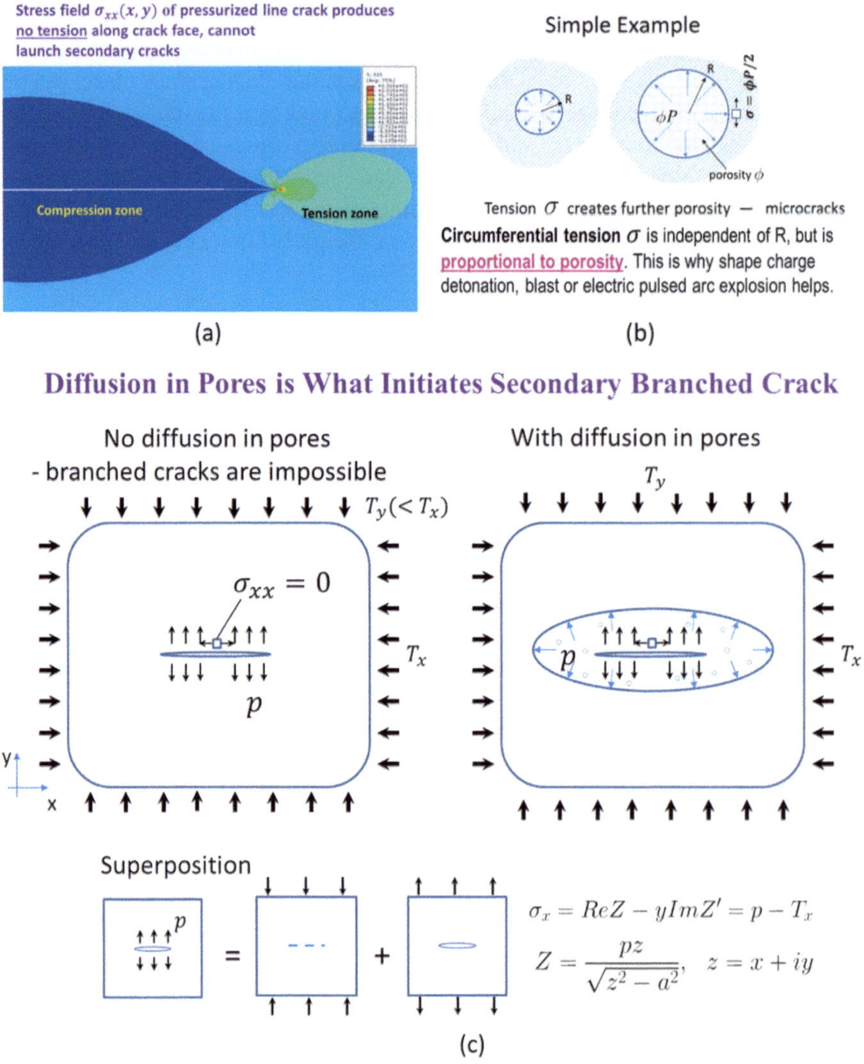

Fig. 13.6 Engendering tension along the wall of crack or cavity. (**a**) Stress field σ_{xx} of pressurized crack. (**b**) Relationship of circumferential tension and porosity. (**c**) Diffusion in pores initiates secondary branched crack

pressure ϕp where p = pore pressure and ϕ = porosity. So the mathematical model must include diffusion of water into the shale surrounding a hydraulic crack. Such diffusion would not be a realistic assumption for granite, for example, but for shale it is, because the typical nano-porosity of shale (e.g., Marcellus shale) is $\phi = 9\,\%$. This assumption is also supported by the fact that the fracking water leak-off is huge, as only about 15 % of the water injected from the wellhead returns to the surface after the fracking ends.

Consequently, one must consider not only the Poiseuille (viscous) flow along the created hydraulic cracks, but also the Darcy diffusion through the pores and flaws in shale.

13.5.3 Water Flow Through Hydraulic Cracks and Pores

The flow along the hydraulically created cracks is assumed to follow the Reynolds equations of classical lubrication theory, which are based on the Poiseuille law for viscous flow; see Fig. 13.7a top left where $q_1.q_2, q_3$ = flux vector components in the vertical (x_1, x_3) and (x_2, x_3) crack planes $(x_1 \equiv x, x_2 \equiv y, x_3 \equiv z)$. where $\tilde{q}_1, [\tilde{q}_3]_{(x_1,x_3)}$ are the volumetric Poiseuille flow rates per unit cross-sectional area (dimension m/s) in plane (x_1, x_3); likewise $[\tilde{q}_3]_{(x_2,x_3)}$ in plane (x_2, x_3); $\tilde{P}$ = water pressure in the cracks; h_1, h_2 = openings of the cracks in (x_2, x_3) and $(x_a.x_3)$ planes (i.e., in x_1 and x_2 directions), and μ' = effective viscosity of fracking water, which is affected by proppant and gellants. The diffusion of water (without proppants) through the pores of shale is governed by the Darcy law [13]; see Fig. 13.7a top right where $k = 1, 2, 3$, $\widehat{q}_k$ = kth component of the vector of volumetric flow rate through the pores, per unit cross-sectional area (dimension m/s); $\widehat{P}$ = water pressure in the pores; μ = water viscosity [1, 2] (taking into account the effect of gellants but not proppants), and κ_h, κ_v = permeabilities of porous shale along the horizontal bedding planes and in the vertical direction (dimension m^2) [53], which are rather different; $\kappa_v \ll \kappa_h$ [36, 46]. In computations κ_v is neglected. Adding the fluxes through the pores and through the cracks gives the equations in Fig. 13.7a (middle left). Water pressure P is, in the global model, common to the pores and the hydraulic cracks.

13.5.4 Combined Diffusion Through Shale Pores and Flow Along the Cracks

In the global model, the water pressures in the cracks and in the pores are considered to be locally the same within each finite or volume element. So the water mass conservation equation must be written for the combined mass of water in the pores and in the cracks. After introducing the Darcy and Poiseuille fluxes, the combined diffusion equation takes the form in Fig. 13.7a (bottom left); in which l = side of cubic element; h_1, h_2 = opening widths of cracks normal to axes x, y (or x_1, x_2); κ_h, κ_v = permeabilities of shale in horizontal and vertical directions (the vertical permeability, normal to the bedding layers, is much smaller).

Although water is considered as incompressible in previous studies, it is in fact about 10–30 times more compressible than rocks. The compressibility of water is considered by the equation for mass density ρ of water in Fig. 13.7a (bottom left), where C_f = compressibility of water with proppant, ρ_0 = mass density at reference pressure, and p_0 = reference pressure taken as gravity pressure of water at the depth

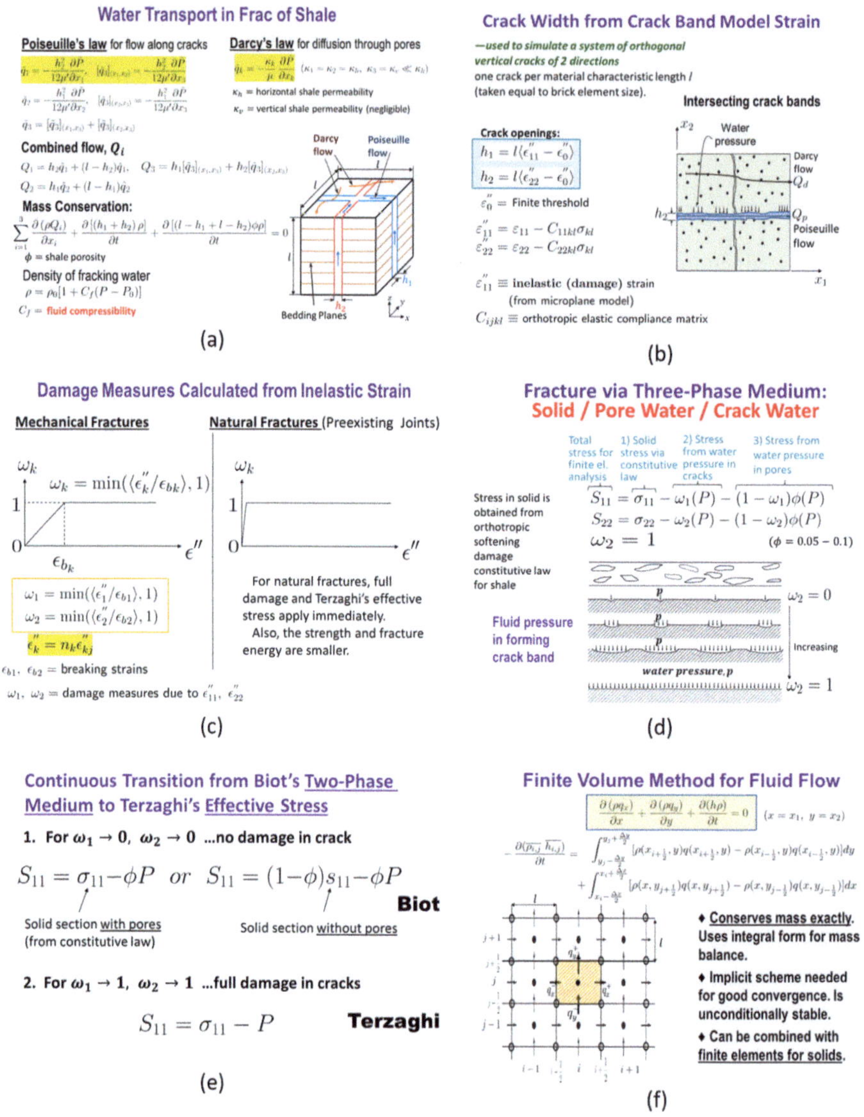

Fig. 13.7 Mathematical modeling for finite element and finite volume analysis. (**a**) Water Transport in Frac of Shale. (**b**) Crack Width from Crack Band Model Strain. (**c**) Damage Measures Calculated from Inleastic Strain. (**d**) Fracture via Three-Phase Medium: Solid, Pore Water and Fracturing Water. (**e**) Continuous Transition from Biot's Two-Phase Medium to Terzaghi's Effective Stress. (**f**) Finite Volume Method for Fluid Flow

of shale layer before fracking. Because the crack width is negligible compared to crack spacing, water compressibility is not important for overall deformations of fracked shale mass, but it matters for the pressure changes in water controlling the flow rates.

13.5.5 Crack Opening Corresponding to Smeared Damage Strain in Crack Band Model

Three potential, mutually orthogonal (or orthorhombic) cracks must be introduced in one finite element. Horizontal cracks are excluded because the vertical overburden pressure (about 80 MPa) generally exceeds the pressure of the fracking water (≤ 70 MPa). The crack band width is taken equal to the material characteristic length l, which is equal to the crack band width or to the element size, and is defined as the minimum possible spacing of stable parallel cracks (note that l must be distinguished from Irwin's material length, which depends on material strength and fracture energy and characterizes the length, rather than width, of the fracture process zone).

The crack openings, h_i $(i = 1, 2)$, which control the rate of water flow, are, in the crack band model, calculated from the continuum damage strain ε_{11}'' or ε_{22}'' across the crack band, which is obtained from strains ε_{11} and ε_{22} calculated from the softening constitutive law; see Fig. 13.7b where n = empirical exponent; subscripts 1 and 2 refer to Cartesian coordinates x_1, x_2 in the horizontal plane; ε_{11}'' = inelastic (or damage) strain, and C_{ijkl} = orthotropic elastic compliance matrix of the shale $(i, j, k = 1, 2, 3)$. The constitutive law for softening damage of the solid (i.e., the shale), i.e., $\sigma = f(\varepsilon)$ must also be orthotropic (this law is another challenge which still awaits proper resolution based on triaxial test data); σ, ε are the stress and strain tensors with Cartesian components σ_{ij} and ε_{ij}; and f is the tensorial function defining the constitutive law.

The distributed microcracking damage is characterized by transversely isotropic damage parameters, ω_1 and ω_2, considered to be functions of the inelastic parts of strains, ε_{11}'' and ε_{22}''. where ε_{b1}, ε_{b2} = specified breaking strain limits (Fig. 13.7c left). For the cracks in intact shale, they were taken as 10^{-5} and, for the cracks along the preexisting cemented joints, as 10^{-7}. The joints, whose volume is assumed to be 0, behave as potential cohesive cracks with reduced strength and reduced fracture energy.

13.5.6 Pore Pressure Effect on Stresses in the Shale

The shale with its cracks and pores is treated as a generalization of both Biot's two-phase medium and Terzaghi's effective stress concept [53]. Three phases may be discerned: (1) the solid shale, of volume fraction $(1 - \omega_1)\phi$, (2) the cracks, of a near-zero volume fraction, and (3) the pores of a significant volume fraction, ϕ. Since no horizontal cracks exist, the equilibrium of the three phases leads to the equations in Fig. 13.7d where ϕ = natural porosity of shale; ω_1, ω_2 = additional porosities for the resultants of fluid pressure in x_1 and x_2 directions (Fig. 13.7d bottom); and S_{ij} = total stress tensor in the solid-fluid system, which are required by finite element analysis to calculate the nodal forces.

There are three contributions to the total stress tensor: (1) σ_{ij} is the stress tensor in the solid including the natural pores, which is calculated from the constitutive law of shale; (2) $\omega_1 p, \omega_2 p$ are the cross-section resultants of water pressure in the cohesive (i.e., bridged) hydraulic cracks; and (3) $(1 - \omega_1)\phi\, p$ and $(1 - \omega_2)\phi\, p$ are the cross-section resultants of the pore water pressure. Alternatively, we could also write $\sigma_{11} = (1 - \phi)s_{11}$ where s_{11} = stress resultant from cross section of solid without the natural pores and preexisting microcracks, whereas σ_{11} = stress that is spread out over the full cross section of solid including the pores (but excluding the hydraulic cracks). Using σ_{11} instead of s_{11} is preferable because σ_{11} is the stress directly measured in material tests.

Note that if there are no cracks with damage, i.e., $\omega_1 = \omega_2 = 0$, the equations from Fig. 13.7e reduce to case 1 in Fig. 13.7e, which characterizes Biot's two-phase medium. In the opposite case of complete damage, i.e., if $\omega_1 = \omega_2 = 1$ (with no bridges across the crack), the same equations in Fig. 13.7e reduce to case 2 in Fig. 13.7e, which gives the classical Terzaghi's effective stress. So we see that the present three-phase medium represents a continuous transition between these two classical (and historically disputed) concepts of soil mechanics.

13.5.7 Numerical Prediction of Evolutions of Hydraulic Crack System

The solid is discretized by finite elements, but this is not the best way for the fluid part of the system. The fluid is better treated by the finite element method [27, 43, 51], which has the advantage that the mass balance of the fluid is enforced exactly (Fig. 13.8a top). This is important if the front of the water infiltration zone is sharp. To combine both methods, the fluid flow elements overlap the finite elements for stress analysis as seen in Fig. 13.8a.

First, to study the behavior of mathematical model, we analyze the fracking of a 2D horizontal square slice of shale mass of unit height and size 5 m × 5 m [21], with the discretization shown in Fig. 13.8a. At each boundary node, we introduce elastic springs that approximately characterize the resistance of the surrounding infinite medium to horizontal expansion. The boundary is, for simplicity, considered impermeable to fluid flow, which means that the simulations are not realistic after the pressurized cracks approach the boundary. The overburden pressure is 80 MPa, corresponding to the depth of 3 km, and the tectonic stresses are $\sigma_h = 30$ MPa and $\sigma_H = 40$ MPa. The horizontal wellbore is considered to run at the bottom of the square in Fig. 13.8b–f. The fracking water first enters conical holes in shale thought to be about 0.5 m deep. These holes are created by explosions of a cluster of three shaped charges, which perforate the high-strength steel pipe. The proppants in water are assumed to suffice to prevent any closing of previously created cracks. The pressure history of water at inlet is a given function of time but here is consider as constant, equal to 70 MPa [21].

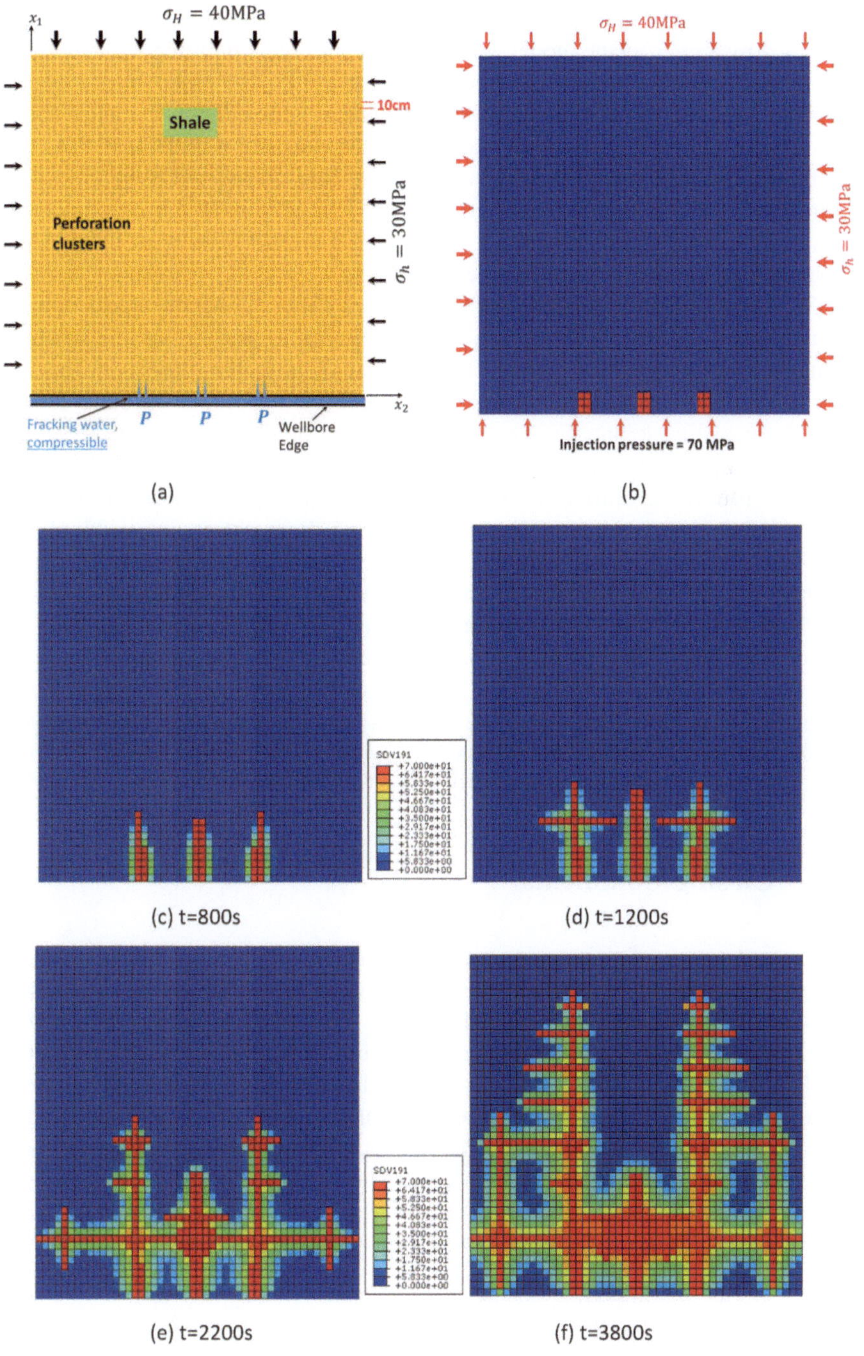

Fig. 13.8 Demonstration of computed evolution of high pressure zones in 2D hydraulic crack system in shale *without* rock joints. (**a**) 2D Hydraulic Fracture Model withou pre-existing rock joints. (**b**) Pressure Field at initial. (**c**) Pressure Field at time t = 800s. (**d**) Pressure Field at time t = 1200s. (**e**) Pressure Field at time t = 2200s. (**f**) Pressure Field at time t = 3800s

The mechanical behavior of shale, which is isotropic in the horizontal plane, is described by a microplane model with postpeak softening, representing an adaptation to shale of the microplane model for concrete [4]. The embedment creep of proppant grains is not yet included in this simulation, whose duration is a few hours. The porosity of shale is $\phi = 0.09$, and the water viscosity is 8.9×10^{-4} Pa s. The tensile strength and fracture energy of shale are here considered to be $\sigma_0 = 4.17$ MPa and $G_f = 0.7$ N/mm in the intact shale, but 3.62 MPa and 0.5 N/mm for cracks running along rock joints (because of overburden and tectonic stresses, these values are higher than those measured in the laboratory).

Two cases have been analyzed (for details, see [21]): (1) All the cracks running through intact shale, and (2) 50 % of cracks running in preexisting joints, whose locations were predefined randomly. Figure 13.8b–f shows the subsequent zones of high pressure for the case of no preexisting joints, and Fig. 13.9 the same for the case of preexisting joints. The finite elements with the highest pressure reveal the locations of open hydraulic cracks. Figure 13.10 from another simulation shows one snapshot from the evolution of hydraulic crack width. The crack width is shown approximately proportional to line thickness and the greatest line width shown corresponds to the width of 10 mm.

Three-dimensional simulations, which are computationally far more demanding, have been demonstrated [21] for a smaller domain—a cubic block of shale with the side of only 0.9 m, shown in Fig. 13.11a. Here only one inlet of water from the horizontal wellbore is considered. All the other parameters are the same. Three subsequent snapshots of the pressure patterns [21] are shown on horizontal and vertical cuts of the block along the planes of symmetry, as shown in Fig. 13.11b, c.

13.6 Closing Comments

The present review of recent researches demonstrates how global fracture analysis with the crack band model, combined with water flow both through hydraulic cracks and through the pores of shale, can be used to simulate the growth of a large system of closely spaced branched hydraulic cracks in shale. Such simulations should allow studying the control parameters that maximize the area of hydraulic cracks and lead to fracking of the largest possible volume of shale with sufficiently close crack spacing. A powerful supercomputer will be needed to scale the analysis to the entire fracking stage.

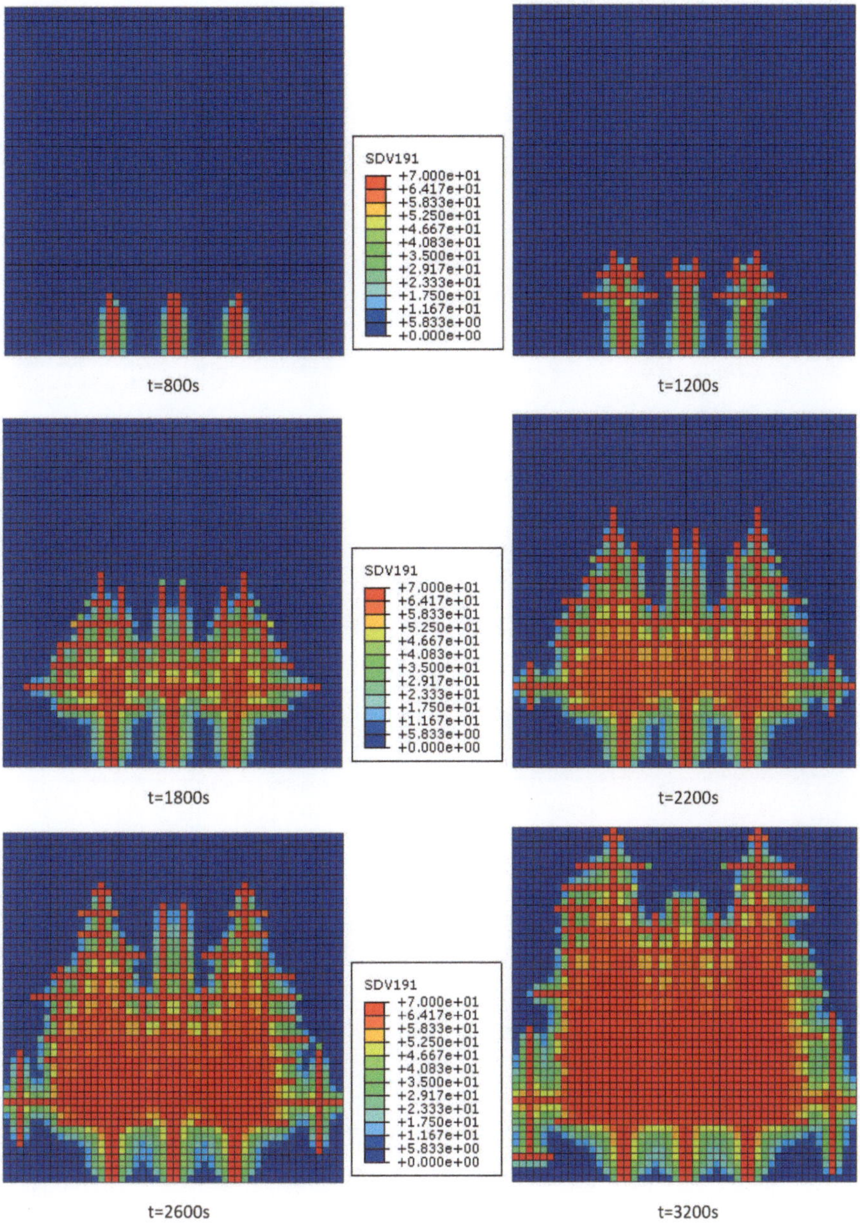

Fig. 13.9 Demonstration of computed evolution of high pressure zones 2D hydraulic crack system in shale *with* rock joints

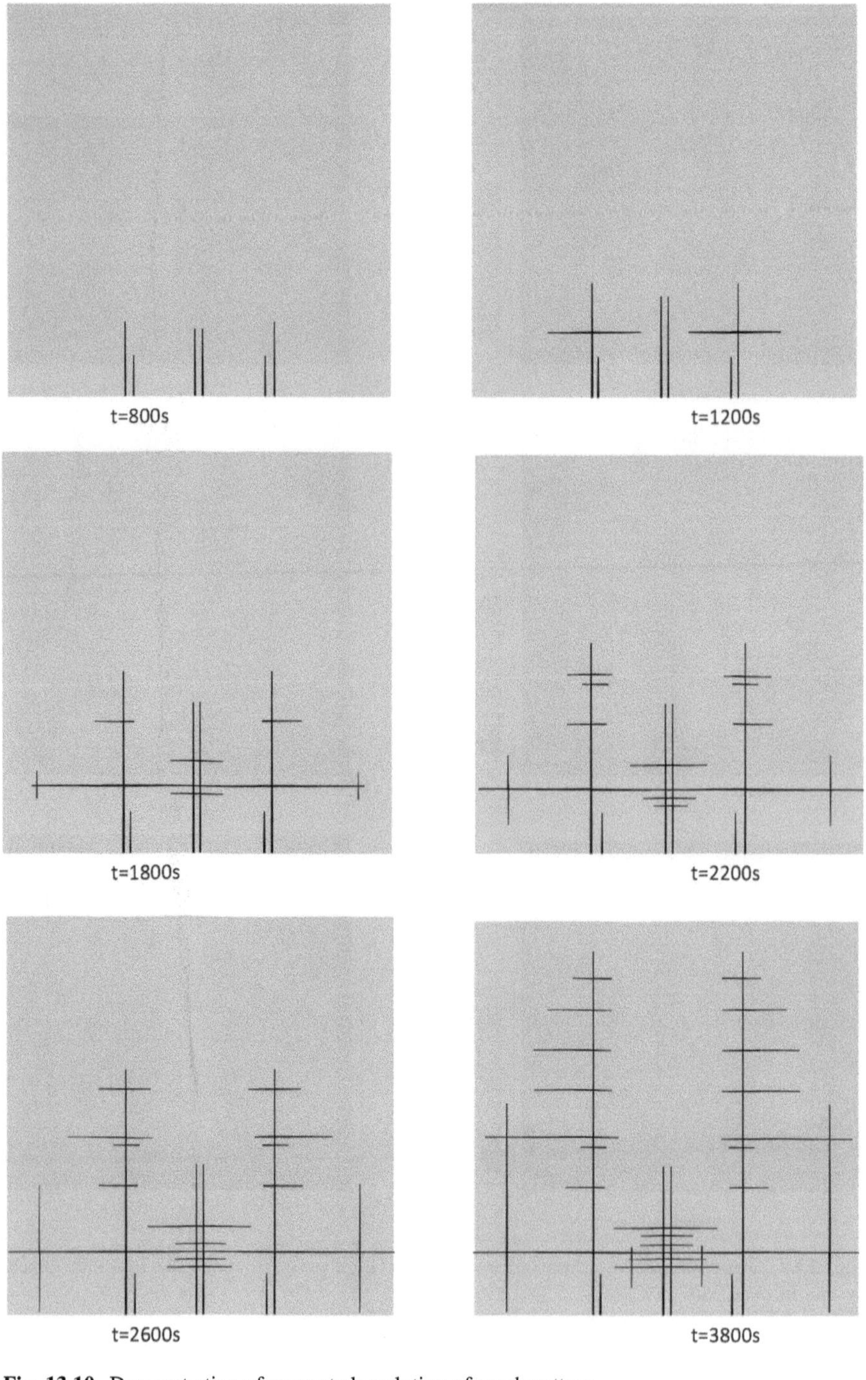

Fig. 13.10 Demonstration of computed evolution of crack pattern

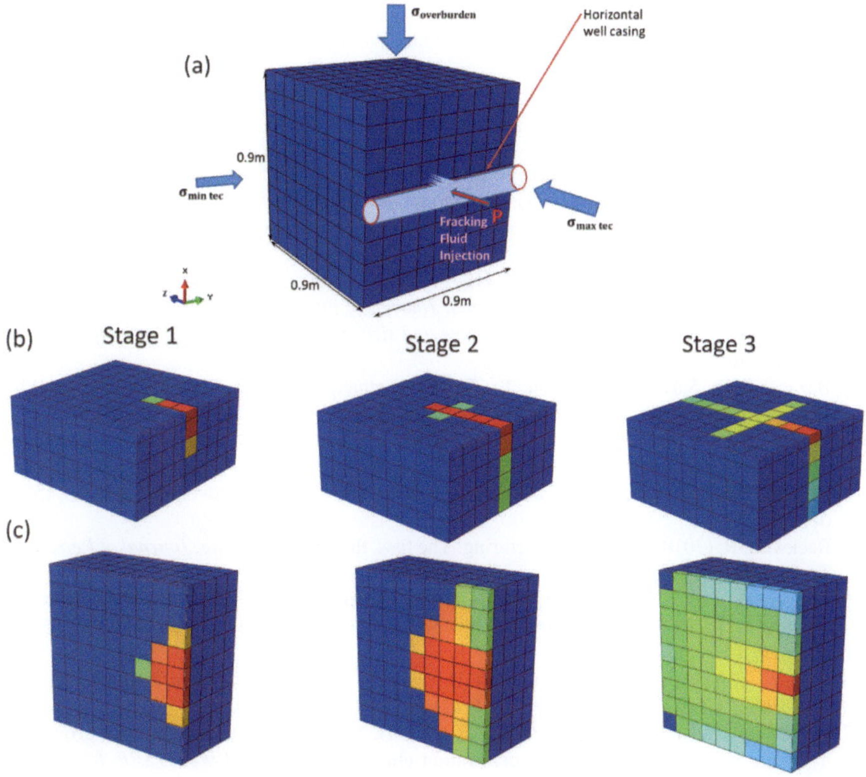

Fig. 13.11 Computed evolution of high pressure zones of 3D hydraulic crack system in a cubic block of shale. (**a**) 3D Hydraulic Fracture Model. (**b**) Pressure Field in YZ Cross Section at Different Time Interval. (**c**) Pressure Field in XZ Cross Section at Different Time Interval

Acknowledgements Financial support from the U.S. Department of Energy through subcontract No. 37008 of Northwestern University with Los Alamos National Laboratory is gratefully acknowledged. Ramifications of fracturing analysis from concrete to shale were partially supported by ARO grant W911NF-15-101240 to Northwestern University.

References

1. Adachi, J., Siebrits, E., Peirce, A., & Desroches, J. (2007). Computer simulation of hydraulic fractures. *International Journal of Rock Mechanics and Mining Sciences, 44*(5), 739–757.
2. Adachi, J. I., & Detournay, E. (2008). Plane strain propagation of a hydraulic fracture in a permeable rock. *Engineering Fracture Mechanics, 75*, 4666–4694.
3. American Petroleum Institute. (1998). Recommended Practices for Core Analysis. API RP40.
4. Bažant, Z. P., Caner, F. C., Adley, M. D., & Akers, S. A. (2000). Microplane model M4 for concrete: I. Formulation with work-conjugate deviatoric stress. *Journal of Engineering Mechanics ASCE, 126*(9), 944–953.

5. Bažant, Z. P., & Cedolin, L. (1991). *Stability of structures: Elastic, inelastic, fracture and damage theories*. New York: Oxford University Press (also 2nd ed. Dover Publ. 2003, 3rd ed. World Scientific 2010).
6. Bažant, Z. P., & Oh, B. H. (1983). Crack band theory for fracture of concrete. *Materials and Structures, 16*, 155–177.
7. Bažant, Z. P., & Ohtsubo, H. (1977). Stability conditions for propagation of a system of cracks in a brittle solid. *Mechanics Research Communications, 4*(5), 353–366.
8. Bažant, Z. P., & Ohtsubo, H. (1978). Geothermal heat extraction by water circulation through a large crack in dry hot rock mass. *International Journal for Numerical and Analytical Methods in Geomechanics*, 2(4), 317–327.
9. Bažant, Z. P., Ohtsubo, R., & Aoh, K. (1979). Stability and post-critical growth of a system of cooling and shrinkage cracks. *International Journal of Fracture, 15*, 443–456.
10. Bažant, Z. P., & Planas, J. (1998). *Fracture and size effect in concrete and other quasibrittle material*. Boca Raton: CRC Press.
11. Bažant, Z. P., Salviato, M., Chau, V. T., Viswanathan, H., & Zubelewicz, A. (2014). Why fracking works. *ASME Journal of Applied Mechanics, 81*(Oct.), 101010-1–101010-10.
12. Bažant, Z. P., & Wahab, A. B. (1979). Instability and spacing of cooling or shrinkage cracks. *ASCE Journal of Engineering Mechanics, 105*, 873–889.
13. Bear, J. (1988). *Dynamics of fluids in porous media*. Mineola, NY: Dover Publications, ISBN 0486656756.
14. Beckwith, R. (2010). Hydraulic fracturing: The fuss, the facts, the future. *Journal of Petroleum Technology, 63*(12), 34–41.
15. Bunger, A. (2013). Analysis of the power input needed to propagate multiple hydraulic fractures. *International Journal of Solids and Structures, 50*. doi:12.1016/j.ijsolstr.2013.01.004.
16. Bunger, A., Detournay, E., & Garagash, D. (2005). Toughness-dominated hydraulic fracture with leak-off. *International Journal of Fracture, 134*(02), 175–190.
17. Bunger, A., Jeffrey, R. G., & Zhang, X. (2014). Constraints on simultaneous growth of hydraulic fractures from multiple perforation clusters in horizontal wells. *SPE Journal, 19* (04), 608–620.
18. Bunger, A., & Pierce, A. (2014). Numerical simulation of simultaneous growth of multiple fracturing interacting hydraulic fractures from horizontal wells. In *Proceedings of ASCE Shale Energy Conference*, Pittsburgh, PA.
19. Bunger, A. P., & Cardella, D. J. (2015). Spatial distribution of production in a Marcellus Shale well: Evidence for hydraulic fracture stress interaction. *Journal of Petroleum Science and Engineering, 133*, 162–166.
20. Červenka, J., Bažant, Z.P., & Wierer, M. (2005). Equivalent localization element for crack band approach to mesh-sensitivity in microplane model. *International Journal for Numerical Methods in Engineering, 62*(5), 700–726.
21. Chau, V. T., Bažant, Z. P. & Su, Y. (2016). Growth model of large branched 3D hydraulic system in gas or oil shale. Philosophical Transactions. Accepted.
22. Cipolla, C. L., Mayerhofer, M. J., & Warpinski, N. R. (2009). Fracture design considerations in horizontal wells drilled in unconventional gas reservoirs. In *SPE Hydraulic Fracturing Technology Conference, 19–21 January, Texas*.
23. Cui, X., Bustin A. M. M., & Bustin R. M. (2009). Measurements of gas permeability and diffusivity of tight reservoir rocks: Different approaches and their applications. *Geofluids, 9*, 208–223.
24. Detournay, E. (2004). Propagation regimes of fluid-driven fractures in impermeable rocks. *International Journal of Geomechanics, 4*, 35. doi:10.1061/(ASCE)1532-3641(2004)4:1(35).
25. Detournay, E. (2016). Mechanics of hydraulic fractures. *Annual Review of Fluid Mechanics, 48*(1), 311–339.
26. Detournay, E., & Peirce, A. (2014). On the moving boundary conditions for a hydraulic fracture. *International Journal of Engineering Science, 84*, 147–155.

27. Eymard, R., Gallouet, T. R., & Herbin, R. (2000). The finite volume method. In P. G. Ciarlet & J. L. Lions (Eds.), *Handbook of numerical analysis* (Vol. VII, pp. 713–1020). Amsterdam: North-Holland
28. Gale, J. F. W. (2002). Specifying lengths of horizontal wells in fractured reservoirs. *SPE Reservoir Evaluation & Engineering, 78600*, 266–272.
29. Gale, J. F. W., Reed, R. M., & Holder, J. (2007). Natural fractures in the Barnett shale and their importance for fracture treatments. *American Association of Petroleum Geologists Memoirs, 91*(4), 603–622.
30. Garagash, D., & Detournay, E. (2000). The tip region of a fluid-driven fracture in an elastic medium. *Journal of Applied Mechanics, 67*(1), 183–192.
31. Guidry, K., Luffel D., & Curtis J. (1995). Development of laboratory and petrophysical techniques for evaluating shale reservoirs - final technical report, Gas Shale Project Area, Restech, Inc., GRI Contract No. 5086-213-1390.
32. Haifeng, Z., Hang, L. Guohua C., Yawei, L., Jun, S., & Peng, R. (2013). New insight into mechanisms of fracture network generation in shale gas reservoir. *Journal of Petroleum Science and Engineering, 110*, 193–198.
33. Javadpour, F. (2009). Nanopores and apparent permeability of gas flow in mudrocks (shales and siltstones). *Journal of Canadian Petroleum Technology, 48*, 16–21.
34. Javadpour, F., Fisher, D., & Unsworth, M. (2007). Nanoscale gas flow in shale gas sediments. *Journal of Canadian Petroleum Technology, 46*, 55–61.
35. Lecampion, B. (2009). An extended finite element method for hydraulic fracture problems. *Communications in Numerical Methods in Engineering, 25*(2), 121–133.
36. Louck, R.G., Reed, R.M., Ruppel, S.C., & Jarvie, D.M. (2009). Morphology, genesis, and distribution of nanometer-scale pores in siliceous mudstones of the Mississippian Barnett shale. *Journal of Sedimentary Research, 79*, 848–861.
37. Mason, J. E. (2011). Well production profiles assess Fayetteville shale gas potential. *Oil & Gas Journal, 109(11)*, 76–76.
38. Maurel, O., Reess, T., Matallah, M., de Ferron, A., Chen, W., La Borderie, C., et al. (2010). Electrohydraulic shock wave generation as a means to increase intrinsic permeability of mortar. *Cement and Concrete Research, 40*, 1631–1638.
39. Metwally, Y. M., & Sondergeld C. H. (2010). Measuring low permeabilities of gas-sands and shales using a pressure transmission technique. *International Journal of Rock Mechanics and Mining, 48*, 1135–1144.
40. Montgomery, C. T., & Smith, M. B. (2010). Hydraulic fracturing: History of an enduring technology. *Journal of Petroleum Technology, 63*(12), 26–32.
41. Nemat-Nasser, S., Keer, L. M., & Parihar, K. S. (1976). Unstable growth of thermally induced interacting cracks in Brittle solids. *International Journal of Solids and Structures***14**, 409–430.
42. Olson, J. E. (2004). Predicting fracture swarms: The influence of subcritical crack growth and the crack-tip process zone on joint spacing in rock. *Journal of the Geological Society of London, 231*, 73–87.
43. Patankar, S. V. (1980). *Numerical heat transfer and fluid flow*. New York: Hemisphere Publishing Corporation.
44. Rijken, P., & Cooke, M. L. (2001). Role of shale thickness on vertical connectivity of fractures: Application of crack-bridging theory to the Austin Chalk, Texas. *Tectonophysics, 337*, 117–133.
45. Society of Petroleum Engineers. (2010). Legends of Hydraulic Fracturing (CDROM), ISBN:978-1-55563-298-4.
46. Soeder, D. J. (1988). Porosity and permeability of eastern Devonian gas shale. *SPE Formation Evaluation, 3*, 116–124.
47. Song, W., Jinzhou, Z., & Yongming, L. (2014). Hydraulic fracturing simulation of complex fractures growth in naturally fractured shale gas reservoir. *Arabian Journal for Science and Engineering, 39*(10), 7411–7419.

48. Stevenson, A. C. (1945). Complex potentials in two-dimensional elasticity. *Proceedings of the Royal Society of London A: Mathematical, Physical and Engineering Sciences* (Vol. 184, No. 997, pp. 129–179). The Royal Society.
49. Tang, Y., Tang, X., & Wang, G. Y. (2011). Summary of hydraulic fracturing technology in shale gas development. *Geological Bulletin of China, 30*, 393–399.
50. Timoshenko, S., & Goodier, J. N. (1970). *Theory of elasticity*. New York: McGraw Hill.
51. Versteeg, H., & Malalasekera, W. (2007). An introduction to computational fluid dynamics: The finite volume method. New York: Pearson Education Limited.
52. Weng, X. (2015). Modeling of complex hydraulic fractures in naturally fractured formation. *Journal of Unconventional Oil and Gas Resources, 9*, 114–135.
53. Zoback, M. D. (2007). *Reservoir geomechanics*. Cambridge: Cambridge University Press.
54. Zou, Y., Zhang, S., Ma, X., Zhou, T., & Zeng, B. (2016). Numerical investigation of hydraulic fracture network propagation in naturally fractured shale formations. *Journal of Structural Geology, 84*, 1–13.

Chapter 14
Fundamentals of the Hydromechanical Behavior of Multiphase Granular Materials

Francesca Casini

Abstract The principal aim of this chapter is to describe the hydromechanical behavior of unsaturated soils based on experimental evidence. The unsaturated soils are media in which the pore space is occupied by more than a fluid, typically liquid and gas. They give rise to very characteristic types of geotechnical problem such as: the loss of strength associated with the increase in water content or degree of saturation and the damage to structure caused by the collapse for saturation induced by wetting. An appropriate description of the behavior of unsaturated soils must incorporate these fundamental effects of wetting on strength and deformation. The experimental evidence in terms of stiffness, compressibility, and strength is presented and discussed.

14.1 Introduction

The fundamental role of suction on unsaturated soil behavior was early recognized [1] and the years between 1950 and 1960 were characterized by an intense experimental activity [2, 3]. This experimental evidence was interpreted using a new concept of effective stress. After that it followed a period of lack of study on the mechanical behavior of partially saturated soils related to the difficulty to apply the effective stress principle [4]. During this period, the partially saturated soils have been labeled as "difficult soils" or "special soils." In principle there is nothing special in a partially saturated soil apart from the simple fact that some of the voids are occupied by air (or another non-wetting fluid). Instead of considering the partially saturated soils as a separate class of materials, they should be treated using the well-known concept at the base of soil mechanics in saturated conditions properly extended to partially saturated conditions.

F. Casini (✉)
Dipartimento di Ingegneria Civile e Ingegneria Informatica (DICII), Università degli Studi di Roma "Tor Vergata", Via del Politecnico 1, Rome 00133, Italy
e-mail: francesca.casini@uniroma2.it

C. Jin, G. Cusatis (eds.), *New Frontiers in Oil and Gas Exploration*,
DOI 10.1007/978-3-319-40124-9_14

Important steps in this direction have been made by considering two stress variables in the definition of the state surface [5], an idea already anticipated by Bishop and Blight [3] and Coleman [6]. The use of the state-surface, the theoretical and experimental assumption in the use of two stress variables, has been further justified by Fredlund and Morgenstern [7] and Fredlund and Rahardjo [8]. An extensive experimental study on the fundamental stress variables to be used in unsaturated conditions was also performed by Tarantino et al. [9].

Since 1980 to date an increasing attention has been paid to the unsaturated soils from the experimental and theoretical point of view. To interpret the experimental data available different constitutive models have been proposed in literature starting from the Barcelona Basic Model (BBM) proposed by Alonso et al. [10]. The constitutive models based on elasto-plasticity proposed in literature (e.g., [11, 12]) have been a successful tool to interpret the hydromechanical behavior of unsaturated soils overcoming the limitations of the state surface approach.

14.1.1 Fundamental Definition in Terms of Volumes and Weights

The unsaturated soils are media in which the pore space is occupied by more than a fluid, typically liquid and gas. To describe properly the behavior of unsaturated soils, it is convenient to make a distinction between phases and species. The porous medium is composed of three species—mineral (–), water (w), and air (a)—distributed in three phases: solid (s), liquid (l), and gas (g) (Fig. 14.1).

A fundamental variable describing the unsaturated soils is the liquid degree of saturation, S_l or the gas degree of saturation, defined respectively as:

$$S_l = V_l/V_v, S_g = 1 - S_l \quad (14.1)$$

where V_l is the liquid volume, V_g is the gas volume, $V_v = V_l + V_g$ is the volume of voids. S_l represents the proportion of pore occupied by liquid.

The void ratio e is defined as:

$$e = V_v/V_s \quad (14.2)$$

where V_s is the volume of solid particles.

The water content w for an unsaturated soil is related to S_l and e through the following equation:

$$w = W_l/W_s = S_l \cdot e/G_s \quad (14.3)$$

where W_l is the weight of liquid (water in this case), W_s is the weight of solid, and $G_s = \gamma_s/\gamma_l$ (typically in the range 2.6–2.75) is the specific gravity and it represents

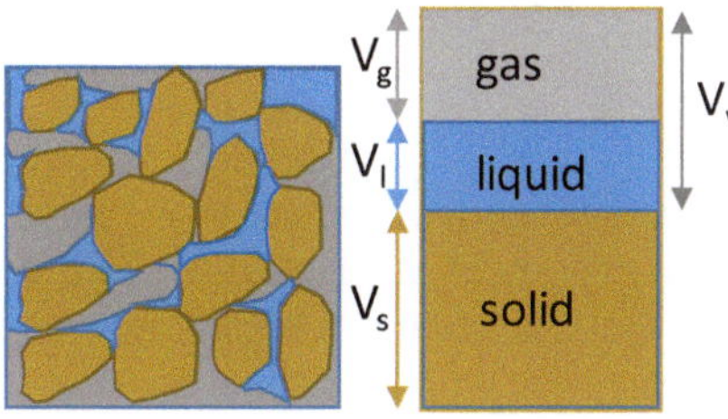

Fig. 14.1 Unsaturated soils volume divisions: solid, liquid, and gas phases in a typical configuration

Table 14.1 A summary of each phase within the soil element

Phase	Volume	Mass	Weight
Solid	V_s	M_s	W_s
Liquid	V_l	M_l	W_l
Gas	V_g	~0	~0

the ratio between the unit weight of solid particles $\gamma_s = W_s/V_s$ and the unit weight of liquid $\gamma_l = W_l/V_l$.

Combining Eqs. (14.2) and (14.3) we obtain $e = w \cdot G_s/S_l$.

The specific volume v is defined as:

$$v = V/V_s = (V_v + V_s)/V_s = 1 + e \tag{14.4}$$

where $V = V_v + V_s$. In unsaturated conditions this could be useful to define the specific volume of liquid as:

$$v_l = (V_l + V_s)/V_s = 1 + e_l \tag{14.5}$$

where $e_l = V_l/V_s$ is the liquid ratio, combining this definition with Eqs. (14.1) and (14.2), $S_l = e_l/e$.

The porosity n is the ratio of volume of void to total soil volume:

$$n = V_v/V = V_v/(V_v + V_s) = e/(1 + e) = (v - 1)/v \tag{14.6}$$

The dry density γ_d, is defined as the ratio of weight of solid particles to total soil volume:

$$\gamma_d = W_s/V = \gamma/(1 + w) = \gamma_s/(1 + e)$$

where $\gamma = W/V$ is the unit weight of soil, $W = W_l + W_s$ is the weight of the soil (W_g is negligible).

A summary of each phase within the soil element is reported in Table 14.1 (where $W = M \cdot g$, gravity acceleration)

14.1.2 Definition of Suction

As highlighted in the introduction, a fundamental rule is played by the suction. To properly describe the suction is useful to recall the definition of soil-water (total potential ψ) defined as "the amount of work that must be done per unit quantity of pure water in order to transport reversibly and isothermally an infinitesimal quantity of water from a pool of pure water at a specified elevation at atmospheric pressure to the soil water at the point of consideration" [13].

The total potential ψ can be divided into four different components (after [14]):

$$\psi = \psi_g + \psi_p + \psi_m + \psi_0$$

where ψ_g is the gravitational potential; ψ_p is the gas pressure potential; ψ_m is the matric potential, and ψ_0 is the osmotic potential. The gravitational potential is driven by difference in elevation; the gas pressure potential is related to the applied gas pressure; osmotic potential is driven by difference in solute concentrations through a semipermeable membrane; matric potential depends on the interaction between solid particles and liquid, and it is composed by the adsorptive and capillary forces (after [15]). If the amount of work is evaluated per unit of volume instead that of unit of mass, the resulting variable is defined as suction with the dimension of a pressure (kN/m^2), with a sign chance to avoid using negative value [14].

The mechanical behavior of unsaturated soils is influenced in different ways by the different components of suction. In particular the most relevant components influencing the behavior of soils are the osmotic and matric suction. Thus the total suction s_t can be defined as:

$$s_t = s + s_0 \tag{14.7}$$

where s is the matric suction and s_0 is the osmotic suction.

The total suction is related to the relative humidity (RH) through the Kelvin's law (see [14] for more details).

Matric suction s is defined usually in terms of capillary forces, e.g., capillary rise, acting within the soils. Capillary rise is caused by the surface tension and interface curvature between the soil particles and the water molecules.

The capillary height, h_c, in a tube can be evaluated imposing the equilibrium at the system reported in Fig. 14.2 composed by the column of water rising in the tube and the surface of separation between liquid and gas (menisci).

$$2\pi r T_s \cos\alpha = \rho_w g h_c \pi r^2 \quad h_c = 2T_s \cos\alpha/(\rho_w g r) \quad s = \rho_w g h_c = 2T_s \cos\alpha/r \tag{14.8}$$

where T_s is the capillary tension (force over a length N/m) and it decreases as the temperature increases. It assumes $T_s(25\ °C) = 72$ mN/m for water, α is the contact angle between liquid and tube and it depends on the surface characteristic of the solid phase and it is different in imbibition and in drying; r is the radius of the tube.

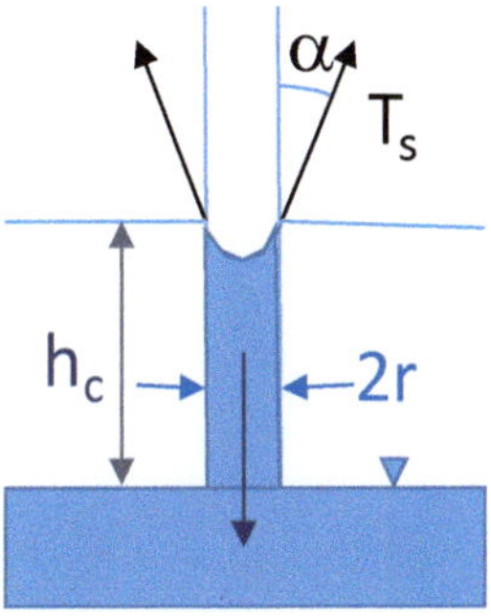

Fig. 14.2 Simplified scheme of capillary rise in a tube

The simplified scheme proposed above can be extended to a more complex geometry of the menisci different from a spherical cap. The menisci can be described through a surface with double curvature characterized by the radius r_1 and r_2, the equilibrium of the system will draw to the following Laplace equation:

$$s = T_s(1/r_1 + 1/r_2) = 2Ts/r^* \tag{14.9}$$

where $2/r^* = (1/r_1 + 1/r_2)$ represents the mean radius of curvature.

In the capillary model the air–water meniscus in a glass tube can be envisaged similar to the air–water interface within a soil pore. The matric suction, as reported in Eqs. (14.8) and (14.9), is proportional to the inverse of the effective pore radius using this model [16, 17].

14.1.3 Soil Water Retention Curve (SWRC)

The soil water retention curve describes the water storage mechanism in an unsaturated soil. The SWRC must be defined in terms of a variable describing the change in volume of water such as gravimetric water content w, degree of saturation S_l, volumetric liquid (water) content $\theta_{l=w} = V_{l=w}/V = n \cdot S_l$ or the liquid (water) ratio $e_{l=w}$ and the stress variable suction $s = u_g - u_l$ defined as the difference between gas pressure and liquid pressure.

The SWRC can be obtained imposing suction and measuring the volume of water at equilibrium (suction controlled technique) or measuring suction at known volume of water with different instrument as tensiometers (e.g., [18, 19]) or psychrometers (e.g., [20]) or filter paper technique.

The SWRC is not unique and it depends on:

- Grain size distribution of soils: clay, silt, and sand have different SWRCs (see Fig. 14.3).
- The direction of the loading path; it is different in wetting and drying (hysteresis Figs. 14.3 and 14.4).
- The void ratio of the sample (Fig. 14.4).
- The microstructural features of the soils [24].

Fig. 14.3 WRC of Jossigny silt (JS silt) and Ruedlingen sand (RS sand) at the same void ratio (after [21–23])

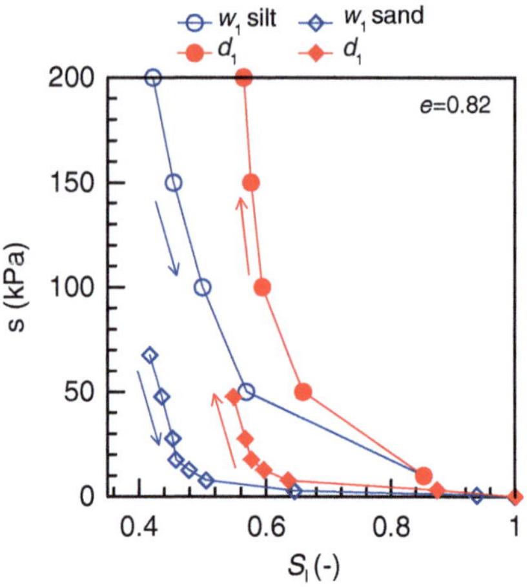

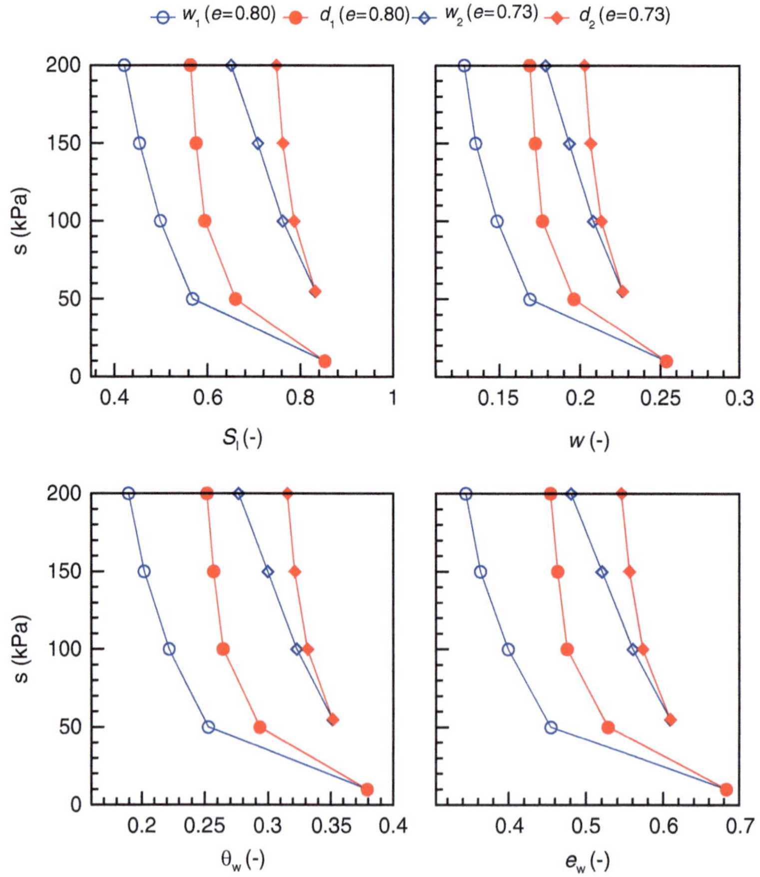

Fig. 14.4 SWRC: void ratio dependencies and hysteresis in different planes for JS silt

In Fig. 14.3 is reported, as an example of typical behaviors, the relation between the degree of saturation S_l and the suction s of two different soils: a silt with clay from Jossigny, east of France (JS silt) and a silty sand from Ruedlingen, north of Switzerland (RS sand). The results have been obtained under suction controlled conditions with a comparable void ratio. The SWRC depends on the direction of the loading path for both soils, fixing $s = 50$ kPa, $S_l = 0.44$ (RS), and $S_l = 0.56$ (JS) along the wetting w_1 while $S_l = 0.58$ (JS) and $S_l = 0.66$ along the drying d_1.

The dependencies of SWRC by loading and void ratios is independent by the variable we choose to represents the volume of water in the sample as reported in Fig. 14.4, where are reported the SWRC of the JS silt at two different void ratio in wetting and drying. The hysteretic effects and the void ratio dependencies are visible in the four plane used to represent the SWRC.

A summary of the different model used to describe the SWRC is reported in Barbour [17] which illustrates the centrality of the SWRC in providing the link between theory, behavior and material properties. In the last decades a lot of work has been done to proper describe the SWRC, see [25–32].

The most common models used are the following:

$$S_e = \frac{\theta_w - \theta_{wres}}{\theta_{wsat} - \theta_{wres}} = \begin{cases} \left(\frac{s}{s_b}\right)^{-\lambda} & (s > s_b) \\ 1 & (s < s_b) \end{cases} \tag{14.10}$$

$$S_e = \frac{\theta_w - \theta_{wres}}{\theta_{wsat} - \theta_{wres}} = \left[\frac{1}{1 + (\alpha s)^n}\right]^m \text{ with } \left(m = 1 - \frac{1}{n}\right) \tag{14.11}$$

where θ_{wres} is the residual water content, θ_{wsat} is the saturated water content. λ is a shape parameter of BC model; m and n are shape parameters of VG model which can be related each other; α related to the inflection point of the WRC (see Eq. 14.15). As reported in Fig. 14.4 the VG model is able to reproduce the typical S-shaped form of the WRC. The comparison between the VG and BC models prediction are reported in Fig. 14.5 for two types of soils JS silt and RS sand and the parameters are reported in Table 14.2.

Different models have been proposed in literature to incorporate the dependencies of the SWRC by void ratio: Mašín [28]) proposed to extend the BC model taking into account the dependencies s_b by void ratio; Salager et al. [33] proposed a method to take into account the dependencies of the inflection point of SWRC by void ratio.

14.1.3.1 Enhanced Models to Describe the WRC Based on Microstructural Features

The dependencies of the SWRC on the microstructural features have been highlighted by Romero and Vaunat [26]. The authors shown that the main wetting and drying paths of the SWRC separating a region of intra-aggregate porosity from

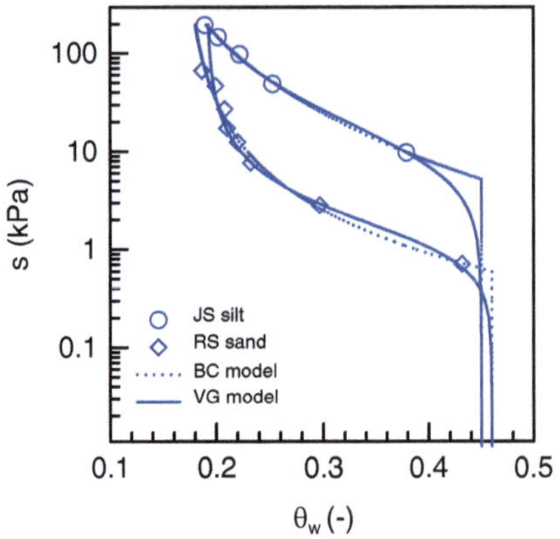

Fig. 14.5 Comparison BC versus VG model prediction for JS and RS soils

Table 14.2 BC and VG model parameters

BC	λ	s_b (kPa)	θ_{wsat}	θ_{wres}
RS	0.55	0.61	0.46	0.17
JS	0.35	5.30	0.45	0.09
VG	n	α (kPa^{-1})	θ_{wsat}	θ_{wres}
RS	2.04	0.76	0.46	0.19
JS	1.59	0.10	0.45	0.14

an inter-aggregate porosity adjoining area for a clayey aggregate. In the intra-aggregate region, the water ratio is not dependent on void ratio and retention curve parameters are mainly dependent on specific surface. However, in the inter-aggregate region, water ratio is dependent on void ratio and strongly coupled to mechanical actions (see Fig. 14.6). They basically extended the model proposed by Fredlund and Xing [25] to take into account the dependencies of the parameters by microstructural features (see [26]).

Recently different models have been proposed in literature to describe the SWRC based on microstructural observations (e.g., [34–36]). Casini et al. [37] proposed an extension to the Van Genuchten model to proper describe the evolution of SWRC of the Jossigny silt based on a wide range of Mercury Intrusion Porosimetry (MIP) results [24, 38].

The MIP test is performed by pushing mercury (a non-wetting fluid) in a soil sample to empty the pore at known pressure. The cumulative volume of mercury intruded in the soil sample is measured as a function of the pressure applied. The following Washburn equation is used to link the pressure applied to the pore entrance diameters of the sample:

$$p = -\frac{n\sigma_{Hg}\cos\theta_{nw}}{x} \tag{14.12}$$

where $\sigma_{Hg} = 0.484$ N/m is the surface tension of mercury at 25 °C, θ_{nw} the contact angle between mercury and the pore wall, and x the entrance or throat pore diameter ($n = 4$) or the entrance width between parallel plates ($n = 2$). The value $n = 4$ is

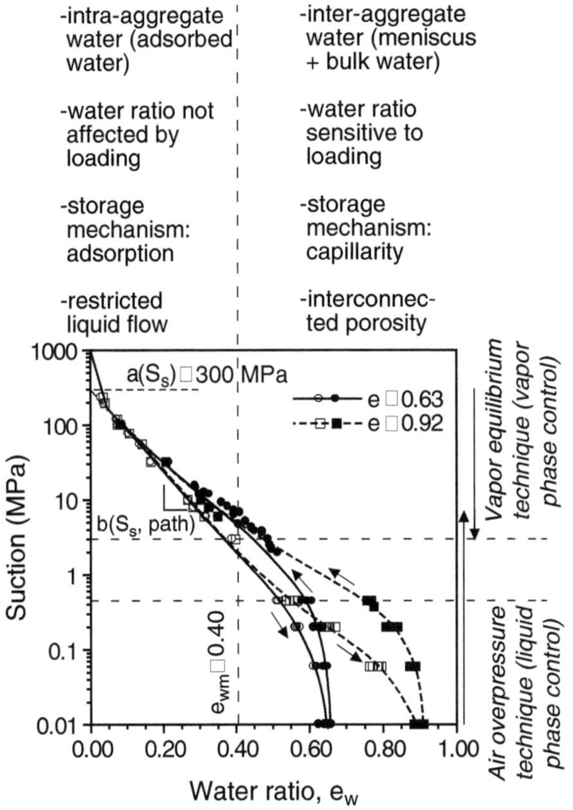

Fig. 14.6 SWRC at constant void ratio for high porosity ($e = 0.92$) and low porosity ($e = 0.63$) specimens. Intra- and inter-aggregate zones (after [26])

often used in MIP [38]. The contact angle ranges between 139° and 150° for clay minerals [39, 40]. The results of MIP tests are usually reported in terms of pore entrance diameter versus cumulative volume of mercury intruded making the analogy between the mercury and the air as a non-wetting fluid [38].

The results of the MIP tests are also reported in terms of cumulative and pore size density function (PSD) versus entrance pore size diameter (e.g., [24, 38]). The cumulative mercury intrusion plot and different normalization results of the pore size distribution (PSD) are reported in Fig. 14.7 for a test of JS silt. The PSD in terms of log diameter has been normalized to the maximum intruded void ratio $e_{nw} = V_{nw}/V_s$ with a non-wetting fluid to obtain a unit area below the curve (Fig. 14.7c). To compare the effect of the initial void ratio of the compacted material, the normalized density function was further scaled by this initial void ratio, as shown in Fig. 14.7d. This last plot PSD* was the one used to compare the effect of the initial state on the pore size distribution.

The MIP results can be used to obtain the relationship between the suction and the degree of saturation or water content at constant void ratio. The mercury intrusion is assimilated to the air-intrusion (a non-wetting fluid) during the drying path of the water retention curve. Thus the injection of mercury with a contact angle

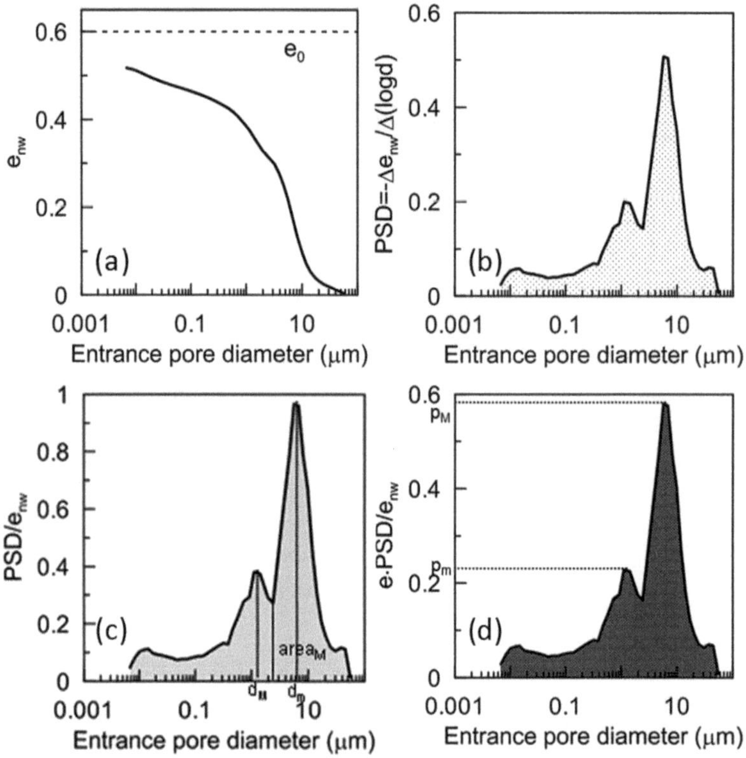

Fig. 14.7 Representation of MIP results: (**a**) Void ratio non-wetting phase; (**b**) Pore size density function (PSD); (**c**) PSD normalized respect to void ratio of non-wetting phase (Mercury) intruded (e_{nw}); (**d**) PSD normalized over e_{nw} and scaled to the void ratio e (after [23])

θ_{nw} is equivalent to the drainage of the water induced by air front advancing for the same diameter of pores intruded. Under the hypothesis of non-deformable soil skeleton, the volume of pores non-intruded by the mercury should be used to evaluate the degree of saturation or the water content corresponding to the equivalent applied air overpressure. Anyway, the non-intruded porosity by the mercury should be taken into account for estimating the residual water content in the evaluation of the water content. The WRC obtained is valid in the range where the capillarity is the predominant physics mechanisms, usually for suction $s < 2$ MPa [37].

The water ratio $e_w = V_w/V_s = S_r \cdot e$ is estimated by the following equations:

$$e_{nw} + e_w = e; \quad e_w = e - e_{nw} = (1 - S_{rnw}) \cdot e \tag{14.13}$$

Where e_{nw} is the non-wetting ratio (mercury), S_r and S_{rnw} the degree of saturation of water and non -wetting fluid respectively ($S_r + S_{rnw} = 1$ for a two fluid mixture).

To properly describe the evolving water retention character, a multimodal retention model has been recently proposed by Casini et al. [37] considering a linear superposition of two subcurves of the van Genuchten type. The model is constructed based on MIP data at specified initial state, and thus, it is representative of this initial pore network configuration. The different model parameters are fitted and their evolution with water content and void ratio studied. The evolving nature along generalized stress paths (with varying void ratio and water content) is taken into account by updating the pore network configuration as the hydraulic paths develop. The multimodal retention model, defined by Durner [41] as a linear superposition of subcurves of the van Genuchten type with the following equation:

$$E_{\mathrm{w}} = \frac{e_{\mathrm{w}} - e_{\mathrm{wres}}}{e - e_{\mathrm{wres}}} = \sum_{i=1}^{k} w_i \left[\frac{1}{1 + (\alpha_i s)^{n_i}} \right]^{m_i} \tag{14.14}$$

has been used by Casini et al. [37], where k is the number of subsystems that assembled together give the global pore size distribution, w_{i} are weighting factors for each subcurve subject to $0 < w_{\mathrm{i}} \leq 1$ and $\sum w_i = 1$. For the parameters of the subcurves (α_{i}, n_{i}, m_{i}) must be imposed the condition $\alpha_{\mathrm{i}} > 0$; $m_{\mathrm{i}} > 0$; $n_{\mathrm{i}} > 1$. Here, the additional constraint $n_{\mathrm{i}} = 1/(1-m_{\mathrm{i}})$ is imposed.

Each subcurve in Eq. (14.14) differentiated two times with respect to the suction s gives the relationship between the suction at the inflection point and the parameters given by the following expression:

$$s_{pi} = \frac{m_i^{1-m_i}}{\alpha_i} \tag{14.15}$$

where s_{pi} is the suction at the inflection point in the E_{w}-s plane. Once α_{i} is obtained, from Eq. (14.15) and substituting in Eq. (14.14), the water storage mechanism in a heterogeneous porous medium becomes:

$$E_{\mathrm{w}} = \frac{e_{\mathrm{w}} - e_{\mathrm{wres}}}{e - e_{\mathrm{wres}}} = \sum_{i=1}^{k} w_i \left[1 + m_i \left(\frac{s}{s_{pi}} \right)^{\frac{1}{1-m_i}} \right]^{-m_i} \tag{14.16}$$

The Eq. (14.16) has been used for model the SWRC data obtained by MIP. The experimental data are modelled with a bimodal water retention model, obtained setting $k = 2$ in the above equation. The pore capillary density function (PCF) is defined as:

$$\mathrm{PCF} = \Delta E_w / \Delta \log(s) \tag{14.17}$$

where E_{w} is the effective degree of saturation as defined in Eq. (14.16).

The model is calibrated best fitting the PCF obtained by experimental results as reported in Fig. 14.8. The area below the peak of lower suction is defined as w_1

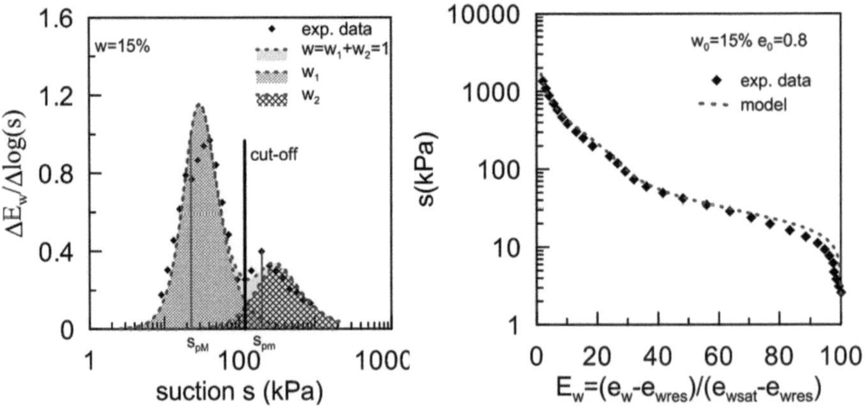

Fig. 14.8 Laboratory versus model prediction: (**a**) Pore capillary function (PCF) macro micro; (**b**) WRC derived from PCF

(macro area below the PCF) while w_2 (micro area) is given by $w_2 = 1 - w_1$ because the area below the curve corresponds to the unity. The suction at inflection points are named respectively s_{pM} and s_{pm} as reported in Fig. 14.8a. A comparison between the model prediction and the experimental data is reported in Fig. 14.8b, the model well fit the experimental results in the entire range of suction. Furthermore, the model is able to well capture the change in shape of the retention curve at the inflection points.

In an unsaturated soil the liquid permeability is function of void ratio and degree of saturation. It is quite difficult to obtain in laboratory (e.g., [42]). As the suction increases the degree of saturation decreases and the permeability of the soils decreases as well. A simple choice to describe the relative permeability is to set the $k_{\text{rel}} = k_{\text{unsat}}/k_{\text{sat}} = S_l^{\lambda}$ with $\lambda = 3$. Different types of models are available in literature to describe the evolution of relative permeability (e.g., [30, 31]).

14.1.4 Stress Variable in Unsaturated Conditions

The effective stress for saturated soils has been defined by Terzaghi [43] as

$$\sigma_{ij}^{'} = \sigma_{ij} - \delta_{ij} u_l \tag{14.18}$$

where u_{l} is the pore water pressure, δ_{ij} is the Kronecker's delta.

"All measurable effects of a change of stress of the soil, that is, compression, distortion and change in shear resistance, are exclusively due to change in effective stress [43]."

The definition is valid under the hypothesis of saturated soils with incompressible grains and the voids completely filled with incompressible fluid. The effects of compressibility of grains and fluid can be taken into account through the definition of a parameter α_l as:

$$\alpha_l = 1 - (1-n)\frac{C}{C_s} \quad \sigma'_{ij} = \sigma_{ij} - \alpha_l \delta_{ij} u_l \tag{14.19}$$

proposed by Suklje [44], which represents a combination of the Biot [45] and Skempton [46] parameters [32, 47].

In saturated conditions, where two phases are considered: solid particle and fluid, the Terzaghi effective stress represents the only stress governing the mechanical behavior of soils. In unsaturated conditions the choice of appropriate stress variable is still an intensive debated issue.

As pointed out in Jommi [48]: "in fact, no single stress variable has ever been found which, substituted for effective stress, allows for a description of all the aspects of the mechanical behavior of a given soil in the unsaturated range." A second variable is generally required to represent the stabilizing influence of suction on intergranular forces and the volumetric effects of its removal or weakening, by wetting [11].

The most common choices of two variables to describe the behavior of unsaturated soils are:

- net stress $\sigma_{ij\text{net}} = \sigma_{ij} - u_g \delta_{ij}$ and suction $s = u_g - u_l$ (e.g., [7, 10]);
- generalized stress approach $\sigma'_{ij} = \sigma_{ij} - u_g \delta_{ij} + \chi(u_g - u_l)\delta_{ij}$ [49] and suction $s = u_g - u_l$; (e.g., [48, 50]).

A summary of the possible choice in unsaturated conditions is reported in Gens et al. [11]; Nuth and Laloui [32, 47] and Sheng [12].

The definition of two stress variables, net or generalized, is not able alone to explain the collapse for saturation during wetting. In addition, we must define an additional yielding curve (Loading-Collapse) which increases with suction (e.g., [10]) or with degree of saturation (e.g., [48, 50]) as function of the net or generalized mean stress.

The net stress approach it is the simplest and most practical choice in terms of stress path representation, it poses difficulties when trying to incorporate the SWRC effects as the hysteresis or the void ratio dependencies. While the generalized stress approach allows the coupling with the SWRC and the representation of stress path is more complex and sometimes and it becomes impossible if, as it is often the case, data on water content is not available or unreliable (after [11]).

The choice of the χ parameter in the definition of generalized stress is still an open question:

$$\sigma'_{ij} = \sigma_{ij} - u_g \delta_{ij} + \chi(u_g - u_l)\delta_{ij} \tag{14.20}$$

A usual choice is $\chi = S_l$ [49], other function can be defined to link the χ parameter with suction (e.g., [29]) or with effective degree of saturation (e.g., [51]).

14.1.5 Small Strain Stiffness

The influence of strain on soil stiffness has been extensively investigated over the past decades (e.g., [52]). The soils exhibit a truly elastic behavior within a strain range very small, shear strain $\gamma_s < 10^{-6}$. This type of behavior is common to all type of soils ranging from clay to sand. As the strain amplitude increases the soil stiffness decays nonlinearly, the stiffness reduction curves exhibit a typical S-shaped with applied strains as reported in Fig. 14.9.

In the following some experimental evidence on the small strain shear stiffness in unsaturated conditions is reported and compared with the saturated results. An interpretation of G_0 with the generalized stress is proposed and validated with the experimental data.

In saturated conditions the small strain stiffness depends on the average effective stress p' [52]. Different authors (e.g., [56]), examining data relative to reconstituted and undisturbed materials, showed that an undisturbed material, due to its "structure," has higher values of G_0 than the same material in a reconstituted state (Fig. 14.10). The effect of this structure is lost as an effect of increasing p'. A linear relationship can be assumed between log G_0 and log p' for both normally and overconsolidated states, with a lower slope for overconsolidation.

Using same representations as Rampello et al. [56], G_0 values measured under controlled suction conditions by resonant column tests by Vassallo et al. [58] have been plotted in Fig. 14.10a as a function of the stress $p' = p - u_g + S_r(u_g - u_l)$ defined by Eq. (14.20) with $\chi = S_l$. Moving from complete saturation to partial saturation induces a translation upwards of experimental curves G_0:p', similar to that highlighted by Rampello et al. [56] comparing the behavior of natural and reconstituted clays. Partial saturation seems to provide the solid skeleton with a structure (or cementation) similar to ageing and other natural time effects. In

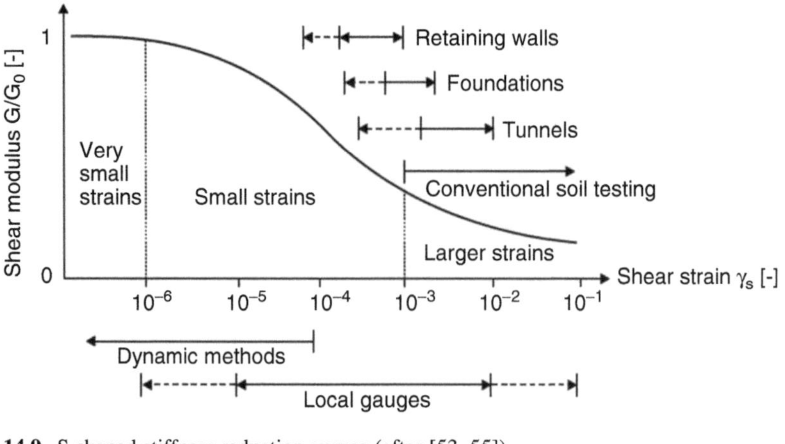

Fig. 14.9 S-shaped stiffness reduction curves (after [53–55])

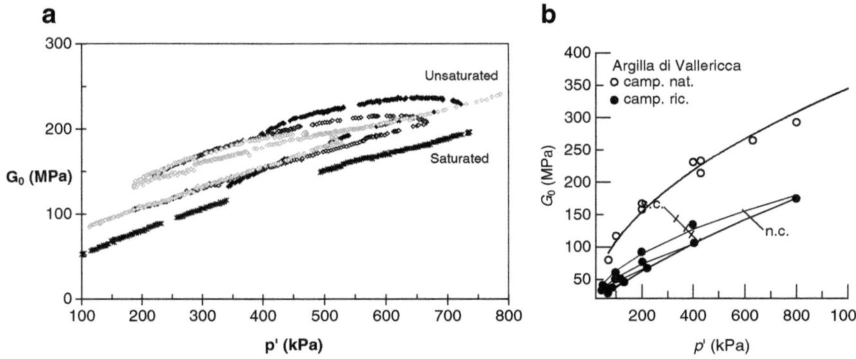

Fig. 14.10 Comparison: (**a**) saturated–unsaturated compacted soil, (**b**) natural–reconstituted saturated soil (after [57])

addition, experimental points relative to measurements of G_0 at beginning of the tests show trends similar to that of an overconsolidated saturated soil [57, 59].

The relationships proposed by Rampello et al. [56] for saturated soils (Eq. 14.21) can be extended by including the effects of S_r on void ratio and on p':

$$G_0 = S^* p_r^{(1-n^*+c)} f(e) p'^{(n^*-c)} \tag{14.21}$$

where

$$f(e) = C/\exp(De) \tag{14.22}$$

and: $c = m/\Lambda$, $\Lambda = (\lambda - k)/\lambda$, $C = exp[c(N(S_r) - 1)/\ \lambda,]$, $D = c/\ \lambda$; p_r is a reference pressure.

Parameters S^*, n^*, and m (linked to the position and the slope of $log\ G_0$: $\log p'$ relationships) were obtained by fitting the saturated soil data (Fig. 14.6). Resulting values are $n^* = 0.61$, $m = 0.2$, $S^* = 3000$.

Equations (14.19) and (14.20) significantly underestimate experimental stiffness of the unsaturated soils. Thus, the extension of saturated relationships to the unsaturated case requires further assumptions. A better agreement can be obtained by supposing that also n^* and S^* depend on S_l. The obtained results are shown in Fig. 14.11 for the same tests of Fig. 14.10. It was found that n^* decreases and S^* increases as S_r decreases. The values used in the figure are $n^* = 0.60$ and $S^* \cong 4300$ for $S_{r,0} \cong 85\,\%$ (Fig. 14.11a), $n^* = 0.59$ and $S^* \cong 4500$ for $S_{r,0} \cong 82\,\%$, $n^* = 0.59$ and $S^* \cong 4800$ for $S_{r,0} \cong 79\,\%$ (Fig. 14.11b).

The effects of partial saturation on the initial shear stiffness G_0 are similar to those observed to the structure of a natural soil, compared to the same soil reconstituted. They cannot be justified by simply taking into account the effect of the partial saturation on the definition of generalized effective and its effects on the void ratio without significantly underestimating the unsaturated soil stiffness. It is necessary to consider explicitly the effect that S_l has on some parameters describing the relationship G_0:p'.

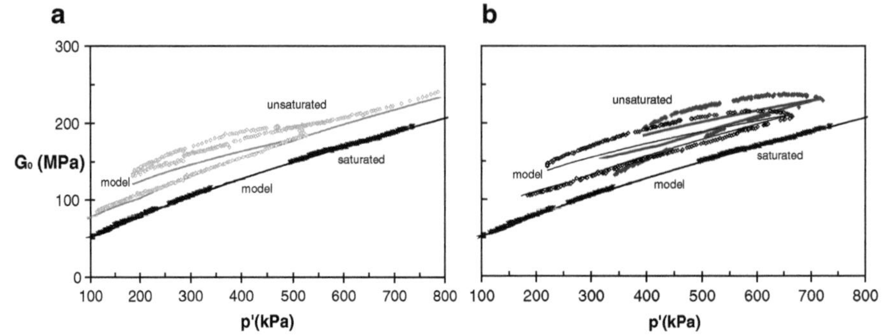

Fig. 14.11 Comparison between experimental results and model predictions of initial shear stiffness G_0. (**a**) compression test carried out at $s = 100$ kPa, (**b**) compression test carried out at $s = 200$ and 400 kPa

14.1.6 Stiffness at Moderate (Larger) Strain: Compressibility

The volumetric change induced by isotropic or anisotropic compression under suction controlled conditions has been extensively investigated in the past decades from experimental and theoretical point of view. This is due to the importance to understand and prevent the decreases in volume during wetting known as "collapse for saturation." The prediction of settlement induced by wetting is of particular importance in ground engineering. To mitigate sudden displacement during the life of a structure or of a natural deposit, it is important to predict the deformation of soils induced by the wetting process as unsaturated soils may either swell or shrink when they experience a wetting path. This is a function of the initial conditions in terms of void ratio and water.

In Fig. 14.12 are reported typical experimental results during 1D compression tests (oedometer) obtained at various level of suction by Casini [21, 22] on JS silt. The experimental results are reported in the plane $e-\log\,\sigma_{\text{vnet}}$ Fig. 14.12a and $e-\log\sigma'_{\text{v}}$ in Fig. 14.12b. As the suction increases the soils become stiffer and the inflection point of the curve moves to the right. At increasing vertical stress the curves tend to converge. Both the representations show similar behavior (Fig. 14.12).

These results show two main common effects of partial saturation on the compressibility of soils:

- The stress at inflection point increases with suction.
- The soil becomes stiffer with suction.

Anyway the curve after the inflection point is characterized by a slope changing with applied vertical stress.

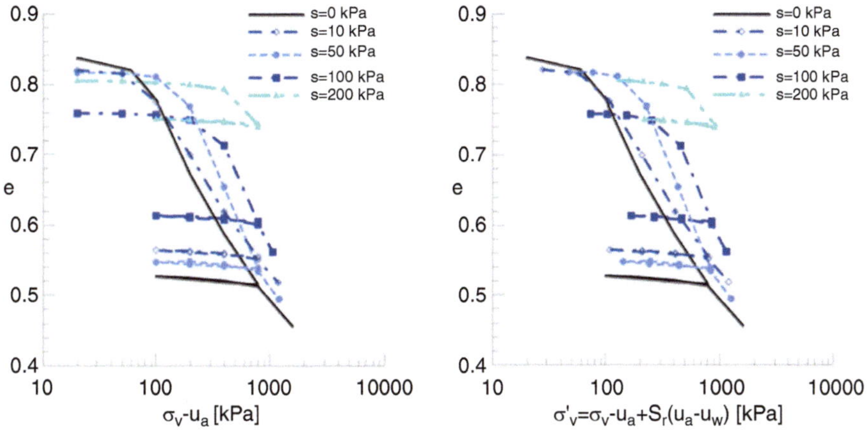

Fig. 14.12 One-dimensional compression at different suction on JS silt: (**a**) $e-\log(\sigma_v-u_a)$ plane; (**b**) $e-\log(\sigma_v-u_a+S_r(u_a-u_w))$

14.1.6.1 Modelling the Compressibility Behavior

In the following a simple model to describe the compressibility behavior of unsaturated soils is presented and validated with experimental results.

The constitutive relationships for the mechanical behavior are defined in terms of Bishop's stress $\sigma'_{ij} = \sigma_{ij} - u_a\delta_{ij} + \chi(u_a - u_w)\delta_{ij}$ and suction $s = u_a - u_w$, where σ_{ij} is the total stress, χ is a weighting parameter, taking into account S_r, and δ_{ij} is the Kronecker delta.

The vertical soil skeleton (effective) stress is defined as:

$$\sigma'_v = \sigma_v - u_a + \chi(u_a - u_w) \tag{14.23}$$

The evolution of the preconsolidation vertical stress σ'_{vc} is characterized by a double mechanism, depends on the rate of plastic strains and by the variation of χ:

$$\dot{\sigma}'_{vc} = \dot{\sigma}'_{vc}\text{sat} - \bar{a}\sigma'_{vc}\dot{\chi} \tag{14.24}$$

where $\bar{a}$ is a constant soil property. Integration of Eq. (14.24) yields:

$$\sigma'_{vc} = \sigma'_{vcsat}e^{\bar{a}(1-\chi)} \tag{14.25}$$

where, $\bar{a}$ controls the rate of change in σ'_{vc} due to the variation of χ.

The hardening is a combination of a reversible component related to change in χ and an irreversible component dependent on the development of plastic strains. The model predicts that a drying process induces some *bonding* (positive hardening) while a wetting process induces some *debonding* (negative hardening).

Considering an oedometric compression test starting from a virgin state, in an analogy with saturated soils, the NCL will be given by:

$$e = B(\chi) - C_c \log\sigma'_v \qquad (14.26)$$

where σ'_v is the vertical effective stress defined by Eq. (14.20) , $B(\chi)$ is the void ratio at 1 kPa and C_c is the compression index.

The dependency of $B(\chi)$ can be obtained easily from Eq. (14.26) and from:

$$e_i = B - C_c \log\sigma'_{vcsat}, \quad e_i = e_p - C_s \log\frac{\sigma'_{vcsat}}{\sigma'_{vc}} \qquad (14.27)$$

where e_i and e_p are the void ratios in saturated and unsaturated conditions respectively on NCL (Fig. 14.13), B is the void ratio at a vertical stress $\sigma'_v = 1$ kPa under saturated conditions, C_s is the swelling index and by substitution:

$$B(\chi) = B + (C_c - C_s)a(1 - \chi) \qquad (14.28)$$

where $a = \frac{\bar{a}}{\ln 10}$.

A variation of χ corresponds to a translation of NCL, parallel to the saturated NCL if χ is constant, by an amount depending on the constitutive parameter a. During a compression phase at constant suction or water content the void ratio change inducing a change in the degree of saturation taking into account by the variation of parameter χ which moves the position on of the NCL approaching the saturated ones as χ increase (Fig. 14.13).

The capability of this model to describe properly the behavior of the unsaturated soils is reported in Fig. 14.14 with the parameter $\chi = \frac{Sl - Slmicro}{1 - Slmicro}$ where Sl and Sl micro are calculated using the SWRC reported in Eqs. (14.13)–(14.16). The tests are performed on Jossigny silt [60] statically compacted at three different void ratio and the same initial water content of at $w = 0.13$ in oedometer. The samples have been compressed up to a vertical loading $\sigma_v = 200$ kPa and finally saturated. The comparison is satisfactory in the two tests where the samples compress during saturation and in the case where the sample swells for the lower void ratio as shown in Fig. 14.14.

14.1.7 Strength of Unsaturated Soils

The strength parameters in unsaturated conditions can be obtained interpreting triaxial tests or shear tests under suction controlled conditions or under constant water content. In the latter case the suction can be measured or can be back-calculated with SWRC.

Typical results of the axial compression phase under suction controlled condition are reported in Fig. 14.15 in terms of deviatoric stress q versus axial

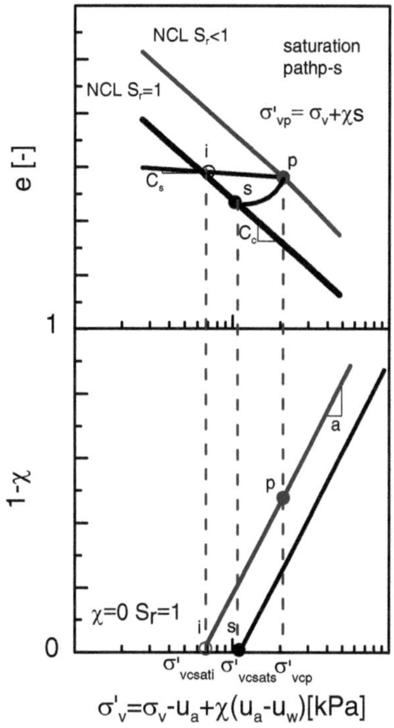

Fig. 14.13 Stress path followed by a point *p* to saturation *s*. Point *i* represents the saturated preconsolidation vertical stress of point *p* (after [23])

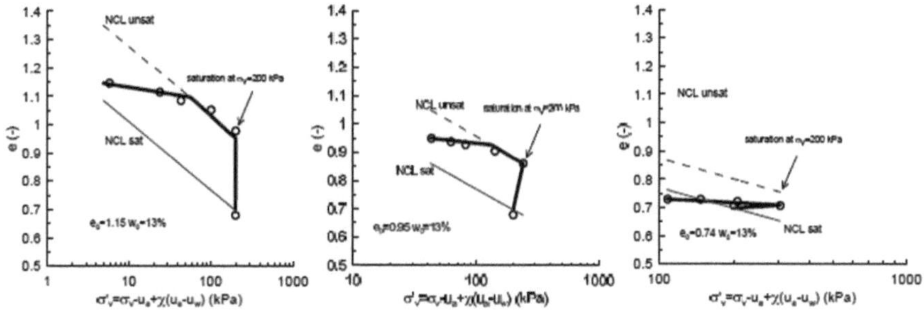

Fig. 14.14 Comparison between model and data (after [60])

strain ε_a. The tests were performed by Vassallo [61] on silt with clay at two different confinement net stresses (Fig. 14.15a $p - u_{a0} = 100$ kPa; Fig. 14.15b $p - u_{a0} = 400$ kPa). The strength increases with suction, the behavior move from ductile to fragile as the suction increases for the lower confinement stress (Fig. 14.15a). While the higher confinement stress the soil shows a ductile behavior over the entire range of suction investigated.

The results of shear tests on a silty sand are reported below. The results are shown in Figs. 14.16, 14.17 and 14.18 in terms of the relation between normalized

Fig. 14.15 Deviatoric stress versus axial strain at different suction: (**a**) $p - u_{a0} = 100$ kPa; (**b**) $p - u_{a0} = 400$ kPa (after [61])

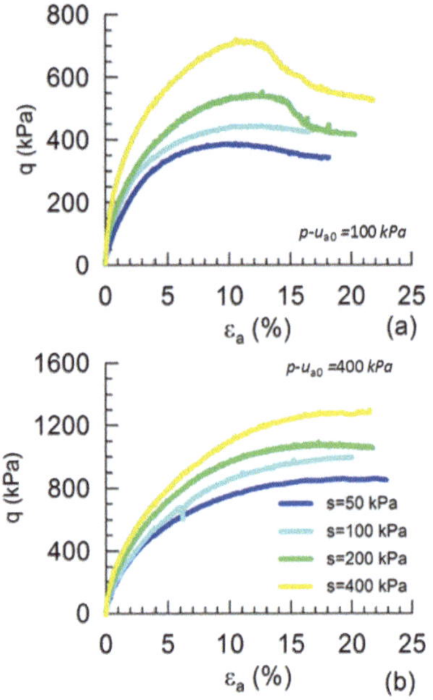

Fig. 14.16 Data at $\sigma_v = 20$ kPa

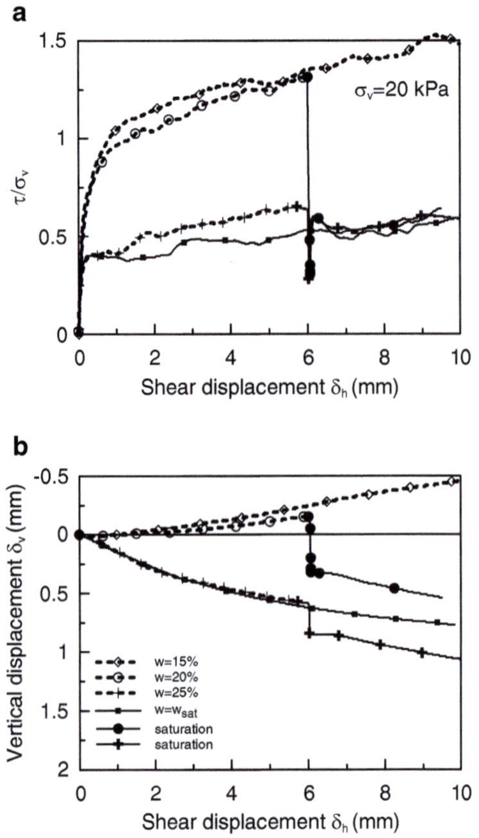

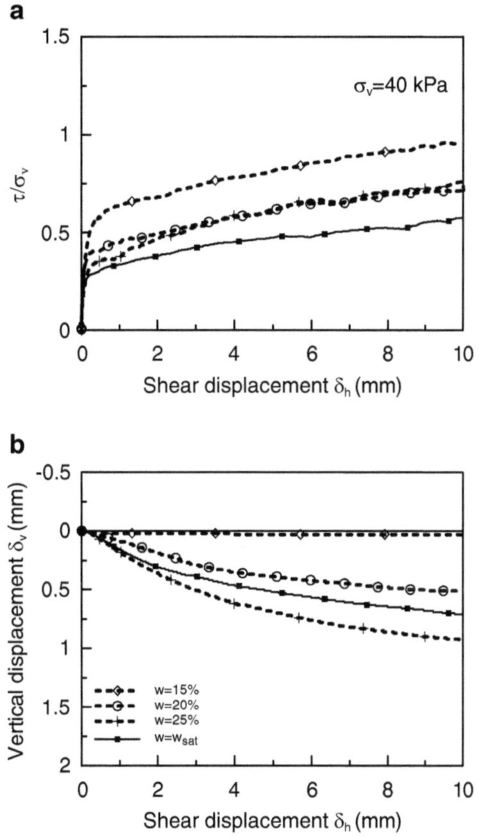

Fig. 14.17 Data at $\sigma_v = 40$ kPa

shear stress (τ/σ_v), vertical displacement δ_v and shear displacement δ_h. A negative vertical displacement indicates an increase in specimen height.

Highest stress ratios are exhibited by two samples at $\sigma_v = 20$ kPa and $w = 0.15$ and 0.20 (Fig. 14.15a). Both specimens show an increase in height (dilatancy) during shearing (Fig. 14.16b). The soil reduces in volume during the saturation of specimens at $w = 0.20$ and 0.25, indicating wetting collapse (Fig. 14.16b). In the subsequent shearing stage, the stress ratios are in good agreement with the saturated sample. The sample under $\sigma_v = 80$ kPa also reduces in volume during the saturation phase (Fig. 14.18).

The stress ratio for the same water content decreases as the vertical stress increases. This is due to the combination of two effects. Firstly, increasing vertical stress leads to a lower void ratio resulting in a higher degree of saturation and therefore less suction. Secondly the overconsolidation ratio (OCR) is decreasing as the vertical stress increases.

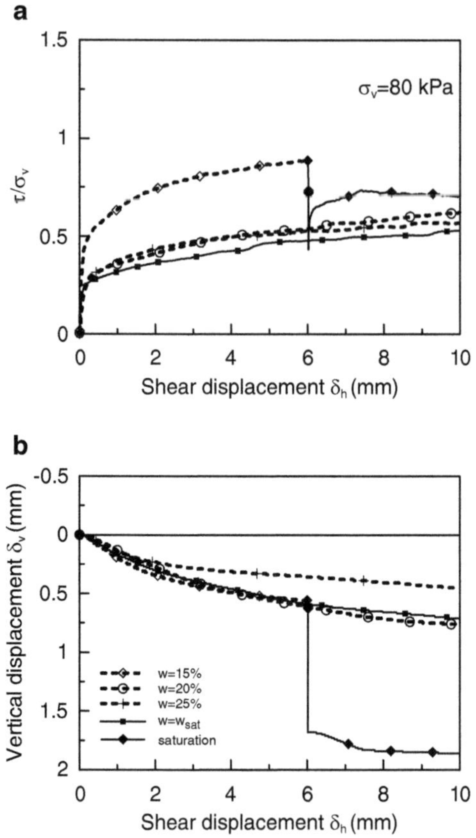

Fig. 14.18 Data at $\sigma_v = 80$ kPa (after [62])

The results of the shear phase can be plotted in the plane $\tau - \sigma_n^{'}$ with the suction obtained through the SWRC. Based on the generalized effective stress definition the Mohr–Coulomb criterion becomes:

$$\tau = \left[\left(\sigma_n - u_g\right) + \chi\left(u_g - u_l\right)\right] \tan\varphi^{'} + c^{\prime} \tag{14.29}$$

Where the terms $\chi\left(u_g - u_l\right)$ can be evaluated by SWRC with $\chi = \chi(S_l)$.

In Fig. 14.19 is reported the comparison between the prediction with Eq. (14.29) with $\chi = S_l$ and the experimental data obtained by Casini et al. [62] for the RS sand. The data are reasonably fitted with a linear envelope with $c^{'} = 0$ and $\varphi^{'} = 31°$.

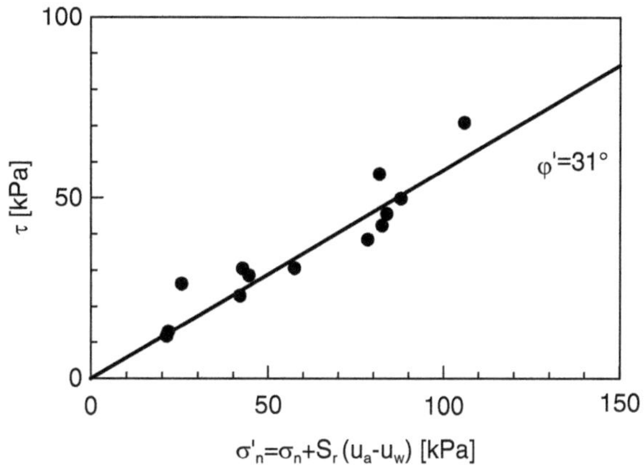

Fig. 14.19 Shear strength envelope using the WRC

References

1. Croney, D. (1952). The movement and distribution of water in soils. *Geotechnique, 3*, 1–16.
2. Bishop, A. W., Alpan, I., Blight, G. E., & Donald, I. B. (1960). Factors controlling the strength of partly saturated cohesive soils. In *Research conference on shear strength of cohesive soils* (pp. 503–532). Boulder: ASCE.
3. Bishop, A. W., & Blight, G. E. (1963). Some aspects of effective stress in saturated and partly saturated soils. *Geotechnique, 13*, 177–197.
4. Jennings, J. E. B., & Burland, J. B. (1962). Limitations to the use of effective stresses in partly saturated soils. *Geotechnique, 12*, 125–144.
5. Matyas, E. L., & Radhakrishna, H. S. (1968). Volume change characteristics of partially saturated soils. *Geotechnique, 18*, 432–448.
6. Coleman, J. D. (1962). Stress strain relations for partly saturated soil. Correspondence. *Geotechnique, 12*, 348–350.
7. Fredlund, D. G., & Morgenstern, N. R. (1977). Stress state variables and unsaturated soils. *Journal of the Geotechnical Engineering Division, ASCE, 103*, 447–466.
8. Fredlund, D. G., & Rahardjo, H. (1993). *Soil mechanics for unsaturated soils*. New York: Wiley.
9. Tarantino, A., Mongiovi, L., & Bosco, G. (2000). An experimental investigation on the isotropic stress variables for unsaturated soils. *Geotechnique, 50*, 275–282.
10. Alonso, E. E., Gens, A., & Josa, A. (1990). A constitutive model for partially saturated soils. *Geotechnique, 40*, 405–430.
11. Gens, A., Sánchez, M., & Sheng, D. (2006). On constitutive modelling of unsaturated soils. *Acta Geotechnica, 1*(3), 137–147.
12. Sheng. (2011). Review of fundamental principles in modelling unsaturated soil behavior. *Computers and Geotechnics 38*(6):757–776.
13. Bolt, G. H. (1976). Soil physics terminology. *Int Soc Soil Sci Bull, 49*, 16–22.
14. Gens, A. (2010). Soil-environment interactions in geotechnical engineering. *Geotechnique, 60* (1), 3–74.
15. Aitchison GD (1965) Soil properties, shear strength and consolidation. Proceedings of the 6th International Conference on Soil Mechanics and Foundation Engineering, Montreal 3, pp. 318–321
16. Taylor, D. W. (1948). *Fundamentals of soil mechanics*. New York: John Wiley & Sons.

17. Barbour, S. L. (1998). The soil-water characteristic curve: A historical perspective. *Canadian Geotechnical Journal, 35*, 873–894.
18. Ridley, A. M., & Burland, J. B. (1993). A new instrument for the measurement of soil moisture suction. *Geotechnique, 43*(2), 321–324.
19. Tarantino, A. (2003). Panel report: Direct measurement of soil water tension. *Proc. 3rd Int. Conf. on Unsaturated Soils*, Recife.
20. Cardoso, R., Romero, E., Lima, A., & Ferrari, A. (2007). A comparative study of soil suction measurement using two different high-range psychrometers. *Experimental Unsaturated Soil Mechanics, Springer Proceedings in Physics* (vol 112, pp. 79–94). Heidelberg, Berlin: Springer.
21. Casini, F. (2008). Effettidelgradodisaturazionesulcomportamentomeccanicodiunlimo. PhD Thesis, UniversitàdiRoma La Sapienza, Roma.
22. Casini, F. (2008). Effetti del grado di saturazione sul comportamento meccanico di un limo. PhD thesis, Universita´ degli Studi di Roma La Sapienza, Italy.
23. Casini, F. (2012). Deformation induced by wetting: A simple model. *Canadian Geotechnical Journal, 49*, 954–960.
24. Romero, E. (2013). A microstructural insight into compacted clayey soils and their hydraulic properties. *Engineering Geology, 165*, 3.
25. Fredlund, D. G., & Xing, A. (1994). Equations for the soil-water characteristic curve. *Canadian Geotechnical Journal, 31*(4), 521–532.
26. Romero, E., & Vaunat, J. (2000). Retention curves of deformable clays. In *Proceedings of international workshop on unsaturated soils: Experimental evidence and theoretical approaches* (pp. 91–106). Trento: Balkema.
27. Gallipoli, D., Wheeler, S., & Karstunen, M. (2003). Modelling the variation of degree of saturation in a deformable unsaturated soil. *Geotechnique, 53*(1), 105–112.
28. Mašín, D. (2010). Predicting the dependency of a degree of saturation on void ratio and suction using effective stress principle for unsaturated soils. *International Journal for Numerical and Analytical Methods in Geomechanics, 34*, 73–90.
29. Khalili, N., & Khabbaz, M. H. (1998). A unique relationship for χ for the determination of the shear strength of unsaturated soils. *Geotechnique, 48*(5), 681–687.
30. Brooks, R. H., & Corey, A. T. (1964). Hydraulic properties of porous media. *Colorado State University Hydrology Paper, 3*, 1–27.
31. van Genuchten, M. T. (1980). A closed-form equation for predicting the hydraulic conductivity of unsaturated soil. *Soil Science Society of America Journal, 44*, 892–898.
32. Nuth, M., & Laloui, L. (2008). Effective stress concept in unsaturated soils: Clarification and validation of a unified framework. *Computers and Geotechnics, 32*, 771–801.
33. Salager, S., Nuth, M., Ferrari, A., & Laloui, L. (2013). Investigation into water retention behaviour of deformable soils. *Canadian Geotechnical Journal, 50*(2), 200–208.
34. Romero, E., Della Vecchia, G., & Jommi, C. (2011). An insight into the water retention properties of compacted clayey soils. *Geotechnique, 61*(4), 313–328.
35. Otalvaro, I. F., Neto, M. P. C., Delage, P., & Caicedo, B. (2016). Relationship between soil structure and water retention properties in a residual compacted soil. *Engineering Geology, 205*, 73–80.
36. Ciervo, F., Casini, F., Papa, M. N., & Rigon, R. (2015). Some remarks on bimodality effects of the hydraulic properties on shear strength of unsaturated soils. *Vadoze Zone Journal, 14*(9).
37. Casini, F., Vaunat, J., Romero, E., & Desideri, A. (2012). Consequences on water retention 32 properties of double porosity features in a compacted silt. *Acta Geotechnica, 7*, 139–150.
38. Romero, E., & Simms, P. H. (2008). Microstructure investigation in unsaturated soils: A review with special attention to contribution of mercury intrusion porosimetry and environmental scanning electron microscopy. *Geotechnical and Geological Engineering, 26*(6), 705–727.
39. Diamond, S. (1970). Pore size distribution in clays. *Clays and Clay Minerals, 18*, 7–23.

40. Penumadu, D., & Dean, J. (2000). Compressibility effect in evaluating the pore-size distribution of kaolin clay using mercury intrusion porosimetry. *Canadian Geotechnical Journal, 37*, 393–405.
41. Durner, W. (1994). Hydraulic conductivity estimation for soils with heterogeneous pore structure. *Water Resources Research, 30*(2), 211–223.
42. Askarinejad, A., Beck, A., Casini, F., & Springman, S. M. (2012). Unsaturated hydraulic conductivity of a silty sand with the instantaneous profile method. *Unsaturated Soils: Research and Applications, 2*, 215–220.
43. Terzaghi, K. (1936). The shearing resistance of saturated soils and the angle between the planes of shear. *International Conference on Soil Mechanics and Foundation Engineering* (pp. 54–55) Cambridge, MA: Harvard University Press.
44. Suklje, L. (1969). *Rheological aspects of soil mechanics*. New York: Wiley.
45. Biot, M. A. (1955). Theory of elasticity and consolidation for a porous anisotropic soil. *Journal of Applied Physics, 26*(2), 182–185.
46. Skempton, A. W. (1960). Effective stress in soils, concrete and rocks. Pore pressure and suction in soils (pp. 4–16). Butterworths: London.
47. Nuth, M., & Laloui, L. (2008). Advances in modelling hysteretic water retention curve in deformable soils. *Computers and Geotechnics, 35*, 835–844.
48. Jommi, C. (2000). Remarks on the constitutive modelling of unsaturated soils. In A. Tarantino & C. Mancuso (Eds.), Experimental evidence and theoretical approaches in unsaturated soils. *Proceedings of the International Workshop on Unsaturated Soils* (pp. 139–153) Trento. Balkema, Rotterdam.
49. Bishop, A. W. (1959). The principle of effective stress. *Tecnisk Ukeblad, 39*, 859–863.
50. Tamagnini, R. (2004). An extended Cam-clay model for unsaturated soils with hydraulic hysteresis. *Geotechnique, 54*(3), 223–228.
51. Alonso, E. E., Pereira, J. M., Vaunat, J., & Olivella, S. (2010). A microstructurally based effective stress for unsaturated soils. *Géotechnique, 60*(12), 913–925. doi:10.1680/geot.8.P.002.
52. Viggiani, G., & Atkinson, J. H. (1995). Stiffness of fine-grained soils at very small strains. *Geotechnique, 45*, 249–265.
53. Atkinson, J. H., & Salfors, G. (1991). Experimental determination of soil properties. In *Proc. 10th ECSMFE* (Vol. 3, pp. 915–956). General report session 1.
54. Mair, M. J. (1993). Developments in geotechnical engineering research: application to tunnels and deep excavations. *Proceedings of Institution of Civil Engineers* (pp. 27–41). Unwin Memorial Lecture 1992.
55. Benz, T. (2007). Small-strain stiffness of soils and its numerical consequences. PhD Thesis Universitat Stuttgart.
56. Rampello, S., Silvestri, F., & Viggiani, G. (1994). The dependence of small strain stiff-ness on stress state and history for fined grained soils: The example of Vallericca clay. In *Proceeding of the First International Symposium on Pre-failure Defomation of Geomaterials* (pp. 273–278). Sapporo.
57. Casini, F., Vassallo, R., Mancuso, C., & Desideri, A. (2007). Interpretation of the Behaviour of Compacted Soils Using Cam-Clay Extended to Unsaturated Conditions. In: Theoretical and numerical unsaturated soil mechanics. Springer Proceedings in Physics (Vol. 113(1), pp. 29-36).
58. Vassallo, R., Mancuso, C., & Vinale, F. (2007). Effects of net stress and suction history on the small strain stiffness of a compacted clayey silt. *Canadian Geotechnical Journal, 44*(4), 447–462.
59. Casini, F., Vassallo, R., Mancuso, C., & Desideri, A. (2008). Application to a compacted soil of a Cam Clay model extended to unsaturated conditions. In D. G. Toll, C. E. Augarde, D. Gallipoli, & S. J. Wheeler (Eds.), *Unsaturated Soils. Advances in Geo-Engineering Proceedings of the 1st European Conference* (pp. 609–615). Durham, UK: E-UNSAT 2008.

60. Casini, F., Vaunat, J., & Romero, E. (2013). A microstructural model on the link between change in pore size distribution and wetting induced deformation in a compacted silt. Poromechanics V. pp. 1309–1313.
61. Vassallo, R. (2003). Comportamento di terreni costipati non saturi a piccole, medie e grandi deformazioni. PhD tesi, University of Napoli Federico II, Naples, Italy.
62. Casini, F., Minder, P., & Springman, S. M. (2011). Shear strength of an unsaturated silty sand. In *Unsaturated Soils—Proceedings of the 5th International Conference on Unsaturated Soils* (Vol. 1, pp. 211–216) Spain: Barcelona.

Chapter 15
Beyond Hydrocarbon Extraction: Enhanced Geothermal Systems

Masami Nakagawa, Kamran Jahan Bakhsh, and Mahmood Arshad

Abstract There are many forms of energy that exist subsurface. Mining typically extracts minerals that are energy rich and processes them to produce electricity. The oil and gas industry extracts hydrocarbons with high energy content that are suitable for energy production. This chapter illustrates the concept of "heat mining" that is a form of mining but has not caught much attention. This term "heat mining" was originally used to describe a general concept of mining heat from deep granitic rocks by injecting cold water and recovering it as steam or water/steam mixture to produce electricity. Beginning with the Fenton Hill Hot Dry Rock project developed by Los Alamos National Laboratory in 1971, there is a long history of development of different types of heat mining projects in Japan, the UK, China, Germany, France, Iceland, and Hungary. This chapter presents a simple conceptual model of a doublet reservoir that can be developed in a sedimentary basin at a depth below where the conventional hydrocarbon resources can be found. In addition, as we move forward with the development of the original heat mining concept from granitic rocks, it is important to understand the consequences of long-term heat extraction. Developing an enhanced geothermal system or engineered geothermal system (EGS) is a complex process and is dependent on a range of geological and operating variables. Stress distribution and re-distribution in and around EGS during the different phases of development may have a significant impact on the reservoir itself as well as surrounding rock masses. Thus, the second part of this paper addresses issues associated with stress redistribution during and after the working cycle of EGS and gives insights in understanding the behavior of stress redistribution in and around basement rock. As the basement rock is thermo-elastically connected to the country rock, newly generated stresses interact with the existing in situ stresses under prevailing conditions of geological, design, and operating variables. Inability to predict the timing and the degree of impact suggests a need of a consorted effort of investigations between oil/gas, mining and geothermal industry, and academic disciplines.

M. Nakagawa (✉) • K.J. Bakhsh • M. Arshad
Department of Mining Engineering, Colorado School of Mines, Golden, CO 80401, USA
e-mail: mnakagaw@mines.edu

C. Jin, G. Cusatis (eds.), *New Frontiers in Oil and Gas Exploration*,
DOI 10.1007/978-3-319-40124-9_15

15.1 Introduction to Sedimentary Enhanced Geothermal Systems (SEGS)

The idea of extracting heat from deep, hot, dry granitic rock where the rock temperature can rise as high as 350 °C at a depth of 5 km was proposed four decades ago during the first oil crisis. In 1971, a Los Alamos National Laboratory experiment demonstrated the successful utilization of heat from the high temperature of the deep crystalline rock to generate electricity [1]. The crystalline formation at elevated temperature has become one of the main targeted geologic formations for geothermal power production.

While prospective areas of crystalline rock for a geothermal system is limited to volcanic and tectonically active setting, deep sedimentary basins for geothermal usage remain largely unexplored even though they comprise an extensive region around the world. The total energy content of sedimentary resources is considerably higher than that of crystalline rock. For example, the estimated energy content of the US sedimentary rock formation to 10 km depth is about 100,000e18 J [2]. The unexplored potential of heat extraction from sedimentary rock presents a significant opportunity to secure the future of the world energy.

Utilization of the heat from sedimentary basins has been limited to direct-use (non-electricity generating) applications. In Szeged in the Carpathian Basin in Hungary, 1200 buildings and a greenhouse area of 400,000 square meters were heated at the end of 1969 [3]. More recently, the number of geothermal wells for direct-use applications in Hungary has grown to 1275 [4]. The Guanzhong sedimentary basin in Shaanxi, China [5], provides 3.5 million square meters of geothermal space heating from 346 geothermal wells as of 2007 [6]. Paris and the Aquitaine Basin in France are other successful direct-use application examples [7].

Harvesting energy for power generation from moderately high- to medium-temperature reservoirs in sedimentary formations was not considered operationally feasible until the development of the binary cycle geothermal power plant. A binary power plant can utilize temperatures in geothermal reservoirs as low as 120 °C. In Germany, they successfully generate 3.36 MWe from the Bavarian Molasses basin, and 3.0 MWe from the Upper Rhine Graben formation exemplifies the promise of sedimentary geothermal reservoirs [8, 9].

Several research papers have been published on different aspects of sedimentary geothermal reservoirs. Gringarten & Sauty [10] developed an analytical model for predicting thermal breakthrough in a horizontal porous reservoir with uniform thickness. They analytically calculated the optimum distance between wells of a doublet system that can keep the temperature of the production well constant. The Dogger aquifer in France has been operated for several decades using the Gringarten scheme [11].

Major sedimentary basins of Colorado, such as the Denver Basin, the Piceance Basin, and the San Luis Basin, have been proposed as candidates for electricity generation in conjunction with direct application [12]. Morgan [13] also highlighted the advantages of using a sedimentary basin as an EGS field laboratory to examine aspects of the EGS system.

De Graaf et al. [14] reviewed the capability of the Coast Geothermal Project within Australia's hot sedimentary basins to generate large amounts of competitively priced, zero-emission, base-load power. The potential of stratigraphic reservoirs of the Great Basin in the Western USA to produce heat energy also has been investigated and results showed the possibility of high thermal recovery without the need for reservoir enhancement [15]. In Germany, the former gas exploration well, Groß Schönebeck, reopened and deepened to access the Lower Permian Rotliegend sedimentary formation. Pairing the extended well with a second well completed in late 2006, a doublet system was constructed in order to generate geothermal electricity [16]. The possibility of using sedimentary reservoirs to build EGS in the lower Triassic sedimentary formations of central Poland area has been investigated. Researchers on the Polish project modeled the performance of a prospective EGS plant using TOUGH2 code. Adjusting the permeability and volume of the fractured zone, they numerically simulated the net power production of an EGS plant [17]. Similarly, two general classes of sedimentary formations in the Great Basin—a multi-layered and single high-permeability layer—have been modeled numerically to investigate the rate of heat extraction. Results showed that lower permeability models have better thermal performance [18].

This chapter provides a simple conceptual model to investigate the behavior of the SEGS system. Thermal performance of the SEGS system is assessed based on advancement of the cold front in a reservoir. This study focuses on the influence of the formation permeability as a site-specific characteristic. In particular, the effect of high-permeability or pre-existing natural fractures on thermal breakthrough and geometry of the heat transfer volume is inspected. Numerical simulation has been conducted using COMSOL Multiphysics. The following section introduces the proposed conceptual model. This conceptual model provides a benchmark for the simulation studies in later sections.

15.2 Description of a Modeled SEGS Reservoir

A segment of a deep, permeable sedimentary basin with a high-temperature gradient of 0.05 K/m is used as a prototype for the reservoir simulation study. The reservoir is assumed to be located in a homogenous sedimentary formation and intersect a vertical fracture at a depth of 3950 m and 3550 m, respectively (Fig. 15.1). Both the injection and production rates are assumed 40 l/s, and the temperature of injection water is 35 °C. Porosity and permeability in the basin are assumed to be uniform, 0.1 and 1e-13 m^2 (100 mD), respectively. It is also assumed that the thermal equilibrium between the injected fluid and the formation was attained instantaneously. The formation is assumed to be incompressible. The study area is restricted to a cubical block with side length of 500 m, and for the sake of simplicity, the gravitational effect is negligible [19]:

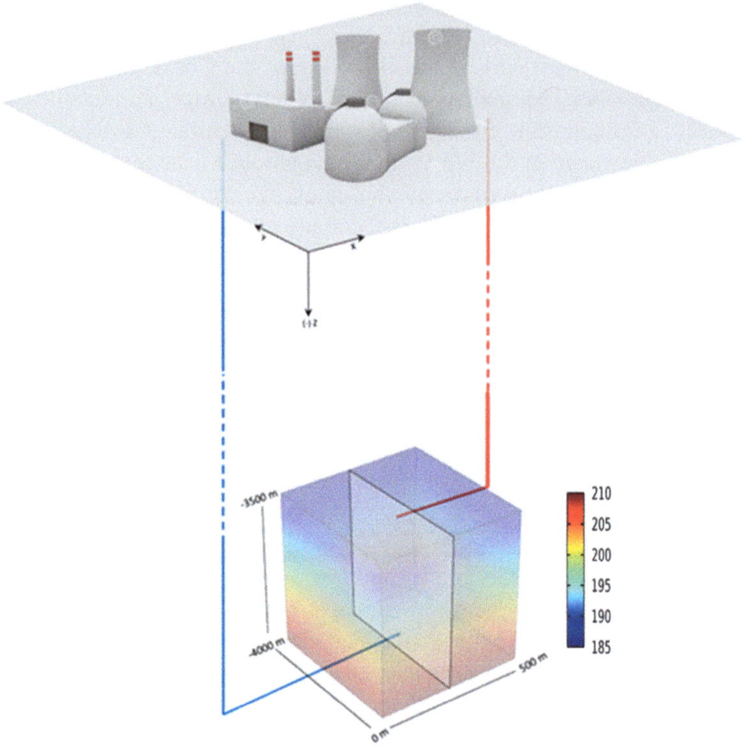

Fig. 15.1 Schematic of the conceptual fractured sedimentary basin located in a region with temperature gradient of 0.05 K/m

$$\Omega_r = \{(x, y, z): 0 < x < 500,\ 0 < y < 500,\ -4000 < z < -3500\}$$

$$\Omega_{inj} = \{(x, y, z): 240 < x < 260, \quad 250 - r_{inj} < y < 250 + r_{inj},$$
$$-3950 - r_{inj} < z < -3950 + r_{inj}\}$$

$$\Omega_{pro} = \{(x, y, z) : 240 < x < 260, \quad 250 - r_{pro} < y < 250 + r_{pro},$$
$$-3550 - r_{pro} < z < -3550 + r_{inj}\}$$

$$\partial\Omega_f = \{(y, z): \ 0 < y < 500, \ -4000 < z < -3500\}$$

The above geometrical regions represent the computational domain for the reservoir, locations of injection and production wells, and fracture, respectively. The numerical model is validated against the Gringarten [11] analytical model. Keeping equal flow rates for injection and production wells leads to a quick steady-state condition.

As the modeled sedimentary basin is assumed to be a homogenous porous media, Darcy's law is used to govern the flow of fluid in the model. Heat transfer-governing equations are also derived from energy balance in the domain.

One way to model a fractured reservoir in which the flow moves freely through the fractures and relatively slowly within the surrounding porous domain is to incorporate Darcy's law for the porous media flow in conjunction with the Navier-Stokes equation in the fracture. Coupling these two equations and solving them numerically are not cumbersome unless the geometrical scale of the incorporating domains is different in the orders of magnitude. In this study, however, a very long and narrow fracture domain with a high aspect ratio ($500 \times 500 \times 0.05$) is confined within a very wide porous block of rock ($500 \times 500 \times 500$), and discretizing the fracture domain explicitly requires a very dense mesh consisting of a huge number of infinitesimally small elements.

An alternative method to avoid an excessive computational cost is a Discrete Fracture Model where the fracture is treated as an interior boundary. The advantage of this approach is in reducing the degrees of freedom (unknowns) and enhancing computational performance particularly for the cases in which the fracture permeability is higher than the surrounding porous medium [20]. In this study, the Discrete Fracture approach is selected.

15.2.1 Flow Equations

Darcy's law is used to describe the behavior of fluid flow in the porous domain of the sedimentary reservoir, and they are

$$\frac{\partial}{\partial t}(\varnothing_r \rho_r) + \nabla \cdot (\rho_r \mathbf{u}) = 0 \qquad x, y,\ z \in \Omega_r \tag{15.1}$$

$$\mathbf{u} = -\frac{\kappa_r}{\mu} \nabla p \qquad x, y,\ z \in \Omega_r \tag{15.2}$$

where $\varnothing_r$ is the rock matrix porosity, ρ_r is the rock matrix density (kg/m^3), $\mathbf{u}$ is the Darcy's velocity (m/s), κ_r is the rock matrix's permeability (m^2), μ is the fluid dynamic viscosity (Pa s), and p is the fluid's pressure in the pore (Pa).

The Neumann boundary condition is applied for mass flow at the top and bottom boundaries, and the Dirichlet condition is applied for sides of the domain:

$\mathbf{n} \cdot \rho_r \mathbf{u} = 0$ at $\partial\Omega_r$ domain top and bottom facets and $\mathbf{n}$ is the vector normal to the boundary.

As mentioned above, in this mode the fracture is considered an interior boundary to avoid computational issues with high-aspect-ratio geometry. Usually at boundaries, flow is defined normal instead of tangent to the boundary plane, and to govern the velocity field along the fracture, Darcy's law equation must be modified by accounting for the fracture thickness d_f:

$$d_{\mathrm{f}}\frac{\partial}{\partial t}(\text{Ø}_{\mathrm{f}}\rho_{\mathrm{f}})+\nabla_{\mathrm{t}}\cdot(d_{\mathrm{f}}\rho_{\mathrm{f}}\mathbf{u})=0 \qquad y,\, z\in\,\partial\Omega_{\mathrm{f}} \tag{15.3}$$

$$\mathbf{u}=-\frac{\kappa_{\mathrm{f}}}{\mu}\,\nabla_{\mathrm{t}}\mathrm{p} \qquad y,\, z\in\,\partial\Omega_{\mathrm{f}} \tag{15.4}$$

where d_{f} is the fracture's thickness (m), Ø_{f} is the fracture porosity, ρ_{f} is the fracture density (kg/m^3), $\mathbf{u}$ is the modified Darcy's velocity on the fracture (m/s), κ_{f} is the permeability of the fracture (m^2), and μ is the fluid dynamic viscosity (Pa s). To restrict the equations to the fracture's plane, ∇_{t}, the tangential gradient operator is applied for flow in the fracture.

It should also be noted that the Neumann boundary condition was applied to impose the no-flow condition on the edge of fracture:

$$n\cdot\mathbf{u}=0 \quad \text{at } \partial^2\Omega_{\mathrm{f}} \quad \text{fracture edges}$$

15.2.2 Heat Transfer Equations

The following equations are used to govern heat transfer in the porous matrix:

$$\left(\rho C_{\mathrm{p}}\right)_{\mathrm{eff}}\frac{\partial T}{\partial t}+\rho C_{\mathrm{p}}\mathbf{u}\cdot\nabla T+\nabla\cdot\mathbf{q}=0 \tag{15.5}$$

$$\mathbf{q}=-k_{\mathrm{eff}}\,\nabla T \tag{15.6}$$

$$\left(\rho C_{\mathrm{p}}\right)_{\mathrm{eff}}=(1-\text{Ø}_{\mathrm{r}})\,\rho_{\mathrm{r}}C_{\mathrm{p_r}}+\text{Ø}_{\mathrm{r}}\,\rho_{\mathrm{w}}C_{\mathrm{p_w}} \tag{15.7}$$

$$k_{\mathrm{eff}}=(1-\text{Ø}_{\mathrm{r}})\,k_{\mathrm{r}}+\text{Ø}_{\mathrm{r}}\,k_{\mathrm{w}} \tag{15.8}$$

where $(\rho C_{\mathrm{p}})_{\mathrm{eff}}$ is the effective volumetric heat capacity at constant pressure (J/m^3 °C), k_{eff} is the effective thermal conductivity of the rock matrix (J/m s °C), and $\rho_{\mathrm{w}}C_{\mathrm{p_w}}$ is the volumetric heat capacity of the water. In this study, thermal conductivity of rock matrix is assumed isotropic.

Neumann boundary condition is applied for the heat flow on the top and bottom boundaries. Hence, there is no heat flux across the top and bottom faces of the rock matrix:

$$-\mathbf{n}\cdot\mathbf{q}=0 \text{ at } \partial\Omega_{\mathrm{r}} \text{ rock block facets}$$

Side boundaries of the domain are considered open and the heat can flow in and out of the domain with a specified exterior temperature defined by the temperature gradient of the region.

The heat transfer equations in a porous matrix also need to be modified to account for the fracture thickness d_f:

$$d_f(\rho C_p)_{eff}\frac{\partial T}{\partial t} + d_f \rho C_p \mathbf{u}\cdot\nabla_t T + \nabla_t\cdot\mathbf{q_f} = \mathbf{n}\cdot\mathbf{q} \quad (15.9)$$

$$\mathbf{q_f} = -d_f k_{eff}\,\nabla_t T \quad (15.10)$$

$$(\rho C_p)_{eff} = (1-\text{Ø}_f)\,\rho_f C_{p_f} + \text{Ø}_f\,\rho_w C_{p_w} \quad (15.11)$$

$$k_{eff} = (1-\text{Ø}_f)\,k_f + \text{Ø}_f\,k_w \quad (15.12)$$

where $(\rho C_p)_{eff}$ is the effective volumetric heat capacity of the fracture-fluid volume at constant pressure (J/m^3 °C), and k_{eff} is the effective thermal conductivity of fluid-fracture mixture (J/m s °C). The third term on the left-hand side of Eq. (15.9) represents conductive heat flux in the fracture-fluid volume where the computational domain is restricted by the tangential gradient, and the term on the right-hand side gives the heat supply through the fracture walls by conduction.

Parameters used for the benchmark case are summarized in Table 15.1 below.

Initially, the temperature everywhere in the system is defined by the temperature gradient of the region, i.e.,

$$T_r(X,Y,Z,0) = T_f(Y,Z,0) = T_s - 0.05\,z \qquad x,y,z \in (\Omega_r \cap \partial\Omega_f)$$

where T_s is the surface temperature (°K), z is the depth (m), T_r is the porous domain temperature (°K), and T_f is the fracture domain temperature (°K).

The Dirichlet boundary condition is applied to fix the temperature of the injection fluid at 35 °C:

$$T(X,Y,Z,t) = 35\,°\text{C} \qquad \text{at } x,y,z \in \text{circumference of the injection well}$$

15.3 Interpretation of Simulation Results

To understand the performance of the reservoir in response to over 30 years of reinjection of cold water, the evolution of the flow and temperature fields in the domain of interest is simulated. As the fluid moves faster through the fractures and slower through the surrounding porous media, the temperature field of the domain develops as shown in Fig. 15.2. A sharp local temperature gradient develops between the fracture plane and the surrounding rock. This gradient, combined with the temperature gradient of the region, determines the thermal performance of the reservoir, and their contributions in reservoir thermal performance change as time elapses.

Table 15.1 The benchmark case geometrical parameters and material properties

Parameter		Symbol	Value	Unite
Fracture	Aperture	d_f	5	cm
	Length	l	500	m
	Height	h	500	m
	Density	ρ_f	1.2e3	kg/m^3
	Porosity	$Ø_f$	0.6	–
	Permeability	κ_f	1e-9	m^2
	Heat capacity	C_{p_f}	1100	J/(kg K)
	Thermal conductivity	k_f	3	W/(m K)
Rock matrix	Domain length	L	500	m
	Top surface depth	z_T	−3500	m
	Bottom surface depth	z_B	−4000	m
	Density	ρ_r	2587	kg/m^3
	Porosity	$Ø_r$	0.1	–
	Permeability	κ_r	1e-13	m^2
	Heat capacity	C_{p_r}	920	J/(kg K)
	Thermal conductivity	k_r	2	W/(m K)
Injection and production wells	Borehole radius	r_b	10.8	cm
	Injection length	l_{inj}	20	m
	Production length	l_{pro}	20	m
	Doublet distance	H	400	m
Fracture fluid (water)	Dynamic viscosity	μ	Function of temperature	Pa s
	Density	ρ_w		kg/m^3
	Heat capacity	C_{p_w}		J/(kg K)
	Thermal conductivity	k_w		W/(m K)
Other	Surface temperature	T_s	10	°C
	Geothermal gradient	$\mathcal{G}$	0.05	K/m
	Time period of modeling	a	30	Year

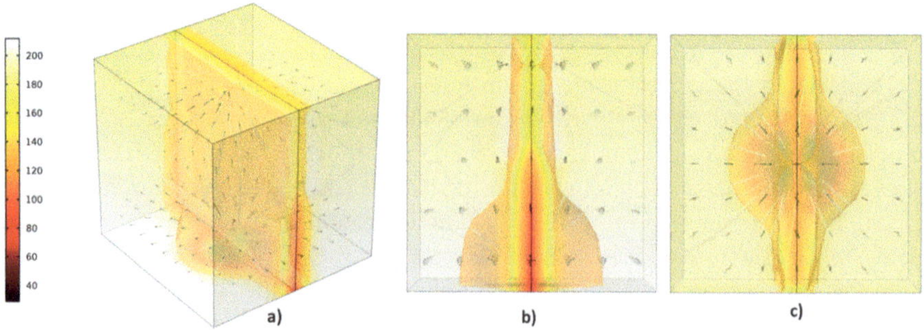

Fig. 15.2 Reservoir performance of the benchmark model over 30 years of heat extraction

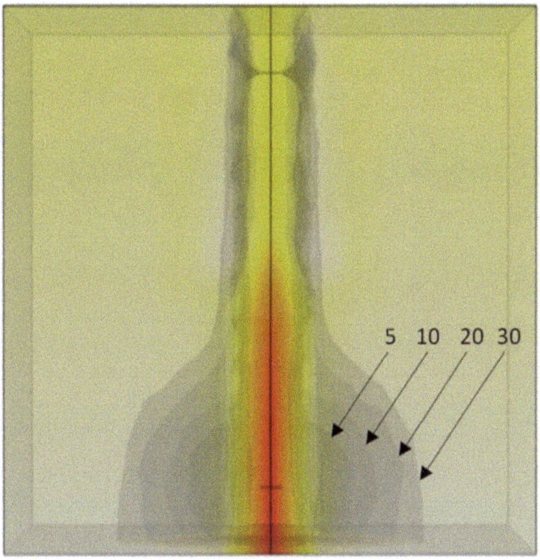

Fig. 15.3 Temperature evolution over 30 years of reservoir operation. Iso-temperature surface of 179 °C for years 5, 10, 20, and 30 is labeled to demonstrate the evolution of the heat transfer volume

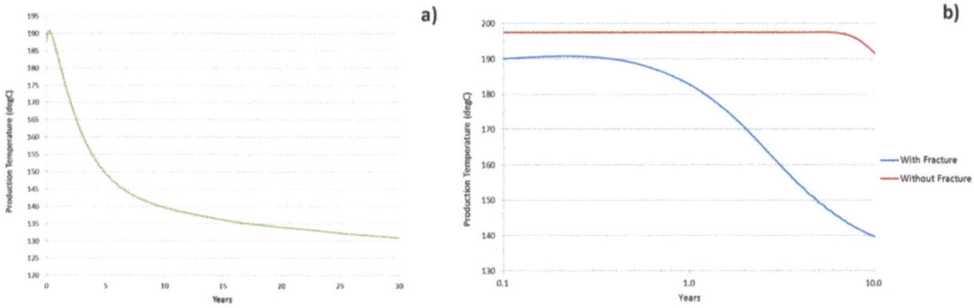

Fig. 15.4 Fluid average temperature verses time: case with high-permeability fracture (**a**) and comparison of production temperature of case with and without high-permeability fracture, logarithmic scale (**b**)

Figure 15.3 shows the 179 °C iso-temperature surface and its evolution over the course of 30 years of reservoir operation. The volume under 179 °C iso-temperature surface is called the "heat transfer volume" in which the temperature drops by 15 % of the maximum value. The size of the heat transfer volume is affected by the significant horizontal temperature gradient imposed by cold-water injection. The contrast between these two temperature gradients is greater near the injection well, resulting in the localized bulging of the iso-temperature surface. As time elapses, the heat transfer area around the line of injection grows faster than the heat transfer area around the production end, forming a bell shape as shown in Fig. 15.3.

The evolving fluid temperature at the production well shows the effect of the high-permeability fracture at the time of thermal breakthrough. Figure 15.4a shows

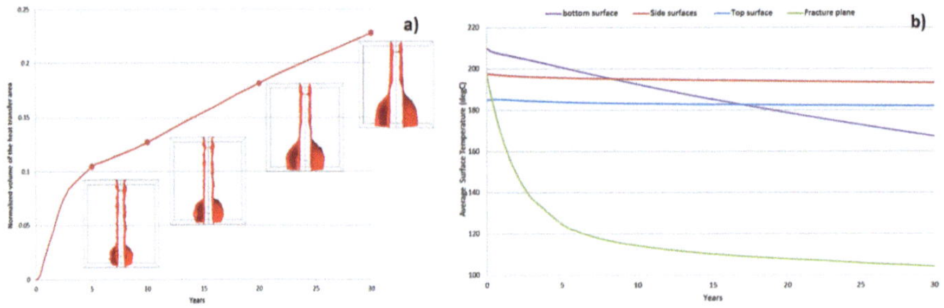

Fig. 15.5 Evolution of heat transfer region that contributes in the reservoir thermal performance (**a**), and average temperature of the domain boundaries over 30 years of heat extraction (**b**)

a sharp decline in the produced fluid temperature at an early stage of the heat extraction due to the relatively low resistance of the fracture to the flow. The slight temperature increase in the first year (see the enlarged view in Fig. 15.4b) is due to the configuration of the doublet system where the injection well is located below the production well. The lower injection well allows the fluid to access higher temperatures (>185 °C) before it migrates toward the higher, cooler production well horizon (185 °C).

To illustrate the role of a high-permeability fracture in thermal breakthrough, Fig. 15.4b shows the production temperature over a logarithmic time scale for a reservoir with and without a fracture, respectively. A well-designed geothermal reservoir requires a well-balanced output of hydraulic and thermal performance. Figure 15.4b partially illustrate this. When there is a fracture, it enhances the hydraulic performance but shortens the thermal breakthrough time. On the contrary, when there is no fracture in the reservoir, the reservoir can extend the thermal breakthrough time at the expense of low hydraulic performance.

Figure 15.5a shows the growth of the normalized heat transfer volume. It appears that contribution grows as time elapses. After 30 years, however, less than 25 % of the domain contributes to heat extraction form the reservoir. This result seems reasonable as at an early stage of the injection the majority of the fluid moves freely through the fracture and just a small portion percolates into the surrounding porous domain.

To address the thermal circulation pattern in the basin and to assess its evolution during the 30 years of heat extraction, the average temperature of the fracture surface and the domain boundaries are monitored (Fig. 15.5b). One can observe that during the first few years of production, longitudinal heat transfer is dominant, and after about 8 years, lateral heat transfer becomes more dominant. This behavior might be important to understand in situ stress redistribution in the area. For now, the authors term this phenomenon as "thermal flip."

15.3.1 Effect of Reservoir Permeability on Thermal Breakthrough Time and Reservoir Thermal Performance

Deep sedimentary basins located at the depth between 2.0 and 6 km with high-temperature gradients would be a good class of candidate for generating electricity through an SEGS system. To investigate the effect of permeability on the reservoir's thermal performance, different permeability values are prescribed for the proposed fractured reservoir model. The permeability values of 1, 10, and 100 mD were used. In addition, the permeability of 500 mD was included for the sake of comparison, even though it represent an unrealistic sedimentary structure.

Figure 15.6a shows the production temperature over 30 years for all four cases. All curves show that permeability has little effect on the thermal breakthrough time. This is likely due to the presence of the high-permeability fracture, which serves as the main conduit for fluid. The average temperature of produced fluid, particularly that of 5 years into operation, is positively correlated with permeability. The 1 and 10 mD cases produced almost the same temperature that is 5 °C lower than benchmark case (100 mD). The rare case of the reservoir with 500 mD permeability produced temperatures about 15 °C higher than the benchmark case. The observed outcome is reasonable since the volume of the reservoir contributing its thermal performance is larger when higher permeability allows the fluid more access to the hot rock. Figure 15.6b confirms this explanation, presenting the normalized volume of the heat transfer area versus production years. The growth rate of volume of the heat transfer area is highest in the case of 500 mD permeability.

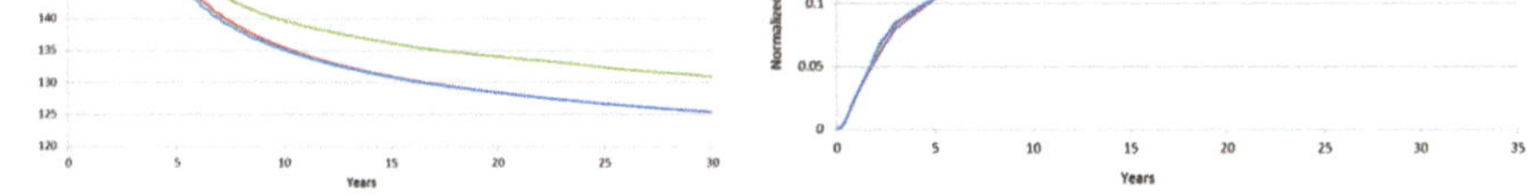

Fig. 15.6 Production temperature over 30 years for cases with permeability of 1, 10, 100, and 500 mD (**a**) and comparison of the normalized volume of the heat transfer area for cases with different permeability during heat extraction (**b**)

15.3.2 *Effect of Boundaries on Reservoir Thermal Performance*

In this section, the effect of the boundary conditions on reservoir thermal performance is investigated. For the first case, it is assumed that the top and bottom allow no mass and heat flux; however, the side boundaries are assumed open, so mass and heat can flux through the domain's sides. A model with this configuration can be a good representation of an SEGS for relatively undeformed horizontally stratified beds deposited from a marine environment, which may extend several kilometers laterally but bounded vertically by gradation or clay layers such as in the Midland Basin.

To address the effect of different boundary conditions on the model, two more configurations are studied: a model with open boundaries where heat and mass can flux within the all boundaries, and a model with closed boundaries where the domain is well insulated from surrounding. The former model could be representative of a SEGS for rounded, equidimensional, tens of kilometers sedimentary basins such as Intracratonic basins. The latter model might be applicable for Lagunal, Fluvial, or Estuarine structures in the stratigraphy, surrounded by impermeable beds such as some Pembina Field deposits.

Figure 15.7 depicts the temperature field in the reservoir at year 30 for the three aforementioned cases. Figure 15.7a–c show the side views of the temperature field for models with open, confined from top and bottom, and closed boundaries, respectively. The top views of these cases are also given in Fig. 15.7d–f. The shapes

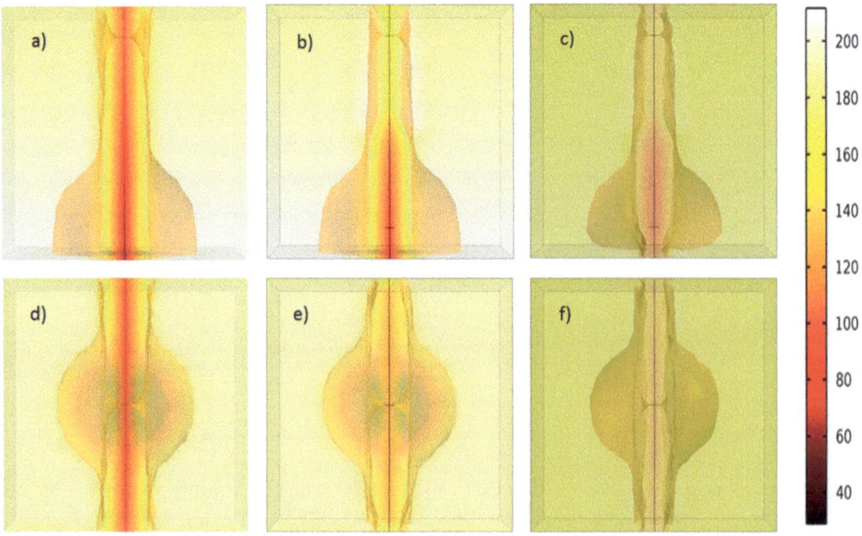

Fig. 15.7 Side views of the temperature field for models with open (**a**), confined from *top* and *bottom* (**b**), and with closed boundaries (**c**). Top views of the models are depicted, respectively (**d–f**)

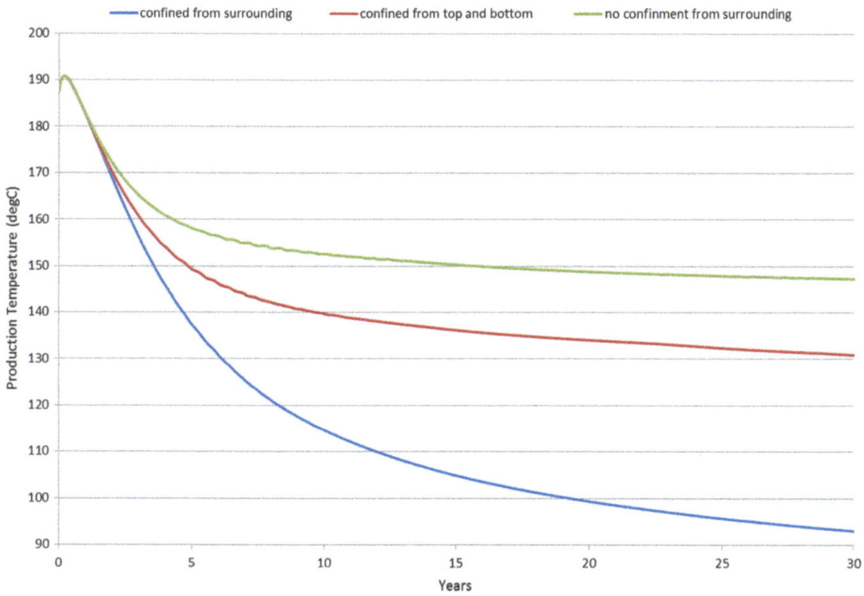

Fig. 15.8 Comparison of production temperature for cases with different boundary system: reservoir with well-insulated boundaries from surrounding (*blue*), with confinement from *top* and *bottom* (*red*) and with no confinement from surrounding (*green*)

of the heat transfer volume for all three cases are almost the same, except the bottom part of the open boundary case becomes more pear shaped. In case with open boundaries, fluid, as an energy carrier, has access to more heat than other two cases in which heat cannot come through the boundaries.

Figure 15.7 also shows that the cooling zone (shown in red) is extended vertically along the fracture plane. This also can be proven by showing produced fluid temperature during heat extraction. Figure 15.8 shows that the production temperature drops to almost 90 °C for the well-insulated case while for the case with open boundaries production temperature only get to 148 °C.

15.4 Issues of Long-Term Heat Extraction

The first part of this chapter described what would happen inside an SEGS reservoir once the reservoir becomes operational. The second part of this paper attempts to demonstrate how heat transfer around the reservoir and stress re-distribution in the surrounding rock mass are related. Enhanced geothermal system (EGS) is one of the innovative technologies and has been under the radar of the geothermal research community for almost 40 years. As EGS is still a developing technology and a complex phenomenon involving a range of geological and operating variables, there are many unanswered questions about how EGS works and what changes take place within and surrounding the EGS reservoir, during and after the working cycle of EGS.

EGS can be developed in distinct geological conditions, and they can be referred to as HWR (hot wet rock), HDR (hot dry rock), or HSR (hot sedimentary rock) systems. In all cases, EGS is a subsurface heat exchange system in which a network of fractured rock at a certain depth with suitable temperatures is stimulated and a fluid flow is established through injection and production wells to recover the earth's heat [21]. Heat in EGS is sourced by a combination of natural radioactivity, earth's heat of formation, and combined flow of heat by conduction, advection, and radiative transport [22]. Even for a small-scale EGS system, radioactive heat production will require millions of years to produce enough temperatures to be useful for EGS, so earth's radioactivity as a source of heat for EGS is not significant [23]. Flow of heat for distances over tens of kilometers by conduction is also insignificant even for earth's thermal diffusion constant of 1 mm^2/s. Radiative transport of heat is negligible for the earth's physical state [24]. Most effective source for a hot crust is by advection involving earth's heat of formation [25]. EGS is just one of the many ways of using earth's heat or geothermal energy for a range of applications. The current capacity of electrical generation using geothermal energy sources in the USA with a capacity factor of 70 % is around 3.4 GW, and was estimated to be 9 GW by USGS in 2008 with the development of known geothermal resources [22].

The US Department of Energy plans to fully implement its Frontier Observatory for Research in Geothermal Energy, commonly known as FORGE, by 2020 [26] and is currently focusing on five different active EGS demonstration projects in four different states, including Nevada, Idaho, Oregon, and California [27, 28]. Although water regulation authorities in the states have different water management regulations, water boards and the Environmental Protection Agency (EPA) have not shown any major reservations regarding this breakthrough technology [29]. The cost of power generation from EGS tends to be higher in early startup days but drops rapidly as the system progresses in action, ranging from USD 2500 to 5000 per installed kW, and USD 0.01 to 0.03 per KW as maintenance and operation costs [27, 28]. Power generation is not the only proposed use of energy recovered by EGS. Many other industrial (agriculture, aquaculture, greenhouse, drying, and process heat) and domestic uses (space heating, snow melting, and bathing and swimming parlors) have taken advantage of heat from the earth. Cascading use of recovered heat is also one of the available options for fully utilizing what EGS has to offer [30].

There are some issues that need monitoring and remedial measures in order for EGS to function properly for the designed span of life. Flow through fractures and heat removal may cause dilation of fracture apertures, and as fluid flows through the path of least resistance, it can lead to fluid short circuiting [31, 32]. In such a case, EGS may not be able to recover the estimated amount of heat.

Chemical or mechanical alteration of basement rock will occur as a result of chemically active or continuous fluid flow. Although water is supposedly the only fluid medium of EGS, flow network, fluid chemistry, and flow rates may be designed in accordance with the properties of base rock [33].

EGS is considered responsible for induced seismicity during its development and production phase. Each fracture-stimulating technique used in EGS, such as hydrofracturing, fluid injection, and acidization, has reportedly contributed to induced seismicity but predominantly only microearthquakes [33, 34].

Heat recovery from earth's crust, being the primary objective of EGS, may be responsible for the changes in the big picture of the project. All EGS factors, directly or indirectly related to EGS operations, ultimately play their role towards changes in stress in and around the basement rocks.

15.5 In Situ Stresses and Their Re-distribution in EGS

The EGS reservoir itself usually attracts most of the scientific and research work, but the focus of this chapter is the bigger picture outside of the EGS reservoir. The question is this: What happens outside the EGS reservoir when heat flows start to cause stress re-distribution in the country rock during and after working life of EGS? For simplicity, the country rock is assumed to be homogeneous.

Once both injection and production wells are completed, fractures stimulated, flow through fractures established, and the heat recovery process started in an EGS reservoir, the basement rock is subjected to slow but continuous thermal, and in turn, structural changes. As the geology over different areas and depths is factually different and EGS is, ideally, supposed to be set up anywhere with supporting ground conditions, there is not much to say with absolute certainty about what is going to happen regarding the new stress state of the area. But knowing the basic behavior of the rock in reaction to thermal changes and flow through fractures, one can provide an educated guess about what is next in structural and stress transition.

Naturally occurring stresses in the earth simply relate to the height of overburden and after a certain depth vertical and lateral stresses are similar. Cooling down of the rocks causes surrounding rocks to shrink. The rocks at usual EGS depths of 3–5 km are thought to exhibit thermoelastic behavior. This is why this shrinking as a result of cooling down generates tensile stresses in the surrounding rock. Newly generated tensile stresses interact with the existing state of stress to develop a whole new pattern of stresses in the EGS zone. It is the new stress pattern that decides the next resulting events. A slight variation in the resulting stress state can define the mode, time and occurrence of the next impactful event and one event can lead to another. This whole chain or cycle of the EGS process during its full life is illustrated in Fig. 15.9 and depends on different sets of variables involved in EGS.

Estimating the span of the changes, timeline of occurrence of eventual events, magnitude of such changes, and extent to which it can affect the surrounding settings is not trivial [35]. For an efficient EGS operation, there must be a balance between heat outflow (heat recovery) and heat inflow towards the basement rock. This means a continuous heat flow must exist from the country rock towards basement rock. Heat flows in the EGS lead to occurrence of different thermal zones in the area. Thermal zones and cooling fronts do not just exist along a straight line between inlets and outlets, but they also exist and expand in the vicinity in more of a distorted spherical shape. Figure 15.10 illustrates different thermal zones

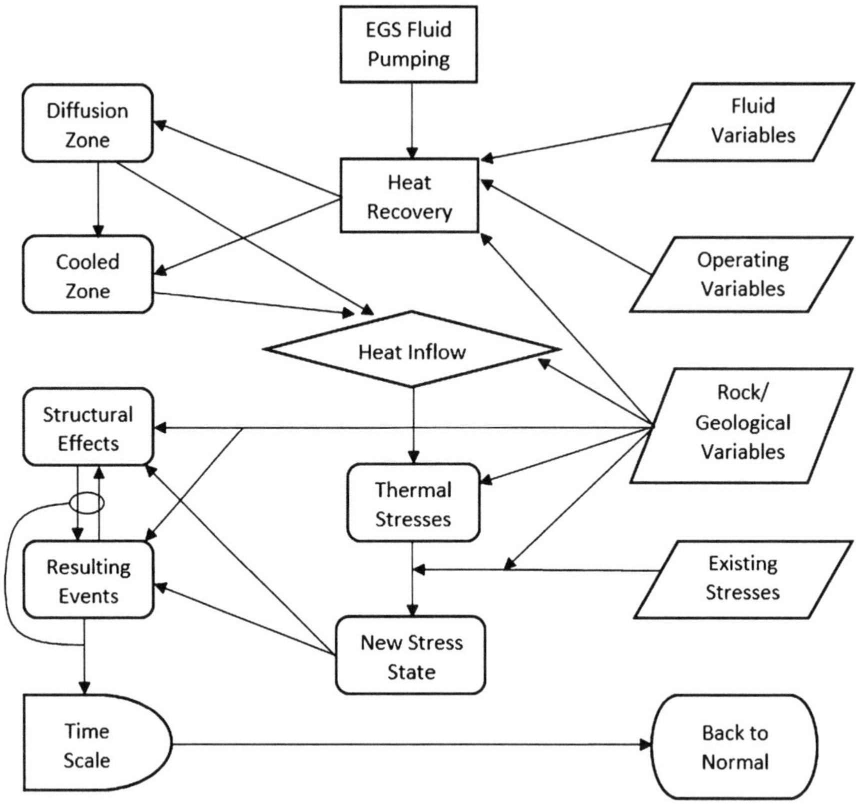

Fig. 15.9 EGS flow chart [21]

during the EGS cycle. These thermal zones have been explained in a DOE report on EGS [22]. The diffusion front separates the rock that is still at ambient temperature from the diffusion zone, which is where rocks start to cool down and the temperature is still between ambient temperature and injection water temperature. The cooling front separates the diffusion zone and cooled zone where the rock temperature has dropped down to injection fluid temperature and does not add much heat to water.

There are two basic facts associated with heat flows in EGS. First, a continuous heat inflow exists from the surrounding to EGS basement rock. Second, there exists earth's natural thermal gradient, implying that rock temperatures above the basement are already less than the temperatures being dealt with. These facts suggest that most of the heat inflows are from everywhere but the rock above the EGS basement rock. This enlightens a fact that most of the heat inflows, thermal changes, and stress and structural effects have to occur in a lateral direction and in the rock under EGS basement and also in the surrounding of the injection well compared to the outlet. Figure 15.11a is the illustrative explanation of the fact and Fig. 15.11b is the illustration of what ultimate diffusion or impact zone may look like.

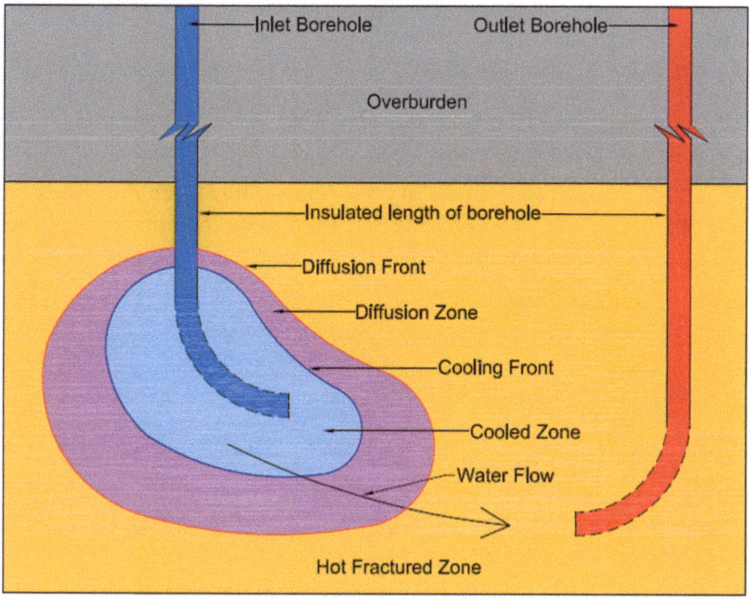

Fig. 15.10 Heat zones in EGS basement rock. In actual EGS designs, the actual placement of the injection well is below the depth of that of the production well

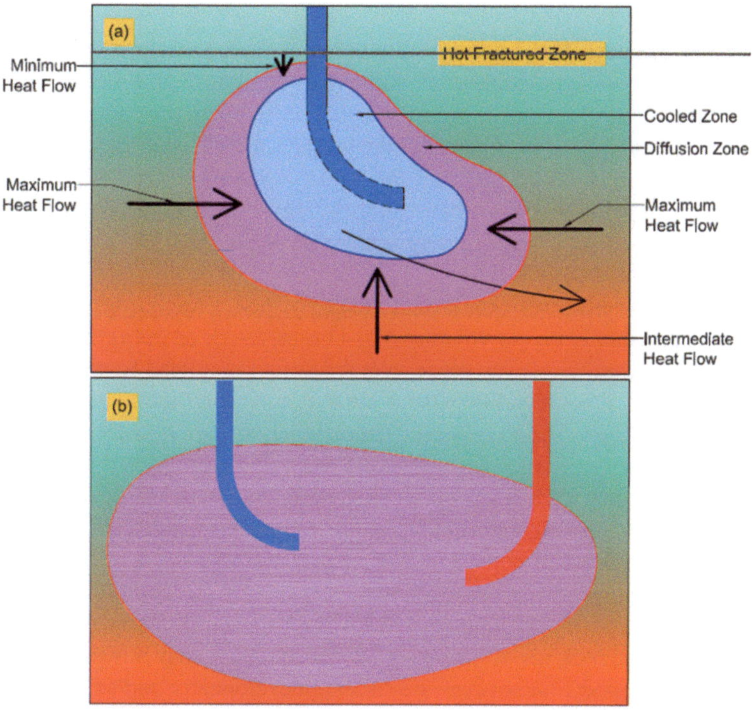

Fig. 15.11 (**a**) Heat flows in EGS basement rock. (**b**) Ultimate diffusion/impact zone

The occurrence of diffusion or ultimate impact zone in a more lateral plane suggests that the impacts of the EGS are majorly in a lateral dimension. Another factor to be considered is the timeline of the whole process. Considering the life of EGS to be 25–50 years, it will take even longer with earth's thermal diffusion constant for the impact zone to nullify the temperature changes and regain the ambient temperature. The slow process may need to be modified to accommodate our energy needs in the future.

15.6 Concluding Remarks

This study used a simple conceptual model of an SEGS with the hope to close the knowledge gap between the oil/gas and geothermal industry. Due to the temperature at the depth a SEGS requires, many forms of hydrocarbons break down, the sedimentary formation is tight and may represents different in situ stress conditions. The model analyzed in this chapter showed that the presence of induced or pre-existing fractures significantly shortens the time to achieve thermal breakthrough. During the heat extraction, the heat transfer region grows to become an elongating semi-spherical dome, influenced by the horizontal temperature gradient imposed by cold-water injection through a highly permeable fracture. The results indicate that the volume of the heat transfer area grows faster near the injection point.

Next, the influence of the permeability of the host rock was investigated. While permeability had no significant effect on thermal breakthrough time, flow through higher permeability resulted in higher production temperature. Sensitivity of the production temperature was reduced for the formation with lower permeability. The heat transfer volume reached 32 % of the total volume of the domain for a model configuration with highest basin permeability (500 mD). This result indicates that even for the case with unrealistically highest permeability, just a portion of the domain contributes in the heat transfer process.

Based on these observations, a multi-stage hydraulic fracturing, which allows access to incrementally bigger volumes of the basin, can be an option to enhance the heat transfer process. However, since over-fracturing may lead to premature thermal breakthrough, it is crucial to balance the hydraulic and thermal performance of the SEGS according to the site-specific characteristics.

Finally, the behavior and effects of stress redistribution due to thermal changes in the surrounding of EGS reservoir were studied. Due to the directionality of the heat transfer from the homogeneous surrounding rock to the homogeneous host rock, there is a long-term consistent trend of in situ stress redistribution. Actual rocks are nothing but being homogeneous, and one of the key questions for the future research is how this long-term, consistent heat transfer in geologic scale influence fractured rock masses. Findings to this question may advance our understanding of the connection between geothermal development and seismic events.

References

1. Armstead, H., & Tester, J. (1987). *Heat mining*. New Fetter Lane, London: E. & F. N. Spon Ltd.
2. Tester, J., et al. (2006). *The future of geothermal energy—impact of enhanced geothermal systems (EGS) on the US in the 21st,* Idaho Falls, Idaho.: Idaho National Laboratory.
3. Boldizsar, T. (1970). Geothermal energy production from porous sediments in Hungary. *Geothermics, 2*.
4. Árpási, M. (2003). Geothermal development in Hungary—country update report 2000–2002. *Geothermics,* 371–377.
5. Erlingsson, T., Jóhannesson, T., Olafsson, E. & Axelsson, G., (2010). Geothermal District Heating System in XianYang, Shaanxi, China. *Proceedings World Geothermal Congress.* Bali, Indonesia.
6. Zhonghe, P., Fengtian, Y., Tianming, H., & Zhongfeng, D., (2010). Genesis analysis of geothermal systems in Guanzhong Basin of China with implications on sustainable geothermal resources development. *Proceedings World Geothermal Congress.* Bali, Indonesia.
7. Laplaige, P. et al., (2005). Geothermal resources in France—current situation and prospects. *Proceedings World Geothermal Congress.* Antalya, Turkey.
8. Agemar, T., Weber, J. & Schulz, R., (2014). Deep geothermal energy production in Germany. *Energies,* 4397–4416.
9. Weber, J., et al. (2015). Geothermal energy use in Germany. *Proceedings World Geothermal Congress.* Melbourne, Australia.
10. Gringarten, A. C., & Sauty, J. P. (1975). A theoretical study of heat extraction from aquafires with uniform regional flow. *Journal of Geophysical Research, 80*(35).
11. Alain, C. & Gringarten, A. C., (1978/1979). Reservoir lifetime and heat recovery factor in geothermal aquifers used for Urban heating. *Pure and applied geophysics, 117*, 297–308.
12. Morgan, P., & Sares, M. A., (2011). *New horizons for geothermal energy in sedimentary basins in Colorado.* s.l., Geothermal Resources Council.
13. Morgan, P. (2013). *Advantages of choosing a sedimentary basin as the site for an EGS field laboratory*. Las Vegas: Geothermal Resource Council.
14. de Graaf, L., Palmer, R., & Reid, I., (2010). The Limestone Coast Geothermal Project, South Australia: A Unique Hot Sedimentary Aquifer Development. *Proceedings World Geothermal Congress.* Bali, Indonesia.
15. Allis, R., Moore, J., Blackett, B., & Gwynn, M. (2011). *The potential for basin-centered geothermal resources in the Great Basin.* San Diego, CA: Geothermal Resources Council Annual Meeting.
16. Huenges, E. et al. (2007). Current state of the EGS project Groß Schönebeck—drilling into the deep sedimentary geothermal reservoir. *Proceedings European Geothermal Congress.* Unterhaching, Germany.
17. Bujakowski, W., et al. (2015). Modelling geothermal and operating parameters of EGS installations in the lower triassic sedimentary formations of the central Poland area. *Renewable Energy, 80*, 441e453.
18. Deo, M., Roehner, R., Allis, R. & Moore, J. (2013). *Reservoir modeling of geothermal energy production from stratigraphic reservoirs in The Great Basin.* Stanford, CA: Stanford University, Thirty-Eighth Workshop on Geothermal Reservoir Engineering.
19. Bakhsh, K. J., Nakagawa, M., Arshad, M., & Dunnington, L. (2016). *Modeling thermal breakthrough in sedimentary geothermal system, using COMSOL multiphysics.* Stanford, CA: Stanfrod University.
20. Romano-Perez, C. A., & Diaz-Viera, M. A., (2015). A comparison of discrete fracture models for single phase flow in Boston, COMSOL conference.
21. Arshad, M., Nakagawa, M., Jahanbakhsh, K., & Dunnington, L. (2016). *An insight in explaining the stress distribution in and around EGS*. Standford, CA: Standford University.

22. Jeanloz, R., Stone, H., et al. (2013). *Enhanced geothermal systems*. Washington, DC 20585: US Department of Energy (DOE), Energy Efficiency and Renewable Energy.
23. Furlong, K. P., & Chapman, D. S. (2013). Heat flow, heat generation, and the thermal state of the lithosphere. *Annual Review of Earth and Planetary Sciences, 41*, 385–410.
24. Donald, L., & Turcotte, G. S., (1982). *Geodynmaics*. s.l.: Cambridge University Press.
25. Stolpher, E., Walker, D., Hager, B. H., & Hays, J. F. (1981). Melt sedregation from partially molten source regions, the importance of melt density and source region size. *Journal of Geophysics Research, 86*, 6261–6271.
26. Lester, P., (2015). *US Department of Energy*. Retrieved October 11, 2015. [Online] http://energy.gov/articles/top-10-things-you-didnt-know-about-enhanced-geothermal-systems.
27. DOE. (2015). *Energy.gov; Office of energy efficiency and renewable energy*. US Department of Energy. Retrieved October 11, 2015. [Online] http://energy.gov/eere/geothermal/enhanced-geothermal-systems-demonstration-projects.
28. DOE. (2015). *Energy.gov; Office of energy efficiency and renewable energy*. US Department of Energy. Retrieved October 11, 2015 [Online] http://energy.gov/eere/geothermal/geothermal-faqs.
29. Cichon, M. (2013). *Renewableenergyworld.com*. Retrieved October 11, 2015 [Online] http://www.renewableenergyworld.com/articles/print/volume-16/issue-4/geothermal-energy/is-fracking-for-enhanced-geothermal-systems-the-same-as-fracking-for-natural-gas.html.
30. Lund, J. W. (2010). Direct utilization of geothermal energy. *Energies, 3*, 1443–1471.
31. Ghassemi, A., & Kumar, G. S. (2007). Changes in fracture aperture and fluid pressure due to thermal stress and silica dissolution/precipitation induced by heat extraction from subsurface rocks. *Geothermics, 36*, 115–140.
32. Ghassemi, A., Nygren, A., & Cheng, A. (2008). Effects of heat extraction on fracture aperture: A poro–thermoelastic analysis. *Geothermics, 37*, 525–539.
33. MIT. (2006). *The future of geothermal*. s.l.: Massachusetts Institute of Technology.
34. Majer, E. L., et al. (2007). Induced seismicity associated with Enhanced Geothermal Systems. *Geothermics, 36*, 185–222.
35. Khademian, Z., Shahriar, K., & Nik, M. G. (2012). Developing an algorithm to estimate in situ stresses using a hybrid numerical method based on local stress measurement. *55*.

Chapter 16
Some Economic Issues in the Exploration for Oil and Gas

Charles F. Mason

Abstract In this chapter I present a simple economic model of exploration, and then discuss some predictions stemming from the model. I also describe some empirical phenomena relevant to exploration: trends in the probability of dry holes, the relation between oil prices and exploratory drilling, and developments in the deep water Gulf of Mexico.

16.1 Introduction

At the end of the twentieth century, two technological innovations were developed that greatly increased the volume of economically recoverable oil reserves in North America. The first of these, hydraulic fracturing, or fracking, was originally developed to enhance the production of natural gas. But over the next 5–10 years, this technique was adopted for production of crude oil, leading to substantial increases in production.[1] At about the same time, developments in the use of sophisticated imaging, such as 3-D imaging, increased the accuracy of exploratory ventures [20]. Together, these techniques made exploration and development of new deposits more efficient, and contributed to the rapid increase in US oil and gas production that has occurred in the past 10 years or so. These efforts led to the development of new hydrocarbon sources from formations that had previously been regarded as uneconomic, particularly shale oil formations in Texas and North Dakota, and offshore resources located in the deep waters of the Gulf of Mexico. Both new sources of oil turned out to be quite prolific.

In this chapter, I describe a simple economic model of exploration, and then discuss the predictions that come from that model. Two aspects are key here: the

[1]Gold [6] and Zuckerman [22] provide engaging and accessible accounts of the development of fracking in the USA since the turn of the century.

C.F. Mason (✉)
H.A. True Chair in Petroleum and Natural Gas Economics, Department of Economics & Finance, University of Wyoming, 1000 E. University Ave., Laramie, WY 82071, USA
e-mail: bambuzlr@uwyo.edu

C. Jin, G. Cusatis (eds.), *New Frontiers in Oil and Gas Exploration*,
DOI 10.1007/978-3-319-40124-9_16

probability that exploration will fail to find sufficient volumes of hydrocarbons (i.e., that the venture will result in a "dry hole"), and the anticipated price at which discovered resources can sold. I then discuss some empirical phenomenon relevant to exploration. First, I discuss the sharp decline in the probability of failure over the past 15 years (a phenomenon likely the result of the technological innovations described above). I then discuss the relation between oil prices and exploratory drilling. Finally, I discuss developments in the Gulf of Mexico, particularly in deep water.

16.2 Modeling Exploration

Early economic models of exploration typically involve a deterministic approach, the idea being that exploration translates into finds in a known way [13]. These models then gave way to analyses that paid explicit attention to uncertainty, particularly in terms of the results of exploration; examples include, but are by no means limited to Devarajan, & Fisher [4], Isaac [8], Lasserre [10], Mason [11], Pindyck [13, 14]. For the purposes of fixing ideas this approach has its merits, but it surely cuts important practical corners. To resolve this shortcoming, one could treat the result of exploratory efforts as a random variable [2] or the evolution of the stock of reserves as unknown [14]. This refinement of the original modeling variants has the attractive feature of analytic convenience and greater modeling realism, but again omits the key feature related to learning—that is, the notion that agents start off with a limited understanding of the underlying probabilistic process, which ignorance they might partially resolve by observing the outcomes from current efforts. Allowing for that sort of learning is a main theme of this monograph in general, and this chapter in particular.

A straightforward way to model exploration is by analogy to investment: the firm spends a certain amount up-front in anticipation of a reward in the event it is successful. To flesh out this idea, consider a firm extracting an oil deposit of size R. The firm values any finds at $V(R)$, which is an affine transformation of R.[2] To see this, suppose that the costs of extraction are $cq + F$, where q is the extraction rate, c is the average variable cost, and F is the fixed cost associated with actively extracting, then the present value of the stream of profits, discounted at rate r, is:

$$\int_0^\infty e^{-rt}\big[(p-c)q - F\big]\mathrm{d}t.$$

[2]That the value of the deposit is proportional to the amount of oil found is sometimes called the "Hotelling valuation principle," after Hotelling [7].

A central result from the Hotelling [7] analysis is that the discounted value of the difference between price and marginal extraction costs is equal for all points in time at which extraction occurs. As a result, the present discounted value of the profit stream is

$$(p-c)\int_0^{\infty} e^{-rt}q\mathrm{d}t - \int_0^{T} e^{-rt}F\mathrm{d}t = (p-c)R - \int_0^{T} e^{-rt}F\mathrm{d}t,$$

as the integral of extraction rate equals economically recoverable reserves, and where T is the date at which extraction ceases (i.e., the deposit is shut in). If T is large, $\int_0^T e^{-rt}F\mathrm{d}t$ is approximately F/r; this leads to the approximate characterization of the value associated with a deposit of size R:

$$V(R) = (p-c)R - Fr.$$

To procure deposits, the firm must first undertake exploratory activities, which cost an amount C_e. These efforts will lead to a discovery R, which is a random variable. Let the probability density function over R be denoted by $\phi(R)$, and the associated cumulative density function by $\Phi(R)$. It is conceivable the firm will drill a "dry hole," which really means the amount discovered is too small to be economically viable.[3] In other words, there is a critical value of discovery, call it $\underline{R}$, with the property that the firm will only develop the oil deposit if the recoverable amount found by exploration exceeds $\underline{R}$.[4] The probability of a dry hole can then be expressed as $\Phi(\underline{R})$. The expected net benefits from exploration are then

$$\int_{\underline{R}}^{\infty} \left[(p-c)s - \frac{F}{r}\right]\phi(s)\mathrm{d}s - C_e.$$

Putting this all together, then, we see that larger levels of exploration will result from lower chances of suffering a dry hole or higher spot prices. I consider each of these elements in the next section, where I discuss some relevant empirical issues.

[3]While it is conceivable that the efforts literally unearth zero oil, a more likely outcome is that the finds are relatively small. For example, Shell's recent failure in the Chukchi sea did not come up completely empty; rather, the find was much smaller than hoped—to the point it did not merit paying the large development costs that would be required to extract and deliver that oil to market.

[4]If one also takes uncertainty about prices into account, then this cutoff must also include an "option value," which can be interpreted as the potential increase in profit associated with waiting a small period of time in the hope that price will rise [5, 12].

16.3 Some Empirical Evidence

In this section, I discuss some empirical considerations that are relevant to exploration decisions. I start by reviewing the pattern of dry hole probabilities over the past 40 years.

16.3.1 Trends in the Probability of a Dry Hole

In the summer of 2015, Shell Oil made the difficult decision to abandon its efforts to find and develop hydrocarbon resources in the Chukchi Sea. Shell walked away from this venture despite having spent billions of dollars on leases and costs related to exploration in the Arctic. The key development that lead to their abrupt departure was the disappointing result from the sole exploratory well they managed to drill in these waters. That well indicated insufficient hydrocarbons were present to motivate further development. While this unfortunate "dry hole" was certainly unwelcome, it was far from unexpected: exploratory ventures come up empty with some regularity. But over the past 15 years or so, this rather gloomy result has become far less likely.

To shed light on the probability an exploratory venture yields a dry hole, I gathered data from the US Energy Information Administration [16]. The information at this website includes the number of exploratory wells drilled and the number of dry holes resulting from those efforts, from which the fraction of wells resulting in dry holes is readily calculated. Figure 16.1 plots this fraction, for every month from January 2007 to December 2011.[5] For a long period of time, from the earliest date at which data are available late into the twentieth century, the probability an exploratory well would come up dry was in the 70–80 % range. As late as the end of the century, the dry hole probability remained well over 50 %; it was about 63 % in January of 2000. But thereafter, this probability began to drop: the fraction of dry holes was smaller than 40 % for a lengthy period, from early 2005 to mid-2008.

It is conceivable that the drop in dry hole probability partly reflected the improving conditions in the global oil market—as price rises, more wells become economic, and so one would expect the fraction of dry holes to fall. But this effect alone cannot explain the pattern displayed in Fig. 16.1. Indeed, the probability of a dry hole was virtually identical in June 2006, when oil was selling for slightly less than $71/barrel, and June of 2008, when oil was selling for nearly $134/barrel. A second explanation for the pattern evidenced in the diagram is that technological

[5]The EIA stopped updating this information after 2011. The available data report the combined number of dry holes associated with exploration for oil dry holes associated with exploration for natural gas.

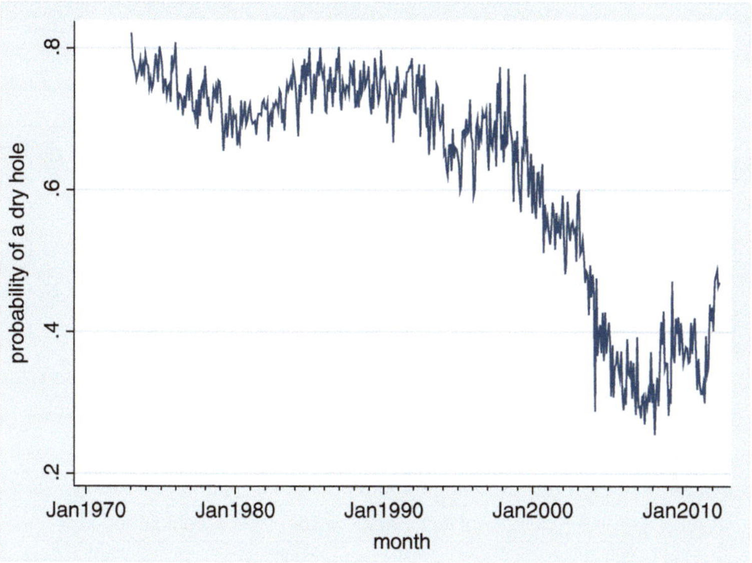

Fig. 16.1 Fraction of exploratory wells that were dry holes, 1973–2011

improvements, such as the development of sophisticated 3-D imaging techniques and the emergence of hydrological fracturing as a method for extracting oil from shale deposits, made successful exploration more likely. This explanation seems a more plausible explanation than the explanation associated with increasing prices.

16.3.2 Trends in Price and Drilling

The model I sketched out in Sect. 16.2 points to the importance of anticipated future revenues in motivating exploration. These expected future revenues are tied to anticipated prices. There is substantial evidence that oil prices evolve slowly over time, so that a good predictor of future price is current price [21]. Thus, one might expect to see the level of exploration moving in the same direction as the oil price.

Figure 16.2 provides some visual evidence corroborating this conjecture. Here, I plot the number of exploratory wells drilled, for each month from January 1986 to December 2011; these values are measured on the left-most *y*-axis.[6] On this graph, I superimpose the real price of crude oil for the same time period, measured against the right-most *y*-axis. This time series is constructed by dividing the monthly average West Texas Intermediate (WTI) spot price—widely regarded as the appropriate benchmark price for crude oil for most of this time frame—by the cost of

[6]As I noted in footnote 5, the EIA stopped reporting these data after 2011.

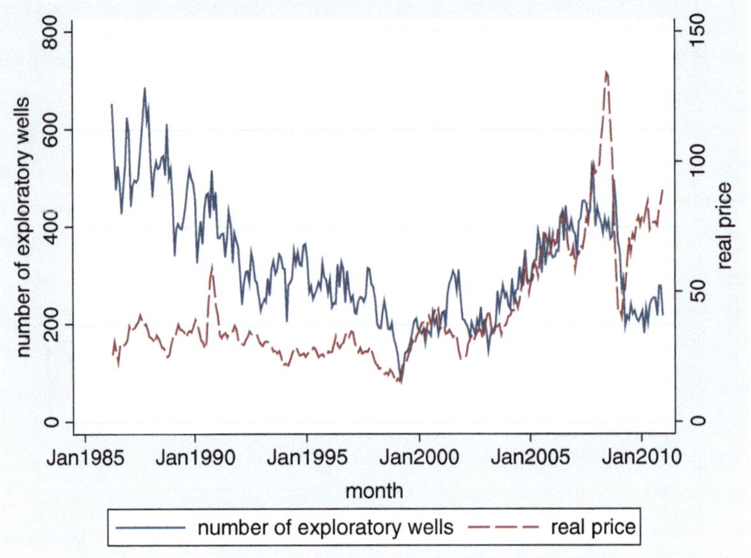

Fig. 16.2 Exploration and real oil prices, 1986–2010

living, as measured by the Consumer Price Index.[7] While the correlation between these two time series is relatively weak through much of the twentieth century, it strengthened considerably after 1999. From that point until 2009, during the depths of the oil price crash, the movements between the two series are strikingly similar. On balance, then, this data is supportive of the conjecture that the oil price is an important driver of exploratory efforts.

16.4 Developments in the Gulf of Mexico

In this section, I look more deeply at factors influencing exploration in a key geographic sector in the oil and gas industry—the Gulf of Mexico. Oil and gas operations have been active in the Gulf of Mexico since the 1940s [15]. In the intervening years, exploration steadily increased; as indicated in Fig. 16.3, crude oil production in the Gulf had reached 300 million barrels by the early 1980s. This pattern accelerated over the next few decades, with output levels exceeding 500 million barrels by the turn of the century. This rise in production was echoed by the increasing role played by Gulf oil production, as a share of total US output, with shares exceeding 20 % by the late 1990s.

[7]Data on the WTI spot price are available at U.S. Energy Information Administration [17].

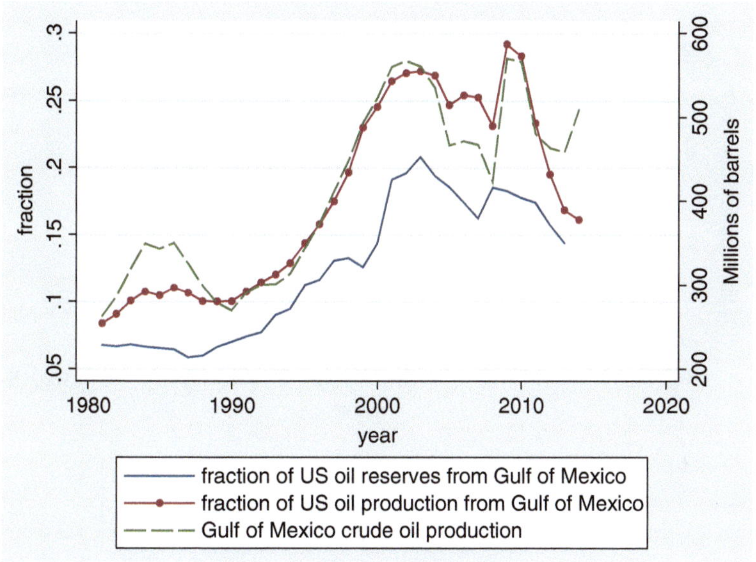

Fig. 16.3 Gulf of Mexico as a fraction of total US: crude oil

During this period, there was a clear migration towards deeper waters. As illustrated by Fig. 16.4, crude oil production from waters less than 600 ft in depth has fallen by about half over the past 20 years or so, with production from deeper waters rising steadily. This increase in production from deeper waters has been sufficient to outweigh the falloff in shallower water production, with the net effect that total production has remained roughly constant at about 40 million barrels per month—with the notable exceptions in August 2005, in the aftermath of Hurricane Rita, and in September 2008, following Hurricanes Gustav and Ike [3, 9].

The initial preference for shallower deposits can be readily explained by differences in exploration costs: while drilling in shallower waters can be readily accomplished at relatively low cost, operations in deeper waters were initially far more expensive [18, p. 12]. However, technological gains lowered drilling costs in deeper waters to the point that these deposits became economic to exploit over the past 15 years [19]. These observations point to deep water resources as increasingly important.

Drilling operations in the Gulf of Mexico are undertaken by six types of rigs: drillship, inland barges, jackups, platform rigs, semisubmersible, and submersibles. These rigs work in different parts of the Gulf. Inland barges, jackups, and submersibles are limited to shallower waters, while drillship, platform rigs, and semisubmersible ply deeper waters. Descriptive statistics concerning these various rigs are listed in Table 16.1. Here, I tabulate average number of days contracted,

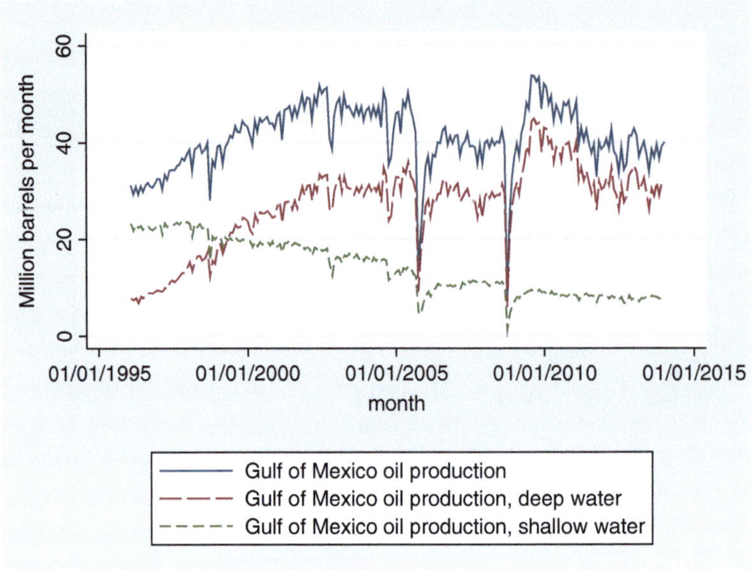

Fig. 16.4 Gulf of Mexico crude oil production

Table 16.1 Characteristics of different drill rigs in the Gulf of Mexico

Rig type (no. of obs'ns)	Number of days contracted	Depth (feet)	Day rate
Drillship (177)	411.3	6432.5	$519,462
	(682.7)	(2133.5)	(85638)
Inland barge (225)	274.2	15.80	$36,539
	(291.2)	(3.0)	(10956)
Jackup (1098)	156.1	122.4	$79,704
	(226.1)	(84.1)	(36638)
Platform rig (72)	144.6	1689.3	$37,847
	(247.2)	(1382.5)	(10417)
Semisubmersible (420)	234.6	4559.0	$409,729
	(332.2)	(3334.6)	(110225)
Submersible (25)	373.68	39.2	$69,620
	257.1829	(19.5)	(20480)

depth, and day rate for each of the six types of drill rigs.[8] Immobile rigs (jackups, platform rigs) tended to be contracted for relatively shorter periods of time, while more mobile rigs (drillship, semisubmersibles, submersibles) were contracted for

[8]These data are available from RigLogix (http://www.riglogix.com/). That information shows the starting and ending dates for each contract, the depth at which the rig operates, and the day rate (price per day), for 2547 drilling contracts under which drilling was undertaken between March 2002 and December 2014.

longer periods of time. This difference may reflect a desire on the part of operators to move contracted rigs between potential drilling sites. I note also that mobile rigs plying deep waters exhibit significantly higher day rates, presumably reflecting the greater difficulty associated with getting the rigs to the desired drilling location, along with the enhanced technical capabilities of those rigs. In light of the trend towards deeper waters that I noted above, the number of contracted drilling days per month in the deep water segment of the Gulf of Mexico merits closer investigation. I next turn to such an examination, in which I analyze the number of contracted drilling days per month associated with the three rig types that are used in deeper waters.[9]

In this analysis, I restrict attention to contracts undertaken between January 2010 and July 2014. The first month in this sample is near the bottom of the great recession; by that time oil markets had shaken off the doldrums associated with the financial collapse. At the end of the sample period, July 2014, oil markets were just about to collapse. Between these two dates, markets were relatively stable—making this a natural time frame in which to analyze drilling efforts.

The number of contracted drilling days can be thought of as a marker for the demand for exploration, and so one expects it to be positively impacted by elements that contribute to the benefits associated with exploration. As I noted above, the key element here is the oil price; one expects firms to contract for a larger number of days when prices are high than when prices are low. The price in question could be the current spot price or it could be an estimated future price. The relevant spot price for the Gulf of Mexico is Louisiana Light Sweet crude (LLS), as the trading hub for LLS is located on the gulf; the alternative spot price, West Texas Intermediate, is located several hundred miles inland.[10] To measure expected future prices, I look to the New York Mercantile Exchange (NYMEX) price; the US Energy Information Administration tabulates these futures prices for four time frames, reflecting 1, 2, 3, or 4 months ahead of the trading date U.S. Energy Information Administration [17]. Finally, one expects that water depth impacts the number of contracted drilling days, for two reasons. On the one hand, deeper waters are likely to be more costly to explore, which would serve to reduce drilling days. On the other hand, to the extent that deeper deposits are more remote there is likely a larger upfront cost associated with contracting for a drill rig, which might induce firms to write longer contracts. *Ex ante*, it is not obvious which of these two opposing effects dominate.

Table 16.2 presents results for two regressions. Both regressions include water depth (measured in feet) as an explanatory variable. In regression 1, I use the NYMEX 4-month ahead futures price to capture expected future prices, while in

[9]Focusing on these three rig types limits observation to operations in waters exceeding 500 m of depth. This cohort lies comfortably within the range the Bureau of Ocean Management interprets as deep water (drilling depths in excess of 1000 ft).

[10]Moreover, during the period I analyze, 2010 to mid-2014, there was a glut of oil in storage near Cushing, OK—the location of the WTI trading hub—which depressed the WTI spot price [1]. This unusual effect did not manifest at the LLS trading hub.

Table 16.2 Number of contracted drilling days in the Gulf of Mexico

Right-side variable	(1)	(2)
Depth	0.044**	0.042*
	(0.021)	(0.022)
NYMEX 4 month-	6.32**	–
ahead futures price	(2.60)	
LLS spot price	–	4.72***
		(1.54)
Constant	−614.7**	−488.2***
	(250.2)	(174.3)
R^2	0.15	0.17

Number of observations: 55 Standard errors in parentheses
$^*p < 0.10$; $^{**}p < 0.05$; $^{***}p < 0.01$

regression 2 I use the LLS spot price (also measured in US Dollars per barrel) to reflect firms anticipated rewards from a successful venture. In both regressions, the coefficient on water depth is positive, though small. The magnitude, roughly 0.04, suggests an increase of 4 contracted days for each 100 ft increase in water depth. This coefficient is statistically important at the 5 % level for regression 1; it is only statistically important at the 10 % level in regression 2. Regarding price, both regressions indicate that anticipated financial rewards exert a substantial influence on contracted drilling. Every three dollar rise in the 4-month ahead was associated with a roughly 20 day increase in the number of contracted drilling days. Similarly, a four dollar rise in the LLS spot price was associated with about a 17-day increase in the number of contracted drilling days. Relatedly, here was considerable variation in prices during the sample period, with spot prices ranging from about $75 (May 2010) to about $120 (April 2012) and futures prices ranging from about $78 (July 2010) to about $111 (April 2011). The range in contracted drilling days was also quite variable, ranging from a low of 7 (November 2012) to a high of 773 (July 2014).

16.5 Discussion

This chapter presented a simple model of exploration, and considered some intriguing related empirical phenomenon in oil and gas markets—the sharp decline in the probability of failure over the past 15 years and the relation between oil prices and exploratory drilling, along with a discussion of developments in the deep water Gulf of Mexico. I close the chapter by discussing the relation between this material and events since July 2014.

When crude oil markets started to collapse in the Summer of 2014, many pundits expected US oil producers to start cutting back on drilling. When drill rig counts did not respond as anticipated, a host of explanations were offered: perhaps firms were desperate to obtain revenues so as to service their debt, or perhaps they had locked in (now) higher prices by undertaking hedges prior to the collapse. No doubt these explanations have some merit, but there are competing explanations. For example, there was talk in late 2014 of oil prices rebounding in 2015, suggesting that key players in the industry anticipated price increases going forward into 2015.[11] When those optimistic conjectures were proved wrong in middle and late 2015, drilling operations persisted, albeit in focused "hot spots," locations where firms anticipated lower drilling costs of hitting oil, combined with significant extraction levels at low production costs.

These two alternative explanations are consistent with the model and empirical results in this paper. Firms explore when they expect the future stream of payoffs to adequately cover the up-front costs. Those anticipated payoffs are larger when firms are optimistic about future prices, or when they are optimistic about the success of the drilling venture. The former conjecture is consistent with the casual evidence in early 2015, as well as the conjecture that some firms had fortuitously locked in higher prices by hedging their bets prior to the oil price collapse. And the results in Sect. 16.3.1 suggest the latter conjecture is also reasonable: in the past 15 years or so, exploration has become more accurate, with the probability of a dry hole falling, thereby making exploration a more profitable gamble. To the extent that firms can focus on better prospects, in hot spots, this effect would be enhanced.

References

1. Brown, S. P., Mason, C. F., Krupnick, A., & Mares, J. (2014). Crude behavior: How lifting the export ban reduces gasoline prices in the united states. Technical Report, Resources for the Future, Washington, DC. Issue Brief 14-03-REV.
2. Cairns, R. D., & van Quyen, N. (1998). Optimal exploration for and exploitation of heterogeneous mineral deposits. *Journal of Environmental Economics and Management, 35*, 164–189.
3. Cruz, A. M., & Krausmann, E. (2008). Damage to offshore oil and gas facilities following hurricanes Katrina and Rita: An overview. *Journal of Loss Prevention in the Process Industries, 21*, 620–626.
4. Devarajan, S., & Fisher, A. C. (1982). Exploration and scarcity. *Journal of Political Economy, 90*, 1279–1290.
5. Dixit, A., & Pindyck, R. (1993). *Investment under uncertainty*. Princeton, NJ: Princeton University Press.
6. Gold, R. (2014). *The boom: How fracking ignited the American energy revolution and changed the world.* New York: Simon and Schuster Paperbacks.
7. Hotelling, H. (1931). The economics of exhaustible resources. *Journal of Political Economy, 39*(2), 137–175.
8. Isaac, R. M. (1987). The value of information in resource exploration: The interaction of strategic plays and institutional rules. *Journal of Environmental Economics and Management, 14*, 313–322.
9. Kaiser, M. J., & Yu, Y. (2010). The impact of hurricanes Gustav and Ike on offshore oil and gas production in the Gulf of Mexico. *Applied Energy, 87*(1), 284–297.

[11] That such predictions turned out to be wrong does not undercut their potential impact on drilling decisions made in the context of those predictions.

10. Lasserre, P. (1991). *Long-term control of exhaustible resources*. Chur, Switzerland: Harcourt Academic Publishers.
11. Mason, C. F. (1986). Exploration, information, and regulation in an exhaustible mineral industry. *Journal of Environmental Economics and Management, 13*, 153–166.
12. Mason, C. F. (2001). Nonrenewable resources with switching costs. *Journal of Environmental Economics and Management, 42*, 65–81.
13. Pindyck, R. S. (1978). The optimal exploration and production of nonrenewable resources. *Journal of Political Economy, 86*, 841–861.
14. Pindyck, R. S. (1980). Uncertainty and exhaustible resource markets. *Journal of Political Economy, 88*, 1203–1225.
15. U.S. Energy Information Administration. (2005). Overview of U.S. legislation and regulations affecting offshore natural gas and oil activity. Technical Report, Office of Oil and Gas.
16. U.S. Energy Information Administration. (2015). Crude oil and natural gas exploratory and development wells. http://www.eia.gov/dnav/pet/pet_crd_wellend_s1_m.htm
17. U.S. Energy Information Administration. (2015). Crude oil spot prices. http://www.eia.gov/petroleum
18. U.S. Minerals Management Service. (1997). Deepwater in the Gulf of Mexico: America's new energy frontier. Technical Report, U.S. Department of the Interior. OCS Report MMS 97-0004.
19. U.S. Minerals Management Service. (2001). The promise of deep gas in the Gulf of Mexico. Technical Report, U.S. Department of the Interior. OCS Report MMS 2001-00374.
20. Wang, Z., & Krupnick, A. (2015). A retrospective review of shale gas development in the united states: What led to the boom? *Economics of Energy & Environmental Policy, 4*, 5–17.
21. Wilmot, N. A., & Mason, C. F. (2013). Jump processes in the market for crude oil. *The Energy Journal, 34*, 33–48.
22. Zuckerman, G. (2013). *The frackers: The outrageous inside story of the new billionaire wildcatters*. New York, NY: Portfolio/Penguin Publishers.

ERRATUM TO

Nano-Scale Characterization of Organic-Rich Shale via Indentation Methods

Ange-Therese Akono and Pooyan Kabir

C. Jin, G. Cusatis (eds.), *New Frontiers in Oil and Gas Exploration*,
DOI 10.1007/978-3-319-40124-9_6

DOI 10.1007/978-3-319-40124-9_6

This chapter was written by two authors namely and Pooyan Kabir who has been missed out to be listed as the author. The affiliation details for Dr. Kabir are as follows;

Department of Civil and Environmental Engineering,
University of Illinois at Urbana-Champaign,
Urbana, IL

The updated original online version for this chapter can be found at
DOI 10.1007/978-3-319-40124-9_6

C. Jin, G. Cusatis (eds.), *New Frontiers in Oil and Gas Exploration*,
DOI 10.1007/978-3-319-40124-9_17

Index

C. Jin, G. Cusatis (eds.), *New Frontiers in Oil and Gas Exploration*,
DOI 10.1007/978-3-319-40124-9

U

V

W

MIX
Papier aus verantwortungsvollen Quellen
Paper from responsible sources
FSC® C105338

If you have any concerns about our products,
you can contact us on
ProductSafety@springernature.com

In case Publisher is established outside the EU,
the EU authorized representative is:
Springer Nature Customer Service Center GmbH
Europaplatz 3, 69115 Heidelberg, Germany

Printed by Libri Plureos GmbH
in Hamburg, Germany